18
VIIIA

on configuration
g point, K

ent

ur distance in Å

	13 IIIA	14 IVA	15 VA	16 VIA	17 VIIA	18 VIIIA
						2 4.00 $1s^2$ 0.95 4.215 **He** 0.205 at 37atm
	5 10.81 $1s^22s^22p^1$ 2300 4275 **B** 2.47/13.0 rhomb.	**6** 12.01 $1s^22s^22p^2$ 4100 4470 **C** 3.52/17.6/1.54 diamond 5.567	**7** 14.01 $1s^22s^22p^3$ 63.14 77.35 **N** 1.03 cubic 5.66/(N_2)	**8** 16.00 $1s^22s^22p^4$ 50.35 90.18 **O** Complex (O_2)	**9** 19.00 $1s^22s^22p^5$ 53.48 84.95 **F**	**10** 20.18 $1s^22s^22p^6$ 24.553 27.096 **Ne** 1.51/4.36/3.16 fcc 4.46
	13 26.98 $(Ne)3s^23p^1$ 933.25 2793 **Al** 2.70/6.02/2.86 fcc 4.05	**14** 28.09 $(Ne)3s^23p^2$ 1685 3540 **Si** 2.33/5.00/2.35 diamond 5.430	**15** 30.98 $(Ne)3s^23p^3$ 317.30 550 **P** complex	**16** 32.07 $(Ne)3s^23p^4$ 388.36 717.75 **S** complex	**17** 35.45 $(Ne)3s^23p^5$ 172.16 239.1 **Cl** 2.03 2.02 complex (Cl_2)	**18** 39.95 $(Ne)3s^23p^6$ 83.81 87.30 **Ar** 1.77/2.66/3.76 fcc 5.31

10 VIII **11** IB **12** IIB

10 VIII	11 IB	12 IIB	13	14	15	16	17	18
58.70 $8 4s^2$ 3187 **Ni** .14/2.49 3.52 fcc	**29** 63.55 $(Ar)3d^{10}4s^1$ 1357.6 2836 **Cu** 8.93/8.45/2.56 3.61 fcc	**30** 65.38 $(Ar)3d^{10}4s^2$ 692.73 1180 **Zn** 7.13/6.55/2.66 2.66/4.95 hcp	**31** 69.72 $(Ar)3d^{10}4s^24p^1$ 302.90 2478 **Ga** 5.91/5.10/2.44 complex	**32** 72.64 $(Ar)3d^{10}4s^24p^2$ 1210.4 3107 **Ge** 5.32/4.42/2.45 diamond 5.658	**33** 74.92 $(Ar)3d^{10}4s^24p^3$ 876 **As** 5.77/4.65/3.16 rhomb.	**34** 78.96 $(Ar)3d^{10}4s^24p^4$ 494 958 **Se** 4.81/3.67/2.32 hex. chains	**35** 79.90 $(Ar)3d^{10}4s^24p^5$ 265.90 332.25 **Br** 4.05/2.36 complex (Br_2)	**36** 83.80 $(Ar)3d^{10}4s^24p^6$ 115.78 119.80 **Kr** 3.09/2.17/4.00 fcc 5.64
106.42 ^{10} 3237 **Pd** .80/2.75 3.89 fcc	**47** 107.87 $(Kr)4d^{10}5s^1$ 1234 2436 **Ag** 10.5/5.85/2.89 4.09 fcc	**48** 112.41 $(Kr)4d^{10}5s^2$ 594.18 1040 **Cd** 8.65/4.64/2.98 2.98/5.62 hcp	**49** 114.82 $(Kr)4d^{10}5s^24p^1$ 429.76 2346 **In** 7.29/3.83/3.25 3.25/4.95 tetr.	**50** 118.71 $(Kr)4d^{10}5s^25p^2$ 505.06 2876 **Sn** 5.76/2.91/2.81 diamond 6.49	**51** 121.76 $(Kr)4d^{10}5s^25p^3$ 904 1860 **Sb** 6.69/3.31/2.91 rhomb.	**52** 127.60 $(Kr)4d^{10}5s^25p^4$ 722.65 1261 **Te** 6.25/2.94/2.86 hex. chains	**53** 126.90 $(Kr)4d^{10}5s^25p^5$ 386.7 458.4 **I** 4.95/2.36/3.54 complex (I_2)	**54** 131.29 $(Kr)4d^{10}5s^25p^6$ 161.36 165.03 **Xe** 3.78/1.64/4.34 fcc 6.13
195.09 $^{14}5d^96s^1$ 4100 **Pt** .62/2.77 3.92 fcc	**79** 196.97 $(Xe)4f^{14}5d^{10}6s^1$ 1337.58 3130 **Au** 19.3/5.90/2.88 4.08 fcc	**80** 200.59 $(Xe)4f^{14}5d^{10}6s^2$ 234.28 630 **Hg** 14.3/4.26/3.01 rhomb.	**81** 204.37 $(Xe)4f^{14}5d^{10}6s^26p^1$ 577 1746 **Tl** 11.9/3.50/3.46 3.46/5.52 hcp	**82** 207.2 $(Xe)4f^{14}5d^{10}6s^26p^2$ 600.6 2023 **Pb** 11.3/3.30/3.50 fcc 4.95	**83** 208.98 $(Xe)4f^{14}5d^{10}6s^26p^3$ 544.52 1837 **Bi** 9.80/2.82/3.07 rhomb.	**84** (209) $(Xe)4f^{14}5d^{10}6s^26p^4$ 527 1235 **Po** 9.31/2.67/3.34	**85** (210) $(Xe)4f^{14}5d^{10}6s^26p^5$ 575 610 **At**	**86** (222) (Rn) 202 211 **Rn**

Lanthanides:

	65	66	67	68	69	70	71
157.25 $f^75d^16s^2$ 3539 **Gd** 3.02/3.58 3.63/5.78 hcp	**65** 158.93 $(Xe)4f^96s^2$ 1630 3496 **Tb** 8.27/3.22/3.52 3.60/5.70 hcp	**66** 162.50 $(Xe)4f^{10}6s^2$ 1682 2835 **Dy** 8.53/3.17/3.51 3.59/5.65 hcp	**67** 164.93 $(Xe)4f^{11}6s^2$ 1743 2968 **Ho** 8.80/3.22/3.49 3.58/5.62 hcp	**68** 167.26 $(Xe)4f^{12}6s^2$ 1795 3136 **Er** 9.04/3.26/3.47 3.56/5.59 hcp	**69** 168.93 $(Xe)4f^{13}6s^2$ 1818 2220 **Tm** 9.32/3.32/3.54 3.54/5.56 hcp	**70** 173.04 $(Xe)4f^{14}6s^2$ 1097 1467 **Yb** 6.97/3.02/3.88 fcc 5.48	**71** 74.97 $(Xe)4f^{14}5d^16s^2$ 1936 3668 **Lu** 9.84/3.39/3.43 3.50/5.55 hcp

Actinides:

	97	98	99	100	101	102	103
(247) $f^76d^17s^2$ **Cm**	**97** (247) $(Rn)5f^97s^2$ **Bk**	**98** (251) $(Rn)5f^{10}7s^2$ 900 **Cf**	**99** (252) $(Rn)5f^{11}7s^2$ **Es**	**100** (257) $(Rn)5f^{12}7s^2$ **Fm**	**101** (258) $(Rn)5f^{13}7s^2$ **Md**	**102** (259) $(Rn)5f^{14}7s^2$ **No**	**103** (260) $(Rn)5f^{14}6d^17s^2$ **Lr**

Elements and Their Compounds in the Environment

Edited by
E. Merian (†), M. Anke, M. Ihnat and
M. Stoeppler

Volume 1:
General Aspects

Related Titles

Joachim Nölte
ICP Emission Spectrometry
A Practical Guide

ISBN: 3-527-30672-2
2002, 279 pp

Bernhard Welz, Michael Sperling
Atomic Absorption Spectrometry

ISBN: 3-527-28571-7
1998, 943 pp

Rita Cornelis, Joe Caruso, Helen Crews, Klaus Heumann (eds)
Handbook of Elemental Speciation
Techniques and Methodology

ISBN: 0-471-49214-0
2003, 670 pp

Markus Stoeppler, Wayne R. Wolf, Peter J. Jenks (eds)
Reference Materials for Chemical Analysis
Certification, Availability, and Proper Usage

ISBN: 3-527-30162-3
200, 322 pp

Elements and Their Compounds in the Environment

Occurrence, Analysis and Biological Relevance
2nd, completely revised and enlarged edition

Edited by
E. Merian (†), M. Anke, M. Ihnat and M. Stoeppler

Volume 1:
General Aspects

WILEY-VCH

WITHDRAWN

WILEY-VCH Verlag GmbH & Co. KGaA

Editors:

Manfred Anke
Am Steiger 12
07743 Jena
Germany

Milan Ihnat
Pacific Agri-Food Research Centre
Agriculture and Agri-Food Canada
Summerland, BC V0H 1Z0
Canada

Markus Stoeppler
Mariengartenstrasse 1a,
52428 Jülich
Germany

Library of Congress Card No: applied for

British Library Cataloging-in-Publication Data:
A catalogue record for this book is available from the British Library.

**Bibliographic information published by
Die Deutsche Bibliothek**
Die Deutsche Bibliothek lists this publication in the Deutsche Nationalbibliografie; detailed bibliographic data is available in the Internet at <http://dnb.ddb.de>.

© 2004 WILEY-VCH Verlag GmbH & Co. KGaA, Weinheim

Printed in the Federal Republic of Germany
Printed on acid-free paper.

Composition, Printing and Bookbinding:
Konrad Triltsch,
Print und digitale Medien GmbH
Ochsenfurt-Hohestadt

ISBN   3-527-30459-2

Preface

This book is the third in an unique line of handbooks, initiated in the early 1980s by Ernest Merian in cooperation with several of his colleagues leading in 1984 to a first book – **Metalle in der Umwelt** – published by Verlag Chemie. The design of the book, which became fondly known as the 'MERIAN', was from the beginning highly appreciated for its clear organization excellently and interdisciplinary covering the broad range from general information, basic elemental data, industrial uses, environmental distribution to biological and medicinal aspects. Since many readers from various scientific branches could benefit from it, an updated and extended English edition titled **Metals and Their Compounds in the Environment** followed relatively quickly in 1991. It was not unexpected that this edition sold very well over many years until it went out-of-print in 2001.

Ernest Merian passed away in 1995 at the age of 75 while traveling to one of his many scientific conferences, but already realized at that time the great success and admiration his work and particularly the voluminous English edition had received worldwide. The book is frequently quoted in the literature as it has been accepted as a major multielement source handbook. Progress, in the intervening years, in many scientific areas called for an update

rather than a simple reprint of the book and the publisher was seeking for someone or some group who that might be willing to produce a new edition. Dr. Steffen Pauly, in charge of the section of WILEY-VCH responsible for a possible update of the 'MERIAN', participated in April 1997 in the 7th International Symposium on Biological and Environmental Reference Materials (BERM-7) in Antwerp, Belgium. BERM-7 was co-organized by Dr. Markus Stoeppler, who took part as an author in the German Edition of the 'MERIAN' and as an author and a member of the scientific advisory board for the English Edition as well. Thus it was natural that Dr. Pauly asked Dr. Stoeppler his opinion about an update. Fortunately there were two other colleagues at the symposium, Prof. Manfred Anke, Germany, and Dr. Milan Ihnat, Canada, both very experienced as authors and editors of scientific publications and also knowledgeable with the subjects treated in the MERIAN. All three have in addition a long history of co-operation.

Thus at Antwerp general agreement was reached for a co-editorship of Anke, Ihnat and Stoeppler, followed by action by the publisher to ask former members of the Editorial Advisory Board and a few new ones for participation and by the editors to discuss a preliminary outline for the content of the book. In the course of the preliminary planning phase, including the six members of the Editorial Advisory Board (three former and three new), it was jointly decided that the comprehensive character of the book could be significantly strengthened and the basis of the literature sources increased if not only metals and some metalloids previously covered, but also several other metals and nonmetals that play important roles in industry, environment, medicine, nutrition and biota, namely alkali metals, alkaline earths, nitrogen, sulfur, phosphorus and the halogens, were to be included as separate chapters. This was finally accepted by the publisher with a slightly changed title influencing the length of the whole text, the organization of the introductory and the analytical part as well. Subsequently the final working phase started in 2002.

This updated and significantly extended 2nd edition of the 'MERIAN' is a tribute to Ernest Merian and therefore also bears his name in the editorial lineup as the book constitutes a continuation and some broadening of his initial comprehensive concept that was not significantly changed by his successors. This book, presented in three volumes, includes 81 chapters written by 83 experts from 20 countries around the world, based on the available international literature on approximately 1700 text pages. As the authors came from very different research areas it was thus unavoidable that their main interest often influences the style, content and general focus of individual chapters.

The first volume is composed of two parts. Part I deals with Element Distribution in the Environment and consists of twelve chapters ranging from "Composition of the Earth's Upper Crust, Natural Cycles of Elements, Natural Resources" to "From the Biological System of the Elements to Bio-

monitoring". Part II discusses in ten chapters "Effects of Elements in the Food Chain and on Human Health" and ranges from "Essential and Toxic Effects of Elements on Microorganisms" to "Ecogenetics". Here several chapters are new or newly written due to the somewhat changed general concept with more emphasis on element essentiality.

The second volume contains Part III covering all metallic elements, with some chapters carefully updated and/or extended and others new (as e.g. Mercury and the Platinum-Group Elements) from the Alkali Metals to Zirconium in 44 chapters of varying length due to the elements' individual essential, eco-chemical and eco-toxicological relevance. The chapter following organization, in order to maintain comparability with the 1st edition and among chapters, consists of seven sections with some freedom for the authors to add and organize subsections individually: 1) Introduction, 2) Physical and Chemical Properties and Analytical Methods (the latter mainly as a subsection with more details for elements for which chemical speciation is essential), 3) Sources, Production, Important Compounds, Uses, Waste Products and Recycling, 4) Distribution in the Environment, in Foods and Living Organisms, 5) Uptake, Absorption, Transport and Distribution, Metabolism and Elimination in Plants, Animals and Humans, 6) Effects (beneficial and/or adverse) on Plants, Animals and Humans, 7) Hazard Evaluation and Limiting Concentrations, 8) Complete References using the Harvard (Name and Date) System. The reference citation system, regrettably not continuously found in the 1st edition, has been, as far as possible, followed in this edition for the benefit of our readers.

The third volume contains Part IV with some important metalloids and nonmetals from Boron to Tellurium and all Halogens; Part V deals with three chapters: Standards and Regulations Regarding Metals and Their Compounds, Analytical Chemistry of Element Determination (Non Nuclear and Nuclear) and Analytical Chemistry of Speciation (Principles, Main methods). The latter is a new contribution by an expert in this area in order to emphasize the increasing importance of speciation for clarification of many elemental actions. Part VI contains additional information in a Glossary (Acronyms, Abbreviations, Symbols and definitions), some general tabulated information, and an Index.

The editors wish to thank all contributing authors for their careful work and for compliance with the general editorial concepts, and publisher and their colleagues in the editorial advisory board for their always quick and very helpful discussions and expert advice. We thank in particular Prof. Dr. Marika Geldmacher von Mallinckrodt, Prof. Dr. Robert F.M. Herber and Dr. Mathias Seifert for very effective support in the final phase of urgent manuscript control during proof reading, and Dipl.-Ing. Karl-Heinz Schaller, for his steady advice when detailed information on actual national and international limit values was required. Our thanks go also to Dr. Steffen Pauly at WILEY-VCH for his encouragement in the planning and realizing this book and his continuous involvement in all editorial matter concerning the publisher's part. We are highly indebted to Dr. Waltraud Wüst at WILEY-VCH for her invaluable help in all technical and organizational matter concerning manuscripts, corrections and daily advice. Without her active support, hard work, friendly attitude, and always good ideas, the often critical deadlines would never have been so successfully reached. The support of Institutional libraries and staff therein for assistance with literature verification and acquisition is gratefully acknowledged. Finally we are indebted to our families for their support and understanding during this time-consuming undertaking.

Jena, Germany
Summerland, British Columbia, Canada
Jülich, Germany

Manfred Anke
Milan Ihnat
Markus Stoeppler

2003 December

Contents

Overview

Volume 2: Metals and Their Compounds

List of Contributors

Anderson, Carolyn J.
Mallinckrodt Institute of Radiology,
Washington University School of Medicine
510 S. Kingshighway Blvd. Campus Box
St. Louis, MO 63110
USA

Angelow, Ljubomir
Boul. Shipchenski prochod 240 D,
Ap. 01111 Sofia
Bulgaria

Anger, Jean Pierre
Laboratoire de Toxicologie, U F R Pharmacie
2, avenue du Professeur Léon Bernard
35043 Rennes Cedex
France

Anke, Manfred
Am Steiger 12
07743 Jena
Germany

Arnhold, Winfried
Ernst-Mey Strasse 14
04229 Leipzig
Germany

Begerow, Jutta
Hygiene-Institut des Ruhrgebiets,
Institut für Umwelthygiene und
Umweltmedizin
Rotthauser Str. 19
45879 Gelsenkirchen
Germany

Bertram, Cornelia
Institut für Pharmakologie und Toxikologie,
Universität Witten/Herdecke
Stockumer Str. 10
58453 Witten
Germany

Bertram, Hans Peter
Lehrstuhl für Pharmakologie und Toxiko-
logie der Universität Witten-Herdecke
Stockumer Str. 10
58453 Witten
Germany

Burkart, Werner
Head, Dept. of Nuclear Sciences and
International Atomic Energy Agency
P.O. Box 100
01400 Vienna
Austria

Coenen, Manfred
Lothringer Str. 18I
30559 Hannover
Germany

Crößman, Gerd
Im Flothfeld 96
48329 Havixbeck
Germany

D'Haese, Patrick C.
University of Antwerp,
Dept. of Nephrology-Hypertension p/a
University Hospital Antwerp
Wilrijkstraat 10
2650 Edegem/Antwerp
Belgium

De Broe, Marc E.
University of Antwerp,
Dept. of Nephrology-Hypertension
p/a University Hospital Antwerp
Wilrijkstraat 10
2650 Edegem/Antwerp
Belgium

Divine, Kevin K.
Lovelace Respiratory Research Inst.
P.O.Box 5890
Albuquerque, NM 87112
USA

Dobrowolski, Jan W.
Dep. of Environmental Mangement
and Environmental Protection
University of Mining and Metallurgy
al. Mickiewicza 30
30-059 Kraków
Poland

Doherty, Patrick
Clinical Engineering,
University of Liverpool
First Floor Duncan Building
Daulby Street
Liverpool, L69 3GA
UK

Drasch, Gustav
Institut für Rechtsmedizin
Frauenlobstr. 7a
80337 München
Germany

Eder, Klaus
Institut für Ernährungswissenschaften
Emil-Abderhalden-Straße 26
06108 Halle (Saale)
Germany

Elsenhans, Bernd
Walther-Straub-Institut für Pharmakologie
und Toxikologie
Nussbaumstraße 26
80336 München
Germany

Ermakov, Vadim V.
Russian Academy of Sciences, V. L.
Vernadsky Institute of Geochemistry and
Analytical Chemistry
Kosygina str. 19
117975 Moscow
Russia

Ewers, Ulrich
Hygiene-Institut des Ruhrgebiets, Abt.
Umweltmedizin und Umwelttoxikologie
Rotthauser Straße 19
45879 Gelsenkirchen
Germany

Fiedler, H. J.
Donndorfstr. 18
1217 Dresden
Germany

Finck, Markus
Kommission Reinhaltung der Luft im VDI
und DIN Normenausschuss KRdL
Postfach 101139
40002 Düsseldorf
Germany

Förstner, Ulrich
TU Hamburg-Harburg,
Arbeitsbereich Umweltschutztechnik
Eißendorfer Str. 40
21071 Hamburg
Germany

Fomin, Annette
Int. Hochschulinstitut Zittau,
Lehrstuhl Umweltverfahrenstechnik
Markt 23
02763 Zittau
Germany

Fraenzle, Stefan
Int. Hochschulinstitut Zittau,
Lehrstuhl Umweltverfahrenstechnik
Markt 23
02763 Zittau
Germany

Garban, Zeno
Calea Lipovci Nr. 315, Ap. 6
01900 Timisoara
Romania

Gebhart, Erich
Institut für Humangenetik
Schwabachanlage 10
91056 Erlangen
Germany

Geldmacher von Mallinckrodt, Marika
Schlehenstraße 29
91056 Erlangen
Germany

Gerhardsson, Lars
Dept. of Occupational and Environment,
University Hospital
22185 Lund
Sweden

Gerth Joachim
TU Hamburg-Harburg,
Arbeitsbereich Umweltschutztechnik
Eißendorfer Str. 40
21071 Hamburg
Germany

Glei, Michael
Department of Nutritional Toxicology,
Institute for Nutrition, FSU Jena
Dornburger Str. 25
07743 Jena
Germany

Goering, Peter L.
Center f. Devices & Radiolog. Health
Food and Drug Admin. (HFZ-112)
12709 Twinbrook Parkway
Rockville, MD 20852
USA

Hecht, Hermann
Herlas 5
95326 Kulmbach
Germany

Herber, Robert F. M
Tollenslaan 16
03723 DH Bilthoven
The Netherlands

Hildebrand, Hartmut F.
Directeur de Recherches INSERM,
Faculté de Médecine
1, Place de Verdun
05904 Lille Cedex
France

Hoppstock, Klaus
Forschungszentrum Jülich GmbH,
Technologie Transfer-Büro (TTB)
52425 Jülich
Germany

Horvat, Milena
Joef Stefan Institute, Department of
Environmental Sciences
Jamova 39
01001 Ljubljana
Slovenia

Ihnat, Milan
Pacific Agri-Food Research Centre
Agriculture and Agri-Food Canada
Summerland, BC V0H IZ0
Canada

Ishida, Koji
Laboratory of Chemistry,
Nippon Medical School,
2-297-2 Kosugi, Nakahara-ku
Japan

Jaritz, Michael
Anni Siemsenstrasse 25
07745 Jena (Winzerla)
Germany

Kabata-Pendias, Alina
UING, ul. Csartoryskick 8
24-100 Pulawy
Poland

Kobayashi, Ryusuke
Japan Industrial and Health Association
5-35-2 Shiba, Minato-ku
Tokyo 108-0014
Japan

Kowalenko, C. Grant
Pacific Agri-Food Research, Centre Agassiz,
Agriculture and Agri-Food Canada
PO Box 1000
Agassiz, BC V0M 1A0
Canada

Kraus, Thomas
Institut für Arbeitsmedizin
Pauwelsstraße 30
52074 Aachen
Germany

Lamberts, Ludwig V.
University of Antwerp,
Dept. of Nephrology-Hypertension p/a
University Hospital Antwerp
Wilrijkstraat 10
2650 Edegem/Antwerp
Belgium

Limbeck, Andreas
Institut für chemische Technologien
und Analytik,
Technische Universität Wien
Getreidemarkt 9/ 164AC
1060 Vienna
Austria

Madden, Emily F.
CDRH/Food and Drug Adminsitration,
HFZ-112
12709 Twinbrook Parkway
Rockville, MD 20852
USA

Markert, Bernd
Int. Hochschulinstitut Zittau,
Lehrstuhl Umweltverfahrenstechnik
Markt 23
02763 Zittau
Germany

Melo, Dunstana
IRD - Instituto de Radioproteção e Dosimetria
Av. Salvador Allende sin Recreio dos
Bandeirantes
Rio de Janeiro, RJ 22780-160
Brazil

Michalke, Bernhard
GSF-Forschungszentrum für Umwelt
und Gesundheit,
Institut für Ökologische Chemie
Ingolstädter Landstr. 1
85764 Neuherberg
Germany

Momcilovic, Berislav
JMJ, POB 291
Ksaverska C.2
10001, Zagreb
Croatia

Neagoe, Aurora D.
Biologische Pharmazeutische Fakultät
Friedrich-Schiller-Univ. Jena
Am Herrenberg 11
07745 Jena
Germany

Nielson, Forrest H.
US Department of Agriculture,
Agricultural Research Service
2420 2nd Avenue North
Grand Forks, ND 58202-9034
USA

Nies, Dietrich H.
Martin-Luther-Universität Halle-Wittenberg
Institut für Mikrobiologie
06099 Halle (Saale)
Germany

Pavelka, Stanislav
Institute of Physiology,
Czech Academy of Sciences
Videnska 1083
142 20 Prague 4
Czech Republic

Peganova, Svetlana
Institut für Ernährungswissenschaften
Emil-Abderhalden-Straße 26
06108 Halle (Saale)
Germany

Peterlik, Meinrad
Institut für Pathophysiologie, Abt. f.
molekulare u. biochemische Pathologie
Währinger Gürtel 18–20
01090 Vienna
Austria

Puxbaum, Hans
Institute for Analytical Chemistry,
Vienna University of Technology
Getreidemarkt 9/151
1060 Vienna
Austria

Rasmussen, Pat E.
Health Canada,
Environmental Health Sciences Bureau
Tunny's Pasture, Addr. Locator 0803C
Ottawa, Ontario K1A 0L2
Canada

Raspor, Biserka
Ruder Boskovic Institute
P.O. Box 1016
41001 Zagreb
Croatia

Ribas-Ozonas, Bartolome
Departamento de Toxicologia,
Instituto de Salud
Carlos III Carretera Majadahonda
28220, Majadahonda (Madrid)
Spain

Rish, Markus A.
Kommuna Str., House Nr. 44-1, Apt. 310
195030 Saint Petersburg
Russia

Rossman, Milton D.
University of Pennsylvania Medical Center
421 Curie Boulevard, 851 BRB II/III
Philadelphia, PA 19104-6160
USA

Sadurski, Wieslaw
Agricultural Institue, IUNG
ul. Czartoryskcih 8
24-100 Pulawy
Poland

Salomons, Wim
Kromme Elleboog 21
09751 RB Haren
The Netherlands

Sapek, Andrzej
Institute for Land Reclamation and
Grassland
Soil and Water Chemistry Dept
Farming at Falenty
05-090 Raszyn
Poland

Sapek, Barbara
Institute for Land Reclamation and
Grassland
Soil and Water Chemistry Dept
Farming at Falenty
05-090 Raszyn
Poland

Schäfer, Ulrich
Gästehaus der Universität
Am Herrenberge 11-523
07745 Jena
Germany

Schaller, Karl-Heinz
Institut und Poliklinik für Arbeits- Sozial-
und Umweltmedizin der Universität
Erlangen-Nürnberg
Schillerstr. 25 und 29
91054 Erlangen
Germany

Schilling, Günther
Julius-Kühn-Straße 31
06108 Halle (Saale)
Germany

Schrauzer, Gerhard N
175 Alameda Blvd.
Coronado, CA 92118
USA

Schümann, Klaus
Walther-Straub-Institut für Pharmakologie
und Toxikologie
Nussbaumstraße 26
80336 München
Germany

Seifert, Mathias
Hauptverband der gewerbl. Berufsgenos-
senschaft BGAG
Königsbrücker Landstraße 2-4
01109 Dresden
Germany

Stoecker, Barbara
Oklahoma State University
Dep. Nutritional Sciences, 301 HES
Stillwater, OK 74078
USA

Stoeppler, Markus
Mariengartenstrasse 1a,
52428 Jülich
Germany

Sunderman, F. William, Jr
270 Barnes Road
Whiting, VT 05778-4411
USA

Sures, Bernd
Zoological Institute I,
Dept. of Ecology and Parasitology,
University of Karlsruhe
Kornblumenstraße 13
76128 Karlsruhe
Germany

Szilagi, Míhaly
Research Institute of Animal Breeding and
Nutrition
P.O.Box
02053 Herceghalom
Hungary

Vormann, Jürgen
Institut für Prävention und Ernährung
Osterfeldstr. 83
85737 Ismaning
Germany

Wedepohl, Karl Hans
Geochemisches Institut der
Universität Göttingen
Goldschmidtstraße 1
37077 Göttingen
Germany

Yokel, Robert A.
College of Pharmacy, Pharmacy Building,
Medical Center
Rose Street
Lexington, KY 40536-0082
USA

Ziegler, Thomas L.
U.S. Geological Survey,
Denver Federal Center
P.O.Box 25046, Mail Stop 964
Denver, CO 80225
USA

Editorial Board

Part I
Element Distribution in the Environment

Elements and their Compounds in the Environment. 2nd Edition.
Edited by E. Merian, M. Anke, M. Ihnat, M. Stoeppler
Copyright © 2004 WILEY-VCH Verlag GmbH & Co. KGaA, Weinheim
ISBN: 3-527-30459-2

1

The Composition of Earth's Upper Crust, Natural Cycles of Elements, Natural Resources

Karl Hans Wedepohl

1.1

Formation of the Earth's Crust

The present natural abundance of the chemical elements at, and close to, the Earth's surface is a function of a sequence of processes:
1. Syntheses of the nuclei of elements in stars, condensation of primitive compounds from solar nebula, aggregation of particles of primitive compounds to form planets.
2. Separation of the Earth's crust and atmosphere from the Earth's mantle during the geologic history.
3. Transformation of the Earth's crust through reactions between rocks, waters, and atmosphere under internal (radiogenic) and external (solar) influence of heat.

The Earth and planetary system were formed 4.6 billion years ago. Certain meteorites as fragments from small planets have preserved a primitive cosmic composition and contain records of the early history of the solar system. Because of the lack of an atmosphere, the lunar surface has not been reworked and still exhibits the craters from the impact of large planetesimals which were abundant in space at the stage of planet formation. The oldest rocks on earth have an age of 3.5 to 4 billion years.

Their masses represent the nuclei of the continents which grew from magmatic melts during the geologic history. These magmas originate from more than 50 km depth. The crust has undergone transformation due to material exchange with the Earth's mantle, weathering, mass transport and increase of temperature and pressure from its growing thickness and from heat of deeper layers.

The crust is the skin of our planet. In the continents it has an average thickness of 40 km, and underneath the oceans of 7 km. Its mass of 2×10^{19} t contributes only 0.4% of that of the total Earth. The crust covers the Earth's mantle, which represents 68% of the Earth's mass. The mantle consists of magnesium-iron silicates and oxides, and reflects the large cosmic abundance of O, Si, Mg, and Fe. The most abundant isotopes of these elements have even numbers of protons and neutrons (Oddo-Harkin rule), indicating their considerable stability in the stellar synthesis of the nuclei. Fractionation of the elements between the Earth's mantle and crust is controlled by their behavior during partial melting of the mantle. The magmas – which are partial melting products of the mantle – preferentially transport volatile and so-called incompatible elements into the crust

Elements and their Compounds in the Environment. 2nd Edition.
Edited by E. Merian, M. Anke, M. Ihnat, M. Stoeppler
Copyright © 2004 WILEY-VCH Verlag GmbH & Co. KGaA, Weinheim
ISBN: 3-527-30459-2

which do not fit well in the crystal structures of the mantle minerals (because of size or valency). This process of element selection is comparable to zone melting. The flux of magmas occurs only partly in a subaerial volcanism. Submarine basaltic magmatism is quantitatively far more important. It is mainly restricted to large meridional ridges in which the ocean crust grows at a rate of about 3×10^{10} t per year through magmatism. At this rate, the ocean crust would double its mass within $< 10^9$ years if the growth were not compensated by consumption. At the continental margins around the Pacific Ocean, the ocean crust is subducted under the continental crust and finally digested by the Earth's mantle. The growing crust of the Atlantic and Indian Oceans pushes the bordering continents in opposite directions. The engine of this dynamic behavior of fragments (plates) of the Earth's crust is the upwelling of radiogenic heat in the Earth's mantle. Large-scale convective transport of heat and matter has lateral branches on which the crustal plates can float. Under the high pressure of the mantle, solid matter has a ductile behavior which allows a slow plastic flow in a convective system. The continental and ocean crustal plates with their upper mantle base of more than 50 km thickness move at velocities of a few centimeters per year. The continental crust is growing in a narrow magmatic belt close to the deep trenches caused by the subduction of ocean crust. Dehydration of the subducted slab of ocean crust triggers the magmatism from which the continents grow. This source is responsible for the average chemical composition of the continental crust which is close to that of the magmatic rock tonalite (named after a locality in the southern Alps).

Beside the important process in which the ocean and continental crusts grow from basaltic and andesite melts originating in the upper mantle, granitic magmas can be formed by partial melting within the continental crust. This causes a major remobilization and vertical transport of matter.

1.2
Alteration of the Earth's Crust

The focal areas of this book have relatively minor relationships to the primary formation of the Earth's crust which has caused a certain distribution of the chemical elements. They mainly deal with products of the alteration of the crust in geologic processes. We can presently still observe the weathering of solid rocks, the erosion of mountain ridges, and the transport of eroded materials as suspended and dissolved constituents in river and rain water, in ice and wind. In-situ weathering forms soils, and soils are the basis of food production for human nutrition. Therefore, soils need special protection against the impact of toxic substances (see Part I, Chapters 4 and 5).

In connection with magma production from the Earth's mantle and its eruption in volcanic processes, the outer shell of the Earth permanently delivers steam and also the more aggressive gases CO_2, SO_2, H_2S, HCl, HF, etc. to the atmosphere and the surface waters. During the Earth's history, oceans, lakes, ice caps, rivers, groundwater, and interstitial waters of sediments have grown to a reservoir of almost 2×10^{18} t water. The existence of this water reservoir at and close to the surface of the planet causes a difference between the Earth and its planetary neighbors Venus and Mars. After weathering and transport, the reactive constituents of the magmatic degassing become finally fixed in the large masses of hydroxide-bearing sediments, in limestones, gypsum and salt deposits. The mentioned chemical sediments ($CaCO_3$, $CaSO_4 \cdot 2H_2O$, NaCl, etc.) have a proportion of 10 to 20% in the sediment shell of the Earth. The major proportion of the sediments consists of the weathering products of preexisting rocks. This mass contains detrital and newly formed minerals after transport and grain size separation in suspension and sedimentation. After diagenetic reconstitution and consolidation, they form clays (mudstones, shales) and sandstones. Greywackes are special sandstones formed from detrital materials which have undergone only minor chemical alteration and separation. Gravity is the major engine for the transport of weathering products at the Earth's surface. Crustal masses which were folded and fractured into mountains and lifted above sea level by continent moving forces are the object of erosion. After decomposition and transport, their matter will be collected in sea basins, mainly at continental margins. Rates of erosion in mountain ridges and the related rates of deposition in near-shore basins scale in the range of fractions of millimeters to centimeters per year. Shales and greywackes which represent important rock masses decomposed without major chemical fractionation approach the average chemical composition of the upper continental Earth's crust (Table 1.1).

Continuous sedimentation forms a layer of deposition which comes towards its base under increasing pressure and temperature with increasing load. The average gradient of rising pressure and temperature with depth is 0.3 kb km^{-1} and 20 °C km^{-1}, respectively. Only minor chemical reactions occur in the majority of sediments up to temperatures of 200 °C. In the range of 300 to 800 °C, former sediments approach new equilibria by various reactions; this process is called regional metamorphism. Water in the porous volume of rocks activates these material reactions. If the metamorphosed sediments which still contain 1–2% water attain temperatures of more than 700 °C, they form granitic partial melts. These magmas rise diapirically if their proportion in the rock exceeds 15%. They transport the low-melting fraction from the lower into the upper crust. The lower crust rarely attains temperatures of 650–700 °C, except that heat is advected from the upper mantle by convection or intrusion of large masses of mafic magmas.

Ore deposits are rare crustal units. They have accumulated certain elements of economic importance. They are formed from magmatic melts, hot brines or from sea water. The hot brines of geothermally heated sea water or formation water abundantly extract metals from the subsurface country rocks in their conduits and precipitate metal sulfides and other compounds, if they mix with surface waters. The total masses of metals in ore deposits are small compared with their large but highly diluted reservoir in common rock species (cf. Table 1.1). The continental crust contains 1.6×10^{15} t Zn, 3.8×10^{14} t Cu, and 3.1×10^{14} t Pb dispersed in its normal rocks. The estimated reserves of the same metals accumulated in ore deposits are: 1.2×10^8 t Zn, 3.4×10^8 t Cu, and 9.3×10^7 t Pb. Even if the hypothetical and speculated resources of Zn, Cu, and Pb exceed 10- to 100-fold the known reserves, the total amount of these metals in ore deposits is small relative to their mass dispersed in normal rocks. The known reserves depend on the state of ore exploration which, because of its costs, is not very far ahead of the rate of near-future mining. High concentrations of toxic metals in near-surface ore deposits can cause environmental problems because of the solubility of their oxidation products and their potential transport in weathering solutions. Dumps of metal mines and

Tab. 1.1: Concentrations of 25 elements in the continental and oceanic Earth's crust and in abundant rock species. (Values in mg kg^{-1})

	Be	Mg	Ti	V	Cr	Mn	Fe	Co	Ni	Cu	Zn	As	Se	Zr
Shales	3	16000	4600	130	90	850	48000	19	68	45	95	10	0.5	160
Greywackes	3	13000	3800	98	88	750	38000	15	24	24	76	8	0.1	300
Limestones	[0.5]^a	26000	400	20	11	700	15000	2	15	4	23	2.5	0.19	19
Granitic rocks	5.5	6000	3000	94	12	325	20000	4	7	13	50	1.5	0.04	145
Gneisses, mica schists	3.8	13000	3870	60	76	600	33000	13	26	23	65	4.3	0.08	168
Basaltic and gabbroic rocks	0.6	37000	9700	251	168	1390	86000	48	134	90	100	1.5	0.09	137
Granulites	2.1	31000	5000	149	228	930	57000	38	99	37	79	1.3	0.17	165
Continental crust	2.4	22000	4010	98	126	716	43000	24	56	25	65	3.1	0.12	203
Oceanic crust (Ocean ridge basalt)	1	45000	9700	252	317	1200	70000	45	144	81	78	[1.5]^a	0.17	104

	Mo	Ag	Cd	Sn	Te	Pt	Au	Hg	Tl	Pb	Bi
Shales	1.3	0.07	0.13	2.5	?	[0.01]^a	0.0025	0.45	0.68	22	0.13
Greywackes	0.7	0.1	0.09	[3]	?	[0.4]^a	0.003	0.11	0.20	14	0.07
Limestones	0.4	0.0x^b	0.16	[0.x]^b	?	[0.001]^a	0.002	0.03	0.05	5	0.02
Granitic rocks	1.8	0.12	0.09	3.5	0.01	0.005	0.0024	0.03	1.1	32	0.19
Gneisses, mica schists	[1.5]^a	0.08	0.10	2.5	[0.02]^a	[0.01]	0.003	0.02	0.65	16	0.10
Basaltic and gabbroic rocks	1	0.11	0.10	1.5	0.008	0.03	0.004	0.02	0.08	3.5	0.04
Granulites	0.6	0.08	0.10	2.5	[0.02]^a	[0.01]^a	0.0015	[0.02]^a	0.28	12.5	0.04
Continental crust	1.1	0.07	0.10	2.3	[0.005]^a	0.4	0.0025	0.04	0.49	15	0.08
Oceanic crust (Ocean ridge basalt)	[0.8]	0.03	0.13	1.4	[0.01]^a	[0.03]^a	0.002	[0.02]^a	0.013	0.50	0.006

^{a} Estimated concentration. ^{b} x = order of magnitude.

mills can contaminate ground and surface waters so that reservoirs for drinking water have to avoid such mining areas.

The elements beryllium, magnesium, titanium, vanadium, chromium, manganese, iron, cobalt, nickel, zinc, zirconium, molybdenum, silver, cadmium, tin, thallium, lead, and bismuth mainly occur in the crystal structures of rock-forming silicates and oxides of the common rocks in the Earth's crust. These elements – with the exception of magnesium, iron, titanium, chromium, and zirconium – are trace elements in the minerals. They follow certain rules as reported by Goldschmidt (1954) in their ten-dency to enter specific crystal structures. Beside their occurrence in silicates, copper, arsenic and selenium have an affinity to sulfides and might be mobilized under weathering conditions. Platinum and gold can occur as native metals in common rocks. Magnesium, iron, titanium, chromium, and zirconium also form their own minerals in abundant rock species. Black shales – which are often characterized by higher concentrations of iron, vanadium, nickel, zinc, cobalt, molybdenum, silver, cadmium, thallium, lead and bismuth than in normal shales – contain several of the listed elements in their sulfide or carbonaceous fraction. These bituminous shales as potential source rocks of crude oils will probably become objects of future mining which might cause environmental problems.

1.3
Average Abundance of 25 Elements in Sedimentary, Magmatic and Metamorphic Rock Species

Data on average concentrations of 25 elements (covered by this book) in the most important sedimentary, magmatic and metamorphic rocks have been compiled in Table 1.1, mainly from sources and reports listed by Wedepohl (1969–1978, 1968, 1981), Heinrichs et al. (1980), Hofmann (1988), Wedepohl (1995), and Gao et al. (1998). These data can be used to estimate the mean abundances of the considered elements in the continental Earth's crust which consists of 7% sedimentary rocks, 27% granitic rocks, 13% gneisses and mica schists, 6% amphibolites and gabbros, and 47% granulites. Granites, gabbros and basalt are formed by consolidation of magmatic melts. Gneisses, mica schists, amphibolites, and granulites are metamorphic products of former sedimentary and magmatic rocks which had to adjust to different temperature and pressure conditions. Granulites are the most common rock species of the lower continental Earth's crust, whereas the remainder of the listed rock types represent the upper crust. Sediments cover large areas of the metamorphic and magmatic crustal rocks. If spread equally over the whole Earth, they would have a thickness of about 1 km. Shales (and partly greywackes) contain the weathering products of large crustal units. Therefore, typical concentrations of our 25 elements in shales are almost equal to the abundances of the respective elements in the average continen-

tal crust. Major deviations from this balance are restricted to the volatile elements As, Se and Hg which are especially accumulated in certain clay sediments. The fact that the average concentrations in abundant crustal rocks on one side and their major weathering products on the other side are almost equal for more than 20 of the selected elements confirms the representativeness of the compiled data.

Table 1.1 also contains data for average concentrations of the 25 selected elements in the oceanic crust. As a first approximation, we have assumed for this part of the compilation that the oceanic crust mainly consists of the so-called ocean ridge basalt (MORB). Large volumes of this basaltic ocean crust have undergone hydrothermal alteration connected with a gain of H_2O, CO_2, Na, Mg and S from heated sea water and losses of Si, Ca, Fe, Mn, etc. from the altered basalt to the ocean water reservoir.

Data from Table 1.1 can be easily used for estimates of the natural background in processes where elements from natural and anthropogenic sources are mixed. Many soils which produce a large proportion of our food, have developed on shales. Therefore, they contain a natural background in the selected 25 elements which is comparable to the values for shales listed in Table 1.1. Lantzy and Mackenzie (1979) have confirmed the resemblance of many soils with shales. Natural dust, the suspended proportion in river water and the detrital silicate fraction in coal is often comparable to shales (and greywackes) in chemical composition. Certain plants as mosses accumulate volatile elements and involatile metals, the former from the atmosphere and the latter from soils. For chemical data on mosses from Norway, see Reimann and De Caritat (1998). The ratios of the metal concentrations in mosses normalized to those in shales range between 1.4 and 0.3

for the elements Se, Ag, Cd, Pb, Zn, Mn, and Hg. In addition to the silicate fraction, coal has accumulated several elements through the physiological action of the pre-existing plants (Mg, Zn, Mo, etc.) and through diagenetic precipitation of sulfides (Fe, Ag, Cu, Cd, etc.). Average concentrations of the 25 elements in coal are listed in Table 1.3. Brick is mainly produced from clays which are comparable to shales in chemical composition (see also Section 1.5). Readily volatile elements as Hg, Bi, Cd, Tl, Pb, Zn, As, and Se are partly lost during firing in brickworks (Brumsack 1977).

Cement is produced from a mixture of limestone and shale or clay. Therefore, the composition of its starting material can be estimated from data listed in Table 1.1.

1.4

Concentration and Transport of 25 Elements in Natural Waters

The most important agents of transport at the Earth's surface are waters and wind. Beside matter in suspension, rivers and rain carry large amounts of dissolved compounds from continental rock weathering to the oceans. Rain also moves salt spray in the opposite direction, from the ocean to the land. Each year, rivers carry about 3.6×10^{13} t water, 8.9×10^8 t suspended matter, and 3.6×10^9 t of dissolved compounds to the oceans. Wind-transported rain compensates for its more than ten times smaller concentration of dissolved constituents by its higher speed of travel. The amount of Mn, Co, Cr, Ni, Ag, and V extracted yearly from the atmosphere by rain exceeds the rate of river transport, which ranges from 2 to 10 (Lantzy and Mackenzie 1979). For the elements As, Cd, Cu, Mo, and Zn, the respective factor

ranges from 20 to 100, and in the case of the more volatile elements Pb, Hg, and Se it ranges from 110 to 790.

Regional geochemical surveys of metal concentrations dissolved and suspended in river waters or occurring in abundant bedrocks will be parts of a future *Geochemical Atlas of the World* (IGCP Project 259). The part on West Germany contains maps about the areal distribution of seven metals in stream waters and of 14 metals in stream sediments (Fauth et al. 1985). These maps inform for instance about an accumulation of Zn, Cd and Pb in the former mining areas of the Rheno-Hercynian sedimentary belt, about relatively high concentrations of Ni, Co, Cr, and V in areas of predominant basaltic rocks and of higher than normal concentrations of Co and Ni in stream waters of low pH in the peat bogs of North Germany. Koljonen (1992) reports in the *Geochemical Atlas of Finland* of the abundance and areal distribution of 40 chemical elements in tillites of Finland. These rocks are produced through the abrasion of surface rocks by glaciers. Tillites contain the matter of crustal rocks related to their abundance in the area sampled by the glaciers.

The dissolved concentrations of the 25 selected elements in ocean deep water are controlled by natural processes. This is not principally the case for river water and rain. The data on river water listed in Table 1.2 (according to Turekian 1969; Wedepohl 1969–1978; and Martin and Meybeck 1979) are mainly from rivers without major contamination from industrialized areas. Suspended clay materials in the rivers have a high capacity to adsorb organic residues and metals from anthropogenic and natural sources (sewage, industrial immissions, soil extraction by acid rain water, etc.), and in this way they keep the level of dissolved metals reasonably low.

Tab. 1.2: Concentrations of 25 elements in seawater[a] (deep water; values in µg kg⁻¹) and river water (values in µg kg⁻¹)

	Be	Mg	Ti	V	Cr	Mn	Fe	Co	Ni	Cu	Zn	As	Se	Zr	Mo
River water	0.1	4100	3	0.9	0.5–1	4	40	0.2	0.3–0.9	2	7–10	1.7	0.2	1–2.5	1
Sea water	0.0002[b]	1.3×10⁶	1	1.9	0.2	0.01[c]	0.1	0.002[d]	0.6	0.25	0.6	1.6	0.09	0.03	10

	Ag	Cd	Sn	Te	Pt	Au	Hg	Tl	Pb	Bi
River water	0.03–0.3	<0.01–0.1[f]	0.006[h]	?	?	0.002	0.07	0.04	0.02–0.4[f]	0.01
Sea water	0.002[e]	<0.01–0.1[g]	0.0006[h]	<0.0001[i]	?	0.01	<0.0002[j]	0.01	0.003	0.000003[k]

a For several elements, there exists a large difference between surface and deep water. Only deep water concentrations are listed. Elements with biologically caused depletion in surface water are e.g., Ni, Cu, Zn, and Cd; b Measures and Edmond (1982); c Landing and Bruland (1980); d Knauer et al. (1982); e Martin et al. (1983); f Valenta et al. (1986), Golimowski et al. (1990), Dorten et al. (1991), Queirolo et al. (2000); g Mart and Nürnberg (1986); h Byrd and Andreae (1982); i Lee and Edmond (1985); j Gill and Bruland (1987); k Lee et al. (1985/86)

Tab. 1.3: Concentrations of 25 elements in brown coal, hard coal and crude oil (in mg kg⁻¹ for solids and mg L⁻¹ for oils). In the averages reported in this table, equal statistical weight is given to mean values from different authors

	Be	Mg	Ti	V	Cr	Mn	Fe	Co	Ni	Cu	Zn	As
Brown coal	0.7[a,b]	3000[a]	140[a,b]	9.5[a,b]	9[a,b]	92[a,b]	5400[a]	2.6[a,b]	5[a,b]	2.5[a,b]	11[a]	2.2[a]
Hard coal	2.6[a,b]	1300[a]	465[a,b]	32[a,b]	13[a,b]	156[a,b]	9000[a]	8[a,b]	22[a,b]	16[a,b]	48[a,b]	21[a,b]
Crude oil	0.0004[c]	0.1[c]	0.1[c]	39[c,d,e,f]	0.12[c,d,e,f]	0.5[c,d,e]	6.5[c,d,f]	0.5[c,d,e,f]	11[c,d,f]	0.7[c,e]	8.0[c,d,e,f]	0.13[c,d,e]

	Se	Zr	Mo	Ag	Cd	Sn	Te	Pt	Au	Hg	Tl	Pb	Bi
Brown coal	0.47[a]	10[b]	2.8[a]	0.01[a]	0.07[a]	1.0[a,b]	?	?	0.0x	0.26[a]	0.03[a]	9[a,b]	0.02[a]
Hard coal	2[a]	28[b]	4[b]	0.44[a]	1.8[a]	2.6[a,b]	?	?	0.0x	0.36[a]	0.62[a]	46[a,b]	0.15[a]
Crude oil	0.20[c,d,e,f]	?	10[c]	0.05[c,f]	0.1[c]	0.01[c]	?	?	0.0009[d,e]	3.4[c,d,e,f]	?	0.3[c]	?

a Brumsack et al. (1984) reported averages of dried brown coal with 18% ash and hard coal with 8.7 and 13.9% ash mainly fired in West-German power plants. b Yudovich et al. (1972) compiled data on large numbers of samples from industrial countries in Europe, USA and USSR. c Bertine and Goldberg (1971) compiled data on crude oil from worldwide sampling. d Hitchon et al (1975) reported on crude oil from Alberta (Canada). e Shah et al. (1970) reported on crude oils from Libya, California, Louisiana, and Wyoming. f Ellrich et al. (1985) reported on crude oils from South Germany.

The proportion of anthropogenic and natural emissions of 18 of the 25 selected elements in the worldwide atmospheric transport has been estimated by Lantzy and Mackenzie (1979).

Sampling and analysis of trace elements in sea water have been improved tremendously within the past decades. Investigations conducted since the mid-1970 s have demonstrated that the majority of the older data were too high due to contamination and unreliable procedures and had to be discarded. Values in Table 1.2 on Mg, Ti, V, Cr, Se, Zr, Mo, and Au are from Turekian (1969). Newer data on Be, Mn, Fe, Co, Ni, Cu, Zn, As, Ag, Cd, Sn, Hg, Tl, Pb, and Bi have been reported by Measures and Edmond (1982), Bruland (1983), Wong et al. (1983), Li (1991) and by the authors listed in the footnotes of Table 1.2. With the exception of Mg, V, and Mo, concentrations in deep ocean water are lower than in river water due to the consumption by organisms, the precipitation of authigenic minerals (e.g. MnO_2) etc. in the sea. Mg, V, As, and Mo occur in almost equal concentrations in deep and surface sea water. Several elements such as Ni, Zn, Cd, and Ba, are highly depleted in the surface layers of the oceans because of their consumption by the organisms. The factor of depletion relative to deep water ranges from 500 (Cd) to 5 (Ni, Ba). Vertical distribution of Mn, Co, Sn, and Pb are characterized by surface maxima. The higher lead concentration was caused by anthropogenic contamination, for example from fuel additives. An increasing extraction of Mn and Co from soils by acid rain water could have caused an additional transport of these elements from the continents to the oceans. The elements Mn, Co, Sn, and Pb have very low absolute concentrations in sea water, which allows the observation of surface contamination.

Assuming a steady-state system in the continental run-off and the deposition of minerals in the ocean, the residence time of the elements in sea water can be estimated from the yearly rate of transport in the rivers and their average concentration in ocean water. For elements such as Zn, Cu, Ni, and Ti, the mean residence time in sea water before incorporation in solid phases is of the order of 10^3 years.

1.5
Average Abundance of 25 Metals in Natural Raw Materials

The combustion of fossil fuels for the production of energy introduces numerous metals into the atmosphere and subsequently into soils, rivers, and oceans. Coal contains the degraded matter of fossil plants. Crude oil is a thermal product of the kerogen and lipid fraction in residues from microorganisms preserved in sediments. The latter is usually perfectly separated from the silicates of the host rocks whereas coal still contains a minor fraction of the interlayered sediments.

Average trace metal concentration in coal and crude oil are listed in Table 1.3. The ranges of variation of these metals in oil are appreciably larger than in coal. Co and Hg scatter over four orders of magnitude.

A large proportion of the Mg, Ti, Cr, Mn, Fe, and Zr concentration in the hard coal listed in Table 1.3 belongs to the detrital sediment material which is contained in coal. This can be concluded from the respective concentrations of the listed elements in shales and greywackes (cf. Table 1.1). Sulfur in coal is produced by bacterial sulfate reduction and diagenetic precipitation as iron sulfide close to the carbonaceous material. Several metals which form sulfides, selenides, and arsenides of very low solubility (e.g., Cu, Cd, Ag) are accumulated from the waters in which the plant material was deposited. Some metals (e.g., Mg, Ni, Cu, Zn) had a physiological function in the pre-existing plants and were, therefore, concentrated by the living material.

The only elements which are specifically high in crude oil relative to coal are V, Mo, and Hg. The former two elements are known for their special accumulation in black shales, which are black as a result of the high proportion of organic residues. The living organisms of the primary production in the surface layers of the oceans do not contain extraordinary concentrations of Mo and V. Therefore, a diagenetic origin of the accumulation of these elements by a scavenging complex formation has to be assumed (see Part II, Chapters 18 and 27). The vanadium and nickel concentrations in crude oils vary with their origin (Tissot and Welte 1984), and therefore it is possible to estimate their origin from V/Ni analysis. For instance, crude oils from Venezuela, Angola, Columbia, Ecuador, and California are rich in vanadium, and those from Indonesia, Libya, and Western Africa contain very little, whereas crude oils from Angola, Columbia, Ecuador, and California are rich in nickel and those from Libya and Tunisia contain little nickel (Tissot and Welte 1984). Arabian and Canadian crude oils contain medium amounts of $10-50 \, \mu g \, kg^{-1}$ V and $3-20 \, \mu g \, kg^{-1}$ Ni. Mercury, as the most volatile metal, can be easily transported in the thermal gradient of the upper Earth's crust and can be trapped by the organic residues on this way.

Of the elements listed in Table 1.3, Zn, As, Se, Ag, Cd, Tl, Pb, and Bi are accumulated to high levels in the finest fraction of particulate aerosols which leave the stacks of coal firing power plants, whereas Hg is the only metal which is predominantly emitted as a gas during coal firing (Brumsack et al.

1984). According to balance computations on the behavior of numerous metals during the combustion of hard coal in West-German power plants, Brumsack et al. (1984) concluded that the fly-ash from the stacks contains about 1000 to $1400 \, mg \, kg^{-1}$ Ni, 3000 to $6000 \, mg \, kg^{-1}$ Zn, 800 to $2000 \, mg \, kg^{-1}$ As, 200 to $300 \, mg \, kg^{-1}$ Se, 20 to $50 \, mg \, kg^{-1}$ Ag, 100 to $200 \, mg \, kg^{-1}$ Cd, 5 to $10 \, mg \, kg^{-1}$ Hg, 10 to $60 \, mg \, kg^{-1}$ Tl, 3000 to $9000 \, mg \, kg^{-1}$ Pb, and 10 to $20 \, mg \, kg^{-1}$ Bi. The concentration of As, Se, Ag, Pb, and Bi in the finest particulate fraction exceed those in natural dust (cf. shale in Table 1.1) by factors of several hundreds. The level of the very toxic cadmium in the finest fly-ash is more than thousand times higher than the cadmium abundance in natural dust. The volatility of the reported elements is caused by the relatively high concentration of chlorine in the firing process (0.6 to 1.2% Cl in stack fly-ash). Large amounts of volatile elements in particulate aerosols are transported by wind over large distances. Their impact on soils and river water can be estimated from the yearly combustion of coal and oil in the world. The consumption of these raw materials for energy is about 3 to 4×10^9 t each. The yearly firing of coal probably causes the worldwide emission of 2.4×10^4 t Pb, 1.6×10^4 t Zn, 6×10^3 t As, 10^3 t Se, 8×10^2 t Hg, 5×10^2 t Cd, and 1.5×10^2 t Tl, if the data computed by Brumsack et al. (1984) on emissions in West Germany are extrapolated. If the yearly run-off from the continents to the oceans of 3.6×10^{13} t H_2O could extract from soils the total amount of the yearly immissions from coal firing, the following concentrations should be observed: $0.7 \, \mu g \, kg^{-1}$ Pb; $0.4 \, \mu g \, kg^{-1}$ Zn; $0.17 \, \mu g \, kg^{-1}$ As; $0.03 \, \mu g \, kg^{-1}$ Se; $0.02 \, \mu g \, kg^{-1}$ Hg; $0.015 \, \mu g \, kg^{-1}$ Cd and $0.004 \, \mu g \, kg^{-1}$ Tl. The estimated concentrations from coal emissions attain the natural soluble concentrations in rivers (cf. Table 1.2) only in the case of Pb and Cd. Admittedly, the total extraction of the soil immissions from coal firing is an extreme assumption. For data and information on the impact of metals from industrial sources on soils, see Schulte and Blum and other authors in Matschullat et al. (1997).

In Section 1.3, it was mentioned that brick is produced mainly from clays and shales, for which average concentrations of the 25 metals are summarized in Table 1.1. The heating of clay materials up to about 1000 °C causes a partial volatilization of Zn, Pb, Cd, Tl, and Bi which ranges from about 40 to 80% of the original metal concentration (Brumsack 1977). Because of the primitive technology of this process, emission is usually not reduced by the installation of filters. Cement is also produced from natural raw materials. The common starting mixture contains three parts limestone and one part clay or shale, with average compositions as listed in Table 1.1. Trace metal concentrations are usually low in limestone relative to shale. The high temperatures required for partial melting of the material are in the range of 1400 to 1500 °C. Electrostatic filters reduce the emission of volatilized elements. Because of the accumulation of toxic metals such as Tl in the filtered particulates, their recycling must be avoided.

1.6
Natural Resources

Human beings have made systematic and organized use of natural raw materials since the Neolithic Revolution. However, with the onset of the industrialization, the amount of matter mined and extracted from its natural occurrence has increased to a size which is comparable with the masses transported in natural cycles. The volume of the yearly water consumption for irrigation and technical purposes has passed the level of 10% of the river discharges to oceans. The food consumption has increased to a size of one permill ($=1‰$) of the total organic production.

Although the mining of metal minerals began a little over 4000 years ago, the size of the operations was small until fossil energy became available for an exponential growth of mineral mining and processing during the 19th century. For the formation of an exploitable ore deposit metals have to be accumulated locally to a high degree relative to the average crustal abundance of the elements. Factors of accumulation range from hundred to several thousands, with the exception of iron and mercury. As a major constituent of the Earth's crust, iron only requires a 10- to 15-fold higher concentration to form an ore deposit, and therefore small iron ore deposits are abundant. Mercury as a very rare element needs abnormally high degrees of accumulation. However, even processes which accumulate ore metals by factors of hundreds or thousands relative to normal crustal abundances are statistically rare because they need a combination of several not very abundant conditions. In many cases, hot water is the transporting agent of the metals. Because of the rarity of favorable conditions, the geographic distribution of mineral resources on a worldwide basis is very uneven, and no technically advanced country is currently self-sufficient. More than 80% of the world production of Cr and Pt is supplied by one country, and a large industrial producer like the United States has to import more than 70% of its demands of Cr, Pt, Ta, Al, Mn, Sn, and Ni metal or ore.

The exploitation of an ore deposit must be preceded by prospection and exploration. Prospection is guided by experience on ore

Tab. 1.4: Annual consumption, reserves and potential resources of 25 metals in tons on a worldwide basis. The majority of data on yearly, consumption according to Gocht (1974), von Baratta (1999) and former volumes of Fischer Weltalmanach reporting on Mg, V, Cr, Mn, Fe, Ni, Cu, Zn, Mo, Ag, Cd, Sn, Pt, Au, Hg, Pb. Information on reserves from Global 2000 (1980), Gocht (1974), and von Baratta (1999). Data on potential resources according to Global 2000 (1980)

	Be	Mg	Ti	V	Cr	Mn	Fe	Co	Ni
Annual consumption	$> 1.3 \times 10^4$	3.3×10^5	3.8×10^6	2.9×10^4	1.1×10^7	2.2×10^7	1×10^9	2.3×10^4	7.8×10^5
Reserves in ore deposits		large	1.6×10^8	2.2×10^7	7.8×10^8	1.8×10^9	9.3×10^{10}	5.1×10^6	5.4×10^7
Potential resources					6.0×10^9	1.1×10^9	1.4×10^{11}		1.0×10^8

	Cu	Zn	As	Se	Zr	Mo	Ag	Cd	Sn
Annual consumption	9.7×10^6	6.9×10^6	4.0×10^4	1.6×10^3	5.2×10^4	6.2×10^4	1.3×10^4	2.0×10^4	1.9×10^5
Reserves in ore deposits	6.0×10^8	1.6×10^8	1.3×10^7	1.1×10^5	3.7×10^7	8.8×10^6	2.0×10^5	2.2×10^5	1.0×10^7
Potential resources	1.8×10^9	4.0×10^9					2.0×10^6		2.7×10^7

	Te	Pt	Au	Hg	Tl	Pb	Bi
Annual consumption	2.0×10^2	1.0×10^2	2.2×10^3	5.8×10^3	35	2.7×10^6	4.8×10^3
Reserves in ore deposits	2.2×10^4	6.5×10^3	5.5×10^4	2.5×10^5	1.4×10^3	1.9×10^8	1.0×10^5
Potential resources		1.6×10^4		4.0×10^5		1.3×10^9	

genesis and on relations between specific ores and certain geologic structures. Exploration has to investigate the size and exploitability of ore bodies. Both are related to the economic situation on the world market or in a specific country. The easily exploitable ore deposits of the world are almost exhausted; therefore, a steady increase of the expenses of prospection, exploration, and exploitation of metal ore deposits is to be expected. The grade of the mined ores has steadily decreased, and this decrease requires an increase in the energy used for mining and mineral processing. The price of energy certainly controls the minimum grade of an ore to be exploitable. Because of the rising expenses for prospection and exploration mining companies restrict the size of reserves, and consequently knowledge of the actual world resources of several metals is not very good. The required degree of metal accumulation in ore deposits high above the level of the crustal abundance allows one to foresee that the world resources of some metals will be exhausted within the next few generations.

Information on consumption, reserves and potential resources for the 25 metals selected for this chapter is compiled in Table 1.4 on a worldwide basis. An identified resource is called a reserve. Potential resources require new mining technology or processing methods. Iron is by far the most important metal for the human civilization, and therefore has the highest consumption (6.3×10^8 t per annum). This is followed by manganese (2.2×10^7 t per annum). The production of Sn, Mg, Ni, Cr, Ti, Pb, Zn, and Cu is in the range of 10^5 to 10^7 t per annum, followed by the group of Ag, Be, Cd, Co, V, As, Zr, and Mo (10^4 to 10^5 t per annum). The latter two classes contain the metals Ni, Cr, Co, V, and Mo, which are mined principally to be added to iron to give it more desirable properties of strength and resistance against corrosion. Manganese is another essential additive for steels. A large proportion of the Sn, Zn, and Cd production is used as a protective coating for iron. Seven of the 25 metals are produced in quantities of 35 to 5800 t per annum. If the current production is maintained instead of being increased, the lifetime of the reserves listed in Table 1.4 will range from the order of 10 to 800 years. The reserves of Cd, Ag, Bi, Au, Zn, Pb, Ti, Hg, and Sn ores will allow mining operations for a future in the range of tens of years.

Three-quarters of the metal minerals are processed and mostly consumed in the relatively small highly industrialized countries which contain one-quarter of the world's population. The aerial concentration of processing and consumption causes environmental problems, and the risk of contaminating soils, rivers and air with toxic trace elements is high in industrialized countries. Beside the firing of coal and oil, processing of ores and the technical use of several metals is a major source of such contamination.

1.7
Concluding Remarks

Human beings depend on soils uncontaminated by toxic elements to produce their food. They need clean drinking water – which is often derived from river water – and they need clean air to breathe. The anthropogenic contamination of soils, waters, and air can only be discovered if the natural background in the abundance of trace metals is known. Therefore, knowledge of the natural cycles of these elements and of the size of the natural reservoirs is required for a better understanding of numerous environmental problems.

References

BERTINE KK and GOLDBERG ED (1971) *Fossil fuel combustion in the Major Sedimentary Cycle*. Science **173**:233–235.

BRULAND KW (1983) *Trace elements in sea water*. Chem Oceanogr **8**:157–220.

BRUMSACK HJ (1977) *Potential metal pollution in grass and soil samples around brickworks*. Environ Geol **2**:33–41.

BRUMSACK HJ, HEINRICHS H and LANGE H (1984) *West German coal power plants as sources of potentially toxic emissions*. Environ Technol Lett **5**:7–22.

BYRD JT and ANDREAE MO (1982) *Tin and methyltin species in seawater: concentrations and fluxes*. Science **218**:565–569.

DORTEN WS, ELBAZ-POULICHET F, MART LR and MARTIN J-M (1991) *Reassessment of the river input of trace metals into the Mediterranean Sea*. Ambio **10**:1–5.

ELLRICH J, HIRNER A and STÄRK H (1985) *Distribution of trace elements in crude oils from Southern Germany*. Chem Geol **48**:313–323.

FAUTH H, HINDEL R, SIEWERS U and ZINNER J (1985) *Geochemischer Atlas Bundesrepublik Deutschland*. Schweizerbartsche Verlagsbuchhandlung Stuttgart.

GAO S, LUO T-C, ZHANG B-H, ZHANG H-F, HAN Y-W, ZHAO Z-D and HU Y-K (1998) *Chemical composition of the continental crust as revealed by studies in East China*. Geochim Cosmochim Acta **62**:1959–1975.

GLOBAL 2000 (1980) BARNEY GO, ed. *Report to the president*. US Council on Environmental Quality and US Foreign Department. US Government Printing Office, Washington, DC.

GOCHT W (1974) *Handbuch der Metallmärkte*. Springer, Berlin-Heidelberg-New York.

GOLDSCHMIDT VM (1954) *Geochemistry*. Clarendon Press, Oxford.

GOLIMOWSKI J, MERKS AGA and VALENTA P (1990) *Trends in heavy metal levels in the dissolved and particulate phase in the Dutch Rhine-Meuse (Maas) delta*. Sci Total Environ **92**:113–127.

GILL GA and BRULAND KW (1987) *Mercury in the Northeast Pacific*. Eos Trans AGU **68**:1763.

HAEFS H (1986) *Der Fischer-Weltalmanach 1987*. Fischer, Frankfurt am Main.

HEINRICHS H, SCHULZ-DOBRICK B and WEDEPOHL KH (1980) *Terrestrial geochemistry of Cd, Bi, Tl, Pb, Zn and Rb*. Geochim Cosmochim Acta **44**:1519–1533.

HITCHON B, FILBY RH and SHAH KR (1975) In: Yen TF, ed. The role of trace elements in petroleum. Ann Arbor Sci Publ, Ann Arbor, Michigan.

HOFMANN AW (1988) *Chemical differentiation of the Earth: relationship between mantle, continental crust and oceanic crust*. Earth and Planetary Sci Lett **90**:297–314.

KNAUER GA, MARTIN JH and GORDON RM (1982) *Cobalt in North-east Pacific waters*. Nature **297**:49–51.

KOLJONEN T (1992) *The Geochemical Atlas of Finland*. Part 2 Till. Geological Survey of Finland, Espoo.

LANDING WM and BRULAND KW (1980) *Manganese in the North Pacific*. Earth Planet Sci Lett **49**:45–56.

LANTZY RJ and MACKENZIE FT (1979) *Atmospheric trace metals: global cycles and assessment of man's impact*. Geochim Cosmochim Acta **43**:511–525.

LEE DS, EDMOND JM and BRULAND KW (1985/86) *Bismuth in the Atlantic and North Pacific: a natural analogue to plutonium and lead?* Earth Planet Sci Lett **76**:254–262.

LEE DS and EDMOND JM (1985) *Tellurium species in seawater*. Nature **313**:782–785.

LI YH (1991) *Distribution patterns of the elements in the ocean: A synthesis*. Geochim Cosmochim Acta **55**:3223–3240.

MART L and NÜRNBERG HW (1986) *The distribution of cadmium in the sea*. In: Mislin H, Ravera O, eds., Cadmium in the Environment, pp. 28–40. Birkhäuser, Basel-Boston-Stuttgart.

MARTIN JH and MEYBECK M (1979) *Elemental mass-balance of material carried by major world rivers*. Mar Chem **7**:173–206.

MARTIN JR, KNAUER GA and GORDON RM (1983) *Silver distribution and fluxes in North-east Pacific waters*. Nature **305**:306–309.

MATSCHULLAT J, TOBSCHALL HJ and VOIGT HJ (1997) *Geochemie und Umwelt*. Springer, Berlin Heidelberg New York.

MEASURES CL and EDMOND JM (1982) *Beryllium in the water column of the Central North Pacific*. Nature **297**:51–53.

QUEIROLO F, STEGEN S, MONDACA J, CORTÉS R, ROJAS R, CONTRERAS C, MUNOZ L, SCHWUGER MJ and OSTAPCZUK P (2000) *Total arsenic, lead, cadmium, copper and zinc in some salt rivers in the northern Andes of Antofagasta, Chile*. Sci Total Environ **255**:85–95.

REIMANN C and DE CARITAT P (1998) *Chemical elements in the environment*. Springer, Berlin-Heidelberg-New York.

SHAH KR, FILBY RH and HALLER WA (1970) *Determination of trace elements in petroleum by*

atmospheric emission of a metal compound is calculated from emission factors available for the different emitting processes. While most of the national emission inventories are focused on SO_2, NO_x, and total particulate matter, data on the emission of metal compounds are relatively sparse. Pacyna (1986a) has reviewed the available trace element emission factors for natural and anthropogenic sources.

On the global scale, compilations have been made for metal emissions by Nriagu (1979), Lantzy and Mackenzie (1979), and Weisel (1981). The divergence of the data reflects the uncertainties in estimating global emissions from very sparse data sets on natural and anthropogenic sources. Pacyna (1986b) and Salomon (1986) presented the first comparison of estimated global anthropogenic emissions of trace metals with emissions from natural sources. A compilation of recently reported data for metal emissions (Pacyna and Pacyna 2001) is shown in Table 2.1. These data indicate a clear dominance of anthropogenic emissions for the most important trace elements such as Cd, Ni, Pb, V and Zn on the global scale.

The long-range transport of Saharan and Asian dust has been identified as the dominant source of mineralic particles over the Atlantic, the Arctic, and the Pacific (SCOPE 1979; Rahn et al. 1979; Duce et al. 1980; Uematsu et al. 1983; Parrington and Zoller 1984). Other important sources of naturally emitted metal compounds are volcanoes (Zoller 1983), forest fires which may be of natural occurrence as well as originate from anthropogenic activities and exudations from vegetation (Pacyna 1986a, b).

Sea spray and gaseous emissions from the oceans contribute only a minor fraction of trace metal compounds to the atmosphere on a global basis (Pacyna 1986b). Volcanoes are thought to be the main source of naturally emitted As and Cd and are important sources of Pb, Se, Zn, and Hg (Table 2.2). Forest fires are likely important emitters of Hg, whereas plant exudations contribute markedly to the flux of naturally emitted As, Zn, and Cd compounds.

Globally anthropogenic emissions of metals (Table 2.3) already exceed the emissions of several trace elements from natural sources. On a regional scale in densely

Tab. 2.1: A comparison of estimated global anthropogenic emissions of trace metals in the mid1990s with emissions from natural sources (Pacyna and Pacyna 2001) (emissions in 10^3 metric tons year^{-1}).

Trace metal	Anthropogenic emissions	Natural emissions: median values	Anthropogenic/national emission ratios
As	5.0	12.0	0.42
Cd	3.0	1.3	2.3
Cr	14.7	44.0	0.33
Cu	25.9	28.0	0.93
Hg	2.2	2.5	0.88
Mn	11.0	317.0	0.03
Mo	2.6	3.0	0.87
Ni	95.3	30.0	3.2
Pb	119.3	12.0	9.9
Sb	1.6	2.4	0.67
Se	4.6	9.3	0.49
V	240.0	28.0	8.6
Zn	57.0	45.0	1.3

Tab. 2.2: Worldwide annual emissions (in 10^6 kg) of trace elements from natural sources. (After a compilation by Pacyna 1986b.)

Source	As	Cd	Co	Cu	Cr	Mn	Ni	Pb	Se	V	Zn	Hg
Atmospheric dust	0.24	0.25	4	12	50	425	20	10	0.3	50	25	0.03
Volcanoes	7	0.5	1.4	4	3.9	82	3.8	6.4	0.1	6.9	10	0.03
Forest fires	0.16	0.01	–	0.3	–	–	0.6	0.5	–	–	0.5	0.1
Vegetation	0.26	0.2	–	2.5	–	5	1.6	1.6	–	0.2	10	–
Sea spray	0.14	0.002	–	0.1	–	4	0.04	0.1	–	9	0.02	0.003

populated areas, anthropogenic emissions of metals are by far the dominant contributors as compared to natural sources. This is reflected by the fact that ambient concentrations of trace elements in source regions are some orders of magnitude higher than in remote regions. Table 2.3 shows that anthropogenic emissions of Cr, Hg, Mn, Mo, Ni, Se, Sn, Tl, and V are mainly derived from stationary fossil fuel combustion, whereas As, Cd, Cu, and Zn are emitted mainly from nonferrous smelters. Cr and Mn are released in large amounts from the iron and steel industries.

Emission data for trace metals on a national scale have been compiled for European countries by Pacyna (1984) for 1979 and have been revised for several elements later (Pacyna 1987). The major emission areas in Europe are: (1) the former Soviet

Tab. 2.3: Worldwide emissions (metric tons year^{-1}) of trace metals from major anthropogenic source categories to the atmosphere during the mid-1990s (Pacyna and Pacyna 2001)

Source category	As	Cd	Cr	Cu	Hg	In
Stationary fossil fuel combustion	809	691	10145	7081	1475	
Vehicular traffic						
Nonferrous metal production	3457	2171	–	18071	164	45
Iron and steel production	353	64	2825	142	29	–
Cement production	268	17	1335	–	133	–
Waste disposal	124	40	425	621	109	–
Other					325[a]	
Total	5011	2983	14730	25915	2235	45
1983 emission[b]	18820	7570	30480	35370	3560	25

Mn	Mo	Ni	Pb	Sb	Se	Sn	Tl	V	Zn
9417	2642	86110	11690	730	4101	3517	1824	240084	9417
			88739						
59	–	8878	14815	552	466	319	–	77	40872
1060	–	36	2926	7	7	–	–	71	2118
-	–	134	268	–	3	–	–	–	2670
511	–	129	821	272	24	115	–	23	1933
11047	2642	95287	119259	1561	4601	3951	1824	240255	57010
38270	3270	55650	332350	3510	3510	3790	5140	86000	131880

[a] Emission of Hg from gold production. [b] Nriagu and Pacyna (1988).

Union; (2) Poland and the Czech Republic; and (3) the Benelux countries and the Western part of the Federal Republic of Germany. For Zn and V, significant emission sources are also located in the UK, Spain, and Italy. While Zn and V emissions are distributed relatively evenly in Europe, emissions of As, Cd, Cu, Cr, Mn, and Be are rather concentrated in Central and Eastern Europe (Pacyna 1986b). A detailed trace metal emission inventory is also available for the Los Angeles area (Cass and McRae 1983, 1986). Regional and local emission inventories are generally used to assess the impact of sources to air quality, pollutant deposition (Vanderborght et al. 1983), or human exposure (Bennett 1981) and to derive control strategies. The European emission inventory has been used to model long-range transport of trace metals to the Arctic (Pacyna et al. 1985).

In future, the focus for environmental studies will be on traffic emissions, as this air pollution source is the only one whose activity is expected to increase in future. Air pollution through other sources like heavy industry and power plants is decreasing due to better technology in combustion processes, increasing efficiency of air pollution control devices and the fact that heavy industry itself is stagnating in developed countries. Therefore, in urban areas with heavy traffic intensity and limited industrial activities such as mining, smelting, petroleum industry and municipal waste incineration, automotive emissions appear to be a source for certain trace metals. The combustion process, brake disk and lining wear and rubber tire wear are potential sources for metals such as copper, zinc, nickel, barium, and lead (Westerlund et al., 2001). Some studies have reported that antimony concentrations were elevated close to major roads, both in airborne particulate matter and in soils (Dietl et al. 1997; Cal-Pietro

et al. 2001). The road traffic emissions of Sb have been explained by the use of certain organic Sb compounds in greases and motor oils (Huang et al., 1994) and by the use of Sb_2S_3 in brake linings (Garg et al. 2000). Tires, and also motor oils, are commonly considered as important sources of Zn from road traffic (Sörme et al., 2001). Mechanical abrasion of the car body can also emit particles containing zinc, nickel, and other alloy components of steel. Another potential source for metals is the resuspension of soil and road dust. Aluminum, Ca, Mg, Si, and Ti are all typical geological marker elements, suggesting that resuspension controls their abundance in atmospheric aerosols (Sternbeck et al. 2002). With the introduction of the three-way catalyst in the mid-1980s, new elements were added to traffic emissions: the platinum group elements Pt, Pd, and Rh (Hodge and Stallord, 1986; Wei and Morrison 1994). These elements are used in the catalyst as active compounds to facilitate the oxidation of hydrocarbons and other incompletely oxidized components, which leads to a reduction of 90% in the amount of hydrocarbons and nitrogen oxides in the exhaust gas. Oxidation catalysts have been developed even for diesel engines, and these are now ready to be integrated into newly manufactured automobiles. Previous investigations have shown that platinum group metals are emitted by the catalyst in the $ng\,km^{-1}$ range (Moldovan et al. 1999) and that this is the result of mechanical abrasion of platinum group elements containing washcoat particles (Palacios et al. 2000).

2.3
Atmospheric Occurrence

A large amount of data are available concerning atmospheric concentrations of trace elements associated with particulate matter. Comprehensive surveys have been compiled by Schroeder et al. (1987) and by Wiersma and Davidson (1986). Atmospheric concentration ranges for remote, rural, and urban locations are given for 14 elements in Table 2.4. The large differences for the remote sites reflect the different regions such as maritime and continental, northern and southern hemispheres.

Lowest concentrations of trace elements have been found in the Antarctic (Cunningham and Zoller 1981), and very low levels in the maritime atmosphere over the Pacific ocean (Gordon et al. 1978; Duce et al. 1983; Parrington and Zoller 1984). A series of elements (V, Cr, Mn, Cu, Zn, Co, Ag, Cd, Ba, Pb, Bi, U) have been measured by Planchon et al. (2002) in snow samples collected at remote, low accumulation sites in Coats Land, Antarctica. Heavy metal concentrations were found to be extremely low, down to 3 pg kg^{-1}, confirming the high purity of Antarctic snow. A review of the data on heavy metals in aerosols over

the seas of the Russian Arctic is presented by Shevchenko et al. (2003).

Remote sites at some distance to the high emission regions of the East USA, Europe, and the former Soviet Union, such as Central USA (Moyers et al. 1977) or the Arctic (Heidam 1981; Barrie et al. 1981; Heintzenberg et al. 1981), receive polluted air masses in an episodic mode. At such sites trace metal concentrations may rise to levels found in rural areas in densely populated countries. Also well documented is the transport of Eurasian aerosols to the Arctic, this occurring particularly in winter (Rahn 1985). Due to meteorological effects, the Arctic atmosphere behaves as a reservoir for long-range transported pollutants during the winter months and is cleaned during spring and summer (Heidam 1986).

High concentrations of trace elements are observed in densely populated and industrial areas (Table 2.4; Schroeder et al. 1987). Fernandez et al. (2000) reported that resuspended soils, industrial activities and traffic emissions were the main sources for

Tab. 2.4: Concentration ranges of various elements associated with particulate matter in the atmosphere (ng m^{-3}). (After Schroeder et al. 1987.)

Location	As	Cd	Ni	Pb	V	Zn
Remote	0.007–1.9	0.003–1.1	0.01–30	0.007–64	0.001–14	0.03–110
Rural	1.0–28	0.4–1000	0.6–78	2–1700	2.7–97	11–403
Urban						
Canada	7.7–626	2–103	4–371	353–3416	10–130	55–1390
USA	2–2320	0.2–7000	1–328	30 –96270	0.4–1460	15–8328
Europe	5–330	0.4–260	0.3–1400	10–9000	11–73	160–8340
Other	20–85	0.6–177	2.3–158	1.3–11020	1.7–180	110-2700

Co	Cr	Cu	Fe	Mn	Se	Sb
0.001–0.9	0.005–11.2	0.029–12	0.62–4160	0.01–16.7	0.0056–0.19	0.0008–1.19
0.08–10.1	1.1–44	3–280	55–14530	3.7–99	0.01–3.0	0.6–7
1–7.9	4–26	17–500	700–5400	20–270	NA	13– 125
0.2–83	2.2–124	3–5140	130–13800	4–488	0.2–30	0.5–171
0.4–18.3	3.7–227	13–2760	294 –13000	23–850	0.01–127	2–470
0.3–10	tr – 277	2.0–6810	21–32820	1.7–590	NA	7–36

NA = not available, tr = traces

the observed high heavy metal concentrations in the area of Seville. Harrison et al. (2003) reported enriched trace metal concentrations at a roadside location in Birmingham, the calculation of enrichment factors allowed the identification of a number of elements which appear to be related with anthropogenic emissions.

The transport pathways of the polluted air masses can be reconstructed by trajectory analysis and elemental tracer techniques (Husain 1986; Rahn and Lowenthal 1985; Chen and Duce 1983; Lowenthal and Rahn 1985).

Elemental balances for atmospheric particles indicate that trace metals comprise only a small fraction of the total aerosol mass. Even in a highly industrialized city such as Linz (Austria), the relative contribution of trace metal compounds (Cd, Cr, Cu, Mn, V, Zn, and Pb compounds) to the total suspended particles (TSP) was found to be about 1%, while Fe-compounds comprised 1–8% of the TSP mass (Puxbaum et al. 1985). The major part of the TSP is formed by electrolytes (Na^+, K^+, NH_4^+, Cl^-, NO_3^-, SO_4^{2-}) (25–35%), carbonaceous material (8–11%), and mineralic components (Ca, Mg, Si, Al compounds) (16–18%). Similar results were found for aerosols collected at a background site in the South African savanna (Puxbaum et al. 2000).

2.4
Size Distributions of Atmospheric Particles and Trace Metals

Atmospheric particles occur in a wide range of sizes. The size distribution can be expressed as a number-, surface-, or mass-density function. If the mass-size distribution is presented differentially versus the logarithm of the particle size ($dM/dlg\ AD$ vs. lg AD; M, particle mass; AD, aerodynamic diameter) for a typical urban aerosol, a trimodal size distribution function (Figure 2.1) is obtained (Whitby 1978). The finest mode (*nucleation mode*) (0.005– 0.1 μm *AD*) is formed by particles from gas to particle conversion reactions as well as by particles from high-energy combustion processes. The particles found in the *accumulation mode* size range (0.1–2.5 μm *AD*) originate from coagulation and condensation processes between and on nucleation mode particles. The interrelations between nucleation and accumulation mode particles can be described by growth laws derived from particle dynamics (McMurry and Wilson 1982, 1983).

The particles in the *coarse mode* (2.5– 100 μm *AD*) are of entirely different origin, and are formed during mechanical processes such as erosion or abrasion and during combustion of ash-containing fuels. Fly ash particles are generally found in the lower size range of the coarse mode. Particle size is a governing factor for the deposition in the human respiratory system. According to the deposition model of the Task Group on Lung Dynamics (1966), particles larger than 10 μm preferentially deposit in the nasopharyngeal compartment, whereas smaller particles penetrate into the bronchial and alveolar compartments (*thoracic fraction*) according to the deposition functions given in Figure 2.2. For ambient air quality control, collection methods for particles < 10 μm *AD* have been promulgated as "PM_{10} standard" (ISO 1981; Purdue 1986). Using the deposition functions of the lung dynamics model or newer deposition data (Stahlhofen 1986) and data on the solubility of some trace metals, the bioavailability of metals via respiratory uptake can be modeled (Bennett 1981; US EPA 1982; Davidson and Osborn 1986). In workplace atmospheres – especially in the case of insoluble fibrogenous particles such as quartz and

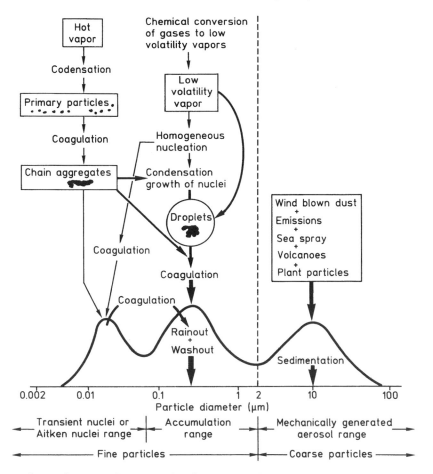

Fig. 2.1 Scheme of an atmospheric aerosol surface area distribution showing the three modes, the main source of mass for each mode, the principal processes involved in inserting mass into each mode, and the principal removal mechanisms. (From Whitby 1978.)

asbestos – the lung-penetrating (alveolar) fraction is of major interest. Currently, several standards exist to define the alveolarly "respirable" fraction. Most European countries follow the Johannesburg convention which defines an *AD* of 5 μm, though in the USA the AD_{50} for respirable particles is 3.5 μm.

Particle size is also a governing factor for the atmospheric lifetime of a particle. Various trace metals are found in different ranges of the size spectra of atmospheric

particles. During combustion or industrial processes, metal compounds may exist in the gaseous state as well as contained in various forms of fly ashes or dusts. After passing control devices such as fabric filters, electrostatic precipitators or wet scrubbers, certain fractions of the compounds will be retained, whereas a remaining part will be released into the atmosphere. Electrostatic precipitators for dust removal do not remove gaseous emissions and have a decreasing collection efficiency for smaller

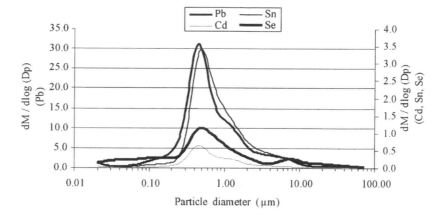

Type(ii)

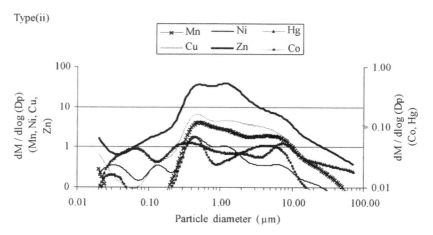

Type(iii)

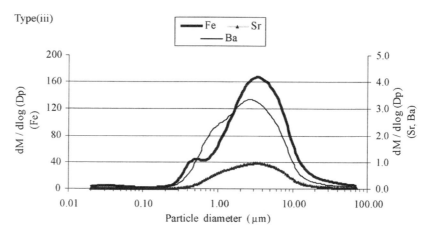

Fig. 2.3   Size distributions of selected trace metals in aerosol samples from central England. (From Allen et al. 2001.)

ways have been shown to contain various forms of Pb and Zn compounds and different minerals (Biggins and Harrison 1979a,b; O'Connor and Jaklevic 1981; Davis 1981; Bloch et al. 1980; reviewed by Harrison 1986). Some work has been performed on speciation of particulate emissions from automobiles, combustion sources, and various industrial processes (see also Section 2.8). Recently, it has been found that a large fraction of the mercury emission from waste incinerators is present in a water-soluble form. It could be shown by mass spectroscopy that $HgCl_2$ and HgO are the major constituents of the emissions (V & F 1988). Halides of Pb, Zn, and Cd could also be identified in the gaseous state (Figure 2.4). For coal fly ashes a differentiation of matrix- versus surface-enriched elements has been performed via leaching techniques (e.g., Hansen and Fisher 1980) or surface analytical methods (e.g. Linton et al. 1976, 1977). In comparative studies of compound forms of elements in oil and coal fly ashes (Henry and Knapp 1980), various forms of V, Fe, Ni, Al, Si, Ca, and Mg have been identified.

Among environmental micropollutants, organometallic compounds are of particular interest because of their toxicity and their increased use. The presence of such compounds in the environment at ultra-trace levels has led to the development of speciation techniques and to the optimization of analytical instrumentation during the past 20 years (Baena et al. 1999; Reuther et al. 1999; Gui-Bin et al. 2000; Cao et al. 2001). The widespread commercial use of organometallic compounds has increased their release and occurrence in the environment, but anthropogenic emissions alone cannot explain their ubiquity, for example, of organotin and organolead compounds in marine and fresh waters, in sediment and biota (Hamasaki et al. 1995). Although controversial, it is assumed that many of these compounds are formed by and interact with natural methylation processes. There is also much controversy about the question whether methylation is chemically or biologically mediated. Therefore, several studies were performed to investigate the occurrence and formation of organometallic compounds in the environment. Jay et al. (2000) reported first results about the chemical speciation and lipid solubility of mercury in the presence of sulfide, and of polysulfides to increase the present knowledge about the methylation of mercury in aquatic systems. A detailed review about the chemistry of

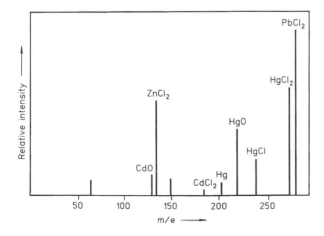

Fig. 2.4 Mass spectrum of gaseous emissions of a municipal incinerator (Vienna). (From V & F 1988)

Rea et al. (2001) showed that dry deposition had the most important influence on Hg, Al, La, Ce, V, As, Cu, Zn, Cd, and Pb fluxes, while foliar leaching strongly influenced Mg, Mn, Rb, Sr, and Ba fluxes in net throughfall.

A strong north to south gradient of Pb, As, and Sb in top soils with concentration differences of a factor of 10 in Norway is indicative for the long-range transport of trace metals from Central Europe to the northern terrestrial ecosystems (Rambaek and Steinnes, 1980). This conclusion has been supported by similar findings in a study of trace element profiles in ombrotrophic bogs from different parts of Norway (Hvatum et al. 1983) as well as from trace element distribution patterns in snow deposits in the Arctic region (Ross and Granat 1986).

Atmospheric deposition is also a major source of metal input into many aquatic ecosystems (Salomons 1986). Helmers and Schrems (1995) reported for the tropical North Atlantic Ocean that wet trace element deposition dominates over dry input. From the increased enrichment factors relative to the Earth's crust, the determined trace metal concentrations were assumed to originate from anthropogenic sources. For atmospheric wet depositional fluxes of selected trace elements at two mid-Atlantic sites, Kim et al. (2000) reported that at least half of the Cr and Mn and more than 90% of the Cd, Zn, Pb, and Ni are from non-crustal (presumably anthropogenic) sources.

For lakes in industrialized areas (e.g., Lake Michigan), the atmospheric load is especially important for lead (60%) and zinc (33%), while for Co, Cd, and Mn the atmospheric flux has been estimated to be 11–13% of the total input. For Al, Fe, and Co the atmosphere is a minor source (Eisenreich 1980). For acid-sensitive lakes, metal concentrations tend to increase with decreasing pH (Dickson 1980; Borg 1983). This effect can be explained by a higher tendency of the metals to remain in solution at lower pH levels and by a possible solubilization of metals from the sediment.

In remote softwater lakes, sediment profiles may be used to evaluate enrichment trends due to anthropogenic activities. Trace metal profiles in sediments of 10 lakes in Ontario (Algonquin Provincial Park) indicate a 2-fold enrichment of Ni, Cu, Zn, and Cd and a 25-fold increase of Pb during the past 100 years (Wong et al. 1984).

Enrichment of lead is even found in sediments from the Atlantic and Pacific oceans (Schaule and Patterson, 1981, 1983). Twenty years ago, fluxes of Pb appeared to be around an order of magnitude higher (68 ng cm^{-2} per year) in the Pacific as compared to pre-industrial levels (from 1 to 7.5 ng cm^{-2}). The present-day fluxes of Pb into the Atlantic are estimated to range from 170 to 330 ng cm^{-2} per year. Migon et al. (1997) presented an assessment of atmospheric inputs to the Ligurian Seas including both wet and dry deposition. For Pb, a wet flux of $1.6 \text{ } \mu\text{g m}^{-2}$ per day and a dry deposition flux of $7.0 \text{ } \mu\text{g m}^{-2}$ per day was determined.

In coastal seawater such as the Western Mediterranean basin, soil-derived particles originated from arid areas (in this case the Sahara). The atmospheric flux of anthropogenic trace metals, however, was dominated by aerosols from industrialized regions of Western Europe. Volcanic activity (Mount Etna) contributes selenium. The atmospheric input of Cr, Hg, Pb, and Zn into the Western Mediterranean basin is of the same order of magnitude as the riverine and coastal inputs of these components (Arnold et al. 1983). For the southern bight of the North Sea, estimates even indicate a predominance of the atmospheric input of

Cu, Zn, Ag, and Pb (which occurs mainly via wet deposition) as compared to the input by the Scheldt river (DeDeurwaerder et al. 1985). First detailed data on aerosol concentrations of trace metals at the Mediterranean coast of Israel are presented by Herut et al. (2001). The extent of the anthropogenic contribution was estimated by the degree of enrichment of these elements compared to the average crustal composition. High values indicating a strong influence from anthropogenic emissions were calculated for Cd, Pb and Zn (median > 100), minor values for Cu, and relatively low values for Fe, Mn and Cr (< 10).

2.7
Historical Trends

The longest historical record about the air chemistry of the atmosphere is found in the 3000 m-deep ice layers in Greenland and the Antarctic. The longest core drilled in the Antarctic ice contains the accumulated material from the past 160,000 years (Legrand et al. 1988). It is highly interesting to assess whether the human emissions of metals have changed the atmospheric aerosol composition on a global or hemispheric scale. Available data compiled by Boutron (1986) indicate that while Greenland is clearly affected by some anthropogenic metals, the Antarctic shows no significant increase in the metal content of the recent snow layers as compared to some hundred-years-old ice layers.

According to data from Murozumi et al. (1969) and Ng and Patterson (1981), the lead concentrations in Greenland ice have increased about 200-fold from prehistoric times to the late 1970s (Figure 2.6). From the beginning of this century up to the late 1970s, the increase has been about 4-fold. A 3- to 4-fold increase during the last cen-

tury has also been derived for zinc (Herron et al. 1977). The concentrations of trace metals in the Arctic ice layers are, however, very low (for Pb and Zn in the range of 1 to $400 \, \text{ng} \, \text{kg}^{-1}$ and 20 to 300 ng kg^{-1}, respectively), so that highly specialized procedures for sample handling and analysis are required. However, for the past three decades a clear decrease in the Pb, Cd, Zn and Cu concentrations has been observed in Greenland snow and ice, documenting the considerable improvements in the control of industrial emissions and the strong decline in the use of Pb alkyl additives (Boutron et al. 1995). From the early 1970s to the present day, a 7.5-fold decrease was observed for Pb, whereas a mean reduction by a factor of 2 was determined for Cd, Zn and Cu.

In recent years, a new class of metals has also been subject to global emission. With the introduction of automobile catalytic converters, noble metals – which originate from the abrasion and deterioration of the surfaces of the catalysts – have been introduced into the environment (Wei and Morrison; 1994; Moldovan et al.; 1999) (see Part III, Chapter 20). The enrichment of Pt, Pd, and Rh has been recently demonstrated in snow from central Greenland (Barbante et al. 2001). The concentrations of Pt, Pd and Rh in snow dated from the mid-1990s are around 40 to 120 times higher than in ice samples dated from 7000 years ago.

Aged ice layers can also be found at high elevation sites at mid latitudes. Briat (1978) analyzed the trace metal trends in an ice core from the Mont Blanc massif covering the years 1948 to 1974. The author concluded that the levels of Pd, V, and Cd have increased by a factor of two during the observation period, whereas no statistically significant trends were found for Mn, Cu, and Zn. Van de Velde et al. (1999) reported an increase of Co, Cr, Mo and Sb in ice core samples collected near the summit of

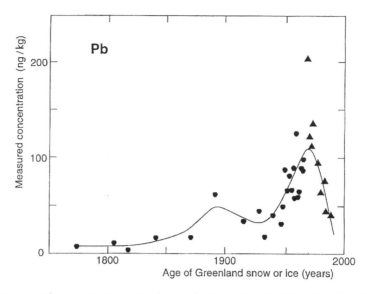

Fig. 2.6 Changes in Pb concentrations in ice deposited at Summit, central Greenland, from 1773 to 1992 compiled from Candelone et al. (1995) (solid circles) and Boutron et al. (1991) (solid triangles). The general time trend is shown with a spline-smoothed curve.

Mont Blanc, compared to pre-industrial levels by factors of 2–16 (Mo×16, Sb×5, Co and Cr×2–3). For the elements Pb, Cd, Cu, Zn, and Ag the concentrations in high alpine snow from Mont Blanc are roughly one order of magnitude higher than those in surface snow from central Greenland, and two orders of magnitude higher than those measured in recent snow in the high central plateau areas of the Antarctic (Batifol and Boutron 1984) (Table 2.5). Heisterkamp et al. (1999) investigated the organolead content of snow and ice samples from the Mont Blanc area. Highest concentrations were observed from 1962 until the late 1980s, with a significant decline during the 1990s.

Several other approaches concerning retrospective studies of trace metal enrichment in lake sediments, ombrotrophic peatlands, and tree-rings have been proposed during the past years. The use of lake sediments for long-term retrospective trace metal monitoring has been questioned because of the strong pH dependence of trace metal mobility in lakes (Nriagu and Wong 1986; Arafat and Nriagu 1986). Lake sediments can, however, be used to study the impact of local metal emission sources on the lake ecology (Nriagu and Rao 1987). For example, Lorey and Driscoll (1999) found a considerable increase in the mercury flux in sediment cores from eight remote lakes in the Adirondack region of New York compared to 1850. From sediment cores from the Central Park Lake in New York, Chillrud et al. (1999) derived temporal trends for Pb, Zn, and Sn resembling the history of solid-waste incineration in New York City, reaching maximum values from the 1930s to the early 1960s.

More promising for long-term retrospective deposition monitoring are ecosystems for which the major source of nutrients is atmospheric deposition. Such conditions are found in ombrotrophic bogs and peatlands (Glooschenko 1986). Bog vegetation has been shown to be a good biomonitoring

Tab. 2.5: Concentration ranges and "typical concentrations" (in parentheses) for trace elements in recent snow from Alpine, Greenland and Antarctic snow

	Pb	*Cd*	*Cu*	*Zn*	*Ag*
Alps (recent snow, > 3000 m altitude)	1–12 (2)	0.1–0.4 (0.2)	0.3–4 (1.5)	2–11 (2.5)	0.01–0.03 (0.02)
Greenland (recent snow)	0.1–0.9 (0.3)	0.001–0.030 (0.01)	0.03–0.1 (0.07)	0.1–0.4 (0.3)	0.007–0.01 (0.008)
Antarctic (recent snow)	0.001–0.005 (0.003)	0.001–0.01 (0.005)	0.015–0.06 (0.03)	– (0.05)	0.002–0.01 (0.005)

Compiled from Batifol and Boutron (1984) and Boutron (1986). Concentration ranges are ranges from averaged results obtained for different sites (in $ng\,kg^{-1}$).

substrate for surveys of regional trace metal deposition (Pakarinen 1981; Glooschenko 1986). Time-resolved studies have been performed using depth profiles in peatlands (Livett et al. 1979; Pakarinen and Tolonen 1977a, b; Shotyk et al. 1996, 2002). A detailed review is provided in Shotyk (1995).

Temporal trends of the ambient concentration and deposition flux of particulate trace metals are now available for various sites (Pirrone et al. 1995; Var et al. 2000; Kemp 2002). From their measurements at seven urban sites across the metropolitan area of Detroit in the 1982–1992 time period, Pirrone et al. (1995) reported downward trends for Fe, Pb, Cr, Cd and Be, and an upward trend for Zn, Ni and Hg. As was reported for lead concentrations in ice layers (Boutron et al. 1995) and peat bogs (Shotyk et al. 1996), a significant reduction of the Pb content in airborne particulate matter was observed in the past two decades at different urban sites in Denmark (Kemp 2002).

2.8
Atmospheric Aerosol Sampling and Analysis (for Biomonitoring, see Part I, Chapter 12)

For the analytical characterization of airborne particles, the combined use of special sampling methods and chemical and physical analytical methods is required (McCrone and Delly 1973; Malissa 1976; Heinrich 1980; Lioy et al. 1980; Liu et al. 1980).

2.8.1
Sampling of Airborne Particles with Not Classifying Methods

The sampling method most widely used for ambient particles is the "high volume sampler". This instrument has been standardized in many countries (FRG: VDI 1972; USA: US EPA 1971) and collects total suspended particles with sampling rates of 40 to 100 m^3h^{-1}. During a sampling interval of 24 h, around 0.01–0.5 g of particles are retained which are evaluated gravimetrically and can be subjected to trace metal analysis.

- Advantages of high-volume samplers are: The large filter size allows samples to be cut to several aliquots that can be used for (a) various types of analysis, (b) replicate analysis, (c) sharing the sample with other groups, and (d) storage of aliquots as "backup" samples.
- Disadvantages of high-volume samplers are: High power consumption, high noise production and in some cases contamination of the particulate samples by abrasion products of the pump have been reported (Countess 1974).

For a number of years, pre-separating inlet systems have been available for high-volume samplers which exclude the particles larger than 2.5 or 10 μm from the filter. Such systems fulfil the PM_{10} standard established by the US EPA in 1984 (McFarland et al. 1984). Pre-separators for PM 2.5 or PM 10 are also available for "mid-" and "low-volume" samplers which are small apparatuses for filter sampling (ASTM 1965; Hrudey 1977; Wedding 1982). In automated air quality measurement stations, semi-continuously registering instruments (irradiation monitors, tape samplers) are used for sampling and measuring the particle concentrations. Under certain precautions the deposits on the filter tapes can be subjected to chemical analysis of selected components.

2.8.2
Sampling with Classifying Methods

The selection of a method for size separating sampling of particulate matter is dependent on the size range and the amount of sampled particles required for the analysis. A survey of size-classifying sampling methods has been provided by IAEA (1978), Hochrainer (1978), LBL (1975), and Hidy (1986). The most widely used principle of size-classifying sampling of particles is the multistage impaction. Cascade impactors with between four and nine impaction stages in the size range of 0.02 to 20 μm *AD* and sampling rates of 0.005 to $70 \, m^3 h^{-1}$ are commercially available and are used for emission and ambient sampling (Marple and Willeke 1976; Hidy 1986; Puxbaum 1979; Lodge and Chan 1987). Very favorable calibration results in the size range of 0.06 to 10 μm *AD* have been obtained for a low-pressure impactor designed by Berner (1978) (Wang and John 1988). For routine sampling of two size frac-

tions of atmospheric particles, the "dichotomous virtual impactor" is used, which can be equipped with a sample changer for up to 36 samples (Goulding et al. 1978). Further possibilities of dichotomous sampling are offered by the "tandem filter" technique (Cahill et al. 1977) and by the use of cyclone preseparators (Lippmann 1976; John and Reischl 1980).

2.8.3
Special Sampling Techniques

For the analytical characterization of single particles, deposition of the particle on very flat surfaces is necessary (e.g., "Nucleopore" filters or organic foils) (McCrone and Delly 1973; Grasserbauer 1978a; Spurny et al. 1979; Chatfield 1984).

2.8.4
Diffusion Controlled Separation of Aerosols

Mixtures of gases and particles can be separated by diffusion denuders combined with filter samplers. Chemically reactive gases are collected on the surface of selectively coated diffusion denuder tubes, whereas particulate components are collected on subsequently placed filters (Stevens et al. 1978; Klockow 1982). Until now, this method has been used exclusively for non-metallic components. Although it can be foreseen that denuder techniques may also be used for the separation of gaseous and particulate metallic components.

2.8.5
Bulk Analysis of the Elements

For the bulk analysis of metallic elements in atmospheric particles, spectroscopic methods (e.g., AAS, ICP-OES, IPC-MS, XRF, PIXE, SSMS) are widely used (Part V, Chapter 2). Here, sample preparation is a crucial step, and the analyst must decide whether the total sample should be digested by a thermal procedure or whether digestion in acids is sufficient. In special cases, electro-chemical methods can be used for the analysis. For large-scale monitoring, nondestructive multielement techniques are preferred such as instrumental neutron activation analysis (INAA) (Ondov et al. 1979; Ragaini 1978) or energy dispersive X-ray fluorescence analysis (EDXRF) (Leyden 1978). For the analysis of very small samples, electron beam microanalysis (EPMA) (Grasserbauer 1978a, b) and PIXE (Cahill 1975) have been used.

2.8.6
Compound-Specific Analysis

Due to the complex composition of ambient particles, the compound-specific analysis of metallic components is a difficult task. As a consequence, most studies of metallic compounds have been performed in emission samples where the compounds are in an enriched state and the composition is generally less complex. Examples for compound specific analysis in source samples include lead compounds in automotive emissions (Habibi 1973), vanadium compounds in fly ashes (Henry and Knapp 1980), or manganese compounds in emissions from turbines (Harker et al. 1975). A speciation of Fe compounds in car exhaust gas and street dust has been performed using Mossbauer spectroscopy (Eymery et al. 1978; Ismail et al. 1985).

Potential methods for the analysis of compounds in ambient particulate samples include X-ray diffraction (McCrone and Delly 1973), infrared spectroscopy (Kellner 1978), AES (Auger electron spectroscopy), ESCA (electron spectroscopy for chemical analysis), SAED (selected area electron diffraction), and SIMS (secondary ion mass

spectroscopy) (Keyser et al. 1978) (see also Section 2.5 on speciation).

2.8.7
Single Particle Analysis and Surface Characterization of Airborne Particles

The determination of size, shape, surface structure, and chemical composition are key steps for the characterization of atmospheric single particles. Table 2.6 provides a survey of the methods used for single particle analysis: Light microscopy (LM), electron beam microanalysis by electron probes (EPMA) or scanning electron microscopy (SEM), ion probe microanalysis (IPMA), transmission electron microscopy (TEM), and scanning transmission electron microscopy (STEM). The selection criteria for the use of the listed methods are the size of the particles and the elements to be analyzed (McCrone and Delly 1973; Grasserbauer 1978a, b; Heinrich 1980; Post and Buseck 1984). Automated single particle analytical methods such as CCSEM (computer controlled scanning electron microscopy) are used for source apportionment studies of aerosols via the PCB (particle class balance) technique (Kim and Hopke 1988).

For particulate emissions (particularly in fly ash samples), it has been found that certain elements – antimony, arsenic, lead, cadmium, chromium, cobalt, manganese, nickel, zinc, sulfur, selenium, thallium, and vanadium – appear to be enriched on the surface of the particles in a water-soluble form (Hansen and Fisher 1980; Keyser et al. 1978).

Tab. 2.6: Survey of analytical methods for the characterization of individual airborne particles. (After Grasserbauer 1978a)

Analytical method	Reagent	Signal	Analytical information	Relative sensitivity	Lower limit of particle diameter (μm)
LM	Light	Reflected, transmitted light	Type of compounds (species, structure), size, shape, morphology	Only pure species can be identified	0.5
EPMA, SEM	Electrons	X ray spectrum	Type of elements and their concentration. Number of particles of a specific composition	> 0.1%	0.1
		SE	Shape, size, morphology. Number of particles of a specific composition		0.01
		BSE, AE	Shape, size, morphology. Number of particles of a specific composition		0.1
IPMA	Ions (O_2^+, O^-, Ar^+),	Secondary ions	Type of elements and their concentration	$mg\ kg^{-1}$	0.5
STEM, TEM	Electrons	X ray spectrum	Type of elements and their concentration	Major and minor compounds	0.02
		Secondary electrons	Shape, size, morphology	> 0.1%	0.005
		Transmitted electrons	Size, shape		0.001
		Diffracted electrons	Structure and lattice parameters	Pure species	0.02
		Energy spectrum	Type of elements	Major components	0.01

2.9
Source Analysis

Results for elements in aerosol samples which are obtained by multielement techniques from data sets from which information about the sources of the components can be extracted (Gordon 1980). Such methods which make use of data obtained at receptor points are called receptor models. The most important receptor models are chemical mass balances (CMB), enrichment factors, time series correlation, multivariate models and spatial models (Cooper and Watson 1980; Gordon 1988). Dispersion modeling has also been used to explain the

occurrence of trace metal concentrations at a remote site in the Arctic (Pacyna et al. 1985).

The two most widely used receptor models in industrialized regions are the CMB and various forms of factor analysis (Hopke 1986). The CMB model requires information about the composition of the contributing aerosol sources in the model region. A great achievement of the CMB method was the identification of road dust as a major contributor to the urban aerosol mass (Cooper 1980) as well as the identification of wood combustion as an important aerosol source (Core et al. 1981). Major drawbacks of the CMB method arise when reactive aerosol components are involved (e.g., nitrates, ammonium compounds, organic constituents), when the regional background has a significant impact on the local aerosol mass concentration, and when several sources with similar elemental profiles contribute to the aerosol composition (Lowenthal et al. 1987). In a simulation study it was shown that for applications on the local scale CMB models yield "acceptable" accuracy and precision (Javitz et al. 1988). Multivariate statistical methods, including factor analysis (FA) and target transformation factor analysis (TTFA), do not require pre-information about the composition of possibly contributing particulate emissions. However, their resolving power to discriminate various source contributions is limited to four to six main factors influencing the aerosol mass (Lowenthal and Rahn 1987). On regional aerosols, satisfactory results are obtained for only one to two factors concerning the pollution elements.

The main advantage of FA and TTFA is to identify unusual sources that may not have previously considered and to find the major contributing source classes (Hopke 1988). More recently, microscopic methods have also been applied to relate individually analyzed particles to suspected sources (Johnson and McIntyre 1983; van Borm and Adams 1988; Kim and Hopke 1988). This approach seems to be applicable especially in regions where emissions of specific components contribute significantly to the aerosol mass.

In remote areas, CMB methods are not applicable due to the mixed influence of numberless sources. FA generally tends to uncover obviously influencing sources such as maritime, crustal, and mixed anthropogenic ones (Heidam 1981). More promising for such regions is the use of enrichment factors (Zoller et al. 1974) and of tracer systems (Rahn 1985). The enrich-

ment factor EF_{crust} relates the concentration of a given element in air to the concentration (X) of a crustal element (Al, Ti, Sc, or Fe) in air, normalized to the ratio of the given element concentration in the crust related to the reference element in the crust:

$$EF_{crust} = \frac{X_{air}/Al_{air}}{X_{crust}/Al_{crust}}$$

EF values near unity suggest that crustal erosion is the primary source of the observed element in the atmosphere, whereas EF values greater than unity indicate that other sources are the main contributors to the concentrations of the observed element. A drawback of this method is the inability to discriminate between anthropogenic contributions and other natural processes which produce an enrichment of the abundance of the observed metal as compared to crustal abundancies. A delicate problem when using EFs, especially in polluted regions, is the similarity of the matrix of major elements in coal fly ash and the crustal composition. A more sophisticated tool is the use of elemental ratios in a multitracer system (Rahn and Lowenthal 1984). With the use of a seven-element tracer system, Lowenthal and Rahn (1985) came to the conclusion that roughly 70% of the tracer elements observed in Alaska came from the former Soviet Union, 25% from Europe, and the remainder from North America. Support for the tracer sets used for Central Europe has been found by Borbely-Kiss et al. (1988).

There is an ongoing search for new tracer sets forming signatures of respective sources ("true" tracers would be emitted exclusively from single source types; while this is rather uncommon, elemental ratios are used as signatures from different sources). However, no simple solutions have arrived until now. Iridium is emitted from volcanoes (Zoller et al. 1983) but is also used as tracer for meteoritic material (Tuncel and Zoller 1987). Selenium is emitted from volcanoes (Tuncel and Zoller 1987) and is also an important tracer for coal emissions (Dutkiewicz and Husain 1988). Gaseous boron has been ascribed to be an important tracer for coal; however, a marine influence has also to be taken into account (Rahn and Fogg 1983). Finally, lanthanides have been used as tracers for petrochemical and oil-refining activities (Olmez and Gordon 1985). With the further development of receptor models it can be expected that their use will be extended to calibration and testing of dispersion models (Hopke 1986).

References

Adams FC, Van Craen MJ and Van Espen PJ (1980) *Enrichment of trace elements in remote aerosols.* Environ Sci Technol 14:1002–1005.

Allen A, Nemitz E, Shi J, Harrison R and Greenwood J (2001) *Size distributions of trace metals in atmospheric aerosols in the United Kingdom.* Atmos Environ 35:4581–4591.

Arafat N and Nriagu JO (1986) *Simulated mobilization of metals from sediments in response to lake acidification.* Water Air Soil Pollut 37:991–998.

Arnold M, Seghaier A, Martin D, Buat-Menard P and Chesselet R (1983) *Geochemistry of Marine Aerosol above the Western Mediterranean Sea.* Journ Etud Pollut Mar Mediterr 6th (Meeting Date 1982), Vol. 6, pp. 27–37.Commission Internationale pour 1'Exploration Scientifique de la Mer Mediterranee, Monaco.

ASTM (American Society for Testing and Materials) (1965) ASTM D 2009 –2065, Philadelphia.

Baena J, Galego M and Valcarel M (1999) *Speciation of inorganic lead and trialkyllead compounds by flame atomic absorption spectrometry following continuous selective preconcentration from aqueous solutions.* Spectrochim Acta B 54:1869–1879.

Baes CF III, and McLaughlin SB (1984) *Trace elements in tree rings: evidence of recent and historical air pollution.* Science 224:494–497.

Barbante C, Veysseyre A, Ferrari C, Van de Velde K, Morel C, Capodaglio G, Cescon P, Scarponi G and Boutron C. (2001) *Greenland Snow evidence of large scale atmospheric contamination for platinum, palladium and rhodium.* Environ Sci Technol 35:835–839.

Barrie LA, Hoff RM and Daggupaty SM (1981) *The influence of mid-latitudinal pollution sources on haze in the Canadian Arctic.* Atmos Environ 15:1407–1419.

Batifol EM and Boutron CE (1984) *Atmospheric heavy metals in the high altitude surface snows from Mt. Blanc French alps.* Atmos Environ 18:2507–2515.

Bennett BG (1981) *Exposure Commitment Assessment of Environmental Pollutants,* Vol. 1, No. 1, MARC Report 23. Monitoring and Assessment Research Center, Chelsea College, London.

Berish CW and Ragsdale HL (1985) *Chronological Sequence of element concentrations in wood of Carya spp. in the Southern Appalachian Mountains.* Can J Forest Res 15:477–483.

Berner A (1978) *A five stage cascade impactor for measurement of mass-size distributions of aerosols (in German).* Chem Ing Tech 50: 399.

Biggins PE and Harrison RM (1979a) *Atmospheric chemistry of automotive lead.* Environ Sci Technol 13:558–565.

Biggins PE and Harrison RM (1979b) *The identification of specific chemical compounds in size-fractionated atmospheric particulates collected at roadside sites.* Atmos Environ 13:1213–1216.

Bloch P, Adams E, Van Landuyt J and Van Goethem L (1980) *Morphological and chemical characterization of individual aerosol particles in the atmosphere.* In: Versino B, ed. Proceedings Symposium Physico-Chemical Behaviour of Atmospheric Pollutants. EUR 6621, ECSC-EEC-EAEC, Brussels, Luxembourg.

Borbely-Kiss I, Haszpra L, Koltay E, Laszlo S, Meszaros A, Meszaros E and Szabo G (1988) *Elemental concentrations and regional signatures in atmospheric aerosols over Hungary.* Phys Scr 37:299–304.

Borg H (1983) *Trace metals in Swedish natural waters.* Hydrobiologia 101:27–34.

Boutron CE (1986) *Atmospheric toxic metals and metal oxides in the snow and ice layers deposited in Greenland and Antarctic from prehistoric times to present.* In: Nriagu JO and Davidson CI, eds. Toxic Metals in the Atmosphere. Wiley, New York.

Boutron C, Görlach U, Candelone J, Bolshov M and Delmas R (1991) *Decrease in anthropogenic lead, cadmium and zinc in Greenland snows since the late 1960s.* Nature **353**:153–156.

BOUTRON CE, CANDELONE J, and HONG S (1995) *Greenland snow and ice cores: unique archives of large-scale pollution of the troposphere of the Northern Hemisphere by lead and other heavy metals.* Sci Total Environ **160/161**:233–241.

BRIAT M (1978) *Evaluation of Levels of Pb, V, Cd, Zn and Cu in the Snow of Mt. Blanc during the last 25 Years.* Atmospheric Pollution, Proceedings 13th Int. Coll. Paris. In: Benarie MM, ed. Studies in Environmental Sciences, Vol. 1. Elsevier, Amsterdam.

CAHILL TA (1975) *Ion excited X-ray analysis of environmental samples.* In: Ziegler J, ed. New Uses of Ion Accelerators, pp. 1–72. Plenum Press, New York.

CAHILL TA, ASHBAUGH LL, BARONE JB, ELDRED RA, FERNEY PJ, FLOCCHINI RG, GOODART CH, SHADDOAN DJ and WOLFE FW (1977) *Analysis of respirable fractions in atmospheric particulates via sequential filtration.* J Air Pollut Control Assoc **27**:675–678.

CAL-PIETRO M, CARLOSENA A, ANDRADE J, MARTINEZ M, MUNIATEGUI S, LOPEZ-MAHAI P and PRADA D (2001) *Antimony as a tracer of the anthropogenic influence on soils and estuarine sediments.* Water Air Soil Pollution **129**:333–348.

CANDELONE J, HONG S, PELLONE C and BOUTRON C (1995) *Post-industrial revolution changes large-scale atmospheric pollution of the northern hemisphere by heavy metals as documented in central Greenland ice and snow.* J Geophys Res **100**:16605–1616.

CASS GR and McRAE GJ (1983) *Source-receptor reconciliation of routine air monitoring data for trace metals: an emission inventory assisted approach.* Environ Sci Technol **17**:129–139.

CASS GR and McRAE GJ (1986) *Emission and air quality relationships for atmospheric trace metals.* In: Nriagu JO and Davidson CI, eds. Toxic Metals in the Atmosphere. Wiley, New York.

CHATFIELD EJ (1984) *Determination of Asbestos Fibers in Air and Water.* ISO-TC-1774 Report.

CAO T, COONEY R, WOZNICHAK M, MAY S and BROWNER R (2001) *Speciation and identification of organoselenium metabolites in human urine using inductively coupled plasma mass spectrometry and tandem mass spectrometry.* Anal Chem **73**:2898–2902.

CHE-JEN L and PEHKONEN S (1999) *The chemistry of atmospheric mercury: a review.* Atmos Environ **33**:2067–2079.

CHEN L and DUCE RA (1983) *The sources of SO_4^{2-} V and mineral matter in aerosol particles over Bermuda.* Atmos Environ **17**:2055–2064.

CHILLRUD S, BOPP R, SIMPSON J, ROSS J, SHUSTER E, CHAKY D, WALSH D, CHIN CHOY C, TOLLEY LR and YARME A (1999) *Twentieth century atmospheric metal fluxes into Central Park Lake, New York City.* Environ Sci Technol **33**:657–662.

COLES DG, RAGAINI RC, ONDOV JM, FISHER GL, SILBERMAN D and PRENTICE B (1979) *Chemical studies of stack fly ash from a coal fired power plant.* Environ Sci Technol **13**:455–459.

COOPER JA (1980) J Air Pollut Control Assoc **30**:855–860.

COOPER JA and WATSON JG JR. (1980) *Receptor oriented methods of air particulate source apportionment.* J Air Pollut Control Assoc **30**:1116–1125.

CORE JE, HANRAHAN PL and COOPER JA (1981) *Air particulate control strategy development – a new approach using chemical mass balance methods.* In: Macias ES and Hopke PK, eds. Atmospheric Aerosols: Source/Air Quality Relationships. ACS Symposium Series 167. Washington, DC.

COUNTESS RJ (1974) J Air Pollut Control Assoc **24**:605.

CRAIG PJ (1980) *Metal cycles and biological methylation.* In: Hutzinger O, ed. The Handbook of Environmental Chemistry, Vol.1, Part A. The Natural Environment and the Biogeochemical Cycles, pp. 169–227. Springer, Berlin-New York.

CUNNINGHAM WC and ZOLLER WH (1981) *The chemical composition of remote area aerosols.* J Aerosol Sci **12**:367–384.

DAVIDSON CI and OSBORN JF (1986) *The sizes of airborne trace metal containing particles.* In: Nriagu JO and Davidson CI, eds. Toxic Metals in the Atmosphere. Wiley, New York.

DAVIS BL (1981) *Quantitative analysis of crystalline and amorphous airborne particulates in the Provo-Orem vicinity, Utah.* Atmos Environ **15**:613–618.

DEDEURWAERDER HL, BAEYENS WF and DEHAIRS FA (1985) *Estimates of dry and wet deposition of several trace metals in the Southern Bight of the North Sea.* In: Lekkas TD, ed. Heavy Metals in the Environment, 5th International Conference, Vol. 1, pp. 135–137. CEP Consultants Ltd., Edinburgh.

DEUTSCH F, HOFFMANN P, and ORTNER H (2001) *Field experimental investigations on the Fe(II)- and Fe(III)-content in cloudwater samples.* J Aerosol Sci **40**:87–105.

DICKSON W (1980) *Properties of acidified waters.* In: Drablos D and Tollan A, eds. Ecological Impact of Acid Precipitation. Proceedings International Conference, Sandefjord, Norway, March 11 –14.

DIETL C, REIFENHÄUSER W and PEICHL L (1997) *Association of antimony with traffic – occurrence in airborne dust, deposition and accumulation in standardized grass cultures.* Sci Total Environ **205**:235–244.

DUCE RA, UNNI CK, RAY BJ, PROSPERO JM and MERRILL JT (1980) *Long range atmospheric transport of soil dust from Asia to the Tropical North Pacific: temporal variability.* Science **209**:1522–1524.

DUCE RA, ARIMOTO R, RAY BJ, UNNI CK and HARDER PJ (1983) *Atmospheric trace elements at Eniwetok Atoll: 1, Concentrations, sources and temporal variability.* J Geophys Res **88**:5321–5342.

DUTKIEWICZ VA and HUSAIN L (1988) *Spatial pattern of non-urban Se concentrations in the Northeastern US and its pollution source implications.* Atmos Environ **22**:2223–2228.

EISENREICH S J (1980) *Atmospheric input of trace metals to Lake Michigan.* Water Air Soil Pollut **13**:287–301.

ELEFTHERIADIS K and COLBECK I (2001) *Coarse atmospheric aerosol: size distributions of trace elements.* Atmos Environ **35**:5321–5320.

EREL Y, PEHKONEN S and HOFFMANN M (1993) *Redox chemistry of iron in for and stratus clouds.* J Geophys Res **98**:18423–18434.

EYMERY JP, RAJN SB and MOINE P (1978) *J Phys Dll:* 2147–2149.

FAUST B and ZEPP R (1993) *Photochemistry of aqueous iron(III)-polycarboxylate complexes: roles in the chemistry of atmospheric and surface waters.* Environ Sci Technol **27**:2517–2522.

FERNANDEZ A, TERNERO M, BARRAGAN F and JIMENEZ J (2000) *An approach to characterization of sources of urban airborne particles through heavy metal speciation.* Chemosphere – Global Change. Science **2**:123–136.

FLAGAN RC and FRIEDLANDER SK (1978) *Particle formation in pulverized coal combustion – a review.* In: Shaw DT, ed. Recent Developments in Aerosol Science. Wiley, New York.

GALLOWAY JN, EISENREICH SJ and SCOTT BC (1980) *Toxic Substances in Atmospheric Deposition: A Review and Assessment,* National Atmospheric Deposition Program Report NC-141. U.S. Environmental Protection Agency Report EPA-5 60/5-80-001, 146 p. Washington, DC.

GARG B, CADLE S, MULAWA P, GROBLICKI P, LAROO C and PARR GA (2000) *Brake wear particulate matter emissions.* Environ Sci Technol **21**:4463–4469.

GERMANI MS, SMALL M, ZOLLER WH and MOYERS JL (1981) *Fractionation of elements during copper smelting.* Environ Sci Technol **75**:299–305.

GLOOSCHENKO WA (1986) *Monitoring the atmospheric deposition of metals by use of bog vegetation and peat profiles.* In: Nriagu JO and Davidson C I, eds. Toxic Metals in the Atmosphere. Wiley, New York.

GORDON GE (1980) *Receptor models.* Environ Sci Technol **14**:792–800.

GORDON GE (1988) *Receptor models.* Environ Sci Technol **22**:1132–1142.

GORDON GE, MOYERS JL, RAHN KA, GATZ DE, DZUBAY TG, ZOLLER WH and CORRIN MH (1978) *Atmospheric Trace Elements: Cycles and Measurements.* Report of the National Science Foundation Atmospheric Chemistry Workshop. Panel on Trace Elements, National Center for Atmospheric Research, Boulder, Colorado.

GOULDING FS, JAKLEVIC JM and LOO BW (1978) US EPA Report No. EPA-600/4/78–034.

GRASSERBAUER M (1978a) *Chapter 8.* In: Malissa H, ed. Analysis of Airborne Particles by Physical Methods. CRC Press, West Palm Beach, Florida.

GRASSERBAUER M. (1978b) *The present state of local analysis.* Mikrochim Acta (Wien) 1978 I:329–350.

GREENBERG RR, GORDON GE, ZOLLER WH, JACKO RB, NEUENDORF D, WANDYOST KJ (1978) *Composition of particles emitted from the Nicosia municipal incinerator.* Environ Sci Technol **12**:1329–1232.

GUI-BIN J, QUN-FANG Z and BIN H (2000) *Speciation of organotin compounds, total tin, and major trace metal elements in poisoned human organs by gas chromatography-flame photometric detector and inductively coupled plasma mass spectrometry.* Environ Sci Technol **34**:2697–2702.

HABIBI K (1973) *Characterization of particulate matter in vehicle exhaust.* Environ Sci Technol **7**:223.

HALES JM (1989) *A generalized multidimensional model for precipitation scavenging and atmospheric chemistry.* Atmos Environ **23**:2017–2031.

HAMASAKI T, NAGASE H, YOSHIOKA Y and SATO T (1995) *Formation, distribution, and ecotoxicity of methylmetals of tin, mercury, and arsenic in the environment.* Critical Rev Environ Sci Technol **25**:45–91.

HANSEN LD and FISHER GL (1980) *Elemental distribution in coal fly ash particles.* Environ Sci Technol **14**:1111–1117.

HARKER AB, PAGNI PJ, NOVAKOV T and HUGHES L (1975) *Manganese emissions from combustors.* Chemosphere **6**:339.

HARRISON RM (1986) *Chemical speciation and reaction pathways of metals in the atmosphere.* In: Nriagu JO and Davidson CI eds. Toxic Metals in the Atmosphere. Wiley, New York.

HARRISON R, TILLING R, ROMERO M, HARRAD S and JARVIS K (2003) *A study of trace metals and polycyclic aromatic hydrocarbons in the roadside environment.* Atmos Environ **37**:2391–2402.

HEIDAM NZ (1981) *On the origin of the Arctic aerosol: a statistical approach.* Atmos Environ **15**:1421–1427.

HEIDAM NZ (1986) *Trace metals in the Arctic.* In: Nriagu JO and Davidson CI, eds. Toxic Metals in the Atmosphere. Wiley, New York.

HEINRICH KFJ, ed. (1980) *Characterization of Particles.* NBS Spec. Pub. 533. US Government Printing Office, Washington, D.C.

HEINTZENBERG J, HANSSON HC and LANNEFORS H (1981) *The chemical composition of Arctic haze at Ny Alesund, Spitsbergen.* Tellus **33**:162–171.

HEISTERKAMP M, VAN DE VELDE K, FERRARI C, BOUTRON C and ADAMS F (1999) *Present century record of organolead pollution in high altitude alpine snow.* Environ Sci Technol **33**:4416–4421.

HELMERS E and SCHREMS O (1995) *Wet deposition of metals to the tropical north and the south Atlantic Ocean.* Atmos Environ **29**:2475–2484.

HENRY WM and KNAPP KT (1980) *Compound forms of fossil fuel fly ash emissions.* Environ Sci Technol **14**:450–456.

HERRON MM, LANGNEAY CC, WEISS HV and CRAGIN JH (1977) *Atmospheric trace metals and sulfate in the Greenland ice sheet.* Geochim Cosmochim Acta **41**:915–920.

HICKS BB (1986) *Differences in wet and dry particle deposition parameters between North America and Europe.* In: Lee SD, Schneiders T, Grant LD and Verkerk PJ, eds. Aerosols. Lewis Publishers, Chelsea, Michigan.

HIDY GM (1986) *Definition and characterization of suspended particles in ambient air.* In: Lee SD, SchneiderS T, Grant LD and Verkerk PJ, eds. Aerosols. Lewis Publishers, Chelsea, Michigan.

HOCHRAINER D (1978) *Chapter 2.* In: Malissa H, ed. Analysis of Airborne Particles by Physical Methods. CRC Press, West Palm Beach, Florida.

HODGE VE and STALLARD MO (1986) *Platinum and palladium in roadside dust.* Environ Sci Technol **20**:1058–1060.

HOPKE PK (1986) *Quantitative source attribution of metals in the air using receptor models.* In: Nriagu JO and Davidson CI, eds. Toxic Metals in the Atmosphere. Wiley, New York.

HOPKE PK (1988) *Target transformation factor analysis as an aerosol mass apportionment method: a review and sensitivity study.* Atmos Environ **22**:1777–1792.

HRUDEY SE (1977) *Chapter 1.* In: Perry R and Young RJ, eds. Handbook of Air Pollution Analysis. Chapman & Hall, London.

HUANG X, OLMEZ I, ARAS K, and GORDON GE (1994) *Emissions of trace elements from motor vehicles: potential marker elements and source composition profile.* Atmos Environ **28**:1385–1391.

HUSAIN L (1986) *Chemical elements as tracers of pollutant transport to a rural area.* In: Nriagu JO and Davidson CI, eds. Toxic Metals in the Atmosphere. Wiley, New York

HVATUM OO, BOLVIKEN B and STEINNES E (1983) *Heavy metals in Norwegian ombrotrophic bogs.* Ecol Bull **35**:351–356.

IAEA (International Atomic Energy Agency) (1978) *Particle Size Analysis in Estimating the Significance of Airborne Contamination.* Tech Rep Ser No. 179. IAEA, Vienna.

ISMAIL SS, GRASS E, WIESINGER G and MOSTAFA AG (1985) *Different Sources of Contamination Detected by Nuclear Methods.* WMO No. 647, pp. 563–589. Proc. TECOMAC.

ISO TC 146 (1981) Am Ind Hyg Assoc J **42**:A64–A68.

JAENICKE R (1986) *Physical characterization of aerosols.* In: Lee SD, Schneider T, Grant LD and Verkerk P J, eds. Aerosols. Lewis Publishers, Chelsea, Michigan.

JAVITZ HS, WATSON JG and ROBINSON N (1988) *Performance of the chemical mass balance model with simulated local scale aerosols.* Atmos Environ **22**:2309–2322.

JAY J, MOREL F and HEMOND H (2000) *Mercury speciation in the presence of polysulfides.* Environ Sci Technol **34**:2196–2200.

JIANG S, ROBBERECHT H and ADAMS E (1983) *Identification and determination of alkylselenide compounds in environmental air.* Atmos Environ **17**:111–114.

JOHN W and REISCHL G (1980) *A cyclone for size selective sampling of ambient air.* J Air Pollut Control Assoc **30**:872–876.

JOHNSON DJ and MCINTYRE BL (1983) *A particle class balance receptor model for aerosol apportionment in Syracuse, N. Y.* In: Dattner SL and Hopke PK, eds. Receptor Models Applied to Contemporary Pollution Problems. Air Pollution Control Association, Pittsburgh, Pennsylvania.

JUNGE CE (1963) *Air Chemistry and Radioactivity.* Academic Press, New York.

KAISER G and TÖLG G (1980) *Mercury.* In: Hutzinger O, ed. The Handbook of Environmental Chemistry, Vol. 3, Part A, Anthropogenic Compounds. Springer, Berlin-New York.

KAZDA M and GLATZEL G (1984) *Heavy metal enrichment and mobility in the infiltration zone of stemflow from beeches in the Vienna Woods (in German).* Z Pflanzenernähr Bodenkd **147**:743–752.

KELLNER R (1978) *Chapter 22.* In: Malissa H, ed. Analysis of Airborne Particles by Physical Methods. CRC Press, West Palm Beach, Florida.

KEMP K (2002) *Trends and sources for heavy metals in urban atmosphere.* Nucl Instrum Methods Phys Res B **189**:227–232.

KEYSER TR, NATUSCH DFS, EVANS CA JR. and LINTON RW (1978) *Characterizing the surfaces of environmental particles.* Environ Sci Technol **12**:768–773.

KIM D and HOPKE PK (1988) *Source apportionment of the El Paso aerosol by particle class balance analysis.* Aerosol Sci Technol **9**:221–235.

KIM G, SCUDLARK J and CHURCH T (2000) *Atmospheric wet deposition of trace elements to Chesapeake and Delaware Bays.* Atmos Environ **34**:3437–3444.

KLOCKOW D (1982) *Analytical chemistry of the atmospheric aerosol.* In: Georgii, HW and Jaeschke W, eds. Chemistry of the Unpolluted and Polluted Troposphere. D. Reidel Publ, Dordrecht.

LANTZY RJ and MACKENZIE FT (1979) *Atmospheric trace metals: global cycles and assessment of man's impact.* Geochim Cosmochim Acta **43**:511–525.

LBL (Lawrence Berkeley Laboratory) (1975) *Instrumentation for Environmental Monitoring – Air.* Lawrence Berkeley Lab. Rep. LBL. 7, Vol. 1, Part 2, 1st Ed. Environmental Instrumentation Group, LBL, University of California, Berkeley.

LEGRAND MR, LORIUS C, BARKOV NI and PETROV VN (1988) *Vostok (Antarctic) ice core: atmospheric chemistry changes over the last climatic cycle (160 000 years).* Atmos Environ **22**:317–331.

LEYDEN DE (1978) *Chapter 3.* In: Malissa H, ed. *Analysis of Airborne Particles by Physical Methods.* CRC Press, West Palm Beach, Florida.

LINDBERG SE (1980) *Mercury partitioning in a power plant plume and its influence on atmospheric removal mechanisms.* Atmos Environ **14**:227–231.

LINDBERG SE and HARRISS RC (1981) *The role of atmospheric deposition of an Eastern US deciduous forest.* Water Air Soil Pollut **16**:13–31.

LINDBERG SE and TURNER RR (1983), *Trace metals in rain at forested sites in the Eastern United States.* In: Proceedings International Conference on Heavy Metals in the Environment, Heidelberg. CEP Consultants Ltd., Edinburgh.

LINDBERG SE, HARRISS RC, TURNER RR, SHRINER DS and HUFF DD (1979) *Mechanisms and Rates of Atmospheric Deposition of Selected Trace Elements and Sulfate to a Deciduous Forest Watershed.* ORNL/TM-6674, 514 p. Oak Ridge National Laboratory, Oak Ridge, Tennessee.

LINDBERG SE, HARRISS RC and TURNER RR (1982) *Reports: atmospheric deposition of metals to forest vegetation.* Science **215**:1609–1611.

LINTON RW, LOH A, NATUSCH DFS, EVANS CA and WILLIAMS P (1976) *Surface predominance of trace elements in airborne particles.* Science **191**:852–854.

LINTON RW, WILLIAMS P, EVANS CA and NATUSCH DFS (1977) *Determination of the surface predominance of toxic elements in airborne particles by IMMS and AES.* Anal Chem **49**:1514–1521.

LIOY PJ, WATSON JG and SPENGLER JD (1980) *APCA Specialty Conference Workshop on baseline data for inhalable particulate matter.* J Air Pollut Control Assoc **30**:1126–1130.

LIPPMANN M (1976) *Size selective sampling for inhalation hazard evaluations.* In: Liu BYH, ed. Fine Particles. Academic Press, New York.

LIU BYH, RAABE OG, SMITH WB, SPENCER HW and KUYKENDAL WB. (1980) *Advances in particle sampling and measurement.* Environ Sci Technol **14**:392–397.

LIVETT EA, LEE JA and TALLIS JH (1979) *Lead, zinc and copper analyses of British blanket peats.* J Ecol **67**:865–891.

LODGE JR and CHAN TL, eds. (1987) *L 87028, Cascade Impactor Sampling and Data Analysis.* American Industrial Hygienists Association, Akron, Ohio.

LOREY P and DRISCOLL C (1999) *Historical trends of mercury deposition in Adirondack Lakes.* Environ Sci Technol **33**:718–722.

LOWENTHAL DH and RAHN KA (1985) *Regional sources of pollution aerosol at Barrow, Alaska, during winter 1979/80 as deduced from elemental tracers.* Atmos Environ **19**(12):2011–2024.

LOWENTHAL DH and RAHN KA (1987) *Application of factor-analysis receptor model to simulated urban- and regional-scale data sets.* Atmos Environ **21**(9):2005–2013.

LOWENTHAL DH, HANUMARA RC, RAHN KA and CURRIE LA (1987) *Effects of systematic error, estimates and uncertainties in chemical mass balance apportionments: Quail Roost II revisited.* Atmos Environ **21(3)**:501–510.

MALISSA H (1976) *Integrated dust analysis.* Angew Chem Int Ed Engl **15**:141–149.

MARPLE VA and WILLEKE K (1976) *Inertial Impactors: Theory,* Design, Use. In: Liu BYH, ed. Fine Particles. Academic Press, New York.

MAYER M (1981) *Contributions for Determining the Size Distribution of Trace Metals in Atmospheric Particles* (in German). Diploma Thesis, Technische Universitat, Wien.

MAYER R (1981) *Natürliche und anthropogene Komponenten des Schwermetallhaushalts von Waldökosystemen.* Gott Bodenkdl Ber **70**:1–152.

MAYER R and ULRICH B (1982) *Calculation of deposition rates from the flux balance and ecological effects of atmospheric deposition upon forest ecosystems.* In: Georgii HW and Pankrath J, eds. Deposition of Atmospheric Pollutants, pp. 195–200. Reidel Publ, Dordrecht, The Netherlands.

MCCRONE WC and DELLY JG (1973) *The Particle Atlas,* Vol. 1–3, 2nd Ed. Ann Arbor Science, Ann Arbor, Michigan.

MCFARLAND AR, ORTIZ CA and BERTCH RW JR (1984) J Air Pollut Control Assoc **34**:544–547.

MCMURRY PH and WILSON JC (1982) *Growth laws for the formation of secondary ambient aerosols: implications for chemical conversion mechanisms.* Atmos Environ **16**:121–134.

MCMURRY PH and WILSON JC (1983) *Droplet phase and gas phase contributions to secondary ambient aerosol formation as a function of relative humidity.* J Geophys Res **88**:5101–5108.

MIGON C, JOURNEL B and NICOLAS E (1997) *Measurement of trace metal wet, dry and total atmospheric fluxes over the Ligurian Sea.* Atmos Environ **31**:889–896.

MILLER FJ, GARDNER DE, GRAHAM JA, LEE RE JR, WILSON WE and BACHMANN JD (1979) *Size considerations for establishing a standard for inhalable particles.* J Air Pollut Control Assoc **29**:610–615.

MOLDOVAN M, GOMEZ M and PALACIOS MA (1999) *Determination of platinum, rhodium and palladium in exhaust fumes.* J Anal At Spectrom **14**:1163–1169.

MOYERS JL, RANWEILER LE, HOPF SB and KORTE NE (1977) *Evaluation of particulate trace species in southwest desert atmosphere.* Environ Sci Technol **11**:789–795.

MUROZUMI M, CHOW TJ and PATTERSON CC (1969) *Chemical concentrations of pollutant lead aerosols,* terrestrial dusts and sea salts in Greenland Antarctic snow strata. Geochim Cosmochim Acta **33**:1247–1294.

NAS (National Academy of Sciences) (1976) *Selenium.* PB 251 Subcommittee on Selenium, Committee on Medical and Biological Effects of Environmental Pollutants. National Research Council, Washington, DC.

NATUSCH DFS and WALLACE JR (1974) *Urban aerosol toxicity: the influence of particle size.* Science **186**:695–699.

NCAR (National Center for Atmospheric Research) (1982) *Regional Acid Deposition: Models and Physical Processes.* NCAR, Boulder, Colorado.

NG A and PATTERSON CC (1981) *Natural concentrations of lead in ancient Arctic and Antarctic Ice.* Geochim Cosmochim Acta **45**:2109–2121.

NOLL K, YUEN P and FANG K (1990), *Atmospheric coarse particulate concentrations and dry deposition fluxes for ten metals in two urban environments.* Atmos Environ **24A**:903–908.

NRIAGU JO (1979) *Global inventory of natural and anthropogenic emissions of trace metals into the atmosphere.* Nature **279**:409–411.

NRIAGU JO and DAVIDSON CI (1986) *Toxic Metals in the Atmosphere.* Wiley, New York.

NRIAGU JO and RAO SS (1987) *Response of lake sediments to changes in trace metal emission from the smelters at Sudbury, Ontario.* Environ Pollut **44(3)**:211–218.

NRIAGU JO and WONG HKT (1986) *What fraction of the total metal flux into lakes is retained in the sediments?* Water Air Soil Pollut **31(3–4)**:999–1006.

NÜRNBERG HW, VALENTA P and NGUYEN VD (1983) *The wet deposition of heavy metals from the atmosphere in the Federal Republic of Germany.* In: Proceedings International Conference on Heavy Metals in the Environment Heidelberg, Vol. I, pp. 115–123. CEP Consultants Ltd., Edinburgh.

O'CONNOR BH and JAKLEVIC JM (1981) *Characterization of ambient aerosol particulate samples from the St. Louis Area by X-ray powder diffractometry.* Atmos Environ **15**:1681–1690.

OLMEZ I and GORDON GE (1985) Science **229**:966–968.

ONDOV JM, RAGAINI RC and BIERMANN AH (1979) Environ. Sci Technol **13**:598–607.

PACYNA JM (1984), *Estimation of the atmospheric emissions of trace elements from anthropogenic sources in Europe.* Atmos Environ **18**:41–50.

PACYNA JM (1986a) *Emission factors of atmospheric elements.* In: Nriagu JO and Davidson CI, eds.

Toxic Metals in the Atmosphere, Wiley, New York.

PACYNA JM (1986b) *Atmospheric trace elements from natural and anthropogenic sources.* In: Nriagu JO and Davidson C I, eds. Toxic Metals in the Atmosphere. Wiley, New York.

PACYNA M (1987) *Long Range Transport of Heavy Metals – Modelling and Measurements.* In: Preprints 16th International Technical Meeting on Air Pollution Modelling and Applications. Committee on Challenges of Modern Society, Bruxelles.

PACYNA JM and PACYNA E (2001) *An assessment of global and regional emissions of trace metals to the atmosphere from anthropogenic sources worldwide.* Environ Rev 9: 269–298.

PACYNA JM, OTTAR B, TOMZA U and MAENHAUT W (1985) *Long-range transport of trace elements to Ny Alesund, Spitsbergen.* Atmos Environ 19: 857–865.

PAKARINEN P (1981) *Metal content of ombrotrophic sphagnum mosses in NW Europe.* Ann Bot Fenn 18: 281–292.

PAKARINEN P and TOLONEN K (1977a) *Distribution of lead in Sphagnum fuscum profile in Finland.* Oikos 28: 69–73.

PAKARINEN P and TOLONEN K (1977b) *Vertical Distributions of N, P, K, Zn and Pb in Sphagnum peat.* Suo 28: 95–102.

PALACIOS MA, GOMEZ MM, MOLDOVAN M, MORRISON G, RAUCH S, McLEOD C, LI R, LASERNA J, LUCENA P, CAROLI S, ALIMONTI A, SCHRAMEL P, LUSTIG S, WASS U, STENBOM B, LUNA M, SAENZ JC, SANTAMARIA J and TORRENTS JM (2000) *Platinum-group elements: quantification in collected exhaust fumes and studies of catalyst surfaces.* Sci Total Environ 207: 1–15.

PAKKANEN T, HILLAMO R, KERONEN P, MAENHAUT W, DUCASTEL G and PACYNA J (1996) *Sources and physico-chemical characteristics of the atmospheric aerosol in southern Norway.* Atmos Environ 30: 1391–1405.

PARRINGTON JR and ZOLLER WH (1984) *Diurnal and longer term temporal changes in the composition of atmospheric particles at Mauna Loa, Hawaii.* J Geophys Res 89: (D2): 2522–2534.

PILINIS C and SEINFELD JH (1987) *Continued development of a general equilibrium model for inorganic multicomponent aerosol.* Atmos Environ 21: 2453–2466.

PIRRONE N, KEELER G and WARNER P (1995) *Trends of ambient concentrations and deposition fluxes of particulate trace metals in Detroit from 1982 to 1992.* Sci Total Environ 162: 43–61.

PLANCHON F, BOUTRON C, BARBANTE C, COZZI G, GASPARI V, WOLFF E, FERRARI C and CESCON P (2002) *Changes in heavy metals in Antarctic snow from Coats Land since the mid-19th to the late-20th century.* Earth and Planetary Science Letters 200: 207–222.

POST JE and BUSECK PR (1984) *Characterization of individual particles in the Phoenix urban aerosol using electron-beam instruments.* Environ Sci Technol 18: 35–42.

PURDUE LJ (1986) *US EPA PM_{10} Methodology Review.* In: Lee SD, Schneider T, Grant LD and Verkerk PJ, eds. Aerosols. Lewis Publishers, Chelsea, Michigan

PUXBAUM H (1979) *Sampling of Inhalable and Lung Penetrating Particles for "Integrated Aerosol Analysis" (in German)* Fresenius Z Anal Chem 298: 110–128.

PUXBAUM H and WOPENKA B (1984) *Chemical composition of nucleation and accumulation mode particles collected in Vienna, Austria.* Atmos Environ 18: 573–580.

PUXBAUM H, QUINTANA E and PIMMINGER M (1985) *Spatial distributions of atmospheric aerosol constituents in Linz (Austria)* Fresenius Z Anal Chem 322: 205–212.

PUXBAUM H, RENDL J, ALLABASHI R, OTTER L and SCHOLES M (2000) *Mass balance of the atmospheric aerosol in a South African subtropical savanna (Nylsvley, May 1997).* J Geophys Res 105: 20697–20706.

RAGAINI RC (1978) *Chapter 7.* In: Malissa H, ed. Analysis of Airborne Particles by Physical Methods. CRC Press, West Palm Beach, Florida.

RAHN KA (1985) *Progress in Arctic air chemistry.* Atmos Environ 19: 1987–1994.

RAHN KA and FOGG TR (1983) *Boron as a Tracer of Aerosol from Combustion of Coal.* Final Technical Report, DOE/PC/51260–4; Order No. DE84004708, 30 p. Avail. NTIS, from: Energy Res Abstr 1984 9(7), Abstr No. 11452

RAHN KA and LOWENTHAL DH (1984) *Elemental tracers of distant regional pollution aerosols.* Science 223: 132–139.

RAHN KA and LOWENTHAL DH (1985) *Pollution aerosol in the Northeast: Northeastern-Midwestern Contributions.* Science 228 (4697): 275–284.

RAHN KA, BORYS RD, SHAW GE, SCHIITZ L and JAENICKE R (1979) *Long range impact of desert aerosol and atmospheric chemistry: two examples.* In: Saharan Dust. SCOPE 14, Wiley, New York.

RAMBAEK JP and STEINNES E (1980) *Atmospheric deposition of heavy metals studied by analysis of moss samples using neutron activation analysis and atomic absorption spectrometry.* Nuclear Methods Environmental Energy Research, pp. 175–180. USDOE CONF-800433.

REA A, LINDBERG S and KEELER G (2001) *Dry deposition and foliar leaching of mercury and selected trace elements in deciduous forest throughfall.* Atmos Environ **35**:3453–3462.

REA A, LINDBERG S and KEELER G (2000) *Assessment of dry deposition and foliar leaching of mercury and selected trace elements based on washed foliar and surrogate surfaces.* Environ Sci Technol **34**:2418–2425.

REUTHER R, JAEGER L and ALLARD B (1999) *Determination of organometallic forms of mercury, tin and lead by in situ derivatization, trapping and gas chromatography – atomic emission detection.* Anal Chim Acta **394**:259–269.

ROSS HB and GRANAT L (1986) *Deposition of atmospheric trace metals in northern Sweden as measured in the snowpack.* Tellus **38B(1)**:27–43.

SALOMONS W (1986) *Impact of atmospheric inputs on the hydrospheric trace metal cycle.* In: Nriagu JO and Davidson CI, eds. Toxic Metals in the Atmosphere. Wiley, New York.

SCHAULE BK and PATTERSON CC (1981) *Lead concentrations in the northeast Pacific: evidence for global anthropogenic perturbations.* Earth Planet Sci Lett **54**:97–116

SCHAULE BK and PATTERSON CC (1983) *Perturbations of the Natural Depth Profile in the Sargasso Sea by Industrial Lead.* Proceedings of NATO Advanced Research Workshop of Trace Metals in Seawater, Erice, Italy, 1981, pp. 407–504. Plenum Press, New York.

SCHROEDER WH, DOBSON M, KANE DM and JOHNSON ND (1987) *Toxic trace elements associated with airborne particulate matter: a review.* J Air Pollut Control Assoc **37**:1267–1285.

SCOPE (1979) *Saharan Dust,* SCOPE 14. Wiley, New York.

SEHMEL GA (1980) *Particle and gas dry deposition: a review.* Atmos Environ **74**:983–1012.

SEINFELD JH (1986) *Atmospheric Chemistry and Physics of Air Pollution.* Wiley, New York.

SLINN WGN (1982) *Prediction of particle deposition to vegetative canopies* Atmos Environ **7**:1785–1794.

SHEVCHENKO V, LISITZIN A, VINOGRADOVA A and STEIN R (2003) *Heavy metals in aerosols over the seas of the Russian Arctic.* Sci Total Environ **306**:11–25.

SHOTYK W (1995) *Peat bog archives of atmospheric metal deposition: geochemical evaluation of peat profiles, natural variations in metal concentrations,* *and metal enrichment factors.* Environ Rev **4**:149–183.

SHOTYK W, CHEBURKIN A, APLEBY P, FANKHAUSER A and KRAMERS J (1996) *Two thousand years of atmospheric arsenic, antimony, and lead deposition recorded in an ombrotrophic peat bog profile, Jura Mountains, Switzerland.* Earth and Planetary Science Letters **145**:E1–E7.

SHOTYK W, WEISS D, HEISTERKAMP M, CHEBURKIN A, APPLEBY P and ADAMS F (2002) *New peat bog record of atmospheric lead pollution in Switzerland: Pb concentrations, enrichment factors, isotopic composition, and organolead species.* Environ Sci Technol **36**:3893–3900.

SMITH RD, CAMPBELL JA and NIELSON KK (1979) *Concentration dependence upon particle size of volatilized elements in fly ash.* Environ. Sci Technol **13**:553–558.

SÖRME L, BERGBÄCK B and LOHM U (2001) *Goods in the anthroposphere as a metal emission source – a case study of Stockholm, Sweden.* Water Air Soil Pollution **129**:213–227.

SPURNY KR, LODGE JP JR, FRANK ER and SHEELSLEY DC (1979) *Aerosol filtration by means of nucleopore filters: structural and filtration properties.* Environ Sci Technol **3**:453.

STAHLHOFEN W (1986) *Regional deposition of inhalable particles in humans.* In: Lee SD, Schneider T, Grant LD and Verkerk PV, eds. Aerosols. Lewis Publishers, Chelsea, Michigan.

STAHLHOFEN W, GEBHART J and HEYDER J (1980) *Experimental determination of the regional deposition of aerosol particles in the human respiratory tract.* Am Ind Hyg Assoc J **41**:385–398.

STELSON AW and SEINFELD JH (1981) *Chemical mass accounting of urban aerosol.* Environ Sci Technol **15**:671–679.

STERNBECK J, SJÖDIN A and ANDREASSON K (2002) *Metal emissions from road traffic and the influence of resuspension – results from two tunnel studies.* Atmos Environ **36**:4735–4744.

STEVENS RK, DZUBAY TG, RUSSWORM G and RICKEL D (1978) *Sampling and analysis of atmospheric sulfates and related species.* Atmos Environ **12**:55–68.

SUZUKI T, KONDO K, UCHIYAMA M and MURAYAMA M (1999) *Chemical species of organotin compounds in sediment at a marina.* J Agric Food Chem **47**:3886–3894.

TASK GROUP ON LUNG DYNAMICS (1966) *Deposition and retention models for internal dosimetry of the human respiratory tract.* Health Phys **12**:173–207.

TUNCEL G and ZOLLER WH (1987) *Atmospheric indium at the South Pole as a measure of the meteoritic component.* Nature (London) **329(6141)**:703–705.

UEMATSU M, DUCE RA, PROSPERO JM, CHEN L, MERRILL J and McDONALD RL (1983) *Transport of mineral aerosol from asia over the North Pacific Ocean.* J Geophys Res **88**:5343–5352.

US EPA (Environmental Protection Agency) (1971) *Reference Methods for the Determination of Suspended Particulates in the Atmosphere (High Volume Method).* US Fed Reg 36, No. 84.

US EPA (Environmental Protection Agency) (1982) *Air Quality Criteria for Particulate Matter and Sulfur Oxides,* Vol. III, EPA 600/8-82-092C.

VALENTA P, NGUYEN VD and NÜRNBERG HW (1986) *Acid and heavy metal pollution by wet deposition.* Sci Total Environ **55**:311–320.

VAN BORM WA and ADAMS FC (1988), *Cluster analysis of electron microprobe analysis data of individual particles for source apportionment of air particulate matter.* Atmos Environ **22**:2297–2307.

VANDERBORGHT B, MERTENS I. and KRETZSCHMAR J (1983) *Comparing the calculated and measured aerosol concentrations and depositions around a metallurgic plant.* Atmos Environ **17**:1687–1701.

VDI (Verein Deutscher Ingenieure) (1972) *Measurement of Particles in Ambient Air (in German).*VDI-Richtlinie 2463. VDI-Verlag, Düsseldorf.

V & F (1988) *Application Notes for the CI-MS 500 Real Time Gas Analyzer.* V & F Analyse- und Meßtechnik, Absams, Tyrol, Austria.

VAN DE VELDE K, FERRARI C, BARBANTE C, MRET I, BELLOMI T, HONG S, and BOUTRON C (1999) *A 200 year record of atmospheric cobalt, chromium, molybdenum, and antimony in high altitude alpine firn and ice.* Environ Sci Technol **33**:3495–3501.

VAR F, NARITA Y and TANAKA S (2000) *The concentration, trend and seasonal variation of metals in the atmosphere in 16 Japanese cities shown by the results of National Air Surveillance Network (NASN) from 1974 to 1996.* Atmos Environ **34**:2755–2770.

VOELKER B, MOREL F and SULZBERGER B (1997) *Iron redox cycling in surface waters: effects of humic substances and light.* Environ Sci Technol **31**:1004–1011.

WALSH PR, DUCE RA and FASCHING JL (1979) *Considerations of the enrichment, sources and flux of arsenic in the troposphere.* J Geophys Res **84**:1719–1726.

WANG HC and JOHN W (1988) *Characteristics of the Berner impactor for sampling inorganic ions.* Aerosol Sci Technol **8**:157–172.

WEATHERS KC, LIKENS GE, BORMANN FH, BICKNELL SH, BORMANN BT, DAUBE BC, EATON JS, GALLOWAY JN, KEENE WC, KIMBALL KD, McDOWELL WH, SICCAMA TG, SMILEY D and TARRANT RA (1988) *Cloudwater chemistry from ten sites in North America.* Environ Sci Technol **22**:1018–1026.

WEDDING JB (1982) *Ambient aerosol sampling. history, present thinking and a proposed inlet for invaluable particles.* Environ Sci Technol **16**:154–161.

WEI C and MORRISON GM (1994) *Platinum analysis and speciation in urban gullypots.* Anal Chim Acta **284**:213–227.

WEISEL CP (1981) *The atmospheric flux of elements from the ocean.* PhD Thesis, University of Rhode Island, R. I. Kingston.

WESTERLUND K (2001) *Metal emissions from Stockholm traffic – wear of brake linings.* Reports from SLB-analysis no. 3:2001, The Stockholm Environment and Health Protection Administration.

WHITBY KT (1978) *The physical characterization of sulfur aerosols.* Atmos Environ **72**:135–159.

WIERSMA GB and DAVIDSON CI (1986) *Trace metals in the atmosphere of rural and remote areas. In:* Nriagu JO and Davidson CI, eds. Toxic Metals in the Atmosphere. Wiley, New York.

WONG HKT, NRIAGU JO and COKER RD (1984) *Atmospheric input of heavy metals chronicled in lake sediments of the Algonquin Provincial Park, Ontario, Canada.* Chem Geol **44(1–3)**:187–201.

WOOD JM and GOLDBERG ED (1977) *Impact of metals on the biosphere. In:* Stumm W, ed. Global Chemical Cycles and their Alteration by Man. Dahlem Konferenzen, Abakon Verlag, Berlin.

ZOLLER WH (1983) *Anthropogenic perturbation of metal fluxes into the atmosphere. In:* Nriagu JO, ed. Changing Metal Cycles and Human Health. Dahlem Konferenzen, Springer, Berlin.

ZOLLER WH, GLADNEY ES and DUCE RA (1974) *Atmospheric concentrations and sources of trace metals at the South Pole.* Science **183**:198.

ZOLLER WH, PARRINGTON JR and KOTRA JMP (1983) *Indium enrichment in airborne particles from Kilauea Volcano: January 1983.* Science **222(4628)**:1118–1121.

ZUO Y, HOIGNE J (1992) *Formation of hydrogen peroxide and depletion of oxalic acid in atmospheric water by photolysis of iron(III)-oxalato complexes.* Environ Sci Technol **26**:1014–1022.

3

Deposition of Acids, Elements, and their Compounds

H. J. Fiedler

3.1

Introduction

In the sequence of emission–transport/conversion–deposition, the last phase has been dealt with in detail (Guderian 2000, 2001). Less emission leads to less deposition. Air pollutants exist as dust particles and in gaseous form, and for the effect of air pollutants their concentration in the air (immission) and their quantity deposited on receptors (deposition) is important. In ecosystems, both nutrients and harmful substances are deposited, and air pollutants may act as either acidic or alkaline, and as reductive or oxidative. The impact of immissions on plants may occur either directly (on plant leaves) or indirectly (through the soil), visible or invisible, latent or acute and chronic. Since they serve as sensitive receptors, plants and ecosystems (e.g., crops, forests, natural vegetation), soils and waters are of major interest. In order to characterize the atmospheric pollutant load of a special site in the landscape, it is first necessary to know the concentration situation and the deposited quantity per time and surface unit – the surface load. Concentrations may be measured over shorter or longer (> 24 h) periods. The concentration of gaseous air pollutants is measured in ppm, ppb or $\mu g\, m^{-3}$,

the deposition in $mg\, m^{-2}\, a^{-1}$, $kg\, ha^{-1}\, a^{-1}$ or $mol\, m^{-2}\, a^{-1}$. ['a' = annum]. In the case of gas concentrations, the transformation of ppbv in $\mu g\, m^{-3}$ is performed by multiplication with the factor 2.86. Exposure is the deposition on an area basis, e.g., $kg\, ha^{-1}\, a^{-1}$.

3.2

Types of Atmospheric Deposition

Total atmospheric deposition includes wet precipitation as well as gaseous (SO_2, NO_x, NH_3) and dry particulate components. Pollutants are removed from the atmosphere by direct adsorption or absorption by the soil, vegetation and water surfaces (dry deposition of sedimentation dust, aerosols, and gases), or by wet deposition. The wet deposition process consists of rainout (scavenging of particles and gases in the cloud, in-cloud scavenging) and washout (uptake of compounds by falling raindrops, below-cloud scavenging). Pollutants in rainfall enter the soil solution directly or after passing the canopy as throughfall or stem flow. Dry deposition is the most important deposition process in polluted areas, wet deposition in rainy remote areas. Whereas the wet deposition of atmospheric pollutants is comparatively uniform within a particular region

Elements and their Compounds in the Environment. 2nd Edition.
Edited by E. Merian, M. Anke, M. Ihnat, M. Stoeppler
Copyright © 2004 WILEY-VCH Verlag GmbH & Co. KGaA, Weinheim
ISBN: 3-527-30459-2

3.4
Deposition of Elements and their Compounds

3.4.1
Dust Deposition

Forests are efficient filters of dust. Dust influences photosynthesis and transpiration, and in part it also corrodes the plant surface. An accumulation of dust takes place near the emittent as well as in forests of higher altitudes after transport.

3.4.1.1
Natural Dusts

Deposition rates of locally generated dust, which is rich in calcium and magnesium, are high in the Limestone Alps of Austria. Alkaline dust particles have the size range of 1 to 20 μm. Dusts containing lime react as alkaline.

3.4.1.2
Industrial Dusts, Deposition of Heavy Metals

For accumulating trace elements in forest ecosystems stemming from the environment, the control parameter is the deposition rate in $kg\,ha^{-1}\,a^{-1}$. In Europe and North America, the atmospheric inputs of heavy metals in the open field and much more in forests reached remarkably high values in the past (Ulrich et al. 1979, Ulrich and Pankrath 1983, Rademacher 2001), and as a consequence several heavy metals have become enriched in the soil. In England, the net annual Cd input by atmospheric deposition, which had averaged $3.2\,g\,ha^{-1}$ over a period of 100 years, increased to $14\,g\,ha^{-1}$ in about 1980 (Johnston and Jones 1992). During the past two decades, the inputs of Cd and Pb have decreased however. In computing the critical loads of heavy metals, one must recognize that no further accumulation of these elements takes place, or their accumulation is lying beneath the critical limit value in the soil or in the soil solution. The accumulation of heavy metals in soils should be regarded as irreversible, and kept at an as low as possible level in order to preserve the agronomic value of soils for the future.

Trace elements and other pollutants transfer via the rainout- or washout-process into the precipitation water (wet deposition). The rainout-process contributes the main part of the concentration of the precipitation water measured on the soil. The washout-process gains importance in the case of fog in heavily polluted atmosphere above urban/industrial agglomerations. In about 1980, in east German urbanized areas, values of $2-16\,\mu g\,dm^{-3}$ Cu, $25-43\,\mu g\,dm^{-3}$ Pb, $1.8-3\,\mu g\,dm^{-3}$ Cd and $145-199\,\mu g\,dm^{-3}$ Zn were determined in rainwater. For Pb, the deposited quantity is ranged from a few $mg\,ha^{-1}\,a^{-1}$ in polar regions to $>10\,kg\,ha^{-1}\,a^{-1}$ in some highly populated and industrialized areas. About 1980, in the Federal Republic of Germany, wet deposition was measured for Pb at between $30\,\mu g\,m^{-2}\,d^{-1}$ and $100-200\,\mu g\,m^{-2}\,d^{-1}$, for Cd between $0.75\,\mu g\,m^{-2}\,d^{-1}$ and $2\,\mu g\,m^{-2}\,d^{-1}$ (Tables 3.1 and 3.2).

Larger parts of the heavy metal emission of smelters and power plants are deposited in the vicinity (up to 10 km distance) of the emitters. Pb is deposited after long-range transport of fine dust in forests of higher altitudes of the mountain area up to $0.5\,kg\,ha^{-1}\,a^{-1}$. In Germany, Pb deposition has been decreasing since 1974, mainly due to reduction of the Pb content in petrol (Ulrich 1991).

As a consequence of radioactive fallout of above-ground nuclear tests (1963–1966) and the reactor damage at Chernobyl, artificial radionuclides were deposited in larger amounts on the Earth's sur-

Tab. 3.1: Field deposition of selected heavy metals in Europe and North America [g ha^{-1} a^{-1}] in the 80 years. From UN/ECE and EK 2001: Der Waldzustand in Europa.

Region	Pb	Cd	Zn	Cu
Northern Europe + North of the USA + Canada	68	1	105	11
Central Europe/low land + shore	110	2	177	27
Central Europe + USA/highland + urban area	150	5	360	45
Central Europe + USA/industrial area	378	17	2018	80
All regions	140	4	242	37

Tab. 3.2: Average concentration (mg L^{-1}) and deposition rate (mg m^{-2} d^{-1}) of soluble inorganic constituents of field deposition and throughfall of spruce stands in 1983. From Brechtel and Sonneborn (1984).

Element	Open field bulk precipitation		Throughfall	
	Concentration	Deposition rate	Concentration	Deposition rate
Cl$^-$	4.2	5.8	9.6	5.8
Cd	0.0008	0.0012	0.0016	0.001
Cu	0.018	0.026	0.057	0.036
Fe	0.076	0.102	0.126	0.075
Mn	0.026	0.035	0.66	0.36
Ni	0.003	0.005	0.007	0.004
Pb	0.022	0.026	0.034	0.019
Zn	0.125	0.161	0.394	0.23

face, e.g., ^{134}Cs, ^{137}Cs, ^{90}Sr, and different Pu-isotopes.

3.4.2

Deposition of Acid Pollutants

Atmospheric deposition of acidifying substances to forests consists of both gases and particles (0.1–1.0 µm). Acid deposition is a mixture of acids and salts (Legge and Krupa 1990). H$^+$ is usually the dominant cation encountered by Ca^{2+}, Mg^{2+}, Na$^+$, K$^+$, and NH$_4^+$. Depending upon the geographical situation, the annual proton (H$^+$ ion) deposition varies between <0.2 and 1 kg ha^{-1} a^{-1}, while SO$_4^{2-}$, NO$_3^-$ and Cl$^-$ act as strong acid anions – the Cl$^-$ anion usually being of marine origin. Inputs of anions of sulfur and nitrogen are high in Central and Western Europe, but lower in

Scandinavia and in the south-western part of Europe (UN/ECE 2001).

Acid deposition interacts with the canopy of forest ecosystems (Lindberg and Lovett 1992), and in this way it may become either more acidic (wash off of accumulated dry deposition) or less acidic (exchanging H$^+$ against basic cations). As an example, in the Polish Swietokrzyski National Park the pH of rainwater (5.1) decreased as throughfall to 4.9 in a beech stand, and to 4.4 in a fir-beech stand. The acidity of stem flow water was even higher, with pH values of 2.9–3.6 for fir trees and 2.9–5.2 for beech trees in 2001. The stemflow water contains leached SO$_2$ sorbed under dry conditions by the bark, as well as organic acids (Kozlowski 2001). High SO$_2$-immissions may damage leaves and needles directly, but buffering of acid in the leaves leads to an acidification of the relevant soils (Table 3.3).

During recent years, Ca deposition has decreased and acid deposition increased in Central Europe. As a result of long-lasting acid input, soils in the lowland and in the highlands with a low buffering capacity under coniferous trees are now in the Al- or even in the Al-Fe-buffer range, at least in the upper soil.

3.4.2.1
Sulfur Deposition

The sulfur throughfall flux is controlled by the atmospheric sulfur concentration. Internal sulfur cycling (net canopy exchange) is a minor contribution to the sulfur flux in the forest floor in heavily sulfur-polluted areas. It is generally less than $0.2 \text{ g S m}^{-2} \text{ a}^{-1}$. SO_2-S is not absorbed irreversibly in the forest canopy, and will leach out as sulfate. The sulfur fluxes in Central Europe caused by filtering of air pollutants in forests are only marginally influenced by canopy exchange. In remote areas, the throughfall fluxes are only slightly higher than bulk precipitation fluxes. At such locations dry deposition to forests is a minor contribution to total atmospheric deposition, and internal cycling of sulfur might be a significant contribution (up to 20%) to the throughfall.

On the high sulfate inputs in the German mountain areas (e.g., Fichtelgebirge, Bayer-ischer Wald, Erzgebirge) the long-range transport of SO_2 is participated. The main source area for SO_2 emission in Europe is also characterized by a high sulfur deposition rate which, before 1900, usually exceeded $5 \text{ g S m}^{-2} \text{ a}^{-1}$. Since 1980, the deposition of sulfur has decreased in Europe, notably in northern Europe. Currently, dry deposition represents about 50–85% of the total atmospheric deposition (about 1.2–$3.0 \text{ g S m}^{-2} \text{ a}^{-1}$).

In the German state of Brandenburg, between 1985 and 2000, an input of 50–$100 \text{ kg } SO_4\text{-S ha}^{-1} \text{ a}^{-1}$ in pine forests was measured. The sulfate input about 1989 in Germany ranged from 20 to $80 \text{ kg } SO_4$ $\text{ha}^{-1} \text{ a}^{-1}$. Sulfur input has strongly decreased in the formerly severely damaged regions of the Erzgebirge (from $> 100 \text{ kg S ha}^{-1} \text{ a}^{-1}$ to about $40 \text{ kg S ha}^{-1} \text{ a}^{-1}$ in 1994–98). In 2001 in Saxony, the annual S-inputs with stand precipitation ranged from 10 to 20 kg ha^{-1}. For some years in this forest region, the S-inputs have been markedly lower than the S-outputs in seepage water. In 2001, areas with a sulfate input of $12.8 \text{ kg } SO_4\text{-S}$ $\text{ha}^{-1} \text{ a}^{-1}$ ($= 800 \text{ mol}_c \text{ ha}^{-1} \text{ a}^{-1}$) may be found in all parts of Europe except for the central and northern regions of Scandinavia. In the Limestone Alps of Austria, the input of sulfate-sulfur in the open field was 8 kg ha^{-1}, and the flux underneath a spruce canopy was 9.6 kg ha^{-1} (Glatzel

Tab. 3.3: Deposition rates (1985–1991, open field and throughfall in kg ha^{-1} a^{-1}) at forest stations in different altitudes of the Eastern Ore Mountains. a) Tharandt Forest: 380 m a.s.l., spruce 109 years. b) Oberbärenburg: 735 m a.s.l., spruce 44 years. From Wienhaus (1996).

Station	SO$_4$-S	NO$_3$-N	NH$_3$-N	Na	K	Ca	Mg
Tharandt:							
Throughfall	129	16	22	9.6	23	78	9.2
open field	31	7.1	8.1	5.9	3.7	36	4.1
Oberbärenburg:							
Throughfall	91	14	15	9.2	28	54	7.2
open field	29	7.1	9.3	7.6	4.1	33	4.2

et al. 1988). The largest fluxes have been measured in forest stands in the South of the Netherlands (up to 17 g S m^{-2} a^{-1}).

3.4.2.2
Nitrogen Deposition

N input in agricultural soils derived from atmospheric deposition of NO$_x$ and NH$_y$ ranged from about 0.5 g N m^{-2} a^{-1} in the central USA to 6 g N m^{-2} a^{-1} in western Europe (Andreae and Schimel 1989, Ulrich 1991) or general 10–60 kg ha^{-1} a^{-1} before 1996.

Throughfall fluxes of N tend to be lower than atmospheric inputs because of uptake of atmospherically deposited N in the canopy. The inorganic nitrogen flux in throughfall will be underestimated in the order of 0.2–0.5 g N m^{-2} a^{-1}. Total deposition of nitrate and also ammonium can be estimated via throughfall monitoring only during winter conditions, when the trees are less active.

Under pristine conditions, atmospheric N deposition is in the order of 1 to 4 kg ha^{-1} a^{-1}. At about 1990, atmospheric N inputs in forests in western Europe ranged from close to pristine background levels in northern Scandinavia to about 140 kg ha^{-1} a^{-1} (throughfall N input) in the most heavily polluted areas in the southern parts of the Netherlands. In the past few years, throughfall fluxes have been of the order of 20–40 kg ha^{-1} a^{-1} over most of west-central Europe. Therefore, strain by nitrogen input on forest ecosystems is still high. In 2001, the annual N-inputs with the stand precipitation in Saxony amounted to 15–30 kg ha^{-1}. High N-inputs of >22 kg N ha^{-1} a^{-1} (1600 mol$_c$ ha^{-1} a^{-1}) take place in Central Europe. In nitrogen-polluted areas, dry deposition of nitrogen to forests will be relatively high and, despite irreversible foliar uptake of nitrogen, the throughfall flux will exceed the bulk precipitation flux.

In past years N-deposition in the open field in the Bavarian Alps was only about 11 kg ha^{-1} a^{-1}, and the input of sulfate-S was also moderate at <5 kg ha^{-1} a^{-1} (Bayer StMLU 2000).

Nitrogen compounds are an important group of acidifying pollutants (Ulrich 1991). High nitrification rates were observed in forest soils even under heavy loadings of acid deposition. For nitrogen, the output of the ecosystem is much more lower than the atmospheric input, also because it is stored strongly in the soil. Therefore, until now sulfate has been regarded as the most important reason for soil acidification.

Nitrogen inputs in ecosystems by acid deposition are transformed in the soil and affect the acid–base relationships of the nitrogen cycle. Forms of nitrogen input most often observed are (NH$_4$)$_2$SO$_4$, HNO$_3$, and NH$_4$NO$_3$. Inputs of these compounds are associated with acidification effects through nitrification, nitrogen uptake by plants in form of NH$_4^+$, cation exchange and leaching.

Ammonium and nitrate deposition

An important part of the nitrogen input of forested ecosystems in Europe is in the form of ammonia and ammonium. Normally, approximately equal amounts of NO$_3$-N and NH$_4$-N are deposited. Farming, by the effect of animal production, causes a rise in the emission of NH$_3$, and in turn the NH$_4$-deposition. In Germany in 1989, the range of land deposition was between 4 and 19 kg NH$_4$ ha^{-1} a^{-1}, whereas in the United Kingdom and North West Europe it was about 10–20 kg N ha^{-1} a^{-1} (Derwent et al. 1988). Close to ammonia source areas, not only high ammonium inputs but also large sulfur inputs occur to the forest soil, this being due mainly to the co-deposition of ammonia and sulfur dioxide on moist tree surfaces and to the deposition of ammonium sulfate particles (Cape 1998). In the higher altitudes of mountain areas, ammonium sulfate particles stemming from long-range transport take part in the NH_4-deposition.

In Europe up to 1990, the average NH_4-throughfall flux was 1.3 g NH_4-N $m^{-2} a^{-1}$ (< 1 to 5 g). In the region of The Netherlands/Belgium/Northern Germany, the contribution of NH_4-N to the total nitrogen flux in stand and bulk precipitation was 70–80%, indicating the high ammonia pollution in this region.

In Europe (1990), the average NO_3-flux in bulk precipitation was 0.60 g NO_3-N $m^{-2} a^{-1}$, and the NO_3-throughfall flux was 1.4 g NO_3-N $m^{-2} a^{-1}$. High nitrate input takes place in the medium and upper altitudes of the mountain area; for example, in 1989 in the Fichtelgebirge it was 45–50 kg NO_3 $ha^{-1} a^{-1}$.

In 1997, in the Northern Limestone Alps of Upper Austria, and between late May and mid October, the annual bulk-deposition of nitrate-nitrogen was 8 kg ha^{-1}, while the throughfall-flux was 9.6 kg ha^{-1} (Katzensteiner 2000). Deposition rates of nitrate were between 3 and 5 μmol_c $m^{-2} h^{-1}$ in periods without precipitation. Depending on the nutritional status of the stands, an enrichment or a depletion of ammonium could be observed in the throughfall. The minimum canopy uptake in the vegetation period was 1.5 to 3 kg ha^{-1} N for spruce stands. Nitrogen saturation with increased output rates of nitrate and subsequent losses of potassium would be critical for the long-term stability of forest ecosystems on leptosols and cambisols.

3.4.3
Deposition of Alkalizing Substances and Basic (Base) Cations

The atmospheric input of basic salts such as carbonates will counteract acid atmospheric inputs and soil acidification by replenishing the basic cations that have leached from the soil. In areas with calcareous soils, e.g., in southern Europe and Lithuania, a relative high atmospheric input of base cations (> 800 mol_c $ha^{-1} a^{-1}$) was observed. In Europe in 1989, the total alkaline deposition was about 77 mEq $m^{-2} a^{-1}$.

The basic cations of interest are calcium, magnesium, and potassium. Besides sea-salts (1–20 μm), sources of Ca, Mg and K are soil dust, fertilizers, fly-ash, and industrial sources. In order to balance the deposition of base cations, correction must be made for the deposition of sea salts, assuming that chloride or sodium originate only from these salts. Furthermore, throughfall data cannot be used directly for estimating the total base cation deposition, due to the internal cycling of base cations in vegetation. The bulk precipitation approach yields a minimum estimate of the atmospheric base cation deposition to forests. The best estimate can be made by combining throughfall measurements of Ca, Mg and K and bulk precipitation data for sodium and making a "filtering" analogy using bulk precipitation data for the other ions. By using this approach, it can be assumed that the forest filtering of Ca, Mg, and K particles is equal to the filtering of sodium-bearing particles.

Calcium is the most important component of the alkaline deposition (about 70%). In Germany in 1985, the Ca^{2+}-input varied between < 10 and 20 kg $ha^{-1} a^{-1}$, whilst in the Bavarian Alps inputs of between 6 and 11 kg $ha^{-1} a^{-1}$ were measured in the open field (Bayer StMLU 2000). *Magnesium* contributes about 8% to total alkaline deposition. The Mg input varied in Germany from < 2 to 6 kg $ha^{-1} a^{-1}$ (1985), and in the Bavarian Alps from 0.1 to 0.8 kg $ha^{-1} a^{-1}$ (Bayer StMLU 2000). *Potassium* is enriched in the throughfall of all stands by an order of magnitude compared to precipitation, with leaching rates in the range of 9 to 11 $\mu mol_c\, L^{-1}\, m^{-2}$. In Europe, potassium contributes about 20% to the total alkaline deposition. Atmospheric input rates of potassium are very low in the Austrian Limestone Alps, while in the Bavarian Alps $3-4$ kg K ha^{-1} a^{-1} was deposited in the open field (Bayer StMLU 2000).

3.4.4
Deposition of Sea-salt Particles and Chloride Deposition

A part of the throughfall flux of sulfur, calcium, magnesium, and potassium is due to the deposition of sea-salt particles. Ion fluxes in bulk and stand precipitation can be corrected for the contribution of sea-salt particles, using sodium or chloride as sea-salt tracers. Sea-salt contributions between 8% (Europe) and 45% (coastal sites) of the total sulfur in stand precipitation were found. Deposited sea-salts are neutral salts, and these do not contribute to either acidic (in the case of sulfur) or alkaline (in the case of base cations) reactions in the forest soil (Büttner et al. 1986).

The atmospheric deposition of Cl^- reflects the maritime influence. In systems of heavy marine influence, the Cl^--inputs can be substantial (about 200 kg Cl ha^{-1} a^{-1}), but over time the output flux is equal to the input. In continental systems, Cl^--concentrations are generally low and mostly unaffected by anthropogenic inputs.

3.5
Deposition and Forest Ecosystems

Of special interest is the impact of the four air pollutants – SO_2, NO_x, H_2SO_4 and HNO_3 – on the health, growth, and mortality of forests, and also the temporal and spatial variability of concentrations of these compounds in the deposition (Forschungsbeirat 1993, Raspe et al. 1998).

Nutritional benefits may often be associated with pollutant inputs. Depending on the amount of atmospheric input involved, the pollutants can be harmful – as seen in the case of N saturation or S eutrophication of forests in Central Europe (Johnson and Lindberg 1992). One of the most important indirect effects of the pollution load of forests is that of accelerated soil acidification by acid rain (Reuss and Johnson 1986, Kreutzer et al. 1998). Acid deposition also affects nutrient cycling processes in forest ecosystems (Ulrich 1991, Matzner 1988, Kazda 1990).

3.5.1
Site and Stand Dependence

3.5.1.1
Elevation
Total rates of deposition to forest canopies are influenced by both topography and elevation. Meteorological factors which increase with elevation, such as wind velocity, precipitation and fog frequency, will increase deposition (Flemming 1993; Langusch 1995, Abyi 1998; Tables 3.3 and 3.4).

3.5.1.2
Soils
Pollutant deposition may increase soil acidity, decrease nutrient availability, and increase the solubility of toxic ions. In central and northern Europe, increased pollutant deposition over several decades has induced soil acidification. In the soil, acid deposition undergoes many reactions, and this leads to a reduced alkalinity and increased aluminum content in the soil solution. The exchange complex of the soil becomes dominated by aluminum, the exchange acidity increases, bases are leached in association with acid anions due

Tab. 3.4: Medium fluxes of water and elements with the open-field bulk precipitation (NF) and throughfall (NB) as well as area output (NA). Element fluxes in kg ha^{-1} a^{-1}. Catchments Schluchsee in the Black Forest (1150–1253 m a.s.l.) and Rotherdbach in the Ore Mountains (675–750 m a.s.l.). From Armbruster et al. (2001).

	Schluchsee (HY 88 – HY 98)			Rotherdbach (HY 95 – HY 99)		
	NF	*NB*	*NA*	*NF*	*NB*	*NA*
mm H$_2$O	1867	1543	1381	989	803	590
H$^+$	0.34	0.26	0.01	0.32	0.91	0.44
Na$^+$	4.1	4.7	21.2	2.3	4.6	27.9
K$^+$	2.1	13.3	7.8	1.0	14.6	14.0
Ca^{2+}	3.9	6.1	13.6	3.5	13.6	43.3
Mg^{2+}	0.7	1.1	2.2	1.3	4.0	15.2
NH$_4^+$-N	5.0	3.6	0.1	6.7	8.7	0.2
NO$_3^-$-N	4.5	5.4	6.9	6.4	11.4	11.4
N$_{ges}$ [a]	9.5	8.9	7.0	13.1	20.1	11.6
SO$_4^{2-}$-S [b]	6.8	8.4	16.3	10.7	34.0	71.3
Cl$^-$	8.4	9.1	9.4	6.1	11.7	58.6
Al$_{ges}$	0.16	0.25	3.3	0.26	0.98	15.6
Mn$_{ges}$	0.06	0.41	0.23	0.05	0.35	1.46
Fe$_{ges}$	0.10	0.14	0.11	0.14	0.47	0.43
DOC	21.4	57.0	18.9			28.0

[a] N$_{tot}$ = NH$_4^+$-N + NO$_3^-$-N [b] Rotherdbach S$_{tot}$ (SO$_4^{2-}$-S + S$_{org}$). HY = hydrological year.

to the charge balance principle, and the chemistry of the surface waters is changed. Soil acidification may set free heavy metals enriched before and dangers by this the quality of ground and surface waters.

The "needle yellowing of spruce in the upper altitudes of the mountain area" has been spread mainly due to air pollution on nutrient-poor soils which have been acidified by acid deposition (Mg deficiency).

3.5.1.3

Forest stands

The direct impact of anthropogenic air pollutants on the needles and leaves of plants may cause a variety of damage. High average loads and especially extreme load pikes of SO$_2$ in the winter months, lead to characteristic damage symptoms in the case of spruce. The relative tolerance of trees to SO$_2$ pollution is an intrinsic factor of morphological and physiological defense mechanisms (thick cuticle, stomatal resistance, presence of free radical scavengers such as ascorbic acid, low sulfate levels).

Forests collect more pollutants than do surrounding surfaces with lower vegetation. For example, the forest edge will disturb the vertical wind profile and induce air turbulence that will in turn increase the dry deposition. The deposition at the front is considerably higher compared to that in the open field (by a factor of between 5 and 20), and also to that within the forest (a factor of 2 to 4). The increased deposition affects the vitality of the trees at the edge. Forest structures (tree species, crown density, stem density) differ widely in terms of aerodynamic roughness and leaf area, and these factors will each influence deposition.

Spruce and fir trees have a larger filtering biomass than pine and deciduous trees. For example, in coniferous stands the dry deposition fluxes of sulfur are about 2.5-fold

Tab. 3.5: Average total atmospheric input (dep.) as well as output of sulfate, nitrogen, base cations and Al in $mol_c ha^{-1} a^{-1}$ 1995–1998. From UN/ECE and EK (2001): Der Waldzustand in Europa

Type of tree	Number of sites	SO_4		N		Ca + Mg + K		Al
		dep.	output	dep.	output	dep.	output	output
Pine	29	517	197	704	7	491	156	138
Spruce	51	685	590	1197	112	448	331	774
Oak	15	637	1025	683	212	519	2184	30
Beech	20	634	604	1327	135	489	717	326

larger than in deciduous stands. Coniferous forest stands are about 1.2- to 2.2-fold more efficient in capturing dry atmospheric nitrogen than are deciduous forest stands The interception deposition is lower in broad-leaved and larch species than in spruce stands, mainly due to a lower leaf index and also to defoliation during winter (Capellato et al. 1993) (Table 3.5).

The exchange of H^+ in precipitation for base cations in the foliage will increase the base cation content of the throughfall and decrease the H^+ concentration (Ibrom 1993). In the Limestone Alps, throughfall of a beech stand had a higher pH-value and a higher alkalinity than throughfall in spruce stands (Katzensteiner 2000).

3.5.2
Forest Decline

Air pollutants act partly by direct fashion, and partly by indirect fashion. Hence, SO_2 can act either directly on spruce needles (wax layer, stomata openings, nutrient content, buffer capacity), or in the form of sulfuric acid in the soil via H- and Al-ions on plant roots. Gaseous pollutants act aboveground, whereas total deposition acts below-ground.

3.5.2.1
Effects of pollutant combinations

Since the late 1970s, the phenomenon of forest decline in stands of Norway spruce has been recorded at the higher elevation sites of Central European mountain regions. The main symptoms were needle yellowing (attributed to Mg-deficiency) and needle loss. At that time in such sites of the north-eastern Bavarian mountains, the following total deposition rates were measured (Schaaf et al. 1991): $1.2–4.5 kmol H^+ ha^{-1} a^{-1}$; $35–189 kg S ha^{-1} a^{-1}$; and $23–46 kg N ha^{-1} a^{-1}$. About $0.3–3.0 kmol H^+ ha^{-1} a^{-1}$ were buffered in the forest canopy, and this resulted in an accelerated leaching of alkali- and earth-alkali ions from the needles. Photo-oxidants, especially ozone, also participated in this process.

3.5.3
Critical Levels and Loads

Within the frame of activities of the UN-ECE has been developed the Critical Levels- and Critical Loads-concept for ecological load limits against different air pollutants (UN-ECE 1988, Nagel and Gregor 1999). Critical levels are defined as concentrations of air pollutants below which no direct damage to receptors (plants, ecosystems, materials) is expected. A critical level is the maximum concentration of a pollutant at which adverse effects will not occur on sensitive

Tab. 3.6: Critical levels of important air pollutants for vegetation (UN-ECE 1990, Wienhaus 1996, Smidt 1997)

Pollutant	Duration	Concentration
Sulfur dioxide	Annual mean	$(15-)20-30 \ \mu g \ m^{-3}$ (7.5 ppb)
Nitrogen oxide	Annual mean	$30 \ \mu g \ m^{-3}$ (15 ppb)
Ammonia	Annual mean	$>8 \ \mu g \ m^{-3}$
Acid rain/fog	Mean value, vegetation period	$>1 \ mol \ H^+ \ m^{-3}$
Ozone	Mean value, growing season	$50 \ \mu g \ m^{-3}$ (25 ppb)

targets. For Europe, the critical levels have been mapped (Hettelingh et al. 1991), and critical levels exist for SO_2, NO_2, and O_3 (Table 3.6). These pollutants may occur in combinations.

3.5.3.1
Critical levels of sulfur

SO_2 concentrations of <5 ppbv cause no plant damage, and standards for the protection of forests are set between 9 and 19 ppb SO_2. The typical range for rural areas is 1 to 30 ppb SO_2. High concentrations of SO_2 in the air may cause direct damage of the needles and leaves of trees, especially in the range of 200 to 2000 ppb SO_2 (Table 3.7).

A zonation of sulfur load areas is further possible on the base of the sulfur content in spruce needles. The contents of total sulfur in the needles of spruce should not exceed the following limits: $<0.11\%$ S in dry matter for 1-year-old; 0.14% S for 2-year-old; and 0.17% S for 3-year-old needles. In the case of beech, the limit in the leaves is 0.08% S.

3.5.3.2
Critical loads

Critical load is a quantitative estimate of an exposure to one or more pollutant. Critical load is the highest deposition of pollutants that will not cause chemical changes leading, for example, to long-term harmful effects on ecosystem structure and function or other sensitive elements of the environment. Harmful effects may be chemical changes in soils and waters or changes in populations.

Critical loads to forest soils have been proposed for the deposition of total nitrogen, total sulfur, and total acidity. Critical loads models assume that indirect effects occur on trees via changes in soil chemistry. However, there exist important direct and indirect impacts of wet and dry deposition on leaves and needles with regard to photosynthesis, nutrient leaching, stomatal function, and leaf surface properties. The values for critical loads are influenced by precipitation, elevation, soil texture, and base cation deposition.

Tab. 3.7: Characteristics of the fume-damage zones in the Ore Mountains of Saxony

Damage zone	Vitality of stands	Long-term SO_2 level [mg m^{-3}]	Growth depression of spruce [%]
Ie	Died	0.120	80
I	Dying	0.100	50
II	Heavily damaged	0.085–0.090	30–40
III	Damaged	0.055–0.065	10–20

3.5.3.3
Critical load of nitrogen

This is normally defined with respect to eutrophication, and not to acidification. The critical load of nitrogen is the maximum deposition of nitrogen compounds that will not cause eutrophication or induce any type of nutrient imbalance in any part of the ecosystem (Table 3.8).

The long-term critical load for nitrogen deposition [mmol N m^{-2} a^{-1}] can be defined as: forest harvesting + denitrification − biological nitrogen fixation.

The long-term nitrogen deposition should not exceed nitrogen removal by forest harvesting. In commercially harvested forests, N-deposition should remain below 10 (15) to 20 kg ha^{-1} a^{-1} in order to prevent N saturation. In the Netherlands, nitrogen deposition in forests is almost 8-fold the critical load.

A critical load for nitrogen as a nutrient (CL$_N$) should be determined as:

$$CL_N = N_u + N_{im} + N_l$$

where N$_u$ is the yearly mean net uptake of nitrogen in tree biomass, N$_{im}$ is the acceptable long-term soil net immobilization, and N$_l$ is the acceptable leaching ($<7-$ 14 mmol m^{-2} a^{-1}) (UN-ECE 1990).

Sometimes, critical loads for nitrogen are defined in order to avoid ecosystem disturbances due to eutrophication, as well as to avoid nitrate leakage (N-saturation) and acidification. Important factors are the initial nutrient status of the ecosystem and its ability to take up nitrogen.

3.5.3.4
Consequences of an increased nitrogen deposition

Nitrogen deposition enters the ecosystem as a slow, steady input rather than in one or a few large doses − this is especially the case with fertilizer input − which is important with respect to nitrification and nitrate leaching (Diese and Wright 1995).

The uptake of gaseous and liquid nitrogen depositions influences the nutritional status of trees (Eilers et al. 1992, Gebauer et al. 1991). Nitrogen input can occur through: (i) stomatal uptake (NO$_x$, NH$_3$, HNO$_3$); (ii) absorption of liquid deposition (NO$_3^-$ and NH$_4^+$ ions) deposited on the needle or bark surfaces; and (iii) root absorption of NO$_3^-$ and NH$_4^+$ ions in the soil solution. For uptake studies of these compounds, ^{15}N or the natural nitrogen isotope ratios in needles, twigs, or roots are used.

Where forest management is intensive, most forests are N-limited. For a certain time, N deposition acts positively on stand growth, the effect being similar to that of nitrogen fertilization. Later on, negative changes become apparent, such as nitrate pollution of drainage water and disturbances in Mg-nutrition, soil acidification by nitrification, and nutrient imbalances. Increased N-loading creates the potential for the dominance of a nitrophilic, acid-tolerant vegetation, and changes from heather to grass in pine forests are an example of this.

Increased N deposition is an important factor in forest decline. The exposure of plants to elevated concentrations of ammo-

Tab. 3.8: Empirical critical nitrogen loads (CL$_N$) for natural vegetation. After Draft Mapping Manual (1990)

Vegetation	kg N ha^{-1} a^{-1}	mmol m^{-2} a^{-1}
Heathlands	7–10	50–70
Raised bogs	5–10	35–70
Coniferous forests	10–12	70–85
Deciduous forests	<15	<110

nia gas and ammonium aerosol in the atmosphere leads to an accumulation of ammonium in the leaves by direct uptake during dry periods. The high nitrogen content causes relative magnesium, calcium, potassium, and phosphorus deficiencies, while ammonium can induce leaching of Ca, Mg and K. An increased deposition rate of ammonia may enhance soil nitrification and Al mobilization.

The mineral N flux density in the soil (throughfall deposition plus net mineralization) is an important parameter, which could predict the degree of saturation in a forest ecosystem (Gundersen 1991). If N deposition in forest ecosystems exceeds $20 \text{ kg N ha}^{-1} \text{ a}^{-1}$, this elevated deposition is a continuous addition to the background flux of mineral N from net mineralization, which normally amounts to $30-50 \text{ kg N ha}^{-1} \text{ a}^{-1}$. In the long term, these additions exceed the capacity of plants and soils. After an excess amount of nitrogen has been built up in the top-soil by high atmospheric input, there exists a potential risk of acidification pushes caused by nitrification during drought periods.

To avoid an aggravation of soil acidification and nutrient imbalances a reduction of NO_x (traffic) and NH_3 (agriculture) emissions is necessary.

3.5.3.5
Critical load of sulfur

Its knowledge may be helpful to avoid acidification. The critical load depends mainly on the chemical weathering ability of the soil, which in turn depends on the soil composition. The long-term critical load of sulfur deposition for forest soils can be defined as (Ivens 1990):

$$Scl = BCw + BCdep$$
$$- BCbr \text{ } [\text{mmol}_c \text{ m}^{-2} \text{ a}^{-1}]$$

where Scl is the critical load of sulfur deposition; BCw is the base cation weathering; BCdep is base cation deposition; and BCbr is net bio-removal of base cations.

For nutrient-poor, noncalcareous forest soils, BCw is in the order of $10-100 \text{ mmol}_c \text{ m}^{-2} \text{ a}^{-1}$. BCbr is generally in the range of 0 to $60 \text{ mmol}_c \text{ m}^{-2} \text{ a}^{-1}$.

3.5.3.6
Consequences of an increased sulfur deposition

In acid precipitation, the dominant strong acid anion is commonly sulfate. SO_2, when entering the soil, is rapidly oxidized to SO_3, and is therefore equivalent to an input of sulfuric acid. Sooner or later, the SO_2 absorbed by plants also becomes SO_4^{2-} and enters the soil. Sulfur deposition therefore increases the sulfate concentration of the soil solution. In most natural acid ecosystems this anion is rare.

Today, the input of nitrogen in European ecosystems is generally larger than that of sulfur. Nevertheless, more sulfate than nitrate is washed out of the soil, because sulfur accumulated in former times now is set free and the ecosystems often are not yet saturated with nitrogen. On a molar charge basis, sulfur deposition induces an equivalent production of acid, leading to an equivalent leaching of base cations or aluminum together with sulfate from the soil.

3.5.3.7
Critical load of acids; long-term effects of acid deposition and acid formation on terrestrial ecosystems

Anthropogenic air pollution is an important source of acidity in ecosystems without carbonates in the soil (Legge and Krupa 1990). The sum of nitrogen and sulfur loads contributes to the critical acidity load. The acidity is linked with the cycles of base cations and phosphorus.

The critical load of soil acidity (CL_a) is calculated by balancing all sources of alkalinity against all sources of acidity in terms of fluxes of equivalents (e.g., $kmol(+)$ $km^{-2} a^{-1}$):

$$CL_a = BC_w - BC_u - L_a - A_N$$

with BC_w the base cation release due to weathering, BC_u the net long-term uptake of base cations by trees and vegetation, L_a the leaching of alkalinity from the soil compartment, and A_N the net acidity produced by nitrogen uptake and ammonium nitrification.

Expressed in terms of the sum of sulfur and nitrogen deposition the critical load of acidity CL_a can be written as:

$$CL_a = CL(SO_4\text{-}D + NO_3\text{-}D - BC\text{-}D)$$

with CL critical load, SO_4-D non-marine sulfate deposition, NO_3-D nitrate deposition, and BC-D non-marine base cation deposition under critical load conditions (UN-ECE 1990).

The pH of the precipitation water in southern Scandinavia fell to pH 4.2 until 1987, this being similar to pH values in the most exposed parts of the European continent. Acid mist contains a series of compounds, including sulfuric acid (H^+) and NH_4^+ (Schemenauer 1986). Cation leaching of the needles may be due to acidic mist and cloud water droplets. Whilst Ca^{2+} and Mg^{2+} are leached in greatest amounts (Ca > Mg) by inorganic acids, solutions of ammonium salts leach K^+. The regulatory ability of the stomata is affected by acid deposition. Leaching of elements from Norway spruce trees treated with ozone plus acid mist is increased with the acidity of the mist applied. Acid deposition may cause reduced base cation uptake (Ca, Mg) in soils where Al is brought into solution.

Acid deposition undergoes many reactions in the soil, and leads to a change in the soil solution composition. The exchange complex of the soil becomes dominated by aluminum, the exchange acidity increases, bases are leached in association with acid anions, and the chemistry of the surface waters is changed. Increased deposition (wet and dry) of acid or potentially acidifying compounds (e.g., ammonia/ammonium) as well as decreased deposition of alkaline or acid-neutralizing compounds may decrease the soil pH.

Depending upon the weathering rate, the critical loads for acids vary in forest soils, from < 20 to > 200 kEq. $H^+ km^{-2} a^{-1}$ (Hettelingh et al. 1991, s. Wienhaus 1996). The acidic depositions should not exceed the weathering rates of soil minerals or the parent rock. Proton input resulting from wet deposition plus organic acid dissociation may exceed the rate of base cation weathering. This results in high aluminum concentrations and a low pH in the soil water. For the pH decreases of deeper soil horizons, the increased flux of deposited strong mineral acids seems to play a major role. Most B horizons are within the aluminum buffer range.

The level of acidification caused by sulfur and nitrogen deposition is affected by several processes, including base cation uptake, nitrogen uptake, and base cation deposition. Acid deposition affects the soil–plant system and, via soil solution, also the surface waters and groundwater. Leaching plus harvesting reduces soil and biomass Ca, and leads to a significant acidification of ecosystems. The soil base cation status decreases.

The main processes which must be considered in a proton budget of ecosystems are: atmospheric deposition of H^+ (including wet deposition and dry deposition of acid precursors such as SO_2 and NO_x); nitrogen transformations; cation accumulation (other than NH_4^+ in perennial biomass and humus); organic acid and carbonic acid dissociation; and anion weathering. Proton sinks must also be considered: H^+ output in the drainage water; nitrogen transformations; anion accumulation other than NO_3^-; and cation weathering (Berdén et al. 1987).

A comparison of element inputs from the atmosphere and element outputs by leaching of the subsoil (matter balances) provides evidence about the storage of elements in the ecosystem. The S- and N-balances provide information about the acidification rate, which is caused by air pollution (Liu et al. 1993). To diminish the impacts of acid inputs, selected forest sites should be limed with calcium carbonate or dolomite at a level of 3 t ha^{-1}, with a replication after 3 to 5 years.

3.5.4
Change with Time

Since the beginning of industrialization, the atmogenic inputs of sulfur dioxide, nitrogen oxides and ammonia have changed in large areas and down to the subsoil the chemical state of soil (Asman et al. 1988). Forest decline by loss and yellowing of needles, growth depression and mortality reached peak values in 1985 and 1986, with especially the higher elevated regions of the mountains being involved in Germany. By contrast, on a regional scale during the past few decades, increased nitrogen inputs, better management and the rising CO_2-concentration has led to faster growth of trees.

From 1983 to 1989, in the German state of Rheinland-Pfalz, between 9 and 12 kg S and 10 to 16 kg N ha^{-1} a^{-1} were deposited in the open field. Under spruce stands, these loads were increased to 24–43 kg S and 28–

41 kg N ha^{-1} a^{-1}, respectively. The maximum input rates were 72 kg S (in 1984) and 45 kg N ha^{-1} a^{-1}. The total acid deposition into spruce stands was calculated as 1.8–3.1 kmol H ha^{-1} a^{-1}, and the S-, N-, and H- deposition rates by far exceeded the critical loads (Forstl. Versuchsanst. 1991).

During the past few years, the nitrogen compounds (NO_x, NH_3) have steadily increased, but then reached a plateau. In the case of nitric oxides, in the eastern Ore Mountains between 1992 and 1998, no trend could be identified relating to air pollution levels. The mean annual level of NO was 2 µg m^{-3}, and that of NO_2 was 10–15 µg m^{-3} (Table 3.9). In contrast, in the foreland of the Ore Mountains the loading was considerably reduced; for example, levels fell from 15 to < 35 kg N ha^{-1} a^{-1} above the critical value in 1995 to 5–15 kg ha^{-1} a^{-1} above the critical value in 2001.

Between 1971 and 1989 in Zinnwald (a town in the crest sites of the Ore Mountains), the medium annual SO_2 load was 71 µg m^{-3}. Subsequently, the annual mean values fell from 50 µg m^{-3} in 1992–93 to 10–20 µg m^{-3} in 1998. In 1999, the annual mean values fell in large areas to less than 10 µg m^{-3} SO_2. Large-scale risks due to sulfur dioxide are no longer to be expected for the forest ecosystems of this region (Zimmermann and Wienhaus 2000). The maximal day values of SO_2-concentrations fell from 170–260 µg m^{-3} in the winters of 1992–93 to 1996–97, and in winter 1997–98 was only 70–80 µg m^{-3}.

The reduction in the emissions of dusts and sulfur dioxide has led during the past few years to an important lowering of the immission and deposition of these substances. The part of nitrogen in total acidification rose generally from 21 to 36%, in the exposed location of Klingenthal (Saxony), from 48–63%. In Saxony, the acidification

Tab. 3.9: Annual concentration (in µg m^{-3}) of trace gasses in Oberbärenburg (735 m a.s.l.) and in the Tharandt forest (380 m a.s.l.), Osterzgebirge. May 1992 to December 1998. From Zimmermann and Wienhaus (2000).

Trace gas/site	1992	1993	1994	1995	1996	1997	1998
SO_2							
Oberbärenburg	59	50	38	35	30	25	15
Tharandt			36	26	37	24	10
NO_2							
Oberbärenburg	14	15	14	11			14
Tharandt		13	13	15	12	8	12

pressure by sulfur and nitrogen immissions has fallen to one-third of its previous values during the last few years. The surplus, as compared to the critical loads, decreased for the total acidity from 5.2 to 1.7 kEq ha^{-1} a^{-1}. This development is overlapped by mobilization of sulfur, which was abundantly in forest soils due to immission in the past. In contrary to the direct impact on needles and leaves, the indirect impacts via the acidified soil last for longer periods of time.

The rates of precipitation deposition are mapped today and allow a comparison to be made with the results of the forest decline inquiry. To get below the critical load limit for the acid input under the present conditions of reduced input of alkaline dust, it is necessary to reduce not only the SO_2-emission but also the emission of NO_x and NH_3. The reduction of the nitrogen compounds is also necessary in view of a reduction of the possible eutrophication of terrestrial ecosystems, the nitrate load of the groundwater and surface waters, as well as a reduction of the ozone load.

References

ABYI M (1998) *Standortskundliche und hydrochemische Untersuchungen in zwei Wassereinzugsgebieten des Osterzgebirges.* Dissertation TU Dresden.

ANDREAE MO and SCHIMEL DS, EDS. (1989) *Exchange of Trace Gases between Terrestrial Ecosystems and the Atmosphere.* John Wiley & Sons, New York.

ARMBRUSTER M, ABYI M, SEEGERT J and FEGER K-H (2003) *Wasser- und Stoffbilanzen kleiner Einzugsgebiete im Schwarzwald und Osterzgebirge – Einflüsse sich verändernder atmosphärischer Einträge und forstlicher Bewirtschaftung.* Schriftenreihe Gesellschaft für Umweltgeowissenschaften. In press.

ASMAN WAH, DRUCKER B and JANSSEN AJ (1988) *Modelled historical concentrations and depositions of ammonia and ammonium in Europe.* Atm Environ 22:725–734.

BAYERISCHES STAATSMINISTERIUM FÜR LANDES-ENTWICKLUNG UND UMWELTFRAGEN (StMLU, Hrsg) (2000) *Monitoring von Schäden in Waldökosystemen des bayerischen Alpenraumes.* Umwelt & Entwicklung, Materialien 155.

BERDÉN M, NILSSON SI, ROSÉN K and TYLER G (1987) *Soil Acidification. Extent, Causes and Consequences.* National Swedish Environmental Protection Board, Solna.

BRECHTEL HM und SONNEBORN M (1984) *Gelöste anorganische Inhaltsstoffe in der Schneedecke unter Fichten- und Buchenaltbeständen und im Freiland der Hessischen Mittelgebirge.* DVWK Mitt 7:527–543.

BREDEMEIER M (1988) *Forest canopy transformation of atmosphere deposition.* Water Air Soil Poll 40:121–138.

BRESSER AHM and MATHY P, eds. *Monitoring Air Pollution and Forest Ecosystem Research.* Report 21 ECE Air Pollution Report Series, 1989.

BÜTTNER G, LAMERSDORF N, SCHULTZ R and ULRICH B (1986) *Deposition und Verteilung chemischer Elemente in küstennahen Waldstandorten.* Ber. Forschungszentrum Waldökosysteme Univ Göttingen B 1:1–172.

CAPE JN (1998) *Enhancement of the dry deposition of sulphur dioxide to a forest in the presence of ammonia.* Atmos Environ **32**: 519–525.

CAPPELLATO R, PETERS NE and RAGSDALE HL (1993) *Acidic atmospheric deposition and canopy interactions of adjacent deciduous and coniferous forests in the Georgia Piedmont.* Can J For Res **23**: 1114–1124.

CONSTANTIN J (1993) *Stoffeinträge in ein Fichtenwaldökosystem durch Deposition luftgetragener Partikel und Nebeltröpfchen.* Ber. d. Forschungszentrums Waldökosysteme, R A, Bd 106.

DERWENT RG, DOLLARD GJ and METCALFE SE (1988) *On the nitrogen budgets for the United Kingdom and North West Europe.* Quart J Royal Met Soc **114**: 1127–1152.

DIESE NB and WRIGHT RF (1995) *Nitrogen leaching from European forests in relation to nitrogen deposition.* Forest Ecology and Management **71**: 153–161.

EILERS G, BRUMME R and MATZNER E (1992) *Aboveground N-uptake from wet deposition by Norway spruce (Picea abies Karst.).* Forest Ecology and Management **51**: 239–249.

FLEMMING G (1993) *Klima und Immissionsgefährdung des Waldes im Osterzgebirge.* Arch f Nat-Lands **32**: 273–284.

FORSCHUNGSBEIRAT WALDSCHÄDEN/LUFTVERUNREINIGUNGEN, ed. (1993) *Dritter Bericht.* Kernforschungszentrum Karlsruhe.

FORSTLICHE VERSUCHSANSTALT RHEINLAND-PFALZ *Immissions-, Wirkungs- und Zustandsuntersuchungen in Waldgebieten von Rheinland-Pfalz.* Trippstadt. Mitteilungen Nr. 16/91.

GEBAUER G, KATZ C and SCHULTZE E-D (1991) *Uptake of gaseous and liquid nitrogen depositions and influence on the nutritional status of Norway spruce.* GSF-Bericht **43**: 83–92. Neuherberg.

GLATZEL G, KATZENSTEINER K, KAZDA M, KÜHNERT M, MARKART G and STÖHR D (1988) *Eintrag atmosphärischer Spurenstoffe in österreichische Wälder; Ergebnisse aus vier Jahren Depositionsmessung.* Bericht FIW Symposium, Univ f Bodenkultur, Wien 60 –72.

GUDERIAN R, ed. *Handbuch der Umweltveränderungen und Ökotoxikologie.* Bd. 1 B. Atmosphäre: Aerosol/Multiphasenchemie, Ausbreitung und Deposition von Spurenstoffen, Auswirkungen auf Strahlung und Klima. Springer, Berlin 2000. Bd. 2A. Terrestrische Ökosysteme. Immissionsökologische Grundlagen – Wirkungen auf Boden – Wirkungen auf Pflanzen. Springer, Berlin, 2001.

HETTELINGH J-P, DOWNING RJ and DE SMET PA (1991) *Mapping Critical Loads for Europe.* CCE Technical Report No 1, Bilthoven.

HICKS BB and MATT DR (1988) *Combining biology, chemistry, and meteorology in modeling and measuring dry deposition.* J Atmos Chem **6**: 117–131.

HÖFKEN KD, MEIXNER F and EHHALT DH (1988) *Dry deposition of NO, NO_2 and HNO_3.* In: van Dop H, ed. Air Pollution Modelling and its Application VI. Plenum Press, New York.

IBROM A (1993) *Die Deposition und die Pflanzenauswaschung (Leaching) von Pflanzennährstoffen in einem Fichtenbestand im Solling.* Ber. d. Forschungszentrums Waldökosysteme, R. A, Bd 105, Göttingen.

IVENS W (1990) *Atmospheric Deposition onto Forests.* Faculty of Geographical Sciences, University of Utrecht, Netherlands. Thesis.

JOHNSTON AE and JONES KC (1992) *The cadmium issue – long term changes in the cadmium content of soils and the crops grown on them.* In: Schultz JJ, ed., Phosphate Fertilizers and the Environment, pp. 255–269. Spec Pub. IFDC-SP-18, Int Fert Develop Ctr, Muscle Shoals, AL.

JOHNSON DW and LINDBERG SE, EDS. (1992) *Atmospheric Deposition and Forest Nutrient Cycling. A Synthesis of the Integrated Forest Study.* Ecological Studies 91, Springer, New York.

KATZENSTEINER K (2000) *Wasser- und Stoffhaushalt von Waldökosystemen in den nördlichen Kalkalpen.* Forstl. Schriftenreihe, Univ. f. Bodenkultur, Wien.

KAZDA M (1990) *Zusammenhang zwischen Stoffeintrag, Bodenwasserchemismus und Baumernährung in drei Fichtenbeständen im Böhmer Wald, Oberösterreich.* Forstl. Schriftenreihe d. Univ. f. Bodenkultur, Wien.

KOZLOWSKI, R (2001) *The rainfall inflow of mineral components to the forest bottom at the Integrated Natural Environment Monitoring Base Station in the Swietokrzyskie Mountains.* In: Joswiak, M. and Kowalkowski, A, eds. The Integrated Monitoring of the Environment in Poland, pp. 207–217. Biblioteka Monitoringu Srodowiska.

KREUTZER K, BEIR C, BREDEMEIER M, et al. (1998) *Atmospheric deposition and soil acidification in five coniferous forest ecosystems: a comparison of the control plots of the EXMAN sites.* Forest Ecology and Management **101**: 125–142.

KROLL G and WINKLER P (1988) *Estimation of wet deposition via fog.* In: Grefen K, Löbel J, eds. Environmental Meteorology, pp. 227–236. Kluwer Acad. Publ., Dordrecht.

LANDESAMT FÜR UMWELT UND GEOLOGIE (1999) *Jahresbericht zur Immissionssituation.* Freiberg, Sachsen.

LANGUSCH J (1995) *Untersuchungen zum Ionen-haushalt zweier Wassereinzugsgebiete in verschiedenen Höhenlagen des Osterzgebirges.* Diss. TU Dresden.

LEGGE A and KRUPA S (1990) *Acidic Deposition. Sulfur and Nitrogen Oxides.* Lewis Publishers, Michigan.

LINDBERG SE and LOVETT GM (1992) *Deposition and forest canopy interactions of airborne sulfur: results from the Integrated Forest Study.* Atmos Environ 26A: 1477–1492.

LIU J-C, KELLER T, RUNKEL K-H und PAYER H-D (1993) *Stoffeinträge und -austräge im Fichten-ökosystem Wank (Kalkalpen) und ihre boden- und ernährungskundliche Bewertung.* GSF-Bericht **39**: 294–310, Neuherberg.

MATZNER E (1988) *Der Stoffumsatz zweier Waldökosysteme im Solling.* Ber. d. Forschungszentrums Waldökosysteme/Waldsterben Reihe A, Bd. 40. Göttingen.

NAGEL H-D und GREGOR H-D, eds (1999) *Ökologische Belastungsgrenzen – Critical Loads & Levels.* Springer, Berlin.

RADEMACHER P (2001) *Atmospheric Heavy Metals and Forest Ecosystems. Current Implementation of ICP Monitoring Systems and Contribution to Risk Assessment.* UN/ECE, Geneva.

RASMUSSEN L (1990) *Study on acid deposition effects by manipulating forest ecosystems.* Commission of the European Communities. Air Pollution Research Report 24.

RASPE S, FEGER KH und ZÖTTL HW, eds (1998) *Ökosystemforschung im Schwarzwald – Auswirkungen von atmogenen Einträgen und Restabilisierungsmaßnahmen in Fichtenwäldern.* Verbundprojekt ARINUS.- Umweltforschung in Baden-Württemberg,- Ecomed-Verlag, Landsberg/Lech.

REUSS JO and JOHNSON DW (1986) *Acid Deposition and the Acidification of Soils and Water.* Ecological Studies 59, Springer, New York.

SÄCHSISCHES STAATSMINISTERIUM FÜR LANDW., ERNÄHR. UND FORSTEN, (ed) *Waldschadensbericht 1996.* Dresden 1996.

SÄCHSISCHES STAATSMINISTERIUM FÜR UMWELT UND LANDWIRTSCHAFT (2001) *Waldzustandsbericht 2001.* Dresden.

SCHEMENAUER R (1986) *Acidic deposition to forests: the 1985 chemistry of high elevation fog (CHEF) project.* Atmos Ocean **24**: 303–328.

SCHWELA D (1977) *Die trockene Deposition gasförmiger Luftverunreinigungen.* Schriftenreihe LIS Nr. **42**, pp. 46–85. Essen.

SÖDERLUND R (1982) *On the difference in chemical composition of precipitation collected in bulk and wet-only collectors.* Univ. of Stockholm, Report CM 57.

ULRICH B (1991) *Deposition of Acids and Metal Compounds.* In: Merian E, ed., Metals and Their Compounds in the Environment, pp. 369–378. VCH Verlag mbH, Weinheim.

ULRICH B, MAYER R and KHANNA P K (1979) *Deposition von Luftverunreinigungen und ihre Auswirkungen in Waldökosystemen im Solling.* Sauerländer, Frankfurt/M..

ULRICH B and PANKRATH J, eds. (1983) *Effects of Accumulation of Air Pollutants in Forest Ecosystems.* D. Reidel Publ. Comp., Dordrecht, Boston.

UN/ECE und EK (2001) *Der Waldzustand in Europa. Kurzbericht 2001.* Bundesforschungsanstalt für Forst- und Holzwirtschaft, Hamburg; Genf und Brüssel.

UN-ECE (1988) *Critical Loads for Sulphur and Nitrogen.* Skokloster, Sweden.

UN-ECE (1990) *Draft Manual. Mapping Critical Levels/Loads.* Convention on Long-range Transboundary Air Pollution.

UN-ECE. Programme Coordinating Centre (1998) *Manual on Methods and Criteria for Harmonized Sampling, Assessment, Monitoring and Analysis of the Effects of Air Pollution on Forests.* 4th edn, Hamburg.

WIENHAUS O (1996) *Wirkungen von Luftverunreinigungen.* In: Fiedler, Grosse, Lehmann, Mittag, Umweltschutz, pp. 202–211. Fischer Verlag, Jena.

WINKLER P, JOBST S and HARDER C (1989) *Meteorologische Prüfung und Beurteilung von Sammelgeräten für die nasse Deposition.* BMFT-Forschungsber., Vorhaben Nr. 07431073, Meteorologisches Observatorium Hamburg.

ZIMMERMANN F and WIENHAUS O (2000) *Ergebnisse von Immissionsmessungen im östlichen Erzgebirge zwischen 1992 und 1998.* Gefahrstoffe-Reinhaltung der Luft **60**: 245–251.

4

Macro Elements in Soil

H. J. Fiedler

In Germany, the chemical state of forest soils is evaluated on the basis of a 4×4 km grid. Additionally, at monitoring stations forest soils are continually investigated in detail. The nutrient state of agricultural soils is analyzed at intervals of 4 years, and continually on permanent research plots. The investigation measures are standardized, and the soil samples are stored for later comparisons. In this way, changes in the chemical properties of the soil or in soil fertility can be monitored and, if necessary, repaired by countermeasures.

4.1
Non-metals

Adsorption of the major anions takes place at low pH values of the soil. Under these conditions, the anions chloride and nitrate are electrostatically and indifferently attracted to positive charges of some soil colloids. Sulfate is more attracted to surfaces, whilst phosphates are specifically and strongly adsorbed. Oxides and hydroxides of Fe and Al, as well as allophane and kaolinite, favor anion adsorption. Hydrous oxides such as goethite (α-FeOOH) or gibbsite (γ-Al(OH)$_3$) generate positive charges below their zero point of charge. Allophane,

which has surfaces rich in hydrous oxide groups, sorbs phosphate most strongly at low pH.

Especially nitrogen and sulfur compounds may change their oxidation state in soil, these changes being catalyzed by microbial enzymes. Nitrate and sulfate are electron acceptors under anaerobic conditions, whilst chloride and phosphate are stable oxidation states in soils.

The oxy anions nitrate, sulfate and phosphate serve as important nutrient sources for plants. Chlorine is essential for plants in trace amounts. Nitrogen and phosphorus, together with potassium, are the most abundantly used nutrients in fertilizers (Bach et al. 1999). Sulfate and chloride salts accumulate in saline soils.

Since cations cannot be leached from soils without an equivalent concentration of anions in solution, soil leaching is highly dependent on the internal balance and atmospheric deposition of mobile anions such as sulfate and nitrate (Fiedler 2001).

4.1.1
Nitrogen

Rocks and their physical weathering products contain almost no nitrogen. This, together with the large requirement of

Elements and their Compounds in the Environment. 2nd Edition.
Edited by E. Merian, M. Anke, M. Ihnat, M. Stoeppler
Copyright © 2004 WILEY-VCH Verlag GmbH & Co. KGaA, Weinheim
ISBN: 3-527-30459-2

sis. Ammonia may partly leave the soil in gas form after surface application of urea in forests, grasslands and rice cultures, or be bound within arable soils as ammonium hydroxide. Manures contain nitrogen as constituent of nonhumin and humin substances besides inorganic forms. The input of nitrogen into the German agricultural area (1985–96) was estimated to be 198 kg N ha^{-1} a^{-1}.

4.1.2
Phosphorus

Phosphorus is bound in igneous rocks in the form of fluorapatite Ca$_5$ F(PO$_4$)$_3$. Dark-colored rocks, such as basalt, have a higher P-content than light-colored ones. For example, quartzporphyry (rhyolite) and its weathering product are both poor in phosphorus. Some phosphorus deposits consist of phosphorite. The mineral apatite is soluble in weak acids and is therefore transformed during weathering in other kinds of phosphates as Al- and Fe- phosphates (variscite AlPO$_4 \cdot$2H$_2$O, strengite FePO$_4 \cdot$2H$_2$O; Al(OH)$_2$H$_2$PO$_4$). Under conditions where the pH is >6, Ca-phosphates are formed (apatite and octocalciumphosphate Ca$_4$H(PO$_4$)$_3$). Vivianite Fe$_3$(PO$_4$)$_2$ is a constituent of peat soils.

The total content of phosphorus in soils ranges from 0.2 to 1.3 g kg^{-1} (see Table 4.3). The phosphorus content of the clay fraction exceeds these values.

In addition to inorganic orthophosphates, phosphorus is bound in organic phosphates (ester linkages) within humus, from which plant available phosphate is set free by microorganisms (mineralization). Examples are nucleic acids, inisitol hexaphosphates as the largest group, and phospholipids. In agricultural soils, the C:P ratio is in the order of 50 : 1.

Phosphates are fixed by Fe-hydroxides and together with these by clay minerals. In tropical soils, phosphorus fixing on Fe-compounds is widespread. Phosphate fixation is appreciable in soils rich in allophane, derived from volcanic rocks.

Soil solution concentrations of phosphate are very low (in the order of 0.1 to 1 mg L^{-1} or about 0.03 μg mL^{-1}). From the soil solution, the element is taken up mainly in the form of H$_2$PO$_4^-$.

Depending upon the distribution of organic and inorganic soil colloids, the content and plant availability of phosphorus changes with soil depth. The pool of insoluble or fixed inorganic phosphorus in the soil is in general much greater than that of organic phosphorus. The smallest pool is that in the soil solution. Leaching losses in mineral soils are minimal (Fiedler et al. 1985a, b). Phosphorus leached from the A-horizon of Podsols is found as Fe- and Al-phosphates in the B-horizon (spodic horizon).

Plant species differ widely in the extent of phosphorus absorption by their roots. Forest trees take up P from insoluble compounds with the help of mycorrhizas, whereas phosphorus is removed from soils by harvested crops (~6 kg ha^{-1} in agriculture), erosion and to a small extent by leaching (~0.1 kg ha^{-1}) and volatilization as phosphine PH$_3$. In the case of erosion, colloids with their surface-bound P are transported into surface waters, and this leads to the eutrophication of aquatic ecosystems, for example in lakes. Leaching losses occur only in sand and peat soils, and in the case of organic phosphates. Under extreme redox situations – as in some paddy soils – phosphate is reduced to PH$_3$ as a gaseous product.

Besides the total content of phosphorus in forest soils, the plant-available form is determined in agricultural soils by extraction with weak sodium bicarbonate, organic acids or their salts (acetate, citric acid, lactic acid) and fluoride.

Rock phosphate is used as fertilizer only for acid and humus-rich soils. Generally, phosphates are fertilized in the form of water- or acid-soluble compounds, derived from rock phosphate (processed mineral phosphate fertilizers). An example is superphosphate, a mixture of monobasic calcium phosphate and calcium sulfate (7–9% P). In double or triple superphosphate the P-content rises to about 20%. Manures contain 0.1 to 0.4% P, partly in inorganic and partly in organic form; both binding forms change with time to insoluble inorganic phosphates. In agriculture, a slight acid pH (about 6.5) warrants the best availability of the mixture of soil phosphates. With an increasing pH value in acid soils, Al- and Fe-phosphates release phosphate in soluble form. Also, reduction of ferric iron releases phosphorus from ferric phosphate. On the other hand, Ca-phosphates dissolve as the pH decreases. Most soils reduce the solubility of fertilizer phosphates, because phosphate interacts rapidly with inorganic soil constituents; therefore, fertilized phosphorus remains near the place of application.

4.1.3
Sulfur

Rock-forming minerals contain sulfur as sulfides of heavy metals and as sulfate in gypsum. Where gypsum is rock-forming in Central Europe – as in the Hartz region – gypsum-rendzina soils developed with their special conditions for plant nutrition and species composition, likewise in deposits of gypsum after mining (Heinze and Fiedler 1984).

Sulfate as anion is not bound to clay minerals but, unlike nitrate, is adsorbed to some extent by soil, the retention increasing with soil acidity. In acid soils, an inorganic form is probably $Al(OH) SO_4$.

Different organic sulfur compounds exist in humus, from which sulfur is set free by soil microorganisms, and normally as sulfate as this is the most stable form in terrestrial ecosystems. The C:S ratio of the organic matter ranges from < 100 (arable soils) to > 400 (forest soils). In the O- and Ah-horizon of forest soils, more than 90% of the total sulfur is bound in organic compounds in the form of carbon-bound S and organic sulfates (Figure 4.2; Table 4.2). Organic sulfates comprise between 10 and 80% of the total organic sulfur in forest soils, whereas in arable soils carbon-bound S comprises between 5 and 30% of the total organic S. The compounds (thiols and

Tab. 4.2: Sulfur content ($\mu g\ g^{-1}$) of sulfur fractions in the humus layer and mineral soil of a brown earth under a spruce stand in the Tharandt Forest. (After Klinger 1995.)

Horizon	Org S (total)	SO₄-S (non-water-soluble)	SO₄-S (water-soluble)
Ol	2107	296	183
Of	2537	264	168
Oh	1639	271	160
Ah	527	183	156
Bv1	130	150	119
Bv2	174	192	79
Bv3	334	176	61

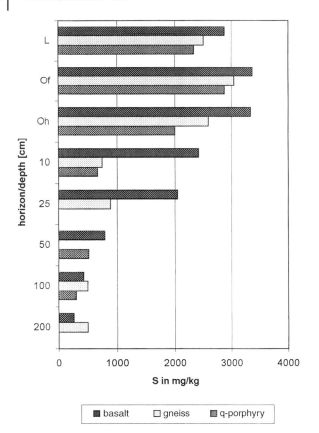

Fig. 4.2 Sulfur in the humus layer and mineral soil above different parent rocks in the Eastern Ore Mountains. After Klinger (1995).

organic disulfides) are relatively resistant against microbial attack. Organic sulfates (mainly estersulfates, thioglucosides and sulfamates) are formed by microorganisms and are easily decomposed.

A low redox potential – as in bog soils – leads to the formation of H_2S, which may be bound to sulfides of iron or may volatilize. Fe-sulfides are the cause for the black color in reduction zones of soils. Sulfide is the stable form under strong reducing conditions, but when changing to aerobic conditions sulfuric acid is formed, and this leads to soil acidification. Both the oxidation and reduction of sulfur compounds involve autotrophic bacteria (sulfur bacteria).

In alkaline soils, acidification is required and realized by fertilization with sulfur or

gypsum. Under Central European conditions, liming of acid soils is necessary; here, together with sulfate ions, basic cations are initially washed out of the soil into the ground water, but later on acid cations such as Al^{3+} are also washed out.

Sulfur is taken up by plants only in the form of sulfate out of the soil solution. Agricultural crops need $10-15$ kg S ha^{-1}a^{-1}, but forests require only $1-2$ kg S ha^{-1} a^{-1}. In areas of clean air, in weathering products of low natural sulfur content and in the case of high harvests in agriculture and fertilizing with S-poor mineral fertilizers, crops need sulfur fertilizers – for example super phosphate that contains gypsum – or potassium sulfates. In industrialized areas – and especially in the northern hemisphere – the problem is one of too-high S-input by immissions, and this leads to S-eutrophication and acidification of terrestrial ecosystems, especially of forests (Fiedler and Klinger 1995). Under these conditions, sulfur is stored in the soil organic matter and as aluminum hydroxy sulfate in the inorganic soil. Now, as the S-input has decreased in spruce and pine forests, the S-output by leaching is far above the input-value. Sulfur is set free from the humus layer as well as the mineral soil.

4.1.4
Chlorine

Chlorine is accumulated in salt sediments of marine origin. The negatively charged chloride ion is not adsorbed in the soil. Whereas in a humid climate the soluble chlorides are washed out, under dry climatic conditions, chlorides remain in the upper soil. They are also accumulated in the soil by irrigation without drainage – an undesirable process in agriculture. Many chlorides are added to the soil via fertilizers containing KCl, and with deicing compounds such as $MgCl_2$ or NaCl which are used on the roads. A further input to ecosystems occurs by natural and artificial immissions. The normal content is < 30 mg Cl/100 g soil, and high values are those > 50 mg Cl/100 g soil. Besides sulfate and nitrate, chloride ions are important in the soil solution. In young marsh soils (salt marsh) on the sea shore, the primary high content of chloride must be washed out by rain before any cultivation of agricultural plants is possible.

4.2
Metals

4.2.1
Exchangeable Bases

Soil minerals such as feldspars, micas, clay minerals, and calcite contain bases in non-exchangeable form. Weathering causes a release of cations from these minerals, after which the ions accumulate on the negatively charged soil colloids (clay and humus fraction of the soil) in exchangeable form. The principal exchangeable bases in soils are the cations Ca, Mg, K and Na (Table 4.3). In a fractional exchange, sodium is released most readily, and Ca least readily. The cation preference of soil clays and silts follows the lyotropic series. Special attention has been given to the K:Ca (plant nutrition), Na:Ca (saline soils) and Ca:Al (acid forest soils) ion pairs. In soil organic compounds, selectivity for multivalent cations such as Ca^{2+} relates mainly to the disposition of their acidic groups, for example carboxylic groups and phenolic hydroxyl groups.

Exchangeable bases constitute a source from which the bases in the soil solution are replenished. The ions serve as nutrients for higher plants and are important for the physical, chemical, and biological properties of soils. The term "total exchangeable bases" refers to the sum of the exchangeable bases Ca, Mg, K, and Na in milligram equivalents (m.e.) per 100 g of soil. The cation-exchange capacity (CEC) is the total amount of exchangeable cations, including acid cations such as H and Al, in $cmol_c kg^{-1}$ soil (see Table 4.3). Depending upon the colloid content of mineral soils, the CEC ranges from 2 to 60 $cmol_c$ kg^{-1} soil, but in raw humus of forest soils the values rise to 150 $cmol_c$ kg^{-1}. The percentage base saturation (BS) is the percentage of the CEC occupied by

Tab. 4.3: Concentrations of macro elements in an acid gneiss podzol-brown earth of the Tharandt Forest, eastern Ore Mountains in Saxony. Total and with NH_4Cl extractable element contents. (After Klinger 1995.)

Total contents

Layer/horizon	Depth	N [g/kg]	P [mg/kg]	S [mg/kg]	Na [g/kg]	K [g/kg]	Mg [g/kg]	Ca [g/kg]	Ca/Al
L		17.5	1121	2152	0.227	0.41	0.374	4.05	1.620
Of		19.2	1104	2670	0.486	0.8	0.82	2.26	0.246
Oh		11.6	642	2019	1.81	3.69	0.589	2.18	0.127
I Aeh	0–3	2.5	295	483	5.847	15.26	1.189	1.64	0.054
I Bv1	15–25	0.7	172	392	6.794	18.54	1.884	1.63	0.028
II Cv1	90–100		220	238	7.476	32.43	1.792	0.81	0.010
II Cv2	130–140		137	100	7.236	43.24	1.521	0.3	0.003
II Cv3	170–180		183	127	10.562	42.5	1.254	0.74	0.008

Exchangeable cations [mg/kg]

Layer/horizon	Depth	H^+	Al^{3+}	Fe^{3+}	Mn^{2+}	Na^+	Mg^{2+}	Ca^{2+}
I Aeh	0–3	50.3	468	159.6	21.2	30.4	23.6	169.2
I Bv1	15–25	5.3	564.8	15.2	94	29.6	9.6	42.8
II Cv1	90–100	1.5	345.2	2	30.4	26.8	4.8	23.6
II Cv2	130–140	0.9	274.8	0.4	26	38.8	6.8	13.6
II Cv3	170–180	1	167.2	0	15.2	26.4	11.6	15.2

mmol ion equivalent/kg

Layer/horizon	Depth	H^+	Al^{3+}	Fe^{3+}	Mn^{2+}	Na^+	Mg^{2+}	Ca^{2+}	CECeff
I Aeh	0–3	49.91	52.04	8.57	0.77	1.32	1.94	8.44	123.0
I Bv1	15–25	5.26	62.80	0.82	3.42	1.29	0.79	2.14	76.5
II Cv1	90–100	1.49	38.38	0.11	1.11	1.17	0.39	1.18	43.8
II Cv2	130–140	0.89	30.55	0.02	0.95	1.69	0.56	0.68	35.3
II Cv2	170–180	0.99	18.59	0.00	0.55	1.15	0.95	0.76	23.0

%

Layer/hoirzon	Depth	H^+	Al^{3+}	Fe^{3+}	Mn^{2+}	Na^+	Mg^{2+}	Ca^{2+}	BS
I Aeh	0–3	40.0	41.7	6.9	0.6	1.1	1.6	6.8	10.8
I Bv1	15–25	6.8	80.9	1.1	4.4	1.7	1.0	2.8	6.9
II Cv1	90–100	3.3	85.0	0.2	2.5	2.6	0.9	2.6	9.0
II Cv2	130–140	2.4	83.3	0.1	2.6	4.6	1.5	1.9	11.8

exchangeable bases (see Table 4.3). Soils of arid regions are usually base-saturated (100%). In humid regions, the relative quantities (in m.e.) of the exchangeable bases in general follow the order Ca > Mg > K > Na, with the content of sodium being very small. The degree of BS is mostly < 100%. Release of bases from the exchange complex occurs by exchange with hydrogen and aluminum ions under the influence of nutrient uptake by plants, nitrification and acid rain (see Table 4.3; Fiedler et al. 1995). During the past 40 years, the BS of forest soils has decreased in large areas due to the high atmogenic acid input (pH < 4.2; BS% < 10).

Between rock, climate, and deposition on the one hand, and the quality of spring water in forests on the other hand, correlations exist. So, springs on basalt sites have higher contents of Ca and Mg and a lower content of K compared with gneiss and rhyolite. Up to 20 mg L^{-1} sulfate-S sulfur correlates with Ca- and Mg-ions, but above 20 mg L^{-1} sulfate-S only with Al-ions (Fiedler and Katzschner 1989, Nebe and Abbiy 2002).

4.2.2
Sodium

In uplifted marine sediments, the sodium content may be high, and in igneous rocks the ratio of potassium to sodium by weight is 0.92. During weathering of sodium-containing rock minerals (e.g., feldspars), sodium is not bound to clay minerals, as it has low bonding energy and so can be found both in the soil solution and in sea water. Further input sources of sodium are similar to those of chlorine (immission, deicing salts).

Soils are said to be "sodic" if they contain an excess of sodium. According to their content of salts and sodium, soils are grouped in saline-nonsodic, saline-sodic, nonsaline-sodic, and normal soils. In arid regions, soils rich in sodium are widely distributed due to salt accumulation as sodium chloride and sodium sulfate. The salt-affected soils are chemically characterized by the specific conductivity (mS·cm^{-1}) [S = Siemens] of the saturation extract and the saturation of the CEC (%) with sodium. The major cationic constituents of the soluble salts in saline soils are sodium, calcium, and magnesium. Precipitation of Ca and Mg as carbonates effects an increase in the proportion

of sodium in soluble and exchangeable form. In saline-sodic soils, sodium comprises more than half of the total soluble cations, and the pH value is below 8.5. In nonsalic-sodic soils, the degree of sodium saturation of the CEC exceeds 15%, and such soils have a pH value between 8.5 and 10, due to a mixture of sodium bicarbonate and sodium carbonate in the soil solution.

A normal soil may be changed by irrigation into a saline-nonsodic or even a saline-sodic soil, because irrigation water always contains soluble salts. The transformation depends on the composition of the irrigation water and the proportion of the water that is removed from the soil by drainage. If excess sodium is present as in saline-sodic soils, the removal of excess salts by leaching with water leads to a reduction of the conductivity of the soil for water. Saturation with sodium causes strong swelling of soil clays, especially montmorillonite.

In the soil, soluble sodium is in equilibrium with the exchangeable sodium, with the equilibrium depending on the water content of the soil. Replacement of exchangeable sodium from sodic soils is achieved by adding soluble calcium salts (gypsum) or acid-forming materials (sulfur, sulfuric acid) and leaching. In fertilizers for normal soils, sodium is not required due to its dispersing effect on soil colloids. In fact, sodium-containing potassium fertilizers may only be used on light soils that support animal nutrition.

4.2.3
Potassium

Potassium makes up 2.6% by weight of the Earth's crust. The element is enriched in acid magmatic rocks such as granite, containing potassium mica (muscovite) and potassium feldspar (orthoclase $KAlSi_3O_8$, 13.9% K). In sediments and weathering tion. Following the reduction of Ca emission during the past years, the humus layer is now once more becoming acid.

Under humid conditions due to the washout of Ca together with anions – either as bicarbonate or as chloride and nitrate – lime or Ca-containing fertilizers must be given from time to time in larger quantities. The content of soil calcium is determined by leaching a sample of soil with ammonium acetate. In agriculture, soils must be especially limed for crops such as sugar beet and wheat. Liming of acid soils causes replacement of exchangeable Al by calcium and also increases the CEC.

References

BACH M, FREDE HG and LANG G (1999) *Nährstoffbilanzen der Landwirtschaft in Deutschland.* AID-Heft 1404, 30 S.

FIEDLER HJ (2001) *Böden und Bodenfunktionen.* expert verlag, Renningen.

FIEDLER HJ and HOFMANN W (1992) *Waldböden auf Thüringer Muschelkalk.* Acta Academiae Scientiarum. Abhandlungen der Akademie gemeinnütziger Wissenschaften zu Erfurt. Bd 1.

FIEDLER HJ and ILGEN G (1989) *Einsatzgebiete, methodische Entwicklung und Verfahren der Boden- und Pflanzenanalyse in der Forstwirtschaft.* Bulletin BGS (Schweiz) **13**:17–35.

FIEDLER HJ, KATZSCHNER W and RICHTER B (1985) *Phosphor in bewaldeten Wassereinzugsgebieten. I. Qualitative Kennziffern.* Wiss Z TU Dresden **34**:155–162.

FIEDLER HJ, KATZSCHNER W and RICHTER B (1985) *Phosphor in bewaldeten Wassereinzugsgebieten. II. Quantitative Kennziffern.* Wiss Z TU Dresden **34**:217–224.

FIEDLER HJ and KATZSCHNER W (1989) *Zur Relation zwischen basischen Kationen und Anionen starker Säuren in Waldgewässern der Mittelgebirge.* Herzynia **26**:94–101.

FIEDLER HJ, KATZSCHNER W and MITSCHICK G (1995) *Gehalt und Austrag des Wernersbaches an Basen-Kationen (Untere Berglagen des Osterzgebirges).* Wiss. Z. TU Dresden **44**:11–16.

FIEDLER HJ and KLINGER T (1995) *Gehalt und Funktion des Schwefels in SO$_2$-belasteten Fichtenforsten.* In: Panstwow inspekcja ochrony srodowiska. Monitoring der anthropogenen Landschaften in Mittel- und Osteuropa S. 23 –42 IV Kolloquium EIPOS-Kielce 29.9– 1.10.1995. Biblioteka Montoringu Srodowiska Warszawa

HEINZE M and FIEDLER HJ (1984) *Chemische Eigenschaften von Gips-Rendzinen und Begleitbodenformen des Kyffhäusergebirges (DDR).* Chemie d. Erde **43**:65–75.

KATZSCHNER W, FIEDLER HJ and PLEISS H (1988) *Zur Nitratstickstoff-Fracht des Wernersbaches im Tharandter Wald.* Wiss Z TU Dresden **37**:233–241.

KATZSCHNER W, FIEDLER HJ and MITSCHICK G (1991) *Ammonium- und Nitritgehalt des Wernersbaches sowie Austrag an NH$_4$-, NO$_2$- und Gesamt-N aus dem Wernersbachgebiet im Tharandter Wald.* Wiss Z TU Dresden **40**:173–178.

KLINGER TH (1995) *Mengen- und Spurenelemente in Waldböden über unterschiedlichen Grundgesteinen des Osterzgebirges.* Diss. TU Dresden, Tharandt.

NEBE W and ABIY M (2002) *Chemie von Quellwässern in bewaldeten Einzugsgebieten des Erzgebirges.* Forstw Cbl **121**:1–14.

5

Trace Elements and Compounds in Soil

Alina Kabata-Pendias and Wiesław Sadurski

5.1
Introduction

Cycling is a fundamental natural process that governs the behavior and distribution of chemical elements in the Earth and its envelope. Soil, as a part of the terrestrial ecosystem, plays a crucial role in elemental cycling. It has important functions as a storage, buffer, filter, and transformation compartment, supporting a homeostatic interrelationship between the biotic and abiotic components.

The chemical composition of soils is diverse and governed by many different factors, of which parent materials and climatic factors usually predominate. Although trace elements (both cationic and anionic forms) are minor components of the soil, they play an important role in soil bioactivity and fertility. Behavior of trace elements in soils depends upon complex reactions between their ionic forms and various components of the various soil phases: solid, aqueous, and gaseous. This relationship is closely related to the main features of the soil biogeochemical system, which are: (i) seasonal and spatial alteration of major soil variables; (ii) heterogeneous distribution of compounds and components; (iii) transformation of element species; (iv) complexa-tion; (v) transfer between phases; and (vi) bioaccumulation.

Today, we are under the "onset" of environmental geochemistry that is concerned mainly with trace elements in all terrestrial compartments. The driving forces of this discipline are: (i) the development of techniques and analytical methods, especially trace analyses; (ii) increased numbers of scholars in this scientific branch; and (iii) rapid transfer of information and possibility of the combination of findings.

In spite of a great accumulation of knowledge on basic geochemical principles that govern the behavior of trace elements in soils, there is still a challenge in studies on these elements for better understanding biogeochemical processes in soils and for fostering sustainable agriculture and environmental and health risk assessment.

5.2
Trace Elements in Soil

Quantitatively, trace elements are negligible chemical constituents of soils, but a number of them are essential as micronutrients for plants and through the food-chain for man and animals. Their unbalanced contents due to either geochemical or anthropogenic

Elements and their Compounds in the Environment. 2nd Edition.
Edited by E. Merian, M. Anke, M. Ihnat, M. Stoeppler
Copyright © 2004 WILEY-VCH Verlag GmbH & Co. KGaA, Weinheim
ISBN: 3-527-30459-2

Tab. 5.3: Trace elements in manganese oxides (mg kg^{-1} dry weight basis)[a]

Element	Mn nodules	Mn minerals from soils[b]	Mn minerals from sediments[c]
Mn (%)	0.4–50	78–60	28–60
Ba	14–2300	32000–54000	110000–128000
Cd	8	–	–
Ce	720	–	–
Co	82–3000	4500–12000	140–12000
Cr	14–120	–	–
Cu	2600	–	130–12600
I	120–900	–	–
Li	–	300–700	2–53400
Mo	410	–	–
Ni	40–4900	100–3400	120–10900
Pb	34–870	–	–
V	88–440	–	–
Zn	30–710	–	50–3800

[a] Data sources: Kabata-Pendias and Pendias 2001, Bartlett 1999. Identified minerals: [b] lithiophorite, birnessite, and hollandite; [c] psylomelane, cryptomelane, and pyrolusite.

Tab. 5.4: Some inorganic ionic species[a] of trace elements and iron occurring in soil solutions

Element	Cations	Anions
Ag	Ag^+,	$AgCl_2^-$, $AgCl_3^{2-}$, $Ag(SO_4)_2^{3-}$
As	$\mathbf{As^{3+}}$	AsO_2^-, $HAsO_4^{2-}$, $H_2AsO_3^-$
B		$\mathbf{BO_3^{3-}}$, $H_2BO_3^-$, HBO_3^{2-}
Be	Be^{2+}, $BeOH^+$	BeO_2^{2-}, $Be(OH)_3^-$, $Be(CO_3)_2^{2-}$
Cd	$CdCl^+$, $CdOH^+$, $CdHCO_3^+$, $\mathbf{CdHS^+}$	$CdCl_3^-$, $Cd(OH)_3^-$, $\mathbf{Cd(OH)_4^{2-}}$, $\mathbf{Cd(HS)_4^{2-}}$
Co	Co^{2+}, Co^{3+}, $CoOH^+$	$Co(OH)_3^-$
Cr	Cr^{3+}, $CrOH^{2+}$	$HCrO_3^{2-}$, CrO_4^{2-}, $Cr(OH)_4^-$, $Cr(CO_3)_3^{3-}$
Cu	Cu^{2+}, $CuOH^+$, $Cu_2(OH)_2^{2+}$	$Cu(OH)_3^-$, $\mathbf{Cu(OH)_4^{2-}}$, $Cu(CO_3)_2^{2-}$
F	AlF^{2+}, AlF_2^+	F^-, AlF_4^-
Fe	Fe^{2+}, $FeCl^+$, $Fe(OH)_2^+$, $FeH_2PO_4^+$	$Fe(OH)_3^-$, $\mathbf{Fe(OH)_4^{2-}}$, $Fe(SO_4)_2^-$
Hg	Hg_2^{2+}, $HgCl^+$, $HgCH_3^+$	$HgCl_3^-$, $\mathbf{HgS_2^{2-}}$
I		I^-, I_3^-, IO_3^-, $\mathbf{H_4IO_6^-}$
Mn	Mn^{2+}, $MnOH^+$, $MnCl^+$, $MnHCO_3^+$	MnO_4^-, $HMnO_2^-$, $Mn(OH)_3^-$, $\mathbf{Mn(OH)_4^{2-}}$
Mo		MoO_4^{2-}, $HMoO_4^-$
Ni	Ni^{2+}, $NiOH^+$, $NiHCO_3^+$	$HNiO_2^-$, $Ni(OH)_3^-$
Pb	Pb^{2+}, $PbCl^+$, $PbOH^+$	$PbCl_3^-$, $Pb(CO_3)_2^{2-}$
Se		$\mathbf{SeO_3^{2-}}$, SeO_4^{2-}, $\mathbf{HSe^-}$, $HSeO_3^-$
V	VO^{2+}	$H_2VO_4^-$, HVO_4^{2-}, VO_3^-
Zn	Zn^{2+}, $ZnCl^+$, $ZnOH^+$, $ZnHCO_3^+$	ZnO_2^{2-}, $Zn(OH)_3^-$, $ZnCl_3^-$,

[a] The symbols given in bold letters indicate the ions occurring only in extreme pH and Eh regimes.

trace element ions. In most soils these minerals have a significant influence on the mobility of trace elements. Clays can form complexes with organic compounds. As Tan (1998) described, due to variable charge of both clays and organic particles, different complexes are formed and influence element mobility and clay dispersion in soil solution. Chelation and complex formation with organic compounds are known to dissolve stable minerals (e.g., $AlPO_4$, $FePO_4$). Tan (1998) has emphasized that various bindings between live organisms, like bacteria and fungus, and soil clays and humic matter seem to be of a great environmental importance.

Phosphates – although negligible constituents of most soils – are known to influence trace metal behavior. Hydrated phosphates can easily fix most of trace elements, and in particular Zn and Pb.

Other minerals such as sulfides, sulfates, and chlorides are not common in soils developed in humid climatic zones. In soils of arid climates, however, they can be the dominant control on the behavior of trace elements. Pyrite (FeS_2) is known to be of environmental concern in soils of specific geochemistry, and chlorides can also be of concern due to their affinity to form soluble metal-chloride complexes with some metals (Table 5.4).

5.2.2
Background Ranges

Contents of trace elements in soil materials from natural as well as from contaminated sites show a high variability in both dimensions, horizontal and vertical. Soil is not homogenous, and the microscale heterogeneity creates a real problem in representative sampling. Thus, the reproducibility and comparability of analytical data for trace elements in soils have been a great concern. The procedure of soil sampling and storage has been broadly described in several publications (Mortvedt et al. 1991, Sparks et al. 1996, Tan 1995).

Several methods were developed to calculate background (pristine) contents of trace elements in soils. There is a great demand for such data as reference values, as today entirely natural contents of trace elements do not exist. In general, these methods are based either on statistical calculations or on the relationship of trace elements to various soil parameters and geologic factors. The best-known methods are described, as compiled by Kabata-Pendias and Pendias (2001) from various sources:

- GB: Geochemical Baseline
- GD: Index of Geochemical Distribution
- IGL: Index of Geochemical Load
- IPD: Index of Pedogenic Distribution
- PEF: Pedochemical Enrichment Factor
- FSP: Factor of Soil Parameters
- SCV: Spatial Concentration Variability
- GIS and MA: Geographical Information System and Multivariate Analyses

The "natural" ("normal") contents of trace elements in soils are of great interest, as background values are needed to assess the degree of soil contamination and to evaluate soil quality – the prerequisites for sustainable land use. The large database of recent surveys on trace elements in soils allows general estimations to be made of the concentration ranges and arithmetic means which are presented for several trace elements in the selected soil units (Table 5.5). These data show a significant role of soil units in the distribution of trace elements and also provide the approximate information on their possible concentrations in uncontaminated soils. The background populations of trace elements in soils seemed to be log-normally distributed, but the contaminated population is also likely to fit a log-normal distribution though with much higher extremes.

5.3
Weathering and Soil Processes

The composition of soils is extremely diverse and, although governed by many different

Tab. 5.5: Mean and maximum values (mg kg^{-1}) of the background ranges reported for trace elements[a] in different soil kinds on the world scale. Adopted from Kabata-Pendias and Pendias (2001) and Tobias et al. (1997)

Elements	Podzols (sandy soils)		Cambisols (silty and loamy soils)		Rendzinas		Various other soils	
	Mean	Maximum	Mean	Maximum	Mean	Maximum	Mean	Maximum
As	4.4	30	8.4	27	–	–	9	67
B	22	134	40	128	40	210	35	100
Ba	330	1500	520	1500	520	1500	300	700
Cd	0.37	2.7	0.45	1.61	0.62	0.84	0.60	2.0
Co	5.5	65	10	58	12	70	5.5	50
Cr	47	530	51	1100	83	500	45	150
Cu	13	70	23	100	23	70	20	120
F	130	1100	385	800	360	840	380	750
Hg	0.05	0.7	0.1	1.1	0.05	0.5	0.18	0.85
I	2.3	10	1.7	8.3	3.4	9.5	3.2	10.5
Li	22	72	46	130	56	105	3.5	90
Mn	270	2000	525	9200	445	7740	470	3900
Mo	1.3	3.7	2.8	7.2	1.5	7.4	1.8	5.2
Ni	13	110	26	110	34	450	16	90
Pb	22	70	28	70	26	50	32	140
Sc	5	30	8	20	8	15	10	20
Se	0.25	1.32	0.34	1.9	0.38	1.4	0.35	1.3
Sr	87	1000	210	1000	195	1000	120	400
V	67	260	76	330	115	500	60	150
Zn	45	220	60	362	100	570	60	480

[a] Related to aqua regia-soluble elements or other forms defined as total contents.

factors, parent material and climatic conditions predominate most commonly. Two main stages are involved in the formation of soil from parent material:

1. Weathering of primary/secondary minerals of parent rocks.
2. Pedogenesis – the formation of the soil profile.

These processes can not be easily distinguished as they take place simultaneously and at the same sites; in fact, most often they are closely interrelated. The principal types of soil-forming processes, including chemical alteration of the substratum are: (i) podzolization; (ii) aluminization; (iii) sialinization; (iv) laterization; (v) alkalization; and (vi) hydromorphic processes.

The mobility of trace elements during weathering is determined, first by the stability of the host minerals, and second by their electrochemical properties. The association of trace elements with minerals in soils reflects often their origin, and this is an important factor in their distribution and behavior. Several chemical and physical processes are involved in weathering of both biotic (living organisms and their decomposition) and abiotic origins. Basic processes can be characterized as follows:

- dissolution/sorption;
- hydrolysis;
- oxidation/reduction; and
- carbonatization.

5.3.1

Dissolution and Sorption

The main soil parameters governing processes of sorption and desorption of trace elements can be presented as follows: (i) pH and Eh values; (ii) fine granulometric fraction < 0.02 mm; (iii) organic matter; (iv) oxides and hydroxides of Fe, Mn and Al; and (v) microorganisms.

The kinetics and mechanisms of chemical reactions in soils have been broadly studied, and comprehensive mathematical models for the particular soil conditions have been presented (Bolt 1979, Huang 2000, Sauvé 2001, Schmitt and Sticher 1991, Sparks 1999, Tan 1998). The diversity of ionic species of trace elements and their various affinities to complex inorganic and organic ligands make possible the dissolution of each element over a relatively wide range of pH and Eh. In most soil conditions the effect of pH on the solubility of trace cations is more significant than that of redox potential (Chuang et al. 1996). However, redox potentials of soils also have a crucial impact on the behavior of trace elements (Bartlett 1999).

Smith and Huyck (1999) described metal mobility under different environmental conditions. Although it is rather difficult to predict trace element mobility in soils and other terrestrial compartments, these authors referred to the capacity of an element to move within fluids after dissolution in surficial environments. The following conditions and behavior of trace elements were distinguished:

- Oxidizing and acid, pH < 3: (i) very mobile – Cd, Co, Cu, Ni, and Zn; (ii) mobile – Hg, Mn, Re, and V; and (iii) somewhat mobile and scarcely mobile – all other metals.
- Oxidizing in the absence of abundant Fe-rich particles, pH > 5: (i) very mobile –

Cd and Zn; (ii) mobile – Mo, Re, Se, Sr, Te, and V; and (iii) somewhat mobile and scarcely mobile – all other metals.
- Oxidizing with abundant Fe-rich particulates, pH > 5: (i) very mobile – none; (ii) mobile – Cd and Zn; and (iii) somewhat mobile and scarcely mobile – all other metals.
- Reducing in the absence of hydrogen sulfide, pH > 5: (i) very mobile – none; (ii) mobile – Cd, Cu, Fe, Mn, Pb, Sr, and Zn; and (iii) somewhat mobile and scarcely mobile – all other metals.
- Reducing with hydrogen sulfide, pH > 5: (i) very mobile – none; (ii) mobile – Mn and Sr; and (iii) scarcely mobile to immobile – all other metals.

It is evident that Fe/Mn-rich particulates and hydrogen sulfide are most significant among abiotic factors in controlling trace metals behavior in the terrestrial environment (Sparks 1995, Bartlett 1999). All metals that are especially susceptible to redox reactions (e.g., Co, Cr, Fe, Mn) play a significant role in dissolution/precipitation reactions in soil, and this is best illustrated by the behavior of Mn:

- The reduction of Mn^{3+} to Mn^{2+} by: Fe^{2+}, Cr^{3+}, Co^{2+} reduces S, phosphate ligands (e.g., $P_2O_7^{4}$), phenols, and other easily oxidizing organic compounds
- The oxidation of Mn^{2+} to Mn^{3+} or Mn^{4+} by: any hydroxyl free radicals or atmospheric oxygen
- Mn^{3+} is an extremely reactive redox species, quickly disappears, either by accepting or by donating an electron.

The impact of soil microorganisms and enzymes on all redox and dissolution and/ or precipitation processes is very crucial, and can exert major control over the behavior of trace elements (Burns and Dick 2002, Naidu et al. 2001, Kostyuk and Bunnerberg 1999). The basic microbial phenomena in cycling processes of trace elements in soils are: (i) transport of an element in/at or out of a cell; (ii) charge alteration of an element; (iii) complexing an element by various produced compounds; (iv) accumulation and immobilization of an element; and (v) methylation. Biological methylation of some elements, such as As, Hg, Se, Te, Tl, Pb, and In, can highly influence their behavior in soils.

5.3.2
Soil Solution

Chemistry of the soil solution provides useful information on soil processes that are important to both agricultural and environmental sciences. Data on the concentrations of trace elements in a "real" soil solution are valuable for predicting the availability of those elements, as well as any toxic effects that they might have on crops and on their biological activities in soils.

Soil solution – the aqueous phase of the soil – is composed of water with colloidal suspension and dissolved substances, which can be various forms of ions (free and complex), and various complex compounds, including bio-inorganic complexes. Ion pairs – specific forms of oppositely charged ions that are thermodynamically stable – can occur in solutions as manifold and of variable compositions. Common ion pairs (mainly hydroxides) in soil solution include $CdOH^+$, $CuOH^+$, $FeCl^+$, $ZnHCO_3^+$, and $PbSO_4^°$. Complex compounds of a relatively high stability constant occurring in soils solution include: $BeF_2^°$, $CdCl_2^°$, $Cd(OH)_2^°$, $CdSO_4^°$, $Co(OH)_2^°$, $Cr(OH)_3^°$, $Cu(OH)_2^°$, $FeCl_2^°$, $FeCl_3^°$, $Hg(OH)_2^°$, $MnSO_4^°$, $Ni(OH)_2^°$, $PbSO_4^°$, $Pb(OH)_2^°$, $VO(OH)_3^°$, and $Zn(OH)_2^°$.

The partitioning of trace elements between the soil and soil solution determines their mobility and bioavailability. However, predicting the properties of soil solution is difficult, and sophisticated techniques must be adopted for the obtaining undiluted soil solution. A major impact of the variation in redox potential of stored soil/sediments samples should be considered.

The concentration of free metal species in soil solution is controlled by several factors, the most significant of which are thermodynamic/kinetic parameters. Mathematical approaches to modeling soil solution – solid-phase equilibria – are broadly described in numerous publications (Lindsay 1979, Sposito et al. 1984, Waite 1991, Wolt 1994, Sparks 1995, Suarez 1999), and several models for calculating activity coefficients for trace metals are overviewed and discussed. Waite (1991) concluded that: "mathematical modeling clearly has a place in extending the information that can be obtained on trace element species distributed by other methods and will be of practical use in systems for which determination of concentrations of all species of interest is impossible because of sensitivity constrains or other analytical difficulties".

Methods used for obtaining soil solution differ widely, and so it is difficult to determine adequately the mean concentrations of trace elements. As Wolt (1994) stated, "No one approach to obtain soil solution is appropriate to all applications". The ranges of some elements measured in the solution obtained by various techniques from uncontaminated soils are, as follows (in $\mu g \ L^{-1}$): B 12 to 800; Cd 0.01 to 5; Co 0.3 to 29; Cr 0.4 to 29; Cu 0.5 to 135; Mn 25 to 8000; Mo 2 to 30; Ni 3 to 150; Pb 0.6 to 63; and Zn 1 to 750 (Kabata-Pendias and Pendias 2001). Wolt (1994) presented mean natural abundance of trace elements in soil solution as follows (in $\mu mol \ L^{-1}$): As 0.01, B 5, Be 0.1, Cd 0.04, Co 0.08, Cr 0.01, Cu 1, Hg 0.0005, I 0.08, Mo 0.0004, Ni 0.17, Pb 0.005, Se 0.06, Sn 0.2, and Zn 0.08.

However, metal content of soil solution can greatly differ under variable water regimes. Du Laing et al. (2003) reported initial data on metals in the pore water in the soil under variably-flooded condition. Especially high variation (by a factor of about 10) was observed for contents of Fe and Mn in the pore water.

In general, the total contents of trace elements in solutions of uncontaminated mineral soils ranges from 1 to 100 $\mu g \ L^{-1}$, while in contaminated soils these values can be much higher. In both types of soil, however, these are negligible portions of the total soil metals. The transfer factor – calculated as a ratio of metals in solution of contaminated soils to their content in solution of the control soils – usually decreases in the following order: Cd > Ni > Zn > Cu > Pb > Cr (Kabata-Pendias and Pendias 2001).

Keller (1997) identified in soil leachates mobile particles not only of clay minerals and iron oxides but also scales of some prosits. The author suggested that it might be worth investigating prosits in soil solutions as pollution indicators.

5.3.3
Speciation (see also Part II, Chapter 3)

As soils consist of heterogeneous mixtures of different organic and organic-mineral substances, crystalline and clay minerals, oxides and hydroxides of Fe, Mn, and Al, and other solid components as well as a variety of soluble substances, the binding mechanisms for trace elements in soils are manifold and vary with the composition of soils and their physical properties. A great deal of confusion persists in the use of the term "speciation" within environmental sciences. According to Ure et al. (1993), the definition of speciation in the context of soils, sediments and sewage sludges refers to the process of identification and quantification of the different defined species, forms and phases in which an element occurs in investigated materials. However, most often "speciation" also means the description of the amounts and types of existing forms of trace elements. Considerable controversy has developed over selective extraction methods to determine the amounts of trace elements associated with various soil phases and compounds.

Metal species resulted from the partitioning of the total metal content that is associated with various solid soil fractions, and usually are estimated using specific operational extraction procedures. Comprehensive reviews of methods applied for the sequential extraction have been provided in several publications (Bourg 1995, Brümmer 1986, Förstner 1986, Kersten and Förstner 1991, Salomons and Förstner 1984, Sauvé 2001, Sheppard and Stephenson 1997, Mortved et al. 1991, Ure and Davidson 1995). Some selected extraction procedures presented in Table 5.6. show a great diversity among reagents used for the determination of commonly distinguished metal species which are, in general: (i) easily exchangeable; (ii) specifically sorbed, e.g., by carbonates, phosphates; (iii) organically bound; (iv) occluded in Fe/Mn oxides and hydroxides; and (v) structurally bound in minerals (residual). Some authors (Boust and Saas 1981, Mathur and Levésque 1988) distinguish also a water-soluble fraction of metals, which actually corresponds to a dilute salt solution extract. Solid-phase fractionation methods allow separate additional fractions of metals, such as silicate clays, sulfides, and specifically sorbed or precipitated. Various concentrations of reagents and different soil/solution ratios are used for buffering, complexing, and reduction processes in soil samples over a broad pH ranges. In consequence, each method gives operational groups of metals that are not the impact of trace elements on plants and soil bioactivity.

5.4
Soil-Plant Transfer

The bioavailability of trace elements has been – and still is – the most crucial problem in agricultural and environmental studies. There is a steady increase of investigations related to both understanding processes of an element being taken up (nutrient and non-nutrient) by plants, and to finding the most reliable methods for predicting the availability of a given element to plants, and in particular to crop plants. All of these topics are broadly discussed in a variety of text books (Mortvedt et al. 1991, Alloway 1995, Brooks 1998, Kabata-Pendias and Pendias 2001, Kamprath 2000, McLaughlin 2001) and various other publications.

During their evolution and course of life (ontogeny and phylogeny), plants have developed a number of biochemical mechanisms which have resulted in adaptation to and tolerance of new or chemically imbalanced growth media. Therefore, plant responses to trace elements in the soil and ambient air are variable in number and should always be investigated for a particular soil-plant system. The prediction of phytoavailability of trace elements is of a crucial importance for both crop production and health risk assessment. The influence of soil parameters and plant absorption ability are the main factors that govern the phytoavailability of an element.

The most important master variables of soils which control element availability can be generalized as follows: pH and redox potential; texture; organic matter (quantity and quality); mineral composition; temperature; and water regimen. Interactions between chemical elements are also known to influence the phytoavailability of some trace elements. It is evident that Ca, P, and Mg are often the main antagonistic elements against the absorption of several micro cations. Some synergistic effects have also been observed for selected pairs of elements (Kabata-Pendias and Pendias 2001).

Plants reveal a great adaptation to variable composition of growth media, and have developed several mechanisms to take up a given nutrient under deficiency conditions in soils; moreover, they can also exclude an element at its high external concentration. However, mechanisms involved in the exclusion processes are much weaker than those developed by roots in the absorption of deficient micronutrients. Thus, an excess of trace metals in soil provides a greater stress to plants than a deficiency of these metals. In general, plants readily take up trace elements that are in the soil solution in either free ionic or complex forms. However, changes in the pH of the root ambient solution and various root exudates can significantly increase availability of certain elements (Mortvedt et al. 1991). On the other hand, the efflux of an excessive amount of zinc (and possibly also other trace elements) from roots seems to be a protective mechanism in metal-contaminated soils (Santa-Maria 1998). The effects of root exudates on the mobility of trace elements are variable, and as Zhao et al. (2001) concluded, root exudates from plants that hyperaccumulate metals (*Thlaspi caerulescens*) are not involved in Zn and Cd accumulation.

Extensive progress has been made in our understanding of the mechanisms and external factors which control the uptake of trace elements by plants under various conditions. However, prediction of the phytoavailability of trace elements – and especially in contaminated environments – is still very difficult. Several models have been used to predict the phytoavailability of trace metals, and in particular that of Cd, Zn, Cu, and Pb (McLaughlin 2001, Mortved et al. 1991); however, these models are rather limited to a given plant and specific growth conditions. Their application to crop plants and field conditions, especially to agricultural landscape, is still uncertain.

In general, the chemical composition of plants closely reflects the chemical properties of whole environments, soils, waters, and air. Using plant chemical status for geochemical prospecting is very old practice (Kabata-Pendias and Pendias 2001), but recently it has been used broadly for the bioindication of contaminated sites and for the environmental biomonitoring (see Part I, Chapter 12).

An overview of methods for assessing bioavailable trace elements – and in particular metals – has been presented in several books (Houba et al. 1999, Mc Laughlin 2001, Mortvedt et al. 1991, Sparks et al. 1996). Methods used to evaluate the pool of soluble (available) trace elements in soils are based mainly on extractions by various solutions:

- Acids: mineral acids at various concentrations.
- Chelating agents: e.g., EDTA, DTPA [+ TEA].
- Buffered salts: e.g., AAAc.
- Neutral salts $CaCl_2$, $MgCl_2$, $Sr(NO_3)_2$, NH_4NO_3.
- Other extractants, like *Coca Cola*, which has proposed for use in routine soil testing.

The use of some other techniques, including electrodialysis, diffusion through membrane, diffusive gradient in thin-film (DGT) (Hooda et al. 2001), and bioindication has also been proposed. However,

since a number of soil parameters and climatic factors have a significant impact on the absorption of trace elements by roots, any method which is applied must be related to a given soil and plant conditions.

Desirable reactions of these extractands are: (i) relatively weak reactions with soil components; (ii) possible relation to the amount taken up by most crop plants; and (iii) possible independence of soil properties. In a view of all discussions and evaluations of recent research results, the earlier concept of Barber (1984) has been renewed for the use of extractands which simulate natural soil solution. These are solutions of neutral salts, but mainly $CaCl_2$ in various concentrations (most frequently 0.01 M). Häni and Gupta (1985) were also some of the first investigators to identify a reason to use neutral salt solution to assess the actual metal bioavailabilities. Recently, Houba et al. (1999) showed 0.01 M $CaCl_2$ to be suitable for the extraction of bioavailable pools of metals. However, there is still a need to evaluate analytical and sampling errors over a wide range of soils. This is especially important when soils are contaminated by metals originating from various sources, whereupon these methods will be recommended for regulatory purposes.

5.5
Contamination and Remediation

Metal contamination of soils (based on the definition presented by Knox et al. 1999, soils are not considered to be polluted unless a threshold concentration exists that begins to affect biological processes, though both terms are used synonymously) is as old as man's ability to smelt and process ores, and goes back as far as the Bronze Age (2500 BC). At present, all of man's activities – for example, mining, various industries, either adsorb or occlude several trace metals.

- Phytoremediation: here, the phytoextraction technique involves growing reasonable yields of plants that hyperaccumulate metals. The method shows promise in practical terms, but the technology needs to be developed. It is a relatively low cost method, however.

5.5.2
Non-In-Situ Methods

The most common of these include:

- Removal: the contaminated soil is exposed to chemical extraction and/or thermal treatment with volatile elements or compounds, and to other leaching or immobilizing processes. It is a high-cost procedure.
- Excavation: the contaminated soil is removed and disposed of elsewhere (e.g., in prepared landfills). This method is very costly, and has possible problems with groundwater contamination.

The problem of concern in soil remediation actions is the cost. Phytoremediation techniques are likely to be less costly than those based on conventional technologies. At present, a real demand for phytoextraction is to increase the yield of plants that hyperaccumulate metals from soils, and to develop adequate technologies for the utilization of plant materials.

Agricultural practices have also been applied to soil remediation. Most commonly, the uptake of metals by plants is diminished by keeping a neutral soil pH and by amendments with materials having a high capacity to bind metals in possibly slightly mobile fractions. Various materials are used for soil amendments and remediation; these are mainly lime, phosphate fertilizers, zeolites, montmorillonite clays, humic

(organic) matter, and biosolids. The beneficial effects of these treatments have been broadly discussed (Cunningham and Berti 2000, Iskandar 2001, Pierzynski et al. 2000).

5.6
Environmental Quality Criteria

Soil acts as a natural buffer controlling the transport of trace elements to the atmosphere, hydrosphere, and biota. There is a strong relationship between the quality of plant food and trace element status of soils. The mobility and phytoavailability of trace elements are crucial parameters for protecting the sanitation of agricultural landscape, for maintenance of the sustainable functioning of the soil, and for the production of healthy food. There is an assumption that a distinction between mobile and non-mobile trace elements is a key for the assessment of ecological safety of soils contaminated by trace elements. However, due to great variability of available species of trace elements, and multifunctional dependence on soil properties, acceptable levels of trace elements in agricultural soils are related mainly to so-called total (most commonly aqua regia-soluble) contents of elements (see Table 5.7).

Assessment criteria and application guidelines for metals in soils are now objectives of legislative actions in most countries. Common practices of disposal and land application of wastes require regulations, and most countries have already installed legislation controlling maximum safe metal levels in biosolids and loading rates for land. Amounts of metals allowed in annual loading to arable soils differ considerable among countries (Table 5.8). Calculated total input of metals due to agricultural practice exceed by some-fold the amounts of metals that are accepted as annual loads

Tab. 5.8: Maximum allowable (or calculated)[c] loading (kg ha^{-1}·year) of trace metals to arable soils in some countries

Metal	USA[a]	EU[b]	Denmark[c]	Poland[d]	UK[e]
As	2	–	–	–	0.7
Cd	1.9	0.15	0.01	0.2	0.15
Cu	75	12	1.68	5	7.5
Cr	150	–	0.23	15	15
Hg	0.85	–	0.004	0.2	0.1
Mo	–	–	–	–	0.2
Ni	21	3	0.32	3	3
Pb	15	15	0.27	10	15
Zn	140	30	4.93	10	15

[a] USEPA (1993); after Miller and Miller (2000). [b] Miller and Miller (2000). [c] Kjølholt (1997) – calculated typical annual loads from various products applied on Danish agricultural soils. [d] Kabata-Pendias and Piotrowska (1987). [e] Smith (1996); after Miller and Miller (2000).

(Kjølholt 1997). Is was estimated, however, that at the global scale the input of most metals is higher from industrial sources than from the application of fertilizers and biosolids in farming (Kabata-Pendias and Pendias 2001). Calculated budgets for Cd, Pb and Zn in soils in Poland clearly indicate that typical input of these metals (mainly from biosolids and atmospheric deposition) leads to their accumulation in arable soils (Kabata-Pendias et al. 1989).

The limits for metal loading rates to soils should consider general ecotoxicity, phytotoxicity, transfer to animals, and risk to human population based on pathways such as direct soil ingestion, contamination of food, and pollution of waters. These limits should also concern the long-term phytoavailability of metals in soils amended with biosolids and compost (Kabata-Pendias and Piotrowska 1987, Chaney et al. 2001, van Hersteren et al. 1999, Huber and Freudenschuss 2001, Logan 2000, Miller and Miller 2000)

Requirements for good/minimum soil quality are based on soil categories and land use (e.g., allotments, gardens, parks, arable soils, waste ground). Plant species (even varieties and genotypes) and environmental conditions also influence the divergent impact of soil metals on plants. It is most important, however, to evaluate acceptable application rates in relation to: (i) initial metal contents of soil; (ii) total amount added of one metal and of all trace elements; (iii) annual and cumulative loading; (iv) relative ratio between interacting elements; (v) input-output balance; (vi) soil properties; and (vii) plant characteristics.

In general, in heavy neutral soils most trace elements would be less mobile and less phytoavailable than in light acid soils. Trace element mobility in soils is also known to be related to land use. Usually, metals in forest soils are more easily mobile, and therefore easier bioavailable and leached, than are trace elements in agricultural soils. This phenomenon will be an environmental concern in the future due to programs of forestation of poor agricultural quality soils, especially in Central and Eastern Europe.

5.7
Final Remarks

Soil not only forms part of the ecosystems, but also occupies a basic role for humans, because the survival of man is tied to the maintenance of its productivity. The history of food supply is an essential part of the history of mankind. Good quantity and quality of food are closely related to chemical balance of agricultural soils, and in particular to proper and safe contents of trace elements.

Some main objectives related to trace elements in soil that require further research include:

- Distribution and quantitative relation of trace elements in soils in various geologic formations or regions.
- Rules of chemical association of elements that form minerals and bio-inorganics.
- Balance and cycling of chemical elements in soils of specific environments.
- Hyperaccumulation of metals by plants.
- Health risk affected by the imbalance of trace elements in environments.
- Interactions of trace elements as multivariant reactions affected by concentration factors.
- Phase distribution (speciation) of elements in soils and bioavailability of various species.
- Methods for determination and prediction of the bioavailability of trace elements.
- Geochemical processes by which the soil regenerates itself, for example, dispersion, transformation, and degradation (e.g., oxidation, mineralization).
- Biomonitoring and bioindication.
- Assessment limits for safe trace element contents of soils protecting soil biological properties, ground waters and food chain.

- Prevention and remediation.

Further research into these highlighted problems – and the data subsequently acquired – should provide valuable information to help our understanding of the complex interactions that exist between trace elements in the soil and the health of plants, animals, and humans. Moreover, these new findings should provide decision makers with the information needed to solve national and international problems, and to ensure a sound environment for future generations.

Acknowledgments

The authors thank Dr. G. Siebielec for reading the text and helping to collect reference material.

References

Adriano DC, Bollag J-M, Frankenberger, JRWT and Sims RC, eds. (1999) *Bioremediation of contaminated soils*, Am. Soc. of Agronomy, Madison, WI.

Alloway BJ (1995) *Heavy metals in soils.* 2nd edn. Blackie Academic and Professional, London.

Baldock IA and Nelson PN (2000) *Soil organic matter.* In: Somner ME, ed. Handbook of Soil Chemistry, pp. B25–B84, CRC Press, Boca Raton, FL.

Barber SA (1984) *Soil nutrient bioavailability.* John Wiley & Sons, New York.

Bartlett RJ (1999) *Characterizing soil redox behavior.* In: Sparks DL, ed. Soil Physical Chemistry, pp. 371–391, CRC Press, Boca Raton, FL.

Berg T and Steinnes E (1997) *Recent trends in atmospheric deposition of trace elements in Norway as evident from the 1995 moss survey.* Sci Total Environ **208**:197–206.

Bolt GM, ed. (1979) *Soil chemistry. B. Physicochemical models.* Elsevier Science Publishers, Amsterdam.

Bourg ACM (1995) *Speciation of heavy metals in soils and groundwater and implications for their natural and provoked mobility.* In: Salomons W,

Förstner U, and Mader P, eds. Heavy metals. Problems and solutions, pp. 19–31. Springer-Verlag, Berlin.

Boust D and Saas A (1981) *A selective chemical extraction procedure applied to trace metals composition; comparison between several reagents on two types of sediments (Seine and Gironde estuaries).* In: Ernst WHO, ed. Proc. Intern. Heavy metals in the environment, pp. 709–711. CEP Consultants, Edinburgh.

Brooks RR, ed. (1998) *Plants that hyperaccumulate heavy metals.* CAB International, Cambridge.

Brümmer GW (1986) *Heavy metal species, mobility and availability in soils.* In: Bernhard M, Brinckman FE and Sadler PJ, eds. The importance of chemical "speciation" in environmental processes. pp. 169–192, Springer-Verlag, Berlin.

Brümmer GW, Gerth J and Tiller KG (1988) *Reaction kinetics of the adsorption and desorption of nickel, zinc, cadmium by goethite.* Soil Sci **39**:37–52.

Burns RG and Dick RP, eds. (2002) *Enzymes in the environment.* Marcel Dekker, Inc. New York.

Chaney RL, Ryan JA, Kukier U, Brown SL, Siebielec G, Malik M and Angle JS (2001) *Heavy metal aspects of compost use.* In: Stoffella PJ and Kahn BA, eds. Compost utilisation in horticultural croppings systems, pp. 323–359. Lewis Publishers, Boca Raton, FL.

Chuang MC, Shu GY and Liu JC (1996) *Solubility of heavy metals in a contaminated soil: effects of redox potential and pH.* Water Air Soil Pollut **90**:543–556.

Cunningham SD and Berti WR (2000) *Phytoextraction and phytoremediation: technical, economic, and regulatory consideration of the soil-lead issue.* In: Terry N and Banuelos G, eds. Phytoremediation of contaminated soils and water, pp. 359–376. CRC Press, Boca Raton, FL.

Dodd RT (1981) *Meteorites a petrologic-chemical synthesis.* University Press. Translated into Russian by Pietayeva MI and Ulyanova AA (1986) Mir, Moscow.

DuLaing G, Vanthuyne D, Vandecasteele B, Tack FMG and Verloo MG (2003) *Metal release from a contaminated soil as affected by hydrological management.* In: 6th Intern. Symposium on Environmental Geochemistry. Final Programme and Book of Abstracts, p 29.

Förstner U (1986) *Chemical forms and environmental effects of critical elements in solid-waste materials-combustion residue.* In: Bernhard M, Brinckman FE and Sadler PJ, eds. The importance of chemical speciation in environmental processes, pp. 465–491. Springer-Verlag, Berlin.

Gerth J, Brümmer GW and Tiller KG (1993) *Retention Ni, Zn and Cd by Si-associated goethite.* Z Pflanzenernähr Bodenk **156**:123–129.

Goldberg SP and Smith KA (1984) *Soil Manganese: E values, distribution of manganese-54 among soil fractions, and effects of drying.* Soil Sci Am J **48**:559–564.

Häni H and Gupta S (1985) *Reason to use neutral salt solutions to assess the metal impact on plants and soils.* In: Leschber R, Davis RD and L'Hermite R, eds. Chemical method for assessing bioavailable metals in sludges and soils, pp. 42–48. Elsevier, New York.

Herbert RB (1997) *Partitioning of heavy metals in podzol soils contaminated by mine drainage water, Dalarna, Sweden.* Water Air Soil Poll **96**:39–59.

Hooda SP, Davison B, Zhang H and Edwards T (2001) *DGT for assessing bioavailable metals in soils.* Proceedings, 6th International Conference on the Biogeochemistry of Trace Elements, Guelph, July 29–August 2, p. 354.

Houba VJ, Temminghof EJM and Van Vark W (1999) *Soil analysis procedure extraction with 0.01 M CaCl₂.* Wageningen Agric. Univ., Wageningen.

Huang MP, ed. (2000) *Soil chemistry.* In: Sumner ME, ed. Handbook of soil science, pp. B1–B352. CRC Press, Boca Raton, FL.

Huber S and Freudenschuss A (2001) *Development of indicators for different soil contamination by heavy metals.* Proceedings of the Second European Soil Forum, pp. 1–7. Napoli, October 23–25.

Iskandar IK, ed. (2001) *Environmental restoration of metals-contaminated soils.* Lewis Publisher, Boca Raton, FL.

Kabata-Pendias A and Adriano DC (1995) *Trace metals.* In: Rechcigl JE, ed. Soil amendments and environmental quality, pp. 139–167. Lewis Publishers, Boca Raton, FL.

Kabata-Pendias A, Dudka S and Galczynska B (1989) *Baseline zinc content of soils and plants in Poland.* Environ. Geochem Health **11**:19–24.

Kabata-Pendias A and Pendias H (2001) *Trace elements in soils and plants,* 3rd edn, CRC Press, Boca Raton, FL.

Kabata-Pendias A and Piotrowska M (1987) *Trace elements a criteria for waste use in agriculture.* P33, IUNG. Pulawy, pp. 46 (in Polish).

Kamprath EJ, ed. (2000) *Soil fertility and plant nutrition.* In: Sumner ME, ed. Handbook of Soil Science, pp. D1–D189. CRC Press, Boca Raton, FL.

KELLER C (1997) *Some microscopic and mineral particles of biological origin in soil solutions.* Eur J Soil Sci 48:193–199.

KERSTEN M and FÖRSTNER U (1991) *Speciation of trace elements in sediments.* In: Batley GE, ed. Trace elements speciation: analytical methods and problems, pp. 246–317. CRC Press, Boca Raton, FL.

KJØLHOLT J (1997) *Sources of agricultural soil contamination with organic micropollutants and heavy metals.* Specially Conference on Management and Fate of Toxic Organics in Sludge Applied to Land. Copenhagen, 30 April–2 May 1997.

KNOX AS (formerly CHLOPECKA A), GAMERDINGER AP, ADRIANO DC, KOLKA RK and KAPLAN DI (1999) *Source and practices contributing to soil contamination.* In: Adriano DC, Bollag J-M, Frankenberger WT and SIMS RC, eds. Bioremediation of contaminated soils, pp. 53–87. Am Soc Agronomy, Inc. Madison, WI.

KNOX AS, SEAMAN J and ADRIANO DC (2000) *Chemophytostabilization of metals in contaminated soils.* In: Wise DL, Trantolo DJ, Cichon EJ, Inyang HI and Stottmeister U, eds. Bioremediation of contaminated soils, pp. 811–836. Marcel Dekker, Inc., New York.

KORZH VD (1991) *Geochemistry of elemental composition of the hydrosphere.* Nauka, Moscow (in Russian).

KOSTYUK O and BUNNERBERG C (1999) *The role of microbiota in the behavior of radionuclides in semi-natural ecosystems.* In: Wenzel WW, Adriano DC, Alloway B, Doner HE, Keller C, Lepp NW, Mench M, Naidu R and Pierzynski GM, eds., 5th International Conference on Biogeochemistry of Trace Elements, pp. 44–45. Vienna.

KUO S, HELMAN PE and BAKER AS (1983) *Distribution and forms of copper, zinc, cadmium, iron, and manganese in soils near a copper smelter.* Soil Sci 135:101–109.

LANTZY RJ and MCKENZIE FT (1979) *Atmospheric trace elements: global cycles and assessment of man's impact,* Geochim Cosmochim Acta 43:511–525.

LI Y-M, CHANEY RL, SIEBIELEC G, and KERSHNER BA (2000) *Response of four turfgrass cultivars to limestone and biosolids compost amendments of a zinc and cadmium contaminated soil at Palmerton, PA.* J Environ Quality 29:1440–1447.

LINDSAY WL (1979) *Chemical equilibria in soils.* Wiley-Interscience, New York.

LOGAN TJ (2000) *Soils and environmental quality.* In: Sumner ME, ed. Handbook of Soil Science, pp. G155–G169. CRC Press, Boca Raton, FL.

MATHUR SP and LEVESQUE M (1988) *Soil test for copper, iron, manganese, and zinc in histosols.* Soil Sci 145:102–110.

MC LAUGHLIN MJ (2001) *Bioavailability of metals to terrestrial plants.* In: Allen HE, ed. Bioavailability of metals in terrestrial ecosystems. Importance of partitioning for bioavailability to invertebrates, microbes and plants, pp. 39–68. SETAC Press, Pensacola, FL.

MORTVEDT JJ, COX FR, SHUMAN LM and WELCH RM, eds. (1991) *Micronutrients in agriculture,* 2nd edn, Chemical methods. Soil Sci Soc Am, Inc., Madison, WI.

MILLER DM and MILLER WP (2000) *Land application of wastes.* In: Sumner ME, ed. Handbook of Soil Science, pp. G-217–G-245. CRC Press, Boca Raton, FL.

MUKHERJEE AB (2001) *Behavior of heavy metals and their remediation in metalliferous soils.* In: Prasad MN, ed. Metals in the environment: analysis by biodiversity, pp. 433–471. Marcel Dekker Inc., New York.

NAIDU R, KRISHNAMURTI GSR, BOLAN NS, WENZEL W and MEGHARAJ M (2001) *Heavy metal interactions in soils and implications to soil microbial biodiversity.* In: Prasad MN, ed. Metals in the environment: analysis by biodiversity, pp. 401–431, Marcel Dekker Inc., New York.

PIERZYNSKI GM, SIMS JT and VANCE GF (2000) *Soils and environmental quality,* 2nd edn, CRC Press, Boca Raton, FL.

RULE JH (1999) *Trace metal cation adsorption in soils: selective chemical extractions and biological availability.* In: Dabrowski A, ed. Adsorption and its applications in industry and environmental protection, pp. 319–349. Elsevier, Amsterdam.

SALOMONS W and FÖRSTNER U (1984) *Metals in the hydrocycle.* Springer-Verlag, Berlin.

SANTA-MARIA GE and COGLIATTI DH (1998) *The regulation of zinc uptake in wheat plants.* Plant Sci 137:1–12.

SAUVÉ S (2001) *Speciation of metals in soils.* In: Allen HE, ed. Bioavailability of metals in terrestrial ecosystems. Importance of partitioning for bioavailability to invertebrates, microbes and plants, pp. 7–38. SETAC Press, Pensacola, FL.

SEA – Swedish Environmental Agency, 2001, http:/www.intrnat.environ.se/estart.htm.

SCHMITT HW and STICHER H (1991) *Heavy metal compounds in the soil.* In: Merian E, ed. Metals and their compounds in the environment, pp. 311–331.VCH-Verlag, Weinheim.

SMITH KS and HUYCK HLO (1999) *An overview of the abundance, relative mobility, bioavailability, and human toxicity of metals.* In: Plumlee GS and Logsdon JJ, eds. The environmental geochemistry of mineral deposits. Part A. Processes, Techniques, and Health Issues. Review in Econ Geol 6A:29–70.

SHEPPARD MI and STEPHENSON M (1997) *Critical evaluation of selective extraction methods for soils and sediments.* In: Prost R, ed. Contaminated soils. 3rd International Conference on the Biogeochemistry of Trace Elements, pp. 69–97, INRA, Paris.

SPARKS DL (1995) *Environmental Soil Chemistry.* Academic Press, San Diego.

SPARKS DL (1999) *Kinetics and mechanisms of chemical reactions at the soil mineral/water interface.* In: Sparks DL, ed. Soil Physical Chemistry, 2nd edn, pp. 135–191. CRC Press, Boca Raton, FL.

SPARKS DL, PAGE AL, HELMKE PA, LOEPPERT RH, SOLTANPOUR PN, TABATABAI MA, JOHNSTON CT and SUMNER ME, eds. (1996) *Methods of soil analysis*, Part 3, Chemical methods. Soil Sci Soc Am, Inc., Madison, WI.

SPOSITO G, LECLAIRE JP, LEVESQUE S and SENESI N (1984) *Methodologies to predict the mobility and availability of hazardous metals in sludge-amended soils.* University of California, Davis, CA.

STUCZYNSKI T and MALISZEWSKA-KORDYBACH B (2001) *Current status of information on heavy metals in European soils—contamination and regulation aspects.* In: Proceedings Second European Soil Forum, pp. 1–11. Napoli, October 23–25.

SUAREZ DL (1999) *Thermodynamics of the soil solution.* In: Sparks DL, ed. Soil Physical Chemistry, 2nd edn, pp. 97–134. CRC Press, Boca Raton, FL.

TAN KH (1998) *Principles of Soils Chemistry*, 3rd edn. Marcel Dekker, Inc., New York.

TAN KH (1995) *Soil Sampling, Preparation and Analysis.* Marcel Dekker, Inc., New York.

TESSIER A, CAMPBELL PGC and BISSON M (1979) *Sequential extraction procedure for the speciation of particulate metals.* Anal Chem 51:844–851.

TOBIAS FJ, BECH J and SANCHEZ ALGARRA P (1997) *Statistical approach to discrimination background and anthropogenic input of trace elements in soils of Catalonia, Spain.* Water Air Soil Poll 100:63–78.

URE AM and DAVIDSON CM, eds. (1995) *Chemical speciation in the environment.* Blackie Academic and Professional, London.

URE A, QUEVAUVILLER PH, MUNTAU H and GRIEPINK B (1993) *Improvements in the determination of extractable contents of trace metals in soil and sediment prior to certification.* CEC BCR Information, Chemical Analysis, Report EUR 14763 EN Brussels.

VAN HERSTEREN S, VAN DE LEEMKULE MA and PRUIKSMA MA (1999) *Minimum soil quality. A use-based approach from an ecological perspective.* Part 1: Metals. Technical Soil Protection Committee, The Hague.

WAITE TD (1991) *Mathematical modeling of trace element speciation.* In: Bartley GE, ed. Trace elements speciation: analytical methods and problems, pp. 117–184. CRC Press, Boca Raton, FL.

WOLT JD (1994) *Soil solution chemistry. Applications to environmental science and agriculture.* J Wiley & Sons, New York.

ZHAO FJ, HAMON RE and MCLAUGHLIN MJ (2001) *Root exudates of the hyperaccumulator Thalaspi caerulescens do not enhance metal mobilization.* New Phytologist 151:613–620.

6

Transfer of Macro, Trace and Ultratrace Elements in the Food Chain

Manfred K. Anke

6.1
Introduction

The transfer of inorganic components from soil to plants and into the food chain of animals and man have been a topic of intensive research since the beginning of modern agricultural chemistry, the biological sciences, and investigation into animal and human nutrition and health. Salm-Horstmar (1849) was one of the first researchers to describe how individual nutrients and their combinations affect the growth of oats. Iron deficiency in plants was first recognized by Gris (1844, 1847), who showed that the condition could be alleviated by spraying the foliage with iron salts. This was probably the first nutrient deficiency disease described in plants. Molisch (1892) considered the discovery of its cause to be "one of the greatest discoveries in the history of plant physiology".

Abiotic toxic damage and accumulation of metals and nonmetals in wild and cultivated plants may result from natural geochemical loads in the soil (Kovalskij 1977) caused by macro, trace and ultratrace elements in water used for irrigation, in natural volcanoes and anthropogenic industrial pollution of the atmosphere. Water, aerosols, and dust contain a variety of aluminum, arsenic, cad-

mium, chromium, copper, chlorine, fluorine, iron, lead, mercury, molybdenum, manganese, nickel, rubidium, selenium, thorium, tin, titanium, uranium, vanadium, zinc, etc., and are therefore often responsible for the accumulation of these elements in – and toxicity symptoms shown by – plants (Dässler 1986). As civilization continues to develop, industrialization increases and food production rises, abiological accumulation of elements and damage to plants will become increasingly common due to overfertilization (manure, fertilizers containing many toxic elements, etc.), the use of loaded water for irrigation purposes, the application of pesticides with high amounts of arsenic, copper, mercury, tin, etc., environmental pollution by industrial waste gases and water (stock gases from fossil fuel power stations, road traffic, combustion of coal and oil, incineration of garbage), and incorrect and unchecked deposition of industrial and other waste products, garbage and sewage sludge (Bergmann et al. 1992). The supply of plants with nonmetals and metals is not only important for covering their requirement but also for the healthy nutrition of animals and man. This applies also to elements that are not essential for plants but important for the fauna.

Elements and their Compounds in the Environment. 2nd Edition.
Edited by E. Merian, M. Anke, M. Ihnat, M. Stoeppler
Copyright © 2004 WILEY-VCH Verlag GmbH & Co. KGaA, Weinheim
ISBN: 3-527-30459-2

Elements essential for all higher plants, besides carbon, hydrogen, oxygen and nitrogen, include the macro elements phosphorus, sulfur, potassium, calcium, and magnesium, and the trace elements boron, copper, manganese, molybdenum, zinc, and nickel. Beside these "essential" elements, ultratrace elements, which comprise all the other stable elements of the periodic system, are important for flora, fauna and man as either toxic or essential elements. Elements especially important to plant nutrition are aluminum, cobalt, sodium, silicon, chlorine and vanadium, whereas the ultratrace elements cadmium, chromium, mercury, arsenic, fluorine, and lead are usually toxic (Bergmann 1992). For the nutrition of animals and man, the essential macro elements are phosphorus, sodium, potassium, calcium, magnesium, sulfur, and chlorine, whereas the trace elements iron, copper, zinc, manganese, molybdenum, nickel, iodine, and selenium are essential and toxic at the same time. Vegetable food supplies the molybdenum and nickel requirements of animals and the manganese, molybdenum, and nickel requirements of man. A deficiency of these elements can only occur in case of genetic defects and parenteral nutrition (Fiedler and Rösler 1993, Macholz and Lewerenz 1989). Besides these elements that are essential for human nutrition and health, the elements lithium, rubidium, cesium, strontium, barium, cadmium, mercury, aluminum, thallium, titanium, tin, lead, vanadium, arsenic, bismuth, chromium, tungsten, and uranium may also be of importance. The transfer of all metals and non-metals to animals and man via the terrestrial food chain is the basis for animal and human nutrition.

6.2
Terrestrial Indicator Plants of the Elemental Load

Generally, nutrient deficiencies or excesses are caused by soil properties, fertilizer application, interactions between mineral elements during uptake and metabolism, and intrinsic factors of plants and crops with intensive nutrient responses. The rules of these element-specific influences were given by Bergmann (1992).

The recognition, identification, and differentiation of nutrient-related disorders in plants is important for the normal distribution of the elements in the terrestrial food chain of animals and man.

First, it is interesting to note that a deficiency or excess of one element does not directly induce a particular symptom in cultivated or wild plants. In this respect, it is necessary to examine the macro, trace and ultratrace element transfer from the different polluted and unpolluted soils into the plant. Indicator plants must be easy to identify, grown worldwide, and indicate the mineral transfer to the food chain. On cultivated soils in many parts of the world, these conditions are met by wheat (*Triticum sativum*), rye (*Secale cereale*), and red clover (*Trifolium pratense sativum*) of the field and meadow varieties (*Trifolium pratense spontaneum*). The green plants were harvested when the rye was in blossom, the wheat shooting, the field red clover in bud, and the meadow red clover in blossom.

Uptake and distribution of metal compounds in the plant are influenced by the amount of plant-available elements in soil, so that it is not easy to obtain meaningful results. At present, there are no methods by which the amount of an element in a soil, accessible to plant roots, can be quantitatively established by direct measurement. Analysis of plant tissue can establish uptake after it has taken place, when suitable indicator plants are available worldwide, and the specific influence of age, plant species and their element contents in stem, leaves and flowers is well known (Mitchell and Burridge 1989). The bioavailability of the elements in the soil is influenced by its pH, drainage status, organic matter, water-holding capacity, microbial activity, cation- and anion-exchange capacity, its ability to supply chelating ligands (Berrow and Burridge 1991), and last – but not least – by the plant species. Their element-specific capacity for uptake and accumulation is very impressive; salient cases in point are the contents of nickel in *Alyssum* spp., but also parsley (1 g kg^{-1} dry matter if the soil is enriched with nickel) (Anke et al. 1995a), of rubidium in twigs of spruce (> 100 mg kg^{-1} dry matter) (*Picea excelsa*) and heather (*Erica carnea*) (Anke et al. 1997), of cobalt in *Crotalaria cobalticola* and *Tragopogon pratensis* (Anke 1961), and of selenium in *Astragalus* spp. (Berrow and Burridge 1991). Of more direct importance for element transportation from soil to plants and the use of indicator plants for measuring the bioavailability of elements are the distribution of an element among different plant parts and the variation of content with age or stage of growth and season. An appreciation of these factors is essential for the correct interpretation of diagnostic plant analysis.

6.2.1
The Element Content of Several Plant Species on the Same Site

The macro element content of field red clover and lucerne, both in the bud, grown on 2 m^2 correlated with a correlation coefficient of 0.39 to 0.84 or a coefficient of determination of 0.15 to 0.70 (Table 6.1).

The certainly rate is stricter in comparison to the correlation coefficient *r*. The correlation of the calcium, potassium, phosphorus, and magnesium contents of the two species is very high, and shows the qualification of both species as indicator plants for these elements. The sodium content of both plant species is very low and demonstrates that their sodium concentration is not correlated under these conditions.

In case of sodium-poor and sodium-rich soils, a correlation of the sodium content of both plant species seems possible.

With the exception of phosphorus, the macro element contents of field red clover and lucerne differ significantly, which shows the species-specific concentration of most macro elements. The same is true for their trace element content with the excep-

Tab. 6.1: The macro element content of field red clover and lucerne (g kg^{-1} dry matter) grown together on 2 m^2 and their correlations (Anke 1968)

Parameter	(n) [1]	Ca [g kg^{-1}]	Mg [g kg^{-1}]	P [g kg^{-1}]	K [g kg^{-1}]	Na [g kg^{-1}]
Field red clover	(55)	16	3.5	2.7	34	0.25
Lucerne	(55)	19	2.7	2.8	27	0.41
p [2]		< 0.001	< 0.001	> 0.05	< 0.001	< 0.001
r [3]		0.84	0.67	0.77	0.81	0.39
cd [4]		0.70	0.45	0.60	0.66	0.15

[1] n = Number of samples; [2] p = Significance level, Student's *t*-test; [3] r = Correlation coefficient; [4] cd = Coefficient of determination.

Tab. 6.2: The trace element contents of field red clover and lucerne (mg kg^{-1} dry matter) grown together on 2 m^2, and their correlations (Anke 1968)

Parameter	Fe [mg kg^{-1}]	Mn [mg kg^{-1}]	Ni [mg kg^{-1}]	Co [mg kg^{-1}]	Zn [mg kg^{-1}]	Cu [mg kg^{-1}]	Mo [mg kg^{-1}]	I [mg kg^{-1}]
Field red clover	149	29	0.86	0.13	33	11.4	0.92	0.11
Lucerne	214	24	1.01	0.16	34	9.2	0.49	0.16
p [1]	<0.001	<0.001	<0.001	>0.05	>0.05	<0.001	<0.001	<0.001
r [2]	0.82	0.90	0.60	0.78	0.81	0.52	0.94	0.71
cd [3]	0.68	0.81	0.36	0.61	0.66	0.27	0.89	0.50

[1] p = Significance level, Student's t-test; [2] r = Correlation coefficient; [3] cd = Coefficient of determination.

tion of zinc and cobalt (Table 6.2). The similarity of zinc and cobalt contents does not influence the high correlation of both elements ($r = 0.78$ and 0.81, respectively). With the exception of copper, all elements correlate with an r of 60 to 94. The molybdenum content of field red clover and lucerne correlated best, with an r of 0.94 and a certainly rate of 0.89. Generally, the trace element content of field red clover and lucerne indicates the different supplies of these elements very well (Anke 1968).

The comparison of the macro element contents of shooting wheat, rye in blossom and field red clover in bud shows similar results as field red clover and lucerne (Table 6.3).

Wheat and rye accumulated significantly lower amounts of ash and macro elements than field red clover. Shooting wheat and rye contained significantly different ash and potassium concentrations only. With the exception of their sodium concentration, all macro elements of both cereals correlated

Tab. 6.3: The macro element contents of green wheat, green rye and field red clover (g kg^{-1} dry matter) grown together on 1 m^2, and their correlations (Anke 2003)

Parameter (n) [1]		Ash [g kg^{-1}]	Ca [g kg^{-1}]	Mg [g kg^{-1}]	P [g kg^{-1}]	K [g kg^{-1}]	Na [g kg^{-1}]
Green wheat (10–39)		74	1.1	0.96	2.7	31	0.64
Green rye (15–28)		57	1.2	0.80	2.5	21	0.46
Field red clover (15–28)		105	14	3.3	2.6	3.0	0.62
p [2]	Wheat:rye	<0.05	>0.05	>0.05	>0.05	<0.01	>0.05
	Wheat:red clover	<0.001	<0.001	<0.001	>0.05	>0.05	>0.05
	Rye:red clover	<0.001	<0.001	<0.001	>0.05	<0.05	>0.05
r [3]	Wheat:rye	0.55	0.64	0.84	0.49	0.51	0.28
	Wheat:red clover	0.09	0.41	0.70	0.16	0.23	0.49
	Rye:red clover	0.00	0.91	0.42	0.13	0.14	0.03
cd [4]	Wheat:rye	0.30	0.41	0.70	0.24	0.26	0.08
	Wheat:red clover	0.01	0.17	0.48	0.02	0.05	0.24
	Rye:red clover	0.00	0.82	0.18	0.02	0.00	0.00

[1] n = Number of samples; [2] p = Significance level, Student's t-test; [3] r = Correlation coefficient; [4] cd = Coefficient of determination.

very well, with $r = 0.49$ (P) to 0.84 (Mg) (Table 6.3). The correlation between the macro elements in wheat and field red clover and those in rye and field red clover is weaker than between wheat and rye, with the exception of sodium. On average, only calcium and magnesium correlated significantly. The ash, phosphorus, potassium, and sodium correlated with $r = 0.26$ or 0.09 to 0.11 on average. The reason for this result is fertilization with phosphorus, potassium (and sodium), which level the concentrations of these elements. All sites of wheat, rye and field clover were very well supplied with phosphorus and potassium. Plants on impoverished soils or soils polluted with phosphorus were not analyzed.

The trace element contents (Fe, Mn, Ni, Zn, Cu, and Mo) of green wheat, green rye and field red clover, on the average of the three species, correlated uniformly with $r = 0.46$ (Ni) and $r = 0.80$ (Zn), or $r = 0.28$ to 0.66 (Table 6.4).

The concentrations of the ultratrace elements cadmium, lead and lithium, on the average of the three species, also correlated very well, with $r = 0.78$, 0.78, and 0.81. The geological origins (Li) and the cadmium and lead pollution of the soils influenced the content in the plant very strongly and accounts for the very high correlation of these ultratrace elements. The same applies to the trace elements, and especially to zinc, manganese and molybdenum. The bioavailability of these elements is strongly influenced by the pH of the soil. With the exception of sodium (and partly of phosphorus and potassium), the analyzed macro, trace and ultratrace element contents of wheat, rye and field red clover grown together on 1 or 2 m² correlated significantly and showed the different bioavailability of the tested elements in the soil (Anke et al. 1980, 1984, 1991, Arnhold 1989, Glei 1995, Grün 1984, Kronemann 1982).

Tab. 6.4: The trace element contents of green wheat, green rye and field red clover (mg kg⁻¹ dry matter) grown together on 1 m², and their correlations (Anke 2003)

Parameter (n) [1]	Fe [mg kg⁻¹]	Mn [mg kg⁻¹]	Ni [mg kg⁻¹]	Zn [mg kg⁻¹]	Cu [mg kg⁻¹]	Mo [mg kg⁻¹]	Cd [mg kg⁻¹]	Pb [mg kg⁻¹]	Li [mg kg⁻¹]
Green wheat (10–39)	79	42	0.33	32	4.6	0.41	0.081	0.75	11.3
Green rye (15–28)	73	28	0.94	31	4.4	0.44	0.041	0.91	4.3
Field red clover (15–28)	110	37	0.80	33	7.4	0.73	0.053	1.5	9.2
p [2] Wheat:rye	>0.05	<0.05	>0.05	>0.05	>0.05	>0.05	<0.01	>0.05	<0.01
Wheat:clover	>0.05	>0.05	<0.001	>0.05	<0.05	<0.05	<0.05	<0.01	>0.05
Rye:red clover	<0.05	>0.05	<0.001	>0.05	<0.05	<0.05	>0.05	<0.05	<0.01
r [3] Wheat:rye	0.76	0.65	0.57	0.86	0.32	0.67	0.88	0.81	0.77
Wheat:clover	0.67	0.89	0.66	0.86	0.86	0.60	0.63	0.67	0.75
Rye:red clover	0.43	0.56	0.14	0.67	0.47	0.60	0.82	0.89	0.90
cd [4] Wheat:rye	0.58	0.42	0.32	0.73	0.10	0.45	0.77	0.66	0.60
Wheat:clover	0.45	0.79	0.44	0.74	0.74	0.36	0.39	0.45	0.56
Rye:red clover	0.18	0.31	0.02	0.45	0.22	0.36	0.68	0.80	0.80

[1] n = Number of samples; [2] p = Significance level, Student's *t*-test; [3] r = Correlation coefficient; [4] cd = Coefficient of determination.

6.2.2
Influence of Geological Origin of the Site on the Macro, Trace and Ultratrace Element Contents of Indicator Plants

The magmatic and sedimentary rocks contain highly different amounts of macro, trace and ultratrace elements, which after weathering of the rocks become components of the soil and of the soil waters. During weathering, the elements are released from the primary minerals and usually also fixed by organic matter. Thus, most of the macro, trace and ultratrace element contents are controlled by conditions of soil formation and the initial contents in the parent rocks.

The distribution in soil profiles follows the general trends of soil solution circulation (Kabata-Pendias et al. 1989, Kabata-Pendias and Pendias 1991). The aim of our experiments was to examine the influence of the geological origin and the condition of soil formation on the correlation of the elements in shooting wheat (n = 550), rye in blossom (n = 485), field red clover (n = 3269) and meadow red clover in blossom

(n = 518). The mean element contents of wheat, rye, field red clover and meadow red clover grown on 12 different soils in Germany, the Czech Republic, and Hungary were correlated (Table 6.5); it transpired that in spite of fertilization with phosphorus, potassium and calcium, the geological origin and the conditions of soil formation led to weak correlations between the elements in the four indicator plants, with the exception of sodium and, partly, potassium. On the average of all correlations of the indicator plants, phosphorus is correlated with $r = 0.52$, magnesium with $r = 0.46$, and calcium with $r = 0.30$.

The correlation of the trace elements in the indicator plants is much higher than that of the macro elements contained in the fertilizers (Table 6.6).

The correlation coefficient of the analyzed trace elements varies, on the average of the four indicator plants, between 0.38 for iron and 0.86 for nickel. The correlation coefficients for iron and copper are relatively low, with $r = 0.38$ and $r = 0.48$, whereas these for molybdenum ($r = 0.63$), manga-

Tab. 6.5: The macro element contents of green wheat, green rye, field red clover and meadow red clover (g kg^{-1} dry matter), and their correlations as a function of the geological origin of the site

Parameter		Ca [g kg^{-1}]	Mg [g kg^{-1}]	P [g kg^{-1}]	K [g kg^{-1}]	Na [g kg^{-1}]
Green DM [1] wheat		1.3	1.1	2.9	28	0.46
Green DM [1] rye		1.1	0.88	2.6	21	0.32
Field red clover		16	3.6	2.9	35	0.42
Meadow red clover		17	4.0	2.5	25	0.52
p [2]	Wheat:rye	>0.05	<0.05	<0.001	<0.01	<0.001
	Wheat:field red clover	<0.001	<0.001	>0.05	>0.05	<0.05
	Rye:field red clover	<0.001	<0.001	<0.001	>0.05	<0.001
	Field:meadow red clover	>0.05	>0.05	<0.001	<0.05	>0.05
r [3]	Wheat:rye	0.25	0.59	0.45	0.31	0.02
	Wheat:field red clover	0.17	0.48	0.77	0.14	0.10
	Rye:field red clover	0.42	0.24	0.25	0.06	0.36
	Field:meadow red clover	0.38	0.51	0.62	0.49	0.14

[1] DM = Dry matter; [2] p = Significance level, Student's *t*-test; [3] r = Correlation coefficient.

Tab. 6.6: The trace element contents of green wheat, green rye, field red clover and meadow red clover (mg kg^{-1} dry matter) and their correlations as a function of the geological origin of the site

Parameter	Fe [mg kg^{-1}]	Mn [mg kg^{-1}]	Ni [mg kg^{-1}]	Zn [mg kg^{-1}]	Cu [mg kg^{-1}]	Mo [mg kg^{-1}]	I [mg kg^{-1}]
Green DM [1] wheat	72	33	0.42	26	4.3	0.43	0.072
Green DM [1] rye	78	30	0.40	28	3.9	0.39	0.106
Field red clover	132	41	0.86	38	9.4	0.84	0.190
Meadow red clover	136	55	1.05	48	8.8	0.96	0.153
p [2] Wheat:rye	>0.05	>0.05	>0.05	>0.05	>0.05	>0.05	<0.05
Wheat:field red clover	<0.01	<0.05	<0.001	<0.001	<0.001	<0.001	<0.001
Rye:field red clover	<0.05	<0.01	<0.001	<0.001	<0.001	<0.001	<0.001
Field:meadow red clover	>0.05	<0.01	<0.05	<0.001	>0.05	<0.05	<0.05
r [3] Wheat:rye	0.31	0.83	0.86	0.87	0.50	0.47	0.71
Wheat:field red clover	0.06	0.60	0.84	0.71	0.76	0.62	0.99
Rye:field red clover	0.48	0.55	0.84	0.77	0.11	0.57	0.94
Field:meadow red clover	0.58	0.80	0.89	0.82	0.50	0.86	0.68

[1] DM = Dry matter; [2] p = Significance level, Student's t-test; [3] r = Correlation coefficient.

nese ($r = 0.70$), zinc ($r = 0.79$), iodine ($r = 0.83$) and nickel ($r = 0.83$) are markedly high. The trace element contents of the indicator plants as a function of the geological origin of the soil correlated much better than the macro elements applied with fertilizers (Anke et al. 1993, Anke 2003).

The ultratrace element contents of the indicator plants as a function of the geological origin of the soil correlated also very well, with r between 0.55 (Li) and 0.82 (Pb) (Table 6.7). The ultratrace elements (Li, Rb, Sr, Ba, Al, Cd, Pb, V, Cr, and U) in green wheat, green rye, field red clover and meadow red clover also correlated with a good correlation coefficient.

The macro, trace, and ultratrace element contents of shooting wheat, rye in blossom, field red clover in bud and meadow red clover in blossom provide good information about the bioavailability of the elements in the soil (Anke 2003, Angelow 1994, Arnhold 1989, Anke et al. 1998, Grün 1984). Wheat, rye and red clover in the defined stages of development are generally suitable as indicator plants of cultivated soils.

The advantage of these species are their worldwide presence and knowledge about their cultivation. Beside these species, dandelion (Anke 1961, Djingova et al. 1986, Kabata-Pendias et al. 1989), nettle (Ernst and Leloup 1987), lucerne (Anke 1961) and mushrooms (Gast et al. 1988) have been used as indicator plants of cultivated soils. Their fitness for this purpose is limited, because they are not present worldwide, and their element concentrations change too quickly or are very specific (mushrooms). Mosses and lichens are known to be good indicator plants for atmospheric pollution, especially in northern countries with limited agriculturally useful areas (Rühling et al. 1987, Jenkins 1987).

6.2.3
Influence of Plant Age on the Macro, Trace and Ultratrace Element Contents of Plants

On average, the ash and macro element contents of the tested monocotyledonous and dicotyledonous plant species decreased from the end of April to the middle of

Tab. 6.7: The ultratrace element contents of green wheat, green rye, field red clover and meadow red clover (mg kg^{-1} dry matter) and their correlations as a function of the geological origin of the site (Anke 2003)

Parameter	Li [mg kg^{-1}]	Rb [mg kg^{-1}]	Sr [mg kg^{-1}]	Ba [mg kg^{-1}]	Al [mg kg^{-1}]
Green DM [1] wheat	11.0	6.8	22.1	9.7	36
Green DM [1] rye	7.0	7.8	26.7	11.7	43
Field red clover	9.6	11	91.6	11.1	54
Meadow red clover	8.5	21	87.0	12.1	70
p [2] Wheat:rye	<0.01	>0.05	>0.05	>0.05	>0.05
Wheat:field red clover	>0.05	<0.05	<0.001	>0.05	<0.01
Rye:field red clover	>0.05	<0.05	<0.001	>0.05	<0.05
Field:meadow red clover	>0.05	<0.001	>0.05	>0.05	<0.001
r [3] Wheat:rye	0.53	0.61	0.64	0.48	0.22
Wheat:field red clover	0.48	0.82	0.02	0.22	0.57
Rye:field red clover	0.50	0.66	0.15	0.28	0.50
Field:meadow red clover	0.68	0.66	0.34	0.12	0.32
Parameter	Cd [mg kg^{-1}]	Pb [mg kg^{-1}]	V [mg kg^{-1}]	Cr [mg kg^{-1}]	As [mg kg^{-1}]
Green DM [1] wheat	0.046	0.63	0.073	0.404	0.116
Green DM [1] rye	0.038	0.80	0.087	0.408	0.136
Field red clover	0.029	1.69	0.098	0.363	0.133
Meadow red clover	0.034	2.10	0.143	0.397	0.137
p [2] Wheat:rye	>0.05	>0.05	>0.05	>0.05	<0.01
Wheat:field red clover	<0.05	<0.01	<0.05	>0.05	<0.05
Rye:field red clover	<0.05	<0.01	>0.05	>0.05	>0.05
Field:meadow red clover	>0.05	>0.05	<0.01	>0.05	>0.05
r [3] Wheat:rye	0.75	0.75	0.81	0.76	0.72
Wheat:field red clover	0.73	0.87	0.58	0.21	0.49
Rye:field red clover	0.49	0.87	0.45	0.13	0.63
Field:meadow red clover	0.46	0.77	0.32	0.41	0.44

[1] DM = Dry matter; [2] p = Significance level, Student's t-test; [3] r = Correlation coefficient.

June by one third (Table 6.8). The uptake of the macro elements by the flora goes ahead of the substance growth by assimilation. With increasing age of the plants, the macroelements are diluted, with the changing proportion of leaves to stem and flowers influencing this process (Graupe et al. 1960). The leguminous plants decrease their contents of the alkaline earth elements (Ca, Mg, Ba, Sr) only slowly (Tables 6.8– 6.10), if at all. The phosphorus, sodium, and potassium concentrations of the analyzed species decreased by 40% on average within six to seven weeks. The mineral supply of wild and domestic animals decreased from spring to summer.

Wild animals of the field take in the highest amounts of macro and trace elements during winter time via rye, wheat, and rape. Game of the forest also consume high amounts of macro, trace and ultratrace elements by eating the needles of spruce

Tab. 6.8: The ash and macro element contents of several plant species as a function of plant age (g kg^{-1} dry matter) (n = 32) (Anke et al. 1994)

Parameter		Ash [g kg^{-1}]	Ca [g kg^{-1}]	Mg [g kg^{-1}]	P [g kg^{-1}]	K [g kg^{-1}]	Na [g kg^{-1}]
Green wheat	End of April	115	4.0	0.88	5.6	56	1.062
	Middle of June	84	2.6	0.74	2.9	42	0.368
	%	73	65	84	52	75	35
Green rye	End of April	123	5.8	1.40	6.4	39	1.044
	Middle of June	61	2.2	0.67	2.6	24	0.532
	%	50	38	48	41	62	51
Fescue grass	End of April	104	4.5	1.48	5.4	52	0.951
	Middle of June	67	2.9	1.05	2.5	34	0.596
	%	64	64	71	46	65	63
Lucerne	End of April	117	6.5	1.84	5.2	46	0.374
	Middle of June	85	5.0	1.78	3.6	74	0.283
	%	73	77	97	69	74	76

and pine (Anke and Brückner 1973, Anke et al. 1978, 1979, 2001a,b,c, Partschefeld et al. 1977).

The trace element content (Zn, Mn, Cu, Fe, Ni, Mo, I) of the plants decreased from the end of April to the middle of June by > 50% on average (Table 6.9).

The greatest decreases were found in iodine (down to 17%), iron (30%) and molybdenum (47%), whereas the zinc, copper, nickel, and manganese contents were lowered to half the amount found at the end of April. The age of the plants has a significant influence on the macro and trace element contents of these elements, which are essential for flora and fauna. Generally, herbivorous wild animals eat foods rich in macro and trace elements in early spring.

The same rules are current for the ultratrace elements lithium, rubidium, arsenic, vanadium, aluminum, and uranium (Table 6.10), which are determined later in different species.

The arsenic content of the tested plants decreased to 30% on average, while their vanadium (40%), uranium (47%), aluminum (48%), lithium (54%) and rubidium concentrations (84%) did not diminish as far. In contrast to these ultratrace elements, the chromium, barium and strontium contents (Table 6.10) of the analyzed species did not decrease with increasing age, but the reasons for that phenomenon are unknown. The influence of age on the mineral content of the vegetation needs to be examined for every element (Anke 2003, Anke et al. 1980, 1998, Jaritz 1999, Krause 1987).

6.2.4

The Element Contents of Plant Stems, Leaves, and Flowers

Generally, though varying with species, leaves contain more ash, macro, trace and ultratrace elements than stems. The proportion of leaf to stem, which depends on the type of plant and its age, influences the concentration of the inorganic components in the whole plant (Table 6.11). *Trifolium hybridium* possesses more stems and fewer

Tab. 6.9: Element contents (mg kg⁻¹ dry matter) of several plant species as a function of plant age (n = 32/ species and element) (Anke et al. 1994, 1984, Groppel 1986, Angelow 1994)

Parameter		Zn [mg kg⁻¹]	Mn [mg kg⁻¹]	Cu [mg kg⁻¹]	Fe [mg kg⁻¹]	Ni [mg kg⁻¹]
Green wheat	End of April	31	56	9.0	273	0.54
	Middle of June	21	31	5.4	92	0.39
	%	68	55	60	34	72
Green rye	End of April	43	36	8.7	254	0.64
	Middle of June	20	12	3.0	30	0.25
	%	47	33	34	12	39
Fescue grass	End of April	45	59	11	185	1.82
	Middle of June	22	26	6.3	63	0.85
	%	49	44	57	34	47
Field red clover	End of April	46	46	13	218	1.48
	Middle of June	30	29	8.1	90	0.61
	%	65	64	62	41	41

Parameter		Mo [mg kg⁻¹]	I [mg kg⁻¹]	Li [mg kg⁻¹]	Rb [mg kg⁻¹]	As [mg kg⁻¹]
Green wheat	End of April	0.38	0.215	18	8.1	0.343
	Middle of June	0.12	0.018	7.7	6.8	0.093
	%	32	8	43	84	27
Green rye	End of April	0.40	0.305	15	3.1	0.509
	Middle of June	0.26	0.043	10	2.4	0.129
	%	66	14	67	77	25
Fescue grass	End of April	0.48	0.184	11	5.6	0.315
	Middle of June	0.28	0.020	5.1	2.9	0.092
	%	58	11	46	52	29
Field red clover	End of April	1.15	0.294	10	15.0	0.350
	Middle of June	0.38	0.103	5.8	11.0	0.133
	%	33	35	58	73	38

leaves than *Trifolium pratense*, which grows at the same time and in the same places, the stems of *Trifolium hybridium* are richer in most of the elements than those of *T. pratense*. During the development of the plants, some – but not all – elements were transported to the flowers and seeds. Generally, cereals are rich in phosphorus, and seeds of legumes are rich in molybdenum (Holzinger 1999) and nickel (Kronemann et al.

1980). In winter, the barks of trees and shrubs are poor in sodium, potassium, phosphorus and iron, but rich in zinc, cadmium and the alkaline earth elements calcium, magnesium, strontium, and barium (Anke et al. 2001 a,b,c).

The cell of the plant cannot distinguish between calcium, strontium and barium. Calcium-rich plants also contain high

Tab. 6.10: Ultratrace element contents of several plant species as a function of plant age (mg kg^{-1} dry matter) (n = 24/species and element) (Anke 2003, Anke et al. 1980, 1998b,c, Jaritz 1999, Krause 1987)

Parameter		V [mg kg^{-1}]	Sr [mg kg^{-1}]	Ba [mg kg^{-1}]	Al [mg kg^{-1}]	Cr [mg kg^{-1}]	U [μg kg^{-1}]
Green wheat	May 4th	0.087	12.7	5.4	194	0.526	9.0
	June 14th	0.033	14.1	10.2	57	0.929	4.4
	%	38	111	189	29	177	49
Couch grass	May 4th	0.059	7.2	10.1	138	0.450	–
	June 14th	0.039	10.2	11.6	87	0.687	–
	%	66	142	115	83	153	–
Field red clover	May 4th	0.105	47	5.1	88	0.368	–
	June 14th	0.034	43	6.5	36	0.316	–
	%	32	91	127	41	86	–
Lucerne	May 4th	0.145	44	5.5	112	0.431	0.009
	June 14th	0.058	45	5.7	54	0.444	0.0039
	%	40	102	104	48	103	43

Tab. 6.11: The macro and trace element contents of several parts of *Trifolium hybridium* and *Trifolium pratense* (n = 6) (mg and mg/by dry matter; respectively)

Part of the plant		Portion [g kg^{-1}]	Ash [g kg^{-1}]	P [g kg^{-1}]	Fe [mg kg^{-1}]	Mn [mg kg^{-1}]	Zn [mg kg^{-1}]	Cu [mg kg^{-1}]	Mo [mg kg^{-1}]
Stem	T. hybridium	67.3	102	1.96	102	31	13	8.5	1.48
	T. pratense	56.6	74	1.46	64	17	12	8.7	0.49
Leaf	T. hybridium	22.4	96	2.48	179	110	25	11.2	1.81
	T. pratense	26.4	113	2.16	154	66	41	12.9	1.31
Flower	T. hybridium	10.3	71	3.89	102	82	37	10.4	2.13
	T. pratense	17.0	71	2.99	99	37	31	9.1	1.95

amounts of strontium and barium (Jaritz 1999, Seifert 1998).

6.3
Influence of Geological Origin of Soil on Macro, Trace and Ultratrace Contents of the Terrestrial Food Chain

The geological origin of the soil influences the macroelement (Ca, Mg, P, K, Na), trace element (Fe, Mn, Co, Mo, Ni, Cu, Zn, I, Se) and ultratrace element content (Cd, Al, As, Sr, Ba, Li, Rb, U, Ti, V, Cr) of the indicator plants (wheat, rye, field and meadow red clover significantly (Anke et al. 2000).

The influence of the geological origin on agriculturally and horticulturally used fertilized soils on the macro element content of the vegetation is smaller than on their heavy metal, light metal, and nonmetal contents. The concentrations of the indicator plants from 13 different soils varied by a mere 17% for calcium, by 15% for phosphorus, by 23% for sodium and potassium, and 32% for magnesium.

The magnesium content of the vegetation is significantly influenced by the magnesium concentration of the geological origin of the soil. By way of fertilization, only phosphate and potassium fertilizers deliver magnesium to the soil and the food chain. The geological origin of the site alters chromium transfer to the food chain by one-quarter, the iron and vanadium transfer by one-third (Anke et al. 1984, 1997c, 2003, Glei 1995, Groppel 1986), barium and strontium by 37–40%, and zinc and titanium by >40%. The geological origin of the soil alters the aluminum, lithium, cadmium, copper, molybdenum and iodine (in this case together with the distance to the seaside) by 45 to 50% (Table 6.13). The geological origin alters the nickel, uranium, and manganese contents of the flora by ~55%, the arsenic and selenium contents by 65%, and the amount of rubidium in the indicator plants by 80%. Highest rubidium concentrations were found in the vegetation of gneiss

and granite weathering soils, and lowest contents in the flora of the sediments of the Triassic time (Bunter relative number 30, Muschelkalk 27, Keuper 21). Generally, the geological origin of the soil influences the element concentration of most elements significantly. Based on the average of the light, heavy and nonmetals summarized in Tables 6.12 and 6.13). , plants grown on the weathering soils of the Rotliegende accumulated the highest, and those on Keuper weathering soils of the Triassic time the lowest concentrations.

6.4
Influence of Pollution on the Terrestrial Food Chain

Abiotic toxic damages to cultured and wild plants may result from natural geochemical loads in the soil originating from salt or certain trace elements in water, the vicinity of volcanoes, or natural atmospheric pollution. As a rule, however, the cause is anthropo-

Tab. 6.12: Influence of geological origin of soil on macro and trace element contents of indicator plants (relative number) (Anke 2003)

Geological origin	Relative number [1]				
	P	Ca	Cr	Fe	Zn
Alluvial river-side soils	93	85	74	83	76
Moor, peat	89	94	87	89	89
Loess	89	89	74	87	67
Diluvial sands	89	83	79	93	86
Boulder clay	93	89	77	71	66
Keuper weathering soils	85	97	74	76	57
Muschelkalk weathering soils	94	100	79	78	64
Bunter weathering soils	96	91	79	66	66
Rotliegende weathering soils	100	96	100	100	91
Phyllite weathering soils	96	84	86	83	100
Granite weathering soils	94	83	96	89	92
Gneiss weathering soils	94	89	77	88	82
Slate weathering soils	100	83	81	73	89

[1] Soil with highest concentration = 100

Tab. 6.13: Influence of geological origin of soil on trace and ultratrace element contents of indicator plants (Anke 2003)

Geological origin	Relative number [1]						
	Al	Cd	Cu	Ni	U	As	Se
Alluvial river-side soils	71	71	74	79	71	70	65
Moor, peat	73	52	52	47	50	70	80
Loess	78	60	86	71	72	66	100
Diluvial sands	78	70	70	71	46	59	49
Boulder clay	73	86	70	71	64	56	49
Keuper weathering soils	54	67	85	65	52	46	40
Muschelkalk weathering soils	71	69	93	64	74	50	50
Bunter weathering soils	68	75	80	60	55	49	36
Rotliegende weathering soils	100	90	100	100	75	87	75
Phyllite weathering soils	93	82	93	94	67	38	37
Granite weathering soils	90	100	82	69	100	58	37
Gneiss weathering soils	83	79	93	83	50	100	47
Slate weathering soils	89	81	94	84	64	44	38

[1] Soil with highest concentration = 100.

genic. The main atmospheric, water, and soil pollutants are the gases fluorine and sulfur dioxide, aerosols and airborne dust, which contains a variety of metals and non-metals such as aluminum, arsenic, cadmium, cesium, chromium, copper, iodine, mercury, nickel, selenium, strontium, manganese, and zinc. These are often responsible for toxicity in plants, animals and man, but in some cases (for instance NO_x, sulfur, iodine) they also deliver essential elements to the food chain of plants, animals and man. In the following, the effects of the pollution are demonstrated by the examples of cadmium, chromium, nickel, and strontium.

6.4.1
Cadmium

Because of its nephrotoxicity causing the Itai-Itai disease in humans, its teratogenic effects, its interactions with iron, copper and zinc, and its potentially existing cancerogenic effects, cadmium (Cd) belongs to

the most dangerous anthropogenic environmentally harmful substances. Cadmium exposure of plants results from the weathering of rocks, which may deliver variable amounts of cadmium to the soil. Low pH values in the soil are favorable to cadmium uptake by the flora, whereas slightly acid neutral and alkaline soils are unfavorable. The same is true for a high proportion of humus and fine earth.

Industrially unexposed soils contain 0.1 to 1.0 mg $Ca\,kg^{-1}$ air-dried soil (Müller et al. 1994). The variation range of the cadmium content of soils due to their geology is overlapped by anthropogenic cadmium. It is assumed that 7600 tons of cadmium were emitted worldwide in 1989 (Nriagu 1989). Individual enterprises of the nonferrous metal industry (Freiberg) emitted 1–2 tons of cadmium annually into the environment even in Germany (Fiedler and Klinger 1996), and this led to an excess of the normal cadmium concentrations in the soils and of the so-called Kloke value of 3 mg $Ca\,kg^{-1}$ air-dried soil (sewage sludge introduction) (Ulken 1985, Anonymous 1992). The different forms of cadmium emissions usually result in species-specific cadmium accumulation in the flora. Different amounts of cadmium are stored in the individual parts of plants. As a rule, fruit and tubers incorporate less cadmium than leaves at the time of their usage. Apples and potatoes, which only double or triple their cadmium concentration at the time of harvesting, accumulate little cadmium (Table 6.14). The thick parts of the stalks (onion, kohlrabi) are able to store much greater quantities of cadmium, though this may be exceeded by the thick parts of root vegetables (carrots, turnips). Grains and seeds (rye, barley, oats) may deliver widely differing cadmium concentrations into the food chain. In cadmium-exposed areas, oats have between four and twenty times

the normal cadmium concentration, and carrots and linseeds may also be extremely cadmium-rich (Anke et al. 1991, 1994, 2000, Müller 1993, Müller et al. 1996, Müller and Anke 1994, Erler et al. 1996, Kronemann et al. 1982). Most cadmium accumulates in leafy forage plants (green maize, turnip leaves, meadow grass) and vegetables; these plants are able to store ten to thirty times as much cadmium than plants in unexposed areas. Seifert (1998) registered similar variations of the cadmium content in the flora even in the case of lower cadmium accumulation in the soil.

Even in forestry districts unexposed to anthropogenic cadmium, the winter grazing of ruminating hoofed game is considerably richer in cadmium than plant raw materials used for human food production (Table 6.15. In particular, the barks and the

Tab. 6.14: Cadmium contents of various species and parts of plants from a control and a cadmium-exposed area ($\mu g \ kg^{-1}$ dry matter)

Species	$n^{1)}$ (u) : n (e)	Unexposed (u)		Exposed (e)		$p^{2)}$	Multipli-cation
		$SD^{3)}$	$\bar{x}^{4)}$	$\bar{x}$	SD		
Apple	(12:6)	63	51	113	71	<0.05	2.2
Potato	(13:31)	24	38	114	194	<0.05	3.0
French bean	(6:6)	35	28	105	46	<0.05	3.8
Rye, grain	(4:6)	45	66	258	154	<0.01	3.9
Onion	(12:6)	50	96	403	194	<0.01	4.2
Tomato	(6:6)	15	32	185	176	<0.05	5.8
Wheat, grain	(12:17)	25	51	345	228	<0.001	6.8
Kohlrabi	(22:6)	50	55	425	326	<0.01	7.7
Carrot	(13:6)	64	73	573	488	<0.001	7.8
Parsley	(9:17)	111	136	1091	1325	<0.001	8.0
Barley, grain	(10:25)	10	26	245	313	<0.05	9.4
Cabbage	(6:12)	44	82	1003	965	<0.05	12
Lettuce	(4:11)	135	352	5148	4012	<0.001	15
Meadow grass	(12:25)	23	49	854	529	<0.001	17
Oat, grain	(6:19)	44	45	868	978	<0.05	19
Fodder beet	(13:21)	36	51	1347	1875	<0.05	26
Beet leaf	(11:21)	126	201	5969	7643	<0.05	30
Green maize	(6:18)	23	55	1709	2444	<0.05	31

[1] n = Number of samples; [2] p = Significance level, Student's t-test; [3] SD = Standard deviation; [4] $\bar{x}$ = Arithmetic mean

Tab. 6.15: Cadmium contents in different winter grazings of hoofed game in areas with and without cadmium exposure ($\mu g\ kg^{-1}$ dry matter)

Species	$n^{1)}$ (u):n (e)	Unexposed (u)		Exposed (e)		$p^{2)}$	$\%^{5)}$
		$SD^{3)}$	$\bar{x}^{4)}$	$\bar{x}$	SD		
Pine bark	(20:9)	630	940	1150	700	> 0.05	122
Heather	(40:5)	170	80	100	60	> 0.05	125
Raspberry twigs	(34:10)	320	350	560	540	> 0.05	160
Bilberry bush	(41:10)	140	170	400	140	< 0.001	235
Fir twigs	(48:10)	140	190	440	230	< 0.001	232
Rye	(17:5)	100	150	420	130	< 0.001	280
Rape	(18:5)	150	210	560	210	< 0.001	267
Fir bark	(12:10)	340	490	1600	1310	< 0.001	327
Pine twigs	(46:10)	160	180	780	840	< 0.001	433
Hair grass	(45:55)	80	80	460	420	< 0.001	575

[1] n = Number of samples; [2] p = Significance level, Student's *t*-test; [3] SD = Standard deviation; [4] $\bar{x}$ = Arithmetic mean; [5] unexposed = 100 %, exposed = x %.

tips of twigs in unexposed areas may contain between 200 and 1000 μg Cd kg^{-1} dry matter. Grazing in exposed forest districts accumulated between two- and six-fold more cadmium than in control areas. Perennial plants apparently accumulate cadmium in their barks. Even without cadmium exposure, the rye and rape grazed in winter still contains three- to four-fold the cadmium amount of pasture grass (150-200 $\mu g\ kg^{-1}$ dry matter). The reason for this is that the cadmium content of annual plants decreases with increasing age; moreover, cadmium uptake by the flora occurs more rapidly than the element becomes available.

This leads to a dilution of the cadmium concentration in plants with increasing age. As the green rye and rape which were grazed in winter were very young, their cadmium content was very high. As a rule, the winter grazing of ruminating hoofed game and other herbivores is richer in cadmium than the winter feedstuffs of farm animals (Anke et al. 1976, 1978, 1979, Anke and Brückner 1973, Partschefeld et al. 1977).

6.4.2
Chromium

The annual production of chromium (Cr) ores amounts to 10 million tons. Cr is used in the production of special steels in the metal-processing industry, for chromium coating in the galvanic industry, as a pigment and catalyst in the chemical industry, as a dye in the textile industry, for leather production in tanneries, and for the impregnation of products in the timber industry. The use of Cr-rich products has led to Cr accumulation in the environment. Local Cr exposure of soil and the environment can occur via sludge, water and air (Anke et al. 1998).

Even after decades, the chromium emissions of a former cement plant induced a significant increase in the Cr content of wild and cultivated plants, without triggering phytotoxic nutritional damage in the flora, fauna and humans (Table 8.16).

The influence of plant species on the Cr content of the flora has remained within moderate limits. The barks, tips of twigs and several perennial plant species which

Tab. 6.16: Effects of emissions of a former cement and incandescent phosphate plant on chromium contents in wild and cultivated plants ($\mu g \, kg^{-1}$ dry matter)

Species	$n^{1)}$ (c) : n (p)	Control area (c)		Polluted area (p)		$p^{2)}$	% [5]
		SD [3]	$\bar{x}$ [4]	$\bar{x}$	SD		
Cucumber	(15 : 8)	293	685	1626	2016	> 0.05	238
Tomato	(13 : 5)	238	343	760	597	< 0.05	222
Onion leek	(86 : 7)	108	380	608	215	< 0.05	160
Lettuce	(16 : 8)	521	1035	1406	625	> 0.05	136
Meadow red clover	(5 : 6)	84	216	287	149	> 0.05	133
Sweet clover	(15 : 4)	67	214	277	129	> 0.05	129
Tansy	(14 : 4)	50	298	362	67	< 0.05	121

Footnotes see Table 6.15.

serve as winter grazing for hoofed game have proved to be particularly rich in Cr, but very little Cr is accumulated in leaves. On average, the Cr content of tubers, roots and stem bulges, fruit and seeds was lower than in leaves. Most of the Cr which accumulated in fruits, vegetables and seeds was concentrated into the skins; hence, it follows that foods produced from peeled tubes, fruits or seeds are Cr-poorer (Anke et al. 1997).

6.4.3
Nickel

During the past few decades there has been a rapid growth in the industrial demand for nickel. Nickel (Ni) is used in steel production, in alloys (e.g., for coins and domestic utensils), in electroplating, and in nickel-cadmium batteries. After a prolonged period of relatively level production, nickel output over the past two years has begun to rise as existing producers have expanded their output. Annual nickel production in the western world has risen by almost 4% to 678 000 tons, and a further increase of several percent is expected in the years to come (Seifert and Anke 1999). The prevalence of nickel allergy is about 10% for women, and about 2% for men (Lieden 1994).

Tab. 6.17: Influence of anthropogenic Ni exposures and the geological origin of the site on the nickel contents in the flora of Saxony

Anthropogenic exposures, geological origin	Relative number
Nickel exposure via air	394
Nickel exposure via water	201
Weathering soils of the Rotliegende	100
Phyllite weathering soils	94
Gneiss weathering soils	83
Loess, boulder clay	83

Riverside soils produce a flora which has a nickel content higher than that of the weathering soils of the Rotliegende. On average, the analyzed plant species of the riverside soils ("meadow grass", meadow red clover, field red clover, wheat, rye, parsley, lettuce) contained about double the nickel content of the same plant species growing on the weathering soils of the Rotliegende (Table 6.17).

To demonstrate this influence, the nickel contents of several plants species growing on the permanent grassland of the riverside of a stretch of water contaminated by galvanic baths are detailed in Table 6.18.

Snakeweed and rough crowfoot accumulated 0.04 and 0.02% nickel in the plant dry matter. Although their nickel propor-

Tab. 6.18: Nickel content in several plant species of the permanent grassland of a nickel-exposed habitat (mg kg^{-1} dry matter) (n = 45)

Species	Variation range	x	s
Snakeweed	388–75	197	147
Rough crowfoot	242–3.5	124	183
Meadow red clover	252–4.0	105	120
White clover	348–2.7	96	127

Tab. 6.19: Nickel contents of several vegetables and fruit from a nickel-exposed environment (mg kg^{-1} dry matter) (n = 33)

Species (n)	Variation range	x	s
Lettuce (6)	238–2.9	72	108
Dill (4)	184–4.6	80	77
Onion (7)	424–3.5	96	165
Onion leek (6)	204–1.8	138	243
Strawberry (6)	483–3.2	146	220
Parsley (4)	1025–6.5	572	502

tions remained essentially below the value typical of hyperaccumulators of class II, the anthropogenic nickel exposure of this habitat manifested itself by the flooding of the riverside with the nickel-rich water of the brook. When the samples were collected, the water contained 1 mg Ni L^{-1}.

The nickel contents of several species of vegetables, spices and fruits cultivated in house gardens at the riverside are listed in Table 6.19. Since nickel accumulation in the garden cultures varied depending on the location of the property, both extreme nickel concentrations and normal values were registered. The highest nickel levels were found in parsley, at >1.0 g Ni kg^{-1} dry matter. Surprisingly, strawberries also stored large amounts of nickel.

Nickel emissions occur in the vicinity of two nickel processing plants in Saxony (Germany), and this led to a quadrupling of the nickel content of the flora. Depending on the distance of the plant site from the emis-

Tab. 6.20: Nickel in various plants in Saxony without and with nickel exposure via air (µg kg^{-1} dry matter) (n = 1032)

Plant species	n$^{1)}$ (u):n (e)	Unexposured (u)		Exposured (e)		p $^{2)}$	% $^{5)}$
		SD $^{3)}$	x̄ $^{4)}$	x̄	SD		
Meadow grass	(283:59)	424	971	8506	18300	<0.05	876
Cabbage	(26:6)	736	842	4652	2815	<0.001	552
Turnip	(69:6)	273	495	2199	1874	<0.001	444
Lettuce	(65:26)	621	1236	5396	5607	<0.001	437
Wheat grain	(65:5)	127	301	1259	922	<0.001	418
Rye grain	(28:5)	138	263	1044	472	<0.001	397
Turnip leave	(21:6)	806	1458	5757	2427	<0.001	395
Tomato	(19:5)	342	575	1914	857	<0.001	333
Parsley	(51:28)	689	1365	4478	4826	<0.001	328
Apple	(17:6)	174	429	1268	824	<0.001	296
Carrot	(33:6)	227	504	1410	1208	<0.001	280
Bean	(22:6)	1575	3075	8223	4992	<0.001	267
Corn	(22:4)	501	1054	2806	2268	<0.01	266
Oat grain	(55:4)	380	712	1456	573	<0.001	204
Potato	(22:7)	230	565	1038	596	<0.01	184
Barley grain	(55:6)	153	246	395	238	<0.05	161

Footnotes see Table 6.14.

sion source, the main wind direction, the soil pH value and other influences, the nickel content in the investigated plant species varied widely, as can be seen from the standard deviations of contents measured in the exposed species (Table 6.20).

The leafy meadow grasses, to which most different monocotyledonous and dicotyledonous species belong, had the highest nickel contents, at 8.5 mg kg^{-1} dry matter. On average, this was nine-fold the level found in "normal meadow grass", though the wide variation range made it impossible to register the difference biostatistically.

In principle, all plant species in the habitats exposed to airborne nickel emissions accumulate nickel, and in this respect it was irrelevant whether these were leafy species, fruits, seeds, tubers, or thickened parts of roots (carrots, turnips). Dwarf beans ready for eating proved to be particularly rich in nickel (8 mg kg^{-1} dry matter), even on normal sites (Anke et al. 1993a,b, 2003, Szentmihaly et al. 1980).

6.4.4
Strontium

With a concentration of 370 mg kg^{-1} in the 16 km-thick Earth's crust, strontium (Sr) occupies 18th position in the frequency list of elements. Strontium occurs as four stable isotopes with atomic masses 84, 86, 87, and 88. The latter isotope, with a relative abundance of 83%, is the most widespread. Isotope-pure [87]Sr is found as a daughter product of the [87]Rb isotope in several minerals, and is used to determine the age of rocks. Celestine (SrSO$_4$) and strontianite (SrCO$_3$) are of economic importance, with 250000–300000 tons of celestine being extracted in 1991. Sr has minimal technological importance, but is used as nitrate in the production of fireworks, as a hydroxide for the removal of sugar from molasses, as

a component of alloys, as a catalyst, as a means of deoxygenation in metallurgy, in high-temperature superconductors. It is also used – in the form of its isotopes – as a marker for calcium metabolism and for the treatment of skeletal metastases (Seifert 1998).

Toxic effects may be expected from an Sr content of > 300 mg kg^{-1} in the skeleton of rats (Seifert 1998). Feeding experiments with strontium carbonate in rats reduced the calcification of teeth, bones, and cartilage (Nagayama et al. 1984). Similar symptoms were observed in humans in the case of endemic chondrodystrophy occurring in Tadzhikistan, and this is assumed also to occur in farm animals. Strontium levels > 1 g kg^{-1} in soil and 50 mg L^{-1} in drinking water were recorded in habitats with a high Sr availability in the food chain (Kovalskij 1977).

The flora on the Muschelkalk and Bunter slopes surrounding the Saale valley in Thuringia contains significantly more strontium than do other regions of Thuringia. This statement is true for most different plant species, from the couch grass which dominates on the Muschelkalk slopes (which contain six-fold more Sr than control areas) to the potatoes in the Saale valley, where double the Sr levels of control areas were recorded (Table 6.21).

The Sr accumulation results from the occurrence of fibrous celestine in the lower Muschelkalk and Bunter (Dinger 1929). This Sr mineral was formed from the water-soluble Sr of the Muschelkalk during the course of millions of years when Sr-rich water came into contact with the Bunter layer, which is almost impervious to water. The water of the local spring horizons of this habitat still contains 24-fold more Sr than do the waters of control areas in Thuringia. Based on an average from 21 samples, the Sr content was

Tab. 6.21: The strontium content of various species and of edible parts of seberal species in a control area and an exposed area (mg kg^{-1} dry matter)

>Species	n[1] (c):n (e)	Control area (c)		Exposed are (e)		p [2]	% [5]	r [6] Sr/Ca
		SD [3]	x̄ [4]	x̄	SD			
Cucumber	(15:8)	8.5	27	122	25	<0.001	452	0.835
Onion	(17:13)	17	27	121	93	<0.01	448	0.402
Lettuce	(16:9)	22	48	214	56	<0.001	446	0.836
Meadow red clover	(15:6)	13	73	315	113	<0.001	432	0.669
Parsley	(13:6)	14	48	187	27	<0.001	390	0.827
Chives	(18:3)	30	53	195	72	<0.001	368	0.628
Apple	(17:7)	0.8	1.4	5.0	1.8	<0.01	357	0.393
Carrot	(14:6)	7.2	24	69	19	<0.01	288	0.612
Tomato	(13:5)	3.8	7.8	22	7.5	<0.05	282	0.567
Kohlrabi	(12:4)	13	29	78	52	>0.05	269	0.459
Potato peel	(18:5)	4.2	7.8	20	2.3	<0.001	256	0.734
Potato	(22:4)	2.1	2.7	4.6	1.2	>0.05	170	0.723

Footnotes [1] – [5] see Table 6.14; [6] r = Correlation coefficient.

> 3000 µg L^{-1}, whilst that of control areas was 130 µg L^{-1} (Seifert 1998). This Sr-rich lower Muschelkalk was processed into cement during the first half of the 20th century, and into incandescent phosphate for agricultural use during the second half of the century. The phosphates may contain considerable amounts of Sr (Kola apatite 20 g Sr kg^{-1}) and contribute to strontium emissions.

The amounts of strontium found in the flora correlate positively and species-specifically with species-specific amounts of Ca (Table 6.21). Although both elements were apparently taken up in a similar ratio, Ca-rich species or parts of plants were found also to contain high levels of Sr, whereas Ca-poor species proved to be Sr-poor. The almost constant Ca:Sr ratio may reduce the occurrence of Sr exposure, and may also be the cause of considerable changes in the proportions of both elements in the skeletons of animals and humans. The cells of plants are unable to distinguish between calcium and strontium (Anke et al. 1999).

6.5
The Influence of Conventional and Ecological (Organic) Farming

The macro, trace and ultratrace element contents of the foodstuffs were also influenced by the farming system and the preparation of the raw materials for food production. Conventionally, working farms use fertilizers, herbicides, fungicides, insecticides, growth promoters and other pesticides, whereas organic farmers use only dung, compost and organic waste of the agricultural production as fertilizers and do not apply any pesticides. Ecologically produced raw materials for food production are not refined (sugar), and the cereals are not ground up. Both processes influence the composition of ecologically produced vegetable (and animal) foodstuffs. Under these conditions, the trace element contents of conventionally and ecologically produced foodstuffs contain different element- and food-specific amounts of macro, trace and ultratrace elements (Table 6.22). The brown sugar of ecological production is unrefined sugarcane, and is significantly richer in all elements than refined beet sugar. Generally, cereals of ecological production in the form of flour, semolina, and pearl barley are richer in inorganic components, because the elements that occur in large quantities in the bran are in fact part of these foods. The same situation applies to bread, cakes and pastries, with the exception of crispbread and coarse-grained wholemeal rye bread, where the whole grain is used in both types of production. These types of bread, when produced conventionally, tend to be richer in the essential elements. Pulses, fruits, herbs, and vegetables of ecological production are poorer in almost all of these elements than those of conventional production (Table 6.22), with the exception of copper, as the use of copper sulfate as a fungicide is not prohibited in ecological production. Fertilization with nitrogen, phosphates, and potassium also delivers macro, trace and ultratrace elements to the food chain of plants, animals and man. Animal foodstuffs of ecological production generally contain lower concentrations of macro, trace and ultratrace elements (Anke et al. 2000, Anke 2003, Röhrig et al. 1998, Röhrig 1998).

6.6
Macro, Trace and Ultratrace Elements in Foodstuffs

Concentrations of macro, trace and ultratrace elements in vegetable foodstuffs range from <1 µg kg^{-1} dry matter for uranium in sugar, to 127 g kg^{-1} dry matter for potassium in lettuce (Table 6.23). The concentrations of the inorganic components are both foodstuff- and element-specific.

Tab. 6.22: Trace and ultratrace element contents of several foodstuffs produced by conventional (c) and organic farming (o)

Element		Sugar	Semo-lina	Rusk	Crisp-bread	Pea, dried	Pear	Kohl-rabi	Carrot	White cabbage	Cauli-flower
Fe [mg kg^{-1}DM[3)]]	c[1)]	5.3	8.3	15	30	55	18	48	56	65	72
	o[2)]	74	25	28	31	47	8.3	28	32	50	70
	%	1396	301	187	103	85	46	58	57	77	97
Zn [mg kg^{-1}DM]	c	0.76	6.8	12	30	39	11	38	47	28	59
	o	3.0	24	24	29	29	5.8	26	31	22	43
	%	395	353	200	97	74	53	68	66	79	73
Cu [mg kg^{-1}DM]	c	0.31	1.5	1.5	4.4	6.9	6.9	4.5	5.7	3.5	3.9
	o	2.2	4.8	4.6	5.2	7.3	6.6	4.6	6.6	4.0	6.3
	%	710	320	307	118	106	96	102	116	114	162
Cr [µg kg^{-1}DM]	c	145	91	163	254	328	234	594	383	382	714
	o	377	289	186	194	199	193	429	242	434	603
	%	260	318	114	76	61	82	72	63	114	84
V [µg kg^{-1}DM]	c	8.0	5.4	9.7	10	5.4	23	29	78	20	42
	o	106	11	8.6	4.1	13	7.7	14	33	5.5	40
	%	1325	203	89	41	240	33	48	42	28	105

[1)] Conventional farming = 100%, organic farming = x%; [3)] DM = Dry matter.

Tab. 6.23: Macro, trace and the ultratrace element contents of several vegetable foodstuffs (mg kg^{-1} dry matter) (Anke et al. 2003)

Element	Sugar	Wheat flour	Wheat and rye-bread	Roll	Lentil	Apple	Potato	Asparagus	Lettuce	Mushroom
Ca	31	264	419	597	401	488	288	2556	15329	1104
Mg	2	341	713	295	1191	433	1349	1869	2496	366
P	18	1187	1848	1710	4401	657	2627	5143	11949	2460
K	61	1777	4099	2870	12269	11388	27576	42934	127438	1297
Na	67	95	8488	6493	78	113	153	436	1732	46214
Fe	5.3	16	22	19	8.3	13	35	116	208	270
Mn	0.24	9.9	16	6.9	13	4.3	6.2	24	34	5.1
Ni	0.140	0.173	0.118	0.098	2.142	0.188	0.975	9.183	4.767	1.575
Zn	0.76	10	21	14	48	4.2	18	94	94	–
Cu	0.31	2.1	2.2	1.8	6.3	2.9	3.9	5.8	11	7.8
Mo	0.023	0.156	0.303	0.250	4174	0.037	0.537	0.602	0.665	0.390
I	0.002	0.021	0.023	0.030	0.029	0.031	0.028	0.101	0.150	0.634
Se	<0.002	0.084	0.020	0.034	0.521	0.022	0.027	0.334	0.025	0.476
As	0.010	0.054	0.548	0.190	0.250	0.046	0.024	0.223	0.122	0.342
Li	0.199	0.905	0.474	0.317	0.748	1.449	1.592	2.217	4.502	5.788
Rb	0.11	0.76	1.39	1.23	6.02	5.02	4.084	68.0	21.8	0.57
Sr	0.17	1.6	3.0	2.4	1.9	2.5	1.9	12	58	12
Ba	3.9	0.9	2.8	1.9	5.4	1.5	1.5	2.8	11.8	11.4
Cd	0.005	0.038	0.039	0.034	0.058	0.019	0.124	0.083	0.547	0.040
Hg	0.002	0.006	0.016	0.012	0.020	0.011	0.034	0.050	0.045	0.263
Al	4.4	4.1	8.3	9.0	18	12	30	66	269	148
Ti	0.071	0.115	0.196	0.125	0.285	0.277	0.442	2.181	3.968	4.288
V	0.008	0.018	0.006	0.007	0.041	0.021	0.019	0.097	0.377	0.625
Cr	0.145	0.113	0.160	0.127	0.358	0.202	0.333	0.948	1.260	1.052
U	0.001	0.0015	0.004	0.002	0.002	0.002	0.003	0.053	0.039	0.105

Sugar contains the lowest amounts of macro, trace and ultratrace elements, and starch is also poor in inorganic substance content. Wheat flour, rye and wheat bread and rolls are also relatively poor in essential and toxic elements, though in general the contents are higher in bread, rolls and pastries than in flour and starch. The supplementation of bread, rolls and pastry increases the element concentrations, with the largest increases being registered for levels of sodium. Pulses (as represented by lentils in Table 6.23) are especially rich in nickel, molybdenum, rubidium, and barium, while fruits and vegetables store variable quantities of all elements.

Vegetables which are especially calcium-rich include lettuce (25 g kg^{-1} dry matter) and asparagus (2.6 g kg^{-1} dry matter). Both species also store large amounts of magnesium (2.5 and 1.9 g kg^{-1} dry matter, respectively), phosphorus (12 and 5.1 g kg^{-1}) (Krämer 1993, Anke and Krämer 1995, Anke et al. 1998, 2002), and potassium (127 and 43 g kg^{-1}). The highest natural concentrations of sodium are found in mixed mushrooms (46 g kg^{-1} dry matter) (Bergmann 1995, Schäfer et al. 2001, Müller et al. 2001). These three vegetable foodstuffs also store the highest amounts of iron (116, 208 and 270 mg kg^{-1} dry matter, respectively) and iodine (101, 130 and 634 µg kg^{-1}), whereas asparagus and lettuce are richest in manganese (24 and 34 mg kg^{-1}), nickel and zinc (94 and 94 mg kg^{-1}). Lettuce, mushrooms and lentils deliver large amounts of Cu to the food chain, whereas lentils and asparagus provide much selenium to the food web of humans.

Asparagus, lettuce and wild mushrooms contain the highest concentrations of the 12 ultratrace elements analyzed, and on some occasions these levels may be dangerous – for example, aluminum (269 mg kg^{-1} dry matter) (Anke et al. 1996, 1997a, b, 1999a, Illing 1995, Drobner 1997, Röhrig et al. 1998, Röhrig 1998, Schmidt 2002, Seeber et al. 1998, Seeber 1998, Müller et al. 1998, Krause 1987). The inorganic components of the vegetable foodstuffs vary greatly, and although being both plant- and element-specific are independent of the geological origin of their site, their age, and the farming system employed.

**6.7
Conclusions**

Intake by plants of the macro, trace and ultratrace elements from the soil is both element- and species-specific. The bioavailability of metals and nonmetals in soils of different geological origin can be estimated with shooting wheat, rye in blossom, field red clover in bud, and meadow red clover in blossom. These are good indicator plants, as they are cultivated worldwide and also easy to identify. In general, element concentrations decrease with increasing plant age and, in most cases, the highest concentrations of the inorganic components are concentrated into the leaves.

Subsequently, although the elements are transferred to the flowers and seeds, the use of the indicator plants wheat, rye and red clover is bound to their vegetation location.

The farming system (whether conventional or organic) influences the element concentrations in the food, as do the geological origin of the site, pollution by metals and nonmetals, plant age, species, and the part of the plant utilized. The element concentrations of vegetable foods range from 1 µg kg^{-1} dry matter for uranium to 125 g kg^{-1} dry matter for potassium.

References

Angelow L (1994) *Rubidium in der Nahrungskette.* Thesis for a lectureship. Biol.-Pharm. Fakult., Friedrich-Schiller-University, Jena, Germany.

Anke M (1961) *Der Spurenelementgehalt von Grünland- und Ackerpflanzen verschiedener Böden in Thüringen.* Z Acker Pflanzenbau **12**:113–140.

Anke M (1968) *Der Mengen- und Spurenelementgehalt von Luzerne, Ackerrotklee und Wiesenklee als Anzeiger der Mineralstoffversorgung.* Arch Tierernaehr **18**:121–133.

Anke M and Brückner E (1973) *Der Mengen- und Spurenelementgehalt verschiedener frequentierter Äsungspflanzen des Rotwildes und des Rothirschgeweihes unterschiedlicher Qualität.* Beiträge zur Jagd- und Wildforschung **8**:21–32.

Anke M, Hennig A, Grün M, Groppel B and Lüdke H (1976) *Cadmium and its influence on plants, animals and man with regard to geological and industrial conditions.* In: Hemphill DD, ed. Trace Substance in Environment in Health – X. pp. 105–111, University of Missouri, Columbia, Missouri.

Anke M, Groppel B, Kronemann H, Dittrich G and Briedermann L (1978) *Der Nährstoffgehalt und die Mengen- und Spurenelementkonzentration des Panseninhaltes freilebender Wiederkäuer (Reh, Capreolus capreolus L.; Hirsch, Cervus elaphus L.; Muffelwild, Ovis misumon L. und Damwild, Cervus (Dama) dama L.) in Beziehung zur Winteräsung.* Math. Naturwiss. R. 27, pp. 189–198, Wiss. Z. Karl-Marx-Univ. Leipzig.

ANKE M, GRÜN M, BRIEDERMANN L, KRONEMANN H, MISSBACH K and HENNIG A (1979) *Die Mengen- und Spurenelementversorgung der Wildwiederkäuer. 1. Mitteilung: Der Kadmiumgehalt der Winteräsung und der Kadmiumstatus des Rot-, Dam-, Reh- und Muffelwildes.* Arch Tierernaehr **29**:820–844.

ANKE M, GROPPEL B, RIEDEL E and SCHNEIDER H-J, (1980) *Plant and mammalian as indicators of exposure to nickel.* In: Brown, SS and Sundermann FW. Nickel-Toxicology, pp. 65–68. Academic Press: London-New York.

ANKE M, SZENTMIHÁLYI S, GRÜN M and GROPPEL B (1984) *Molybdängehalt und -versorgung der Flora und Fauna.* Math.-Naturwiss. R. Wiss. Z. Karl-Marx-University Leipzig **33**:135–147.

ANKE M, GROPPEL B, GRÜN M and KRONEMANN H (1991) *Relations between the cadmium content of soil, plant, animals and humans.* In: Momcilovic B, ed. Trace Elements in Man and Animals – 7. pp. 26–10–26–11, University of Zagreb, Zagreb.

ANKE M, LÖSCH E, ANGELOW L, GLEI W, ARNHOLD W and ILLING H (1993) *Die Nickelbelastung der Nahrungskette von Pflanze, Tier und Mensch in Deutschland. 1. Nickelbelastung der Flora.* Mengen- und Spurenelemente **13**:365–381.

ANKE M, LÖSCH E, HÜBSCHMANN H and KRÄMER K (1993a) *Die Nickelbelastung der Nahrungskette von Pflanze, Tier und Mensch in Deutschland. 2. Auswirkung der Nickelbelastung bei der Fauna.* Mengen- und Spurenelemente **13**:382–399.

ANKE M, LÖSCH E, ANGELOW L and KRÄMER K (1993b) *Die Nickelbelastung der Nahrungskette von Pflanze, Tier und Mensch in Deutschland. 3. Der Nickelgehalt der Lebensmittel und Getränke des Menschen.* Mengen- und Spurenelemente **13**:400–414.

ANKE M, GROPPEL B and GLEI M (1994) *Der Einfluß des Nutzungszeitpunktes auf den Mengen- und Spurenelementgehalt des Grünfutters.* Das wirtschaftseigene Futter **40**:304–319.

ANKE M and KRÄMER K (1995) *Der Calciumgehalt der Lebensmittel und Getränke sowie die Calciumaufnahme bzw. Calciumbilanz Erwachsener Deutschlands – ein Vergleich nach der Duplikat- und Marktkorbmethode erzielten Ergebnisse.* In: Holmeier HJ, ed. Magnesium und Calcium. pp. 223–241, Wissenschaftliche Verlagsgesellschaft mbH: Stuttgart.

ANKE M, ANGELOW L, GLEI M, MÜLLER M and ILLING H (1995a) *The biological importance of nickel in the food chain.* Fresenius J Anal Chem **352**:92–96.

ANKE M, ANGELOW L, MÜLLER M, ILLING-GÜNTHER H, LÖSCH E, HARTMANN E, SCHWARZBACH A and SEIFERT M (1996) *Der Titangehalt der Lebensmittel und Getränke in Deutschland (1988 und 1992).* Mengen- und Spurenelemente **16**:929–942.

ANKE M, ANGELOW L, GLEI M, ANKE S, LÖSCH E and GUNSTHEIMER G (1997) *The Biological Essentiality of Rubidium.* In: Pollet S ed. International Symposium on Trace Elements in Human. New Perspectives, pp. 245–263, Athens, Greece.

ANKE M, ANGELOW L, GLEI M, MÜLLER M, GUNSTHEIMER U, RÖHRIG B, ROTHER C and SCHMIDT P (1997a) *Rubidium in the food chain of humans. Origins and intakes.* In: Fischer PWF, L'Abbé MR, Cockell KA and Gibson RS, eds. Trace Elements in Man and Animals – 9: Proceedings of the Ninth International Symposium on Trace Elements in Man and Animals. pp. 186–188, NRC Research Press, Ottawa, Canada.

ANKE M, ARNHOLD W, MÜLLER M, ILLING H, SCHÄFER U and JARITZ M (1997b) *Lithium.* In: O'Dell BL, Sunde RA, eds. Handbook of Nutritionally Essential Mineral Elements. pp. 465–477, Marcel Decker, Inc.: New York, Basel, Hong Kong.

ANKE M, JARITZ M, HOLZINGER S, SEIFERT M, GLEI M, TRÜPSCHUCH A, ANKE S, MOCANU H, GUNSTHEIMER G and GUNSTHEIMER U (1997c) *Der Chromtransfer in der Nahrungskette. 1. Mitteilung: der Einfluß der geologischen Herkunft des Pflanzenstandortes, des Pflanzenalters, der Pflanzenart, des Pflanzenteiles und der Chromemission auf den Chromgehalt der Flora.* Mengen- und Spurenelemente **17**:883–893.

ANKE M, ILLING-GÜNTHER H, HOLZINGER S, JARITZ M, GLEI M, MÜLLER M, ANKE S, TRÜPSCHUCH A, NEAGOE A, ARNHOLD W and SCHÄFER U (1997d) *Chromtransfer in der Nahrungskette. 2. Mitteilung: Der Chromgehalt pflanzlicher Lebensmittel.* Mengen- und Spurenelemente **17**:894–902.

ANKE M, GLEI M, GROPPEL B, ROTHER C and GONZALES D. (1998) *Mengen-, Spuren- und Ultraspurenelemente in der Nahrungskette.* Nova Acta Leopoldina NF 79. Nr. **309**:157–190.

ANKE M, SEIFERT M, JARITZ M, HOLZINGER S, ANKE S, HARTMANN E and LÖSCH E (1999) *Strontium transfer in the food chain of humans.* In: Pais I , ed. 8th International Trace Element Symposium 1998. pp. 1–22, University of Horticulture and Food Sciences, Budapest.

ANKE M, HOLZINGER S, JARITZ M, SCHÄFER U, MÜLLER R, DROBNER C and GUNSTHEIMER U

(1999a) *Mangantransfer in der Nahrungskette der Menschen. 2. Mitteilung: Der Mangangehalt pflanzlicher Lebensmittel.* Mengen- und Spurenelemente **19**:1013–1019.

ANKE M, DORN W, MÜLLER R and SCHÄFER U (2000) *Schwermetalle im Ernährungspfad der Menschen.* In: Fritsche W and Zerling L, eds. Umwelt und Mensch – Langzeitwirkungen und Schlußfolgerungen für die Zukunft. Abhandlungen der Sächsischen Akademie der Wissenschaften zu Leipzig, Mathematisch-naturwissenschaftliche Klasse **59**:45.61

ANKE M, ILLING-GÜNTHER H, GÜRTLER H, HOLZINGER S, JARITZ M, ANKE S and SCHÄFER U (2000a) *Vanadium – an essential element for animals and humans?* In: Roussel et al., eds. pp. 221–225. Trace Elements in Man and Animals 10. Kluwer Academic/Plenum Publishers: New York.

ANKE M, SEEBER O, GLEI M, DORN W and MÜLLER R (2000b) *Uranium in the food chain of humans in central Europe – risks and problems.* In: Gârban Z and Dragan P, eds. pp. 7–22. Metal Elements in Environment, Medicine and Biology, Publishing House "Eurobit": Timisoara, Romania.

ANKE M, MÜLLER R, DORN W, SEIFERT M, MÜLLER M, GONZALES D, KRONEMANN H and SCHÄFER U (2000c) *Toxicity and Essentiality of Cadmium.* In: Ermidou-Pollet S, Pollet S, eds. 2nds. International Symposium on Trace Elements in Human. New Perspectives, pp. 343–361, Athens, Greece.

ANKE M, ARNHOLD W, SCHÄFER U and MÜLLER R (2001) *Nutrients, macro, trace and ultra trace elements in the feed chain of mouflons and their mineral status. First part: Nutrients and Macroelements.* In: Nahlik A and Uloth W, eds. Third International Symposium on Mouflon. pp. 225–242, Lover Print, Sopron Hungary.

ANKE M, ARNHOLD W, MÜLLER R and ANGELOW L (2001a) *Nutrients, macro, trace and ultratrace elements in the food chain of mouflons and their mineral status. Second Part: Trace Elements.* In: Nahlik A and Uloth W, eds. Third International Symposium on Mouflon. pp. 243–261, Lover Print, Sopron Hungary.

ANKE M, JARITZ M, HOLZINGER S, ARNHOLD W, MÜLLER R, ANGELOW L and HOPPE C (2001b) *Nutrients, macro, trace and ultratrace elements in the food chain of mouflons and their mineral status. Third Part: Ultratrace Elements.* In: Nahlik A and Uloth W, eds. Third International Symposium on Mouflon. pp. 263–280, Lover Print, Sopron Hungary.

ANKE M, KRÄMER-BESELIA K, LÖSCH E, MÜLLER R, MÜLLER M and SEIFERT M (2002) *Calcium supply, intake, balance and requirement of man. First information: Calcium content of plant food.* Mengen- und Spurenelemente **21**:1386–1391.

ANKE M, DORN W, SCHÄFER U and MÜLLER R (2003) *The biological and toxicological importance of nickel in the environment and the food chain of humans.* In: Romancík V, Koprda V, eds. 23rd International Symposium "Industrial Toxicology '03". pp. 7–21, Bratislava, Slovak Republic.

ANKE M (2003) Unpublished results.

ANONYMOUS (1992) *Klärschlammverordnung.* Bundesges Bl 912 –934.

ARNHOLD W (1989) *Die Versorgung von Tier und Mensch mit dem lebensnotwendigen Spurenelement Lithium.* Dissertation, Sektion Tierproduktion und Veterinärmedizin, University Leipzig, Germany.

BERGMANN K (1995) *Die Bedeutung tierischer Lebensmittel für die Natrium- und Kaliumversorgung des Menschen.* Thesis, Vet.-Med.-Fac., University Leipzig, Germany.

BERGMANN W (1992) *Nutritional Disorders of Plants.* Gustav Fischer Verlag, Stuttgart-New York.

BERROW ML and BURRIDGE JC (1991) *Uptake, distribution and effects of metal compound on plants.* In: Merian, E, ed. Metals and Their Compounds in the Environment. pp. 399–410, VCH, Weinheim-New York-Basel-Cambridge.

DÄSSLER HG (1986) *Einfluss von Luftverunreinigungen auf die Vegetation*, 3. Auflage, Gustav Fischer, Jena.

DINGER K (1929) *Über die Herkunft des Strontiums in den Schichten des unteren Muschelkalks und des Röt in der Umgebung von Jena.* Chem Erde 167–177.

DJINGOVA R, KULEFF I, PENEV I and SANSONI B (1986) *Bromine, copper, manganese and lead content of the leaves of Taraxacum officinale (dandelion).* Sci Total Environ **50**:197.

DROBNER C (1997) *Die Selenversorgung Erwachsener Deutschlands.* Thesis, Biol.-Pharm. Fakulty, Friedrich-Schiller University, Jena, Germany.

ERLER M, SCHEIDT-ILLING R, ANKE M, GLEI M, MÜLLER M, ARNHOLD W, MOCANU H, NEAGOE A, ANGELOW L, ROTHER C and HARTMANN E (1996) *Cadmium in der Nahrungskette des Menschen eines teerbelasteten Lebensraumes (Rositz, Thüringen)* Mengen- und Spurenelemente **16**:847–856.

ERNST WHO and LELOUP S (1987) *Perennial herbs as monitor for moderate levels of metal fall-out.* Chemosphere **16**:233–238.

FIEDLER HJ and KLINGER T (1996) *Die Spurenelementsituation in den Waldböden des Osterzgebirges.* In: Haase G and Eichler E, eds. Wege und Fortschritte der Wissenschaft. pp. 679–697, Akademie Verlag: Berlin, Germany.

FIEDLER HJ and RÖSLER HJ (1993) *Spurenelemente in der Umwelt.* Gustav Fischer Verlag: Jena, Stuttgart.

GAST CH, JANSEN E, BIERLING J and HAANSTRA L (1988) *Heavy metals in mushrooms and their relationship with soil characteristics.* Chemosphere **17**:789–795.

GLEI M (1995) *Magnesium in der Nahrungskette unter besonderer Berücksichtigung der Magnesiumversorgung des Menschen.* Thesis for lectureship, Biol.-Pharm. Fac., Friedrich-Schiller-University, Jena, Germany.

GRAUPE B, ANKE M and ROTHER A (1960) *Die Verteilung der Mengen- und Spurenelemente in verschiedenen Ackerpflanzen.* Jahrbuch der Arbeitsgemeinschaft für Fütterungsberatung **3**:357–362.

GRIS E (1847) *Addition á une précédente. Note concernant des expériences sur l'application des sels de fer á la végétation, et spécialement au traitement des plantes chlorosées, languissantes et menacées d'une mort prochaine.* Compt Rend Acad SCI Paris **25**:276–778.

GRIS E (1844) *Nouvelles expériences sur l'action des composés ferrugineux soluble, appliqués à la végétation, et spécialement au traitement de la chlorose et de la débilités plantes.* Compt. Rend Acad SCI Paris **19**:1118–1119.

GROPPEL B (1986) *Jodmangelerscheinungen, Jodversorgung und Jodstatus der Wiederkäuer.* Thesis for lectureship University Leipzig, Sec. Tierproduktion und Veterinärmedizin, Germany.

GRÜN M (1984) *Der Einfluss des Bleistatus auf Futterverzehr, Lebendmassezunahme, Mortalität, Reproduktionsleistung und Blutbild von Schafen und Kälbern – Die Bleibelastung der Wiederkäuer in der Deutschen Demokratischen Republik.* Thesis for lectureship University Leipzig, Sec. Tierproduktion und Veterinärmedizin, Germany.

HOLZINGER S (1999) *Die Molybdänversorgung des Menschen unter Berücksichtigung verschiedener Ernährungsformen.* Thesis, Biol.-Pharm. Fac., Friedrich-Schiller-University, Jena, Germany.

ILLING-GÜNTHER H (1995) *Bestimmung, biologische Bedeutung und Versorgung des Menschen mit Vanadium.* Thesis, Biol.- Pharm. Fac., Friedrich-Schiller-University, Jena, Germany.

JARITZ M (1999) *Barium in der Nahrungskette unter besonderer Berücksichtigung der Barium-*aufnahme des Menschen. Thesis, Biol.-Pharm. Fac., Friedrich-Schiller-University, Jena, Germany.

JENKIS DA (1987) *Trace elements in saxicolous lichens.* In: Coughtrey PJ, Martin MH and Unsworth MH, eds. Pollutant Transport and Fate in Ecosystems. pp. 249–253. Blackwell Sci Publ: Oxford.

KABATA-PENDIAS A, GALCZYNSKA B and DUDKA S (1989) *Baseline zinc content of soils and plants in Poland.* Environ Geochem Health **11**:19–24.

KABATA-PENDIAS A and PENDIAS H (1991) *Trace Elements in Soils and Plants.* 2nd edition. CRC Press: Boca Raton, Ann Arbor, London.

KOVALSKIJ VVM (1977) *Geochemische Ökologie Biogeochemie.* VEB Deutscher Landwirtschaftsverlag: Berlin.

KRÄMER K (1993) *Calcium- und Phosphorausscheidung Erwachsener Deutschlands nach der Duplikat- und Marktkorbmethode.* Thesis, Biol.-Pharm. Fac., Friedrich-Schiller-University, Jena, Germany.

KRAUSE M (1987) *Die biologische Bedeutung des Arsens.* Thesis, University of Leipzig, Tierproduktion und Veterinärmedizin, Germany.

KRONEMANN H (1982) *Die Kadmiumbelastung von Pflanze, Tier und Mensch in der DDR und der VR Ungarn.* Thesis, Sec. Tierproduktion und Veterinärmedizin, University of Leipzig, Germany.

KRONEMANN H, ANKE M and GRÜN M (1982) *Der Cadmiumgehalt der Nahrungsmittel in der DDR.* Zentralbl Pharm **171**:556–558.

KRONEMANN H, ANKE M, THOMAS S and RIEDEL E (1980) *The nickel concentration of different food- and feed-stuffs from area with and without nickel exposure.* In: Anke M, Schneider H-F, Brückner CHR, eds. Nickel. 3. Spurenelement – Symposium. pp. 221–228, University of Leipzig and Jena.

LIEDÉN C (1994) *Occupational contact dermatitis due to nickel allergy.* Sci Total Environ **148**:283–285.

MACHOLZ R and LEWERENZ H.J (1998) *Lebensmitteltoxikologie,* pp. 270–316, Akademie – Verlag: Berlin.

MITCHELL RL and BURRIDGE JC (1989) *Trace elements in soils and crops.* In: Environmental Geochemistry and Health. A Society Discussion, March 1978, pp. 15–24, London.

MOLISCH H (1892) *Die Pflanze in ihrer Beziehung zum Eisen.* Gustav Fischer: Jena, Germany

MÜLLER M (1993) *Cadmiumaufnahme und Cadmiumausscheidung Erwachsener nach der Markt- und Duplikatmethode.* Thesis, Biol.-Pharm. Fac., Friedrich-Schiller-University, Jena, Germany.

MÜLLER M and ANKE M (1994) *Distribution of cadmium in the food chain (soil-plant-human) of a cadmium exposed area and the health risks of the general population.* Sci Total Environ **156**: 151–158.

MÜLLER M, ANKE M, HARTMANN E and ILLING-GÜNTHER H (1996) *Oral cadmium exposure of adults in Germany. I. Cadmium content of foodstuffs and beverages.* Food Addit Contam **13**: 359–378.

MÜLLER M, ANKE M and ILLING-GÜNTHER H (1998) *Aluminium in foodstuffs.* Food Chem **61**:419–428.

MÜLLER M, MACHELETT B and ANKE M (1994) *Cadmium in the food chain soil-plant-animal/man and the current exposure in Germany.* In: Pais I, ed. Proceedings, 6th International Trace Element Symposium, pp. 205–225, University of Horticulture, Budapest.

MÜLLER R, ANKE M, BUGDOL G, LÖSCH E and SCHÄFER U (2001) *Der Natriumtransfer in der Nahrungskette des Menschen. 1. Mitteilung: Die biologischen Grundlagen des Natriumtransfers vom Boden über die Flora und Fauna bis zum Menschen.* In: Anke M, Müller R and Schäfer U, eds. Mengen-, Spuren- und Ultraspurenelemente in der Prävention. pp. 208–221, Wissenschaftliche Verlagsgesellschaft mbH: Stuttgart, Germany.

NAGAYAMA M, SABURI T, OKA N, YAMADA S and MATSUMOTO A (1984) *Sudanophilia at the sites of calcification in hard tissues of rats given strontium carbonate.* Shika Kiso Igakkai Zasshi **26**:549–553.

NRIAGA JO (1989) *Heavy metals in the atmosphere.* Nature **338**:47.

PARTSCHEFELD M, GROPPEL B, ANKE M and GRÜN M (1977) *The Cd-exposure of game grazing and of roes, red deer, fallow-dear and muoflons in the GDR.* In: Anke M and Schneider H-J, eds. Kadmiumsymposium. Scientific contributions of the Friedrich-Schiller-University of Jena. 258–265.

RÖHRIG B (1998) *Der Zink- und Kupfergehalt von Lebensmitteln aus ökologischem Landbau und der Zink- und Kupferverzehr Erwachsener Vegetarier.* Thesis, Biol.-Pharm. Fac., Friedrich-Schiller-University of Jena, Germany.

RÖHRIG B, ANKE M, DROBNER C, JARITZ M and HOLZINGER S (1998) *Zinc intake of German adults with mixed and vegetarian diets.* Trace Elem Electrolytes **15**:81–86.

RÜHLING A, RASMUSSEN L, PILEGAARD K, MÄKINEN A and STEINNES E (1987) *Survey of atmospheric heavy metal deposition in Nordic countries in 1985.* Report for Nordic Council of Ministers, Kobenhavn.

SALM-HORSTMAR FÜRST ZU (1849) *Versuche über die nothwendige Aschenbestandteile einiger Pflanzen-Spezies.* J Prakt Chem **46**:193–211.

SCHÄFER U, ANKE M, BERGMANN K, LÖSCH E, MÜLLER R and MÜLLER M (2001) *Der Natriumtransfer in der Nahrungskette des Menschen. 2. Mitteilung: Der Natriumgehalt pflanzlicher Lebensmittel.* In: Anke M, Müller R and Schäfer U, eds. Mineralstoffe. pp. 222–234, Wissenschaftliche Verlagsgesellschaft mbH: Stuttgart, Germany.

SCHMIDT P (2002) *Quecksilberverzehr und –bilanz Erwachsener Deutschlands in Abhängigkeit von Geschlecht, Zeit, Kostform, Alter, Gewicht, Leistung und Messverfahren.* Thesis, Landwirtschaftliche Fakultät Martin-Luther-University Halle–Wittenberg, Germany.

SEEBER O (1998) *Das Angebot, die Aufnahme und die Bedeutung des toxischen Schwermetalls Uran bei erwachsenen Mischköstlern in Deutschland.* Thesis, Friedrich-Schiller-University of Jena, Germany.

SEEBER O, HOLZINGER S, ANKE M, LEITERER M and FRANKE K (1998) *Die Uranaufnahme erwachsener Mischköstler in Deutschland.* Mengen- und Spurenelemente **18**:924–931.

SEIFERT M (1998) *Cadmium und Strontium in der Nahrungskette eines industriell belasteten Lebensraumes im mittleren Saaletal.* Thesis, Biol.-Pharm. Fac., Friedrich-Schiller-University of Jena, Germany.

SEIFERT M and ANKE M (1999) *Alimentary nickel intake of adults in Germany.* Trace Elem Electrolytes **16**:17–21.

SZENTMIHALYI S, REGIUS A, ANKE M, GRÜN M, GROPPEL B, LOKAY D and PAVEL J (1980) *The nickel supply of ruminants in the GDR, Hungary and Czechoslovakia dependent on the origin of the basic material for the formation of soil.* In: Anke M, Schneider H-J and Brückner CHR, eds., 3. Spurenelement-Symposium, Nickel. pp. 229–236, University of Leipzig und Jena, Germany.

ULKEN R (1985) *Nähr- und Schadstoffgehalt in Klär- und Flußschlämmen, Müll und Müllkomposten.* VDLUFA – Schriftenreihe **22**:95–101.

7

Elements and Elemental Compounds in Waters and the Aquatic Food Chain

Biserka Raspor

7.1

Introduction

Understanding the distribution of chemical forms of metals within certain water types, and their uptake into biota, is based on the electronic configuration of elements and the empirical classification of electron acceptors (metals) and donors (ligands) to "hard" and "soft" categories (Morgan and Stumm 1991, Raspor 1991). The relationship between the chemical properties of elements, and their uptake and accumulation – which has implications on detoxification and food chain transfer – will be considered. Classification of trace metals as either essential (Fe, Cu, Mn, Zn, Co) or non-essential (Hg, Cd, Ag, Pb) should be performed with caution, bearing in mind that the former can exert beneficial effects at low concentrations and harmful ones at higher levels.

There are numerous biotic and abiotic parameters which influence metal uptake and accumulation. The abiotic factors (salinity, temperature, light, pH, Eh, and ligand concentration) influence the relationship between metal species distribution and organisms. Using the available speciation techniques (see Part IV, Chapter 3; Raspor 1980), biological responses can often be pre-dicted from a knowledge of the chemical properties of the metal and the complexation capacity of the surrounding media. In many instances, the major features influencing biological responses can be identified despite the difficulties in quantifying the heterogeneous mixtures of ligands present in natural waters (Langston and Bryan 1984).

Except for Hg, other metals are not biomagnified along the food chain. The reason for this is reduced bioavailability of metals. Concentration factors tend to be the highest in the primary producers and in the organisms at lower trophic levels, for which the dominant source of metals is uptake from water. Organisms counter the reactivity of metals and their potential toxicity by ligand binding and compartmentalization. The bound forms include insoluble phosphates and sulfur compounds which are formed and accumulated within membrane-limited vesicles in specialized tissues such as the liver and kidney. These metals are also unavailable to predators that consume the tissues because they are not absorbed by their digestive systems and therefore are not transferred along the food chain (Nott 1998).

Elements and their Compounds in the Environment. 2nd Edition.
Edited by E. Merian, M. Anke, M. Ihnat, M. Stoeppler
Copyright © 2004 WILEY-VCH Verlag GmbH & Co. KGaA, Weinheim
ISBN: 3-527-30459-2

7.1.1
Periodic Table of the Elements

The properties of chemical elements are periodic functions of their atomic number (Masterton et al. 1986). As one moves across a period or down a group of the Periodic Table, the physical properties of elements change in a smooth, regular fashion. Within a given group, the elements show very similar chemical properties, because they have the same outer-electron configuration. Elements may thus be classified as follows:

- Main-group elements in the Periodic Table are confined to the two groups at the far left and the six groups at the right-hand side of the table, assigned as groups IA to VIIIA; the latter includes the noble gases.
- Transition elements are those in the center of the Periodic Table, between the IIA and IIIA main-group elements and are assigned as groups IB to VIIIB.
- Lanthanides refer to elements with atomic numbers from 57 to 71, while actinides to the elements with atomic numbers from 89 to 103.

According to the physical properties, elements are classified as metals, nonmetals, and metalloids.

7.1.1.1
Metals
Of 108 elements known to date, 84 belong to the group of metals, 17 to nonmetals, and seven to the metalloids. The predominance of metals over other classes of elements is also reflected in nature. Of the ten most abundant elements in the Earth's crust, seven are metals: Al, Fe, Ca, Na, K, Mg, and Ti (see Part I, Chapter 1; Giddings 1973).

Metals have low ionization energy; that is, they easily lose the outermost electron(s)

and therefore have relatively free electrons to move about.

Due to the loss of valence electrons, positive ions are smaller than the metal atoms from which they are formed. The sodium atom has a radius of 0.186 nm while the sodium ion has a radius of 0.095 nm. The difference in radii between atom and cation is due to the excess of protons in the ion, which draws the outer electrons closer to the nucleus (Masterton et al. 1986).

Along the same period, the ionization energy increases from left to right, and in the same chemical group decreases down the group. This means that the ability of elements to form cations changes in the opposite manner. The successive alkali metals have a minimal ionization energy, which indicates that these metals form cations very easily. In general, transition metals have somewhat higher ionization energies than the main-group elements, and thus are generally less reactive since they oxidize less readily. Potassium and Ca react vigorously with water, while among the transition metals in the first series only Sc reacts rapidly with water, while Mn reacts slowly.

All transition metals form cations of $+1$, $+2$ and $+3$ oxidation state by loss of successive s and d electrons which have energies of the same order of magnitude. These metals often have more than one oxidation state and hence more that one set of compounds, for example, Cu^+/Cu^{2+}, Fe^{2+}/Fe^{3+}, Co^{2+}/Co^{3+}. In contrast to the transition metals, the metals of the main groups IA and IIA are present in only one oxidation state, $+1$ and $+2$, respectively (Masterton et al. 1986).

7.1.1.2
Nonmetals
Elements on the right-hand side of the diagonal which consists of B, Si, As, Se, Te, and Po are classified as nonmetals. They have high ionization energies and therefore do

not lose electrons in order to achieve the stable electron configuration of the noble gas, but form ions by accepting the electrons. This is the reason why nonmetals have no free electrons which could serve for conducting electricity and heat. Nonmetals are usually present in water as anions, such as O^{2-}, F^-, Cl^-. Due to the gain of electrons and the increased repulsion of the outer electrons, negative ions are larger than nonmetal atoms, from which they are formed. The radius of the chlorine atom is 0.099 nm, while that of the chloride ion is 0.181 nm (Masterton et al. 1986).

7.1.1.3
Metalloids

On the right-hand side of the Periodic Table, between metals and nonmetals, exist seven elements which, according to their physical properties, are difficult to classify as either metals or nonmetals. They have properties in between those of elements in the two other classes. In particular, their electronic configuration is intermediate between that of metals and nonmetals. These elements are B, Si, Ge, As, Sb, Te, and Se. They are often called metalloids (see Part III, Chapters 4 and 10; Part IV, Chapters 1, 3, 6, 7, and 8).

7.1.1.4
Ionic Metal Compounds

Generally speaking, within the aqueous phase metal ions might undergo the following reactions: complexation; precipitation; and changes of the oxidation state. Metal ions in natural water systems can interact with the inorganic and organic types of ligands in the water phase and/or at the surface of the solid phase (Morgan and Stumm 1991). Ionic compounds are formed via ionic bonds, as the result of electrostatic interaction of the oppositely charged ions (Pytkowicz 1983). Ionic compounds evolve

from the reaction of the elements of low ionization energy (usually the metals of the IA and IIA groups) with the elements of high ionization energy (the nonmetals of the VIA and VIIA groups).

In an aqueous electrolytic solution, ions of opposite charge are held together by electrostatic forces within the critical distance, forming ion-pairs. These forces decrease with $1/r^2$, where r is the interionic distance (Pytkowicz 1983). Ion triplets may also occur, as is the case for $[CaMg(CO_3)]^{2+}$ (Pytkowicz and Hawley 1974). When the ion-pair is formed, the metal ion or the ligand or both retain coordination water, so that cation and anion are separated by one or more water molecules (Stumm and Morgan 1981). Ion-pairs are also called outer-sphere complexes.

Estimates of the stability constants of ion-pairs can be made on the basis of a simple electrostatic model which considers coulombic interactions between the ions (Stumm and Morgan 1981). The areas of particular importance of application of the ionic model to aquatic chemistry are the hydration energies of cations and complex formation constants (Whitfield and Turner 1983).

7.1.1.6
Covalent Metal Compounds

Within molecules, atoms are held together by strong forces called covalent bonds (Lewis 1916). These bonds are formed when a metal as an electron-acceptor (Lewis acid) reacts with an electron-pair donor (Lewis base). Metals that form coordinate bonds most readily are small, and highly charged with empty orbitals, such as the transition metals (Pytkowicz 1983).

Atoms of two different elements always differ at least slightly in their affinity for the electrons. Hence, covalent bonds between unlike atoms are always unsymmetrical, respectively polar (Masterton et al. 1986). The greater the electronegativity of an atom, the greater the affinity for bonding electrons. Pauling (1960) used bond energies to calculate relative electronegativity values for the various elements, arbitrarily defining the most electronegative element, that is, fluorine with the value 4.0. The assigned electronegativity values are presented in the Periodic Table of Elements (Table 7.1). It would be helpful to know the electronegativity value for each oxidation state of an element and for each individual valence orbital (Whitfield and Turner 1983).

The greater the difference of electronegativities between two elements, the more ionic is the bond between them. A difference of 1.7 units corresponds to a bond with 50% ionic character. Electronegativity differences less than 1.7 units imply that the bonding is mainly covalent (Masterton et al. 1986). Bonds are stronger and the bond energy is higher for a multiple rather than for a single bond between the same two atoms.

Since the predominant ligands in the aquatic environment are the water molecules, the occurrence and distribution of metal complexes should begin with the consideration of hydration of ions.

7.2
Hydration of Ions

Metal ions dissolved in water are already complexed and present in the form of hydrated ions. Therefore, the explanation of the formation and stability of complex ions in aqueous solution begins with the structure of liquid water and hydrated ions themselves (Cotton and Wilkinson 1980).

The structure of liquid water is the subject of intense study and controversy. The polar nature of the water molecule and its ability to form strong intermolecular hydrogen bonds result in the cooperative association of multimolecular aggregates (Horne 1969).

Tab. 7.1: Electronegativity values of the elements, listed in groups of the Periodic Table of the elements. (After Pauling 1960.)

IA	IIA	IB	IIB	IIIB	IVB	VB	VIB	VIIB	VIIIB	IIIA	IVA	VA	VIA	VIIA	VIIIA
H 2.1															He –
Li 1.0	Be 1.5									B 2.0	C 2.5	N 3.0	O 3.5	F 4.0	Ne –
Na 0.9	Mg 1.2									Al 1.5	Si 1.8	P 2.1	S 2.5	Cl 3.0	Ar –
K 0.8	Ca 1.0	Sc 1.3	Ti 1.5	V 1.6	Cr 1.6	Mn 1.5	Fe 1.8	Co Ni 1.8 1.8	Cu Zn 1.8 1.6	Ga 1.6	Ge 1.8	As 2.0	Se 2.4	Br 2.8	Kr –
Rb 0.8	Sr 1.0	Y 1.2	Zr 1.4	Nb 1.6	Mo 1.8	Tc 1.9	Ru 2.2	Rh Pd 2.2 2.2	Ag Cd 1.9 1.7	In 1.7	Sn 1.8	Sb 1.9	Te 2.1	I 2.5	Xe –
Cs 0.7	Ba 0.9	57–71 1.1–1.2	Hf 1.3	Ta 1.5	W 1.7	Re 1.9	Os 2.2	Ir Pt 2.2 2.2	Au Hg 2.4 1.9	Tl 1.8	Pb 1.8	Bi 1.9	Po 2.0	At 2.2	Rn –

Hydrogen bonding is the specific association of the hydrogen atom of one molecule with the lone pair electrons of another. Hydrogen bonding is responsible for many of the extraordinary physical properties of water. Each water molecule has approximately 4.4 neighbors in the first coordination shell. Thus, liquid water is a highly structured liquid in which the tetrahedral coordination observed in ice is still evident (Westall and Stumm 1980).

An understanding of ionic hydration is a prerequisite for understanding the chemistry of water. The hydration atmosphere of an ion in solution has a complex internal structure and its outer boundary is difficult to establish. An ion in solution is represented as being surrounded by two zones. An inner layer can be equated to what is often called the "primary" hydration shell, which is composed of dense, electrorestricted and immobilized water molecules strongly bound by the coulombic field on the ion. Furthermore, there is a region of comparative randomness, of disrupted water organization, of broken structure. At some further distance from the ion, the water structure is "normal", although the molecules may be slightly polarized by the ubiquitous charge field. Frank and Evans (1945) suggested that the structure-enhanced zone is present and intact in all ions, while the particular characteristics of different types of ions arise from the variability of the structure within the broken zone (Horne 1969).

For a hydrated metal ion, we wish to know the coordination number and the manner in which the water molecules are arranged around the metal ion, or according to Taube (1954) "formulas" of the ion-water complexes. Some experimental methods measure only the most tightly bound water molecules, whereas other methods measure the loosely bound water molecules as well.

Therefore, various methods yield different hydration numbers (Horne 1969).

Generally speaking, cations are more hydrated than the anions of the same negative charge, and the greater the charge of the ion, the more heavily hydrated is the ion. The primary hydration number of Mg^{2+} is higher than that of Li^+, even though these differently charged cations have nearly the same crystal radius. In a given charge type, the smaller the crystal radius of the ion, the heavier is the hydration. The protons in water are hydrated as well, and although they are usually written as H^+ or H_3O^+ (hydronium ion) the best available evidence points strongly to the existence of $H_9O_4^+$, which is the prevailing form (Horne 1969).

Ultrasonic velocity measurements are convenient for measuring hydration numbers from ion compressibilities (Padova 1964). For the di- and trivalent cations of the first transition series, the aqua ions are octahedral $[M(H_2O)_6]^{2+}$ or $[M(H_2O)_6]^{3+}$, although in Cr(II), Mn(II) and Cu(II) definite distortions of the octahedra are present (Cotton and Wilkinson 1980).

The crystals of the first series of the transition metals are colored; for example, Ti(III), V(II), V(III), Mn(II), Fe(II), Fe(III), Co(II), Ni(II). The hexaaquo-salts of these metals dissolve in water without changing the color. The absorption spectra of crystals and solutions for these transition metals are perfectly in agreement, so that there is no doubt about the octahedral coordination of these transition metals in water (Schneider 1968).

The hydration of ions can also be conceptualized in terms of the residence time of water molecules near an ion. If an average water molecule is in a position near an ion for a longer time than it would be at some greater distance from the ion, then the ion is positively hydrated. However, if the water molecule is more mobile near the ion than it would be at some distance from the ion, the term "negative hydration" is used. The residence time concept of hydration is complementary to the concept of the hydration number (Horne 1969).

There are vast differences in the average length of time that a water molecule spends in the coordination sphere. For Cr(III) and Rh(III) the residence time is so long that when a solution of $Cr(H_2O)_6^{3+}$ in ordinary water is mixed with water enriched in ^{18}O, many hours are required for complete equilibration of the enriched solvent water with the coordinated water (Cotton and Wilkinson 1980). Taube (1954) measured the kinetics of water exchange in the solution of Cr(III). The half-time of this reaction is 2×10^6 s (Morgan and Stone, 1985). For Rh(III), the reaction of water exchange is even slower. It can be concluded that Cr(III) and Rh(III) show a clear inert behavior with respect to the exchange of water molecules in their hydration sphere. For most other aqua-ions the exchange of water molecule(s) occurs too rapidly to permit the same type of measurements (Schneider 1968).

7.3
Metal Complex Formation

When covalent or inner-sphere types of complexes are formed, kinetically speaking, a dehydration step must precede the association reaction. The metal cation in a complex is called the central ion, the molecules or anions bound directly to it are called ligands, and the number of bonds formed by the central ion is its coordination number.

Based on the exchange rate of hydration water and comparing the coordination number of covalent complexes with other ligands than water it is obvious that the complex formation represents ligand exchange of an equivalent number of water molecules

from the hydration shell. In many instances the exchange of the first coordinated water molecule controls the overall rate of complexation (see Morgan and Stumm 1991).

Metal complexes can be classified as kinetically labile or kinetically inert. It should be noted that there is no relationship between thermodynamic stability (as determined by a large formation constant) and kinetic inertness.

7.3.1
Labile Complexes

Complex ions that exchange ligands almost instantaneously are regarded as labile. Typically, they exchange ligands in water solution with a half-time of a minute or less. In general, the half-time of the water exchange reactions from the primary shell of the ion covers the range of about fifteen orders of magnitude (Schneider 1968). For additional information on labile complexes, see Raspor (1991).

7.3.2
Inert Complexes

In contrast to the labile complexes, in nonlabile or inert complexes the hydration water is slowly exchanged with the added ligand. An example for the slow rate of water exchange is the aqua-complex $[Cr(H_2O)_6]^{3+}$. An additional illustration of labile and inert type of complexes, which reflects different bonding strength and therefore different electronic configuration is given by the complexes of the same metal atom but at two different oxidation states; for example, $[Co(H_2O)_6]^{3+}$ is inert type of complex while $[Co(H_2O)_6]^{2+}$ is labile; $[Fe(CN)_6]^{4-}$ is inert while $[Fe(CN)_6]^{3-}$ is labile type of complex (Masterton et al. 1986).

For each reaction of complex formation in an aquatic system, the thermodynamic stability has to be defined, besides the kinetic stability of the particular chemical form, which refers to the rate of transformation leading to the attainment of equilibrium. Theoretically, the change of the free energy of complex formation indicates whether the observed reaction is thermodynamically possible to occur. As an additional practical parameter, a kinetic factor defines the height of the energy barrier E_a, which chemical reactants have to overcome. It determines the rate of the reaction and the measurable amount of the product of a chemical reaction.

Most simple ionic equilibria in aqueous solutions tend to be very rapid, their rates often being controlled by diffusion. Further it is probably correct to assume that most equilibria in the dissolved phase are reached rapidly (Horne 1969). Rates of precipitation and even more of dissolution are usually slower (Westall and Stumm 1980).

7.4
Hard and Soft Acceptors and Donors

On the basis of experimental evidence, Ahrland et al. (1958) and Pearson (1963) classified the electron acceptors (metals) and electron donors (ligands) into "hard" and "soft" categories, according to the stability of the complexes of particular type of metal with the particular type of ligand (Morgan and Stumm 1991). The stability of metal complexes formed by any ligand with a series of metals (Irving and Williams 1948) may be expected to increase with electronegativity of the metal concerned (see Table 7.1). According to the Irving-Williams series, the formation constants for a given ligand with a divalent metal ions, are in the order:

$$Ba^{2+} < Sr^{2+} < Ca^{2+} < Mg^{2+} <$$
$$Mn^{2+} < Fe^{2+} < Co^{2+} < Ni^{2+} < Cu^{2+} > Zn^{2+}$$

The separation of metals into distinct classes was based on empirical thermodynamic data, namely, trends in the magnitude of equilibrium constants that describe the formation of metal-ion/ligand complexes. On the basis of these criteria, metal ions can be divided into three groups: hard, soft, and borderline. The partition of a particular ion in each group is shown in Figure 7.1 (Nieboer and Richardson 1980; see also Morgan and Stumm 1991).

Hard cations include ions of the alkali metals, alkaline earth metals, lanthanides, actinides, and aluminum. They bind mainly via electrostatic interactions and form strongest complexes with electron donors from VIA and VIIA main groups in the Periodic Table (see Table 7.1). Their ligand preference has the sequence:

$$N \gg P > As < Sb : O \gg S > Se > Te:$$
$$F^- \gg Cl^- > Br^- > I^- \text{ with } F^- > O > N.$$

Hard or class A cations preferentially bind to hard bases; that is, to oxygen sites of inorganic anions (O_2^{2-}, OH^-, $H_2PO_4^-$, CO_3^{2-}, SO_4^{2-}) and organic molecules provided with oxygen-containing groups (carboxylate, carbonyl, alcohol, ester). Considering organic types of ligands, hard cations will also bind to nitrogen sites, although less strongly.

Soft or class B cations include transition metals from the triangle in the Periodic Table of Elements, with Cu^+ at its apex (note Cu^{2+} is borderline) and Ir^{3+} and Bi^{3+} at its base. They bind mainly via covalent interactions, forming their strongest complexes with electron donors in the following sequence:

$$N \ll P > As > Sb : O \ll S \cong Se \cong Te:$$
$$F^- \ll Cl^- < Br^- < I^- \text{ with } S > N > O > F^-.$$

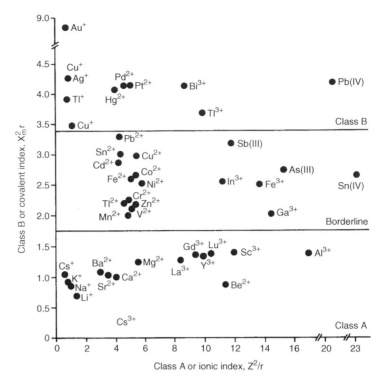

Fig. 7.1 A separation of metal ions and mettaloid ions, As(III) and Sb(III), into three categories: class A, borderline, and class B ions. The class B index $X_m^2 r$ is plotted for each ion against the class A index Z^2/r. In these expression X_m is the metal-ion electronegativity, r its ionic radius, and Z its formal charge. Oxidation states given by Roman numerals imply that simple, hydrated cations do not exist even in acidic solutions. (Reprinted from Nieboer and Richardson (1980), Copyright 2002, with permission from Elsevier Science).

Soft cations preferentially bind to soft bases; that is, with inorganic anions I^- and CN^-, while in organic molecules they preferentially bind to sulfur (sulfhydryl, disulfide, thioether) and nitrogen sites (amino, imidazole, histidine, nucleotide base).

The borderline cations comprise the first row of transition metals, in their common oxidation states, as well as Ga^{3+}, In^{3+}, Cd^{2+}, Sn^{2+}, and Pb^{2+}. The hydrogen ion and the metalloid ions As(III) and Sb(III) are also included in this category (Nieboer and Richardson 1980). They are able to form stable complexes with numerous ligands, and preference for a given donor group will be determined by factors including the degree of hard-character, in part by the soft-character, in part by the relative availability of ligand(s) in the system, and in part by the steric environment of the reaction site(s) themselves (Whitfield and Turner 1983). Thus, according to Figure 7.1, among borderline metal ions, class B (i.e., soft character) increases in the order (Nieboer and Richardson 1980):

$$Mn^{2+} < Zn^{2+} < Ni^{2+} < Fe^{2+}$$
$$\cong Co^{2+} < Cd^{2+} < Cu^{2+} < Pb^{2+}.$$

Since it is not possible to determine an ionic radius for H^+, its position is not indicated in Figure 7.1. However, the chemistry and chemical reactivity calculations clearly show that H^+ should be regarded as a borderline ion (e.g., Evans and Huheey 1970), although this is in contrast to the traditional view that a proton is a pure class A cation (Pearson 1963, 1969). In acid–base equilibria of inorganic and organic types of ligands, pH is an important parameter as the protons compete with metal ions and, depending on the pH of the aqueous solution, displace metals from the binding site or *vice versa*.

An examination of trends in the magnitude of metal-ligand equilibrium constants determined in aqueous solution reveals some interesting features. Soft or class B cations, in spite of their own preference for soft bases, when reacting with hard bases form complexes that are more stable than those with hard cations of comparable Z^2/r values (see Figure 7.1). The same observation holds for borderline cations relative to hard cations. Presumably, this feature signifies that in addition to the largely ionic interactions observed for class A ion, borderline and class B ions of comparable size and charge make significant covalent contributions to the overall interaction energy (Nieboer and Richardson 1980).

Another observation of interest is that ions with values of Z^2/r greater than 8 (Figure 7.1), with few exceptions, tend to hydrolyze and form metal hydroxides and oxo anions in mildly acidic and some even in quite acidic solutions (e.g., Huheey 1978).

Ions with intermediate values of $X_m^2 r$ and concurrent large Z^2/r values, form water-soluble organometallic cations which involve metal-carbon bonds; for example, $(CH_3)_2Pb^{2+}$, $(CH_3)_2Tl^+$, $(CH_3)_3Sn^+$, CH_3Hg^+ and the volatile $(CH_3)_3As$, corresponding to the complexes of Pb^{4+}, Tl^{3+}, Sn^{4+}, Hg^{2+}

and As^{3+}, respectively. Two methylating agents, *S*-adenosylmethionine and methylcobalamin, are capable of converting As and Se and respectively Hg, Sn, Pb and Sb into organic forms (Thayer 1993).

7.5
Bioavailability of Metals

The term "bioavailable" is used to refer to the proportion of a chemical species that might be taken up from the environment into an organism (Sanders and Riedel 1998). It is important to note that the total metal concentration in the surrounding water does not represent that concentration which is available to the biota (Luoma 1983; Brezonik et al. 1991).

Within the aquatic environment, metals may be distributed as the dissolved complexes, bound to organic and inorganic particulate matter and bound on sediment surfaces. Within the water phase, metals exist in equilibrium among hydrated metal ions, metal bound to organic (e.g., amines, humic acid) and inorganic (e.g., OH^-, CO_3^{2-}) type of ligands. The chemical composition of seawater strongly influences the speciation of metals. In turbid estuarine waters, a large proportion of the total metal load is bound in or to organic and inorganic particulate matter (Salomons and Förstner 1984). In estuarine water, metal speciation alters with the ionic strength (i.e., the salinity). Dissolved organic complexes and particulate matter may undergo flocculation and, for some metals, a large proportion of the metal load, when transported in the river water, sinks to the sediments of the estuary. However, metals such as Cd are displaced from particulate matter because they form chloride complexes (Elbay-Poulichet et al. 1987). In estuaries, the speciation of metals that remain in solution is affected by the increasing concentration of anions (particularly chlorides), and for most metals hydrated ions constitute a relatively minor proportion of the total dissolved metal concentration (e.g., Mantoura et al. 1978; Sipos et al. 1980; Ahrland 1988).

Alterations in the physico-chemical parameters of water can strongly influence the relative proportions of the metal species that can be taken up. Changes in the pH, redox potential, salinity, temperature, etc. can all greatly influence the availability of metals for uptake into aquatic organisms (Mantoura et al. 1978). This is an important consideration for biomonitoring studies, which are performed with the aim of defining the amount and the bioavailability of metal contaminants. The fact that certain types of organisms do not contain elevated metal content in their tissues does not preclude the possibility that metals are not present at elevated levels in the aquatic environment; they might be firmly bound in highly stable complexes. As physico-chemical conditions alter – for example, after the resuspension of sediments due to turbulence – rapid conversion of metals to bioavailable ionic forms may result in higher concentrations in biota (Salomons and Förstner 1984; Samiullah 1990).

7.5.1
Metal Uptake into the Organism

The metal uptake refers to its entrance into an organism. Organisms obtain metals by direct uptake from the surroundings across the entire body surface of the organism, across specialized respiratory structures (gills and lungs), across the digestive epithelium if water is imbibed, via ingested food, or by a combination of routes (Brown and Depledge 1998). The most important routes of metal uptake by aquatic organisms are those from solution and from food. The knowledge on the predominant route of metal uptake, either from solution or from food, would be of benefit for estimating the usefulness of various organisms as biomonitors (Phillips and Rainbow 1993).

From solution the metals are taken up by:
- Active uptake (energy-dependent ion pump, such as the cations Na, K and Ca).
- Passive uptake by facilitated diffusion into the cell (e.g., Simkiss and Taylor 1989a).

Once inside the cell, exchange to stronger ligands may take place, thereby preventing back diffusion and forming a kinetic trap for the metal (Langston and Bryan 1984).

The reason for the existence of different uptake mechanisms lies in the different chemical properties of two classes of metals (Nieboer and Richardson 1980; Phillips and Rainbow 1993).

As already explained in Section 7.4, the major ions of Class A (e.g., Na, K, Ca) do not have a high affinity for ligands containing sulfur and nitrogen, and therefore do not bind to the membrane carrier proteins, for transport into the cell. Thus, active ion pumps (Figure 7.2) are required for the movement of these ionic metals against concentration gradients across the hydrophobic membrane (Phillips and Rainbow 1993).

Metal ions falling into the Class B or Borderline categories according to Nieboer and Richardson (1980) (see Figure 7.1) have high affinities for ligands containing sulfur and nitrogen, and therefore bind relatively easily to proteins and other cellular macromolecules (e.g., Nieboer and Richardson 1980). The high affinity of such metal ions for proteins and other cellular constituents provides the basis for their passive uptake, mediated by carrier proteins, as presented in Figure 7.2. The initial binding with the protein is a passive process, and metal transfer across the membrane into the cell occurs

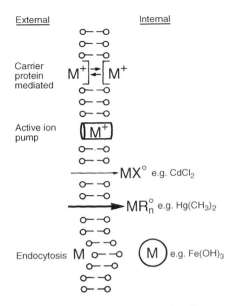

External   Internal

Carrier protein mediated $M^+ \rightleftharpoons M^+$

Active ion pump $\boxed{M^+}$

$\longrightarrow MX^{\circ}$ e.g. CdCl$_2$

$\longrightarrow MR_n^{\circ}$ e.g. Hg(CH$_3$)$_2$

Endocytosis   M   (M) e.g. Fe(OH)$_3$

Fig. 7.2   Possible models of the uptake of trace metals from solution. (From Simkiss and Taylor 1989a reproduced with permission).

along a series of metal-binding ligands of increasing affinity. Thus, the metal is passed down a thermodynamic gradient into the cell, where it binds finally with the ligand of highest metal affinity. This will give rise to either storage of the metal or its transfer out of the cell, perhaps ultimately to ligands in circulating fluids or specific target organs. Thus, metal uptake into the organism continues as a passive process, apparently against a concentration gradient.

Certain trace metals are available for uptake into organisms from solution only as hydrated ions, whereas others are transported across biological membranes as inorganic complexes. In experiments in which the hydrated species of Cu and Cd were either carefully controlled by organic chelators or determined by means of ion-selective electrodes, the toxicity and bioavailability were correlated with the concentration of hydrated metal ions rather than the total dissolved

metal concentration (e.g., Zamuda and Sunda 1982, Sanders et al. 1983). In most studied estuarine organisms, such findings are consistent with the increasing uptake and toxicity of these metals as the salinity decreases, because the hydrated ion concentration also increases (McLusky et al. 1986).

Given that neutral complexes of metals are more lipid-soluble than ionic species, Simkiss (1983) has suggested that such neutral species may be transported by direct diffusion across hydrophobic cell membranes. Simkiss and Taylor (1989a) proposed that uncharged inorganic species (such as CdCl$_2$; see Figure 7.2) and organic derivatives may also diffuse into organisms due to their high lipid solubility. The uncharged Hg(II) complexes such as HgCl$_2$, Hg(CH$_3$)$_2$, are also transported across lipid bilayer membranes by direct diffusion (e.g., Langston and Bryan 1984, Sanders and Riedel 1998).

Some trace metals will become incidentally incorporated into active transport pumps for the major metal ions. The hydrated Cd ion has a similar ionic radius to that of the Ca ion, and Cd will therefore be taken up to some extent through Ca-ion pumps. For example, Cd may enter a variety of crustaceans, mollusks and fish via active transport through Ca-ion pumps. The relative significance of this route of entry into the cytosol as opposed to that of facilitated diffusion varies with the organism concerned and with environmental conditions. Mollusks and malacostracan crustaceans with a high physiological demand for Ca (whether for shell formation or calcification of the exoskeleton) may exhibit atypically high Ca-ion pump activities, particularly in water of low salinity. These ion pumps will incorporate Cd, even to the extent that this becomes a predominant route of Cd entry into the cytosol. Using radiotracer labeling, Markich and Jeffree (1994) demonstrated

branes is due primarily to the diffusion of the uncharged $HgCl_2^°$ complex (Langston and Bryan 1984). Cadmium uptake in *F. vesiculosus* is significantly reduced in the presence of Zn. In "hard" water types the competition of Ca^{2+} ions for uptake sites may influence the toxicity of Cd^{2+} (Langston and Bryan 1984).

In general, the term "accumulation" refers to the amount of chemical that remains in an organism following exposure over a particular period of time (Sanders and Riedel 1998). The accumulation of trace elements by an aquatic organism results from the net balance of the processes of metal uptake and excretion. All aquatic organisms take up metals in significant quantities, but for many species the excretion of accumulated metals may be insignificant. According to accumulator/regulator classification, the metal accumulation strategies for the aquatic organisms fall along a spectrum from high metal uptake (in barnacles, ascidians, mollusks) to low metal uptake (in decapod crustaceans and finfish) (Phillips and Rainbow 1989). Thereby it is important to specify for which metal the "regulatory" ability applies. Moreover, for a particular species, the "regulatory" ability for a given metal may vary with route of uptake (Brown and Depledge 1998). Chemical properties of metals, which influence their bioavailability, also determine their retention and biochemical/toxicological reactions within cells.

The Class A metals (see Figure 7.1) such as Na and K form such weak complexes that their retention within cells is established only by restriction within membranes and is maintained by selective energy pumps. Metals such as Ca, Mg, Sr and Mn show preference for ligands containing the charged oxygen atoms of pyrophosphate ions. Moving across the Periodic Table from Class A to Class B metals, there is an increasing tendency for the latter to be retained by the thermodynamic traps in polymers. Transition metals, such as Cu, Cd, and Hg, show preference for the SH-groups of metallothioneins, which are inducible type of proteins (see Langston and Bryan 1984, Erk and Raspor 2000). Some degree of control is available by regulation of ligand production, although in many cases this is not achieved and leads to accumulation of metals in excess of requirements. When confronted with an excess to normal metabolic requirements, an organism must metabolize, eliminate or otherwise detoxify the burden to prevent harmful complexation to, and inhibition of, its enzyme systems (Langston and Bryan 1984).

Once in the organism, metals may become associated with ligands having a strong binding capacity, resulting in their accumulation. Complexation with a variety of organic ligands can significantly affect the availability and toxicity of trace metals (Langston and Bryan 1984). In animals, metallothioneins are one of the key determinants of the ability to withstand exposure to trace metals such as Cu and Cd (Raspor and Pavičić 1991, Sanders and Riedel 1998). Several mechanisms exist, including the binding of metals to soluble metallothioneins and other metal-binding proteins (see also Roesijadi 1992, Engel and Brouwer 1989) and the sequestration of elements in metal-rich insoluble deposits or granules which may or may not be associated with lysosomes (Simkiss and Taylor 1989b).

7.6
Aquatic Food Chain

Food chains consist of variable numbers of trophic levels linked in successive prey and predator relationships. Networks of these chains form complex food webs that route the supply, transfer and disposal of potentially toxic metals within ecological systems. For further information, the reader is recommend to consult a condensed literature overview (see Table 12.1 in Nott 1998) on a number of food chains that have been investigated, showing uptake routes of different metals from food, water, sediment and their transfer to higher trophic levels. The results in the quoted papers are assessed and marked subjectively for the relative efficiency with which the metals are taken up, transferred and in some cases excreted.

Metals, which enter aquatic organisms directly from the surrounding water, are compartmentalized within tissues in membrane-limited vacuoles and bound to ligands to reduce the toxic reactivity (Roesijadi 1992, Viarengo and Nott 1993). Amiard (1988) suggested that phytoplankton fixes metals and makes them unavailable to oysters. Compartmentalization processes remove metals from tissue fluids, and diffusion gradients inwards from surrounding water are maintained (Nott 1998).

Three Mediterranean species of marine snail which dwell in the same seawater adjacent to a nickel smelting plant accumulate markedly different levels of metals (Nott and Nicolaidou 1989). The highest levels occurred in the sediment feeder *Cerithium vulgatum* and the lowest levels in the predator *Murex trunculus*, which preys on *C. vulgatum*. It has been established that metals in *C. vulgatum* occur in the digestive gland, where they are accumulated within intracellular phosphate granules and residual lysosomes. The metals are unavailable to the animal in the sense that they are insoluble and within membrane-limited compartments at high concentration. When the digestive glands are consumed by the carnivore, the metals remain insoluble and unavailable in the gut, so that the detoxifica-

tion system operating in *C. vulgatum* also protects the carnivore *M. trunculus* (Nott and Nicolaidou 1989, Nott 1998).

As shown by Nott and Nicolaidou (1989), metal-containing granules produced in the tissues of the prey still contained the same metals after passing through the gut of a predator (carnivore). This indicates that the detoxification system of the prey also protects the predator by rendering the metals unavailable to its digestive system.

Of note is the explanation on metal retention in crustaceans (Rainbow 1988, 1998, Rainbow and White 1989), in gastropods (Nott et al. 1993) and in decapods (Nott and Nicolaidou 1994). In the gastropod *L. littorea* the metal is finally accumulated in the digestive gland (Langston and Zhou 1987, Bebianno et al. 1992, Langston et al. 1998), where it is released periodically into the lumen when the digestive cells disintegrate (see Figure 12.5 in Nott 1998). Cadmium occurs as a soluble and labile element in the cytosol and is reabsorbed by the digestive epithelium. Other metals, which are bound as insoluble compounds and enclosed in membrane-bound vesicles, are excreted.

7.6.1
Concentration Factors

The relationship between the concentration of a metal in the surrounding water and the concentration in an organism is defined as concentration factor (Baudo 1981). Concentration factors in excess of a thousandfold can be attained (Martinčić et al. 1984), and are affected by speciation, active and passive uptake, modalities of uptake, and transformation, transport and distribution between and within tissues, and elimination.

Concentration factors represent the net balance of continuous uptake and excretion, which can result in negligible or excessive accumulation according to the metal and its availability and the species of organism (e.g., Baudo 1981, 1985, Fowler 1982, Suedel et al. 1994). Metal bioavailability is a direct result of reactivity and thermodynamic equilibrium, and biomagnification occurs at stages in a food chain when proportionally more metal is retained than energy, in the form of weight gain (Nott 1998).

Concentration factors for metals within organisms and the retention of stable metal species within ecosystems both contribute to the transfer of metals along food chains and the toxic effects. Thus, primary producers can accumulate high concentrations, and these are consumed by organisms on secondary trophic levels in the food chain. If the food type of the secondary consumer contains biochemically reactive metals, they will be absorbed and accumulated; however, if they are insoluble they will pass through the gut and be excreted in the feces.

7.6.2
Trophic Levels

The trophic levels range from phytoplankton and macrophytes to zooplankton, invertebrates, fish and mammals. Food chain transfer is affected by the distribution of metals between different tissues of the prey, and by the marked degree to which this compartmentalization can vary at different trophic levels.

Phytoplankton has a large ratio of surface area to volume, and metals dissolved in the surrounding water have access to the entire surface of each cell (Baudo 1981, Sanders and Riedel 1998). Therefore, primary producers are vulnerable to the effects of metal excess and they show some of the highest levels of accumulation in food

chains (Sanders et al. 1989, 1990, Lindsay and Sanders 1990). The initial accumulation from seawater provides much of the momentum for subsequent transfer along food chains (Preston et al. 1972, Nott 1998).

Arsenic is incorporated by algae and transformed into reduced and methylated forms (see Section 7.5.2) which are nontoxic to phytoplankton but may be toxic to higher animals (Sanders and Riedel 1998, Edmonds and Francesconi 1998). The copepod *Eurytemora affinis*, barnacle *Balanus improvisus* and oyster *Crassostrea virginica* do not take As from water, but they do take it from phytoplankton (Sanders et al. 1989). Phytoplankton is more efficient at taking As from surrounding water than invertebrates are from food (see Table 12.1 in Nott 1998).

Along the food chain, from the primary producers on, the larger zooplankton organisms, macro invertebrates and higher trophic levels have tissues that are differentiated. Metal is taken up by permeable epithelia of the gut and gills and, internally, it is transported, metabolized, stored and excreted by other specialized tissues (Fowler 1982, Viarengo and Nott 1993, Langston et al. 1998, Rainbow 1998, Olsson et al. 1998). Within these tissues, metals are compartmentalized in particular cells and organelles. Highest concentrations can occur in storage tissues that are "glandular", and these range from the digestive gland/hepatopancreas in crustaceans and mollusks (reviewed in Viarengo and Nott 1993) to the pyloric caeca in starfish (Pelletier and Larocque 1987) and the liver in fish (Maage et al. 1991). These compartments can account for widely differing proportions of the total body weight as presented by Nott (1998). The weight of the glandular tissue expressed as a percentage of the total soft body weight is termed the "hepatosomatic index". This index can be higher than 10%

in invertebrates, but much lower in higher animals – especially bony fish, where it can be less than 1%. In wild Atlantic salmon the liver contains much higher levels of Cu and Se than any other tissue, and high levels of Fe are confined to the liver, spleen, and kidney (Maage et al. 1991). This compartmentalization has implications for food chains because, for example, predators that consume salmon flesh without the offal will avoid a dietary intake of metals.

Whole-body analyses of fish and other animals do not reflect high concentrations of metals in the liver and pancreas, which can disrupt normal biochemical processes in these tissues. Also, once saturation of a storage system occurs, spillage into other compartments can disrupt enzyme systems and produce toxic effects, without producing any significant increase in whole-body analyses (Roesijadi 1992, Langston et al. 1998, Olsson et al. 1998).

7.6.3
Biomagnification

According to Bryan (1979), absorption from food is often the most important route for metal bioaccumulation and transfer along food chains, but there is little evidence that predators at high trophic levels will contain the highest concentrations. Cesium in fish is an exception in that a high degree of assimilation from prey results in magnification along the food chain (Pentreath 1977). Of 18 metals considered by Bryan (1976) in various organisms, Hg is one of the few where mean levels in fish exceed those in phytoplankton or seaweed as measured on a dry weight basis.

Mercury, Cu, Zn, Pb, and Cr in two different food chain experiments (seawater-plankton-fish, and seawater-phytoplankton-mussel) all had reduced concentration factors at higher trophic levels (Laumond et al. 1973).

Mercury is rarely amplified between invertebrates and small fish (Knauer and Martin 1972, Leatherland et al. 1973), but it is sometimes amplified in large fish, where there are effects linked to both trophic level (Ratkowsky et al. 1975) and the age of the animals (Jackson 1998). Bacteria are important in the transformation of mercury to methyl mercury. Such conversion is probably the major source of methyl mercury present in the aquatic food chains (see Part III, Chapter 17; Jackson 1998).

The biomagnification of metals along food chains in natural communities has been established as a common feature only in the case of Hg, where magnification results from selective retention of methyl mercury at each trophic level. The major criteria responsible are the extremely high assimilation efficiency of methyl mercury and its long biological half-life, together with the greater longevity of most top predators. The importance of food type in influencing Hg levels is exemplified by the work of MacCrimmon et al. (1983), who showed that a change from low Hg-containing invertebrates to high Hg-containing smelt resulted in a dramatic increase in Hg accumulation in lake trout, *Salvelinus namaycush* (Langston and Bryan 1984). Mercury is biomagnified, principally as lypophilic methyl mercury, presumably because at the cellular level for that type of chemical species the biochemical processes which could counter its reactivity do not exist, contrary to other metals which excess and toxic reactivity could be reduced by means of compartmentalization within tissues (see Sections 7.6 and 7.6.1).

References

AHRLAND S, CHATT J and DAVIES NR (1958) *The relative affinities of ligand atoms for acceptor molecules and ions.* Q Rev Chem Soc **12**:265–276.

AHRLAND S (1988) *Trace metal complexation by inorganic ligands in sea water.* In: West TS and Nürnberg HW, eds. The determination of trace metals in natural waters, Chapter 7, pp. 223–252. Blackwell Scientific Publications, Oxford.

AMIARD JC (1988) *Les méecanismes de transfer des éléments métalliques dans les chaînes alimentaires aboutissant a l'huître et a la moule, mollusques filtreurs; formes chimiques de stockage, conséquences écotoxicologiques.* Océanis **14**:283–287.

BAUDO R (1981) *Is analytically defined chemical speciation the answer we need to understand trace element transfer along a trophic chain?* In: Trace element speciation in surface waters and its ecological implications, pp.275–290. Proceedings of the NATO-AIOL Workshop, Genova-Nervi, Plenum Publishing Corporation, New York.

BAUDO R (1985) *Transfer of trace elements along the aquatic food chain.* Memorie dell Istituto Italiano di Idrobiologia **43**:281–309.

BEBIANNO MJ, LANGSTON WJ and SIMKISS K (1992) *Metallothionein induction in Littorina littorea (Mollusca: Prosobranchia) on exposure to cadmium.* J Mar Biol Ass UK **72**:329–342.

BREZONIK PL, KING SO and MACH CE (1991) *The influence of water chemistry on trace metal bioavailability and toxicity to aquatic organisms.* In: Newman MC and McIntosh AW, eds. Metal ecotoxicology concepts and applications, pp. 1–31. Lewis, Boca Raton.

BROWN MT and DEPLEDGE MH (1998) *Determinations of trace metal concentrations in marine organisms.* In: Langston WJ and Bebianno MJ, eds. Metal metabolism in aquatic environments, Chapter 7, pp. 185–217. Chapman & Hall, London.

BRULAND KW (1983) *Trace elements in seawater.* In: Riley RP and Chester R, eds. Chemical Oceanography, Volume 8, pp. 157–220. Academic Press, London.

BRYAN GW (1976) *Heavy metal contamination in the sea.* In: Johnston R, ed. Marine Pollution, pp. 185–302. Academic Press, London and New York.

BRYAN GW (1979) *Bioaccumulation of marine pollutants.* Phil Trans Roy Soc Lond B**286**:483–505.

BRYAN GW (1984) *Pollution due to heavy metals and their compounds.* In: Kinne O, ed. Marine ecology, pp. 1289–1431. John Wiley, Chichester.

COTTON FA and WILKINSON G (1980) *Advanced inorganic chemistry*, 4th edition, Chapter 3, pp. 61–106. John Wiley & Sons, New York.

DEPLEDGE MH and PHILLIPS DJH (1986) *Circulation, respiration and fluid dynamics in the gastropod mollusc, Hemifusus tuba (Gmelin).* J Exp Mar Biol Ecol **95**:1–13.

EDMONDS JS and FRANCESCONI KA (1998) *Arsenic metabolism in aquatic ecosystems.* In: Langston WJ and Bebianno MJ, eds. Metal metabolism in aquatic environments, Chapter 6, pp. 159–183. Chapman & Hall, London.

ELBAY-POULICHET F, MARTIN J-M, HUANG WW and ZHU JX (1987) *Dissolved Cd behaviour in some selected French and Chinese estuaries. Consequences on Cd supply to the ocean.* Mar Chem **322**:125–136.

ENGEL DW and BROUWER M (1989) *Metallothionein and metallothionein-like proteins: physiological importance.* Adv Comp Environ Physiol **5**:53–75.

ERK M and RASPOR B (2000) *Advantages and disadvantages of voltammetric method in studying cadmium-metallothione interactions.* Cell Mol Biol **46**:269–281.

EVANS RS and HUHEEY JE (1970) *Electronegativity, acids and bases. III. Calculation of energies associated with some hard and soft acid-base interactions.* J Inorg Nucl Chem **32**:777–793.

FOWLER SW (1982) *Biological transfer and transport processes.* In: Kullenbergg, ed. Pollutant transfer and transport processes, Volume II, pp. 1–65. CRC Press.

FRANK HS and EVANS MW (1945) *Free volume and entropy in condensed systems. III. Entropy in binary liquid mixtures; partial molal entropy in dilute solutions; structure and thermodynamics in aqueous electrolytes.* J Chem Phys **13**:507–532.

GEORGE SG, PIRIE BJS and COOMBS TL (1976) *The kinetics of accumulation and excretion of ferric hydroxide in Mytilus edulis (L.) and its distribution in the tissues.* J Exp Mar Biol Ecol **23**:71–84.

GIDDINGS JC (1973) *Chemistry, man and environmental change, an integrated approach.* Canfield Press, San Francisco.

HOBDEN DJ (1967) *Iron metabolism in Mytilus edulis. I. Variation in total content and distribution.* J Mar Biol Ass UK **47**:597–606.

HORNE RA (1969) *Marine chemistry*, pp. 11–53. John Wiley & Sons, New York.

HUHEEY JE (1978) *Inorganic chemistry*, 2nd edn. Harper & Row, New York.

IRVING H and WILLIAMS RJP (1948) *Order of stability of metal complexes.* Nature **162**:746–747.

JACKSON TA (1998) *Mercury in aquatic ecosystems.* In: Langston WJ and Bebianno MJ, eds. Metal metabolism in aquatic environments, Chapter 5, pp. 77–158. Chapman & Hall, London.

JENKINS KD and SANDERS BM (1986) *Relationships between free cadmium ion activity in seawater, cadmium accumulation and subcellular distribution, and growth in polychaetes.* Environ Health Perspect **65**:205–211.

KALK M (1963) *Absorption of vanadium by tunicates.* Nature (London) **198**:1010–1011.

KNAUER GA and MARTIN JH (1972) *Mercury in a marine pelagic food chain.* Limnol Oceanogr **17**:868–876.

LANGSTON WJ and BRYAN GW (1984) *The relationships between metal speciation in the environment and bioaccumulation in aquatic organisms.* In: Kramer CJM and Duinker JC, eds. Complexation of trace metals in natural waters, Part VI Biological response, pp. 375–392. Martinus Nijhoff/Dr W. Junk Publishers, The Hague.

LANGSTON WJ and ZHOU M (1987) *Cadmium accumulation, distribution and metabolism in the gastropod Littorina littorea: the role of metal-binding proteins.* J Mar Biol Ass UK **67**:585–601.

LANGSTON WJ, BEBIANNO MJ and BURT GR (1998) *Metal handling strategies in molluscs.* In: Langston WJ and Bebianno MJ, eds. Metal metabolism in aquatic environments, Chapter 8, pp. 219–283. Chapman & Hall, London.

LAUMOND F, NEUBURGER M, DONNIER B, FOURCY A, BITTEL R and AUBERT M (1973) *Experimental investigations, at laboratory, on the transfer of mercury in marine trophic chains.* Rev Int Oceanogr Med 31–**32**:47–53.

LEATHERLAND TM, BURTIN JD, CULKIN F, MCCARTNEY MJ and MORRIS RJ (1973) *Concentrations of some trace metals in pelagic organisms and of mercury in northeast Atlantic.* Deep-sea Res **20**:679–685.

LEWIS GN (1916) *The atom and the molecule.* J Am Chem Soc **38**:762–785.

LINDSAY DM and SANDERS JG (1990) *Arsenic uptake and transfer in a simplified estuarine food chain.* Environ Toxicol Chem **9**:391–395.

LUOMA SN (1983) *Bioavailability of trace metals to aquatic organisms.* Sci Total Environ **28**:1–22.

MAAGE A, JULSHAMN K and ULGENES Y (1991) *A comparison of tissue levels of four essential trace elements in wild and farmed Atlantic salmon (Salmo salar).* Fiskeridirektoratets Skrifter, Serie Ernring **4**:111–116.

MACCRIMMON HR, WREN CD and GOTS BL (1983) *Mercury uptake by lake trout, Salvelinus namaycush, relative to age, growth, and diet in Tadenac Lake with comparative data from other PreCambrian Shield lakes.* Can J Fish Aquat Sci **40**:114–120.

MANGUM CP (1979) *A note on blood and water mixing in large marine gastropods.* Comp Biochem Physiol **63A**: 389–391.

MANTOURA RFC, DICKSON A and RILEY JP (1978) *The complexation of metals with humic materials in natural waters.* Estuarine Coastal Mar Sci **6**:387–408.

MARKICH SJ and JEFFREE RA (1994) *Absorption of divalent trace metals as analogs of calcium by Australian freshwater bivalves: an explanation of how water hardness reduces metal toxicity.* Aquat Toxicol **29**:257–290.

MARTINČIĆ D, NÜRNBERG HW, STOEPPLER M and BRANICA M (1984) *Bioaccumulation of heavy metals by bivalves from Lim Fjord (North Adriatic Sea)* Mar Biol **81**:177–188.

MASTERTON WL, SLOWINSKI EJ and STANITSKI CL (1986) *Chemical Principles,* 6th edn., pp. 224–287. College Publishing, Philadelphia.

MCLUSKY DS, BRYANT V and CAMPBELL R (1986) *The effects of temperature and salinity on the toxicity of heavy metals to marine and estuarine invertebrates.* Oceanogr Mar Biol Ann Rev **24**:481–520.

MORGAN JJ and STONE AT (1985) *Kinetics of chemical processes of importance in lacustrine environments.* In: Stumm W, ed. Chemical Processes in Lakes, pp. 389–426. John Wiley & Sons, New York.

MORGAN JJ and STUMM W (1991) *Chemical processes in the environment, relevance of chemical speciation.* In: Merian E, ed. Metals and their compounds in the environment, Chapter I.3, pp. 67–103. VCH Weinheim.

NIEBOER E and RICHARDSON DHS (1980) *The replacement of the nondescript term "heavy metals" by a biologically and chemically significant classification of metal ions.* Environ Pollut Ser B **1**:3–26.

NOTT JA (1998) *Metals and marine food chains.* In: Langston WJ and Bebianno MJ, eds. Metal metabolism in aquatic environments, Chapter 12, pp. 387–414. Chapman & Hall, London.

NOTT JA and NICOLAIDOU A (1989) *The cytology of heavy metal accumulations in the digestive glands of three marine gastropods.* Proc Roy Soc Lond B**237**:347–362.

NOTT JA and NICOLAIDOU A (1994) *Variable transfer of detoxified metals from snails to hermit crabs in marine food chains.* Mar Biol **120**:369–377.

8
Elements and Compounds in Sediments

Ulrich Förstner and Wim Salomons

8.1
Introduction

The composition of sediments reflects the natural (geological) conditions of their source, but at the same time it is a reflection of human activities in river basins, estuaries, and the coast. Most of the human population lives along rivers or at the coastal environment, and hence fluvial, estuarine, lake and marine sediments show increased levels of metals and may require management intervention.

Point and diffuse sources contribute to heavy metals in sediments. Point sources have dominated the input of heavy metals in surface waters for most of the past century, and both dated sediment cores and archived sediment samples show the impact of uncontrolled industrialization between 1900 and 1970 on sediment composition. Although these point sources are no longer in existence, or they have reduced their output due to regulations, these contaminated sediments are still present in the environment and pose an important management issue of clean-up. Important in this respect are river flood plains and dredging and disposal of "old" sediments in locks, weirs, and river stretches. In this chapter, the available technologies for clean-up or containment will be discussed, and the potential impact of contaminated sediments on surface quality (remobilization) for those cases were removal is not feasible will be outlined.

During the past 20 years, all major rivers in Europe have shown a decline in metal levels. However, when comparing current metal levels with existing regulations, discrepancies persist which either prevent the reuse of sediments or restrict their disposal. This has led to new management questions of identifying those point and diffuse sources which continue to contribute to the elevated levels, and those which can be regulated by taking into account a cost-benefit analysis. This requires an holistic approach of the sediment issue at the river basin scale to be made, and this will be discussed in the last section of the chapter.

8.2
Remobilization of Metals from Polluted Sediments

Due to the capacity of sediments to store and immobilize toxic chemicals in so-called "chemical sinks", direct effects of pollution may not be directly manifested. This positive function of sediments does not guaran-

Elements and their Compounds in the Environment. 2nd Edition.
Edited by E. Merian, M. Anke, M. Ihnat, M. Stoeppler
Copyright © 2004 WILEY-VCH Verlag GmbH & Co. KGaA, Weinheim
ISBN: 3-527-30459-2

tee, however, that the chemicals are safely stored for ever. Factors influencing the storage capacity of sediments or the bioavailability of the stored chemical can change and indirectly cause sudden and often unexpected mobilization of chemicals in the environment (Stigliani 1988). From the discussions on the "Chemical Time Bomb" (CTB) concept during the early 1990s it became apparent first, that it is imperative to know what sediment properties will control the toxicity levels of a chemical, and how sensitive the chemical toxicity is to changes in these properties. Second, the relevance of a sediment property to a CTB depends on how this property is affected by long-term environmental changes, for example, socioeconomic or climatic changes (Hesterberg et al. 1992).

The solubility, mobility, and bioavailability of sediment-bound metals can be increased by four major factors in terrestrial and aquatic environments:

- lowering of pH, locally from mining effluents, or regionally from acid precipitation;
- increasing salt concentrations, by the effect of competition on sorption sites on solid surfaces and by the formation of soluble chloro-complexes with some trace metals;

- increasing occurrence of natural and synthetic complexing agents, which can form soluble metal complexes with trace metals that are otherwise adsorbed to solid matter; and
- changing redox conditions, e.g., after land deposition of polluted anoxic dredged materials.

In some cases (which will be described here), mobilization is a change in the chemical environment affecting lower rates of precipitation or adsorption – compared to "natural" conditions – rather than active release from contaminated solid materials.

8.2.1
Acidity

Acidity imposes problems in all aspects of metal mobilization in the environment: the toxicity of drinking water; growth and reproduction of aquatic organisms; the increased leaching of nutrients from the soil and the ensuing reduction of soil fertility; the increased availability and toxicity of metals; and the undesirable acceleration of mercury methylation in sediment (Fagerström and Jernelöv 1972).

In Swedish lakes, a pronounced correlation was observed between dissolved metal levels and pH (Figure 8.1). This phenom-

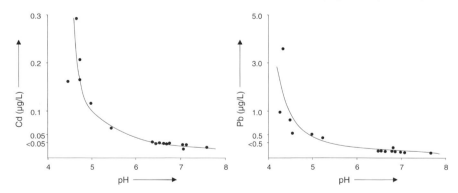

Fig. 8.1   Dissolved metal concentrations ($\mu g\,L^{-1}$) in relation to pH values in 16 lakes at the West coast of Sweden. (After Dickson 1980.)

enon is probably due to the combined effects of:

- changing solid/dissolved equilibria in the atmospheric precipitation;
- washout processes on soils and rocks in the catchment area;
- enhancing groundwater mobility of metals; and
- active mobilization from aquatic sediments.

8.2.2
Salinity

The effect of higher salinity seems to be particularly critical for resuspended cadmium-rich sediments in estuaries (Salomons and Förstner 1984). As a result of biological or biochemical pumping, the tidal flats may act as a source of dissolved metals. Release of trace metals from particulate matter has been reported from several estuaries (Scheldt, Gironde, Elbe/Weser, Savannah/Ogeechee), and has been explained by oxidation processes and by intensive breakdown of organic matter (both mediated by microorganisms), whereafter the released metals become complexed with chloride and/or ligands from the decomposing organic matter in the water. According to experimental data reported by Salomons and Mook (1980), these effects can even be found in salt-polluted inland waters: at chloride contents of 200 mg L^{-1} (e.g., the Lower Rhine river), the "normal" adsorption rate of cadmium would be reduced by approximately 20%; at 1.000 mg L^{-1} Cl^{-} (e.g., the Weser River in Germany), this rate would be only half compared to the sorption of Cd under natural salt concentrations.

8.2.3
Complexing Agents

Significant effects on the mobility of heavy metals can be expected by strong synthetic chelators, such as nitrilotriacetate (NTA), which is used as a substitute for polyphosphate in detergents, and ethylenediaminetetraacetate (EDTA), which is also used for replacing phosphate, but in the metal-processing, galvanotechnology, and photographic industries. The extent of metal mobilization depends on the concentration of the complexing agent, its pH-value, the mode of occurrence of heavy metals in the suspended sediment, and on competition by other cations. Active remobilization seems to show reliable results at NTA concentrations above ~1–2 mg L^{-1}; such concentrations of NTA could rarely be expected in normal river waters, but may occur at even higher levels in sewage treatment plants. "Passive" effects of NTA (where the complexing agent may negatively influence the natural adsorption processes) start at lower NTA concentrations of 200 to 500 μg NTA L^{-1}, and it has been found by Salomons (1983) that zinc adsorption is already significantly affected at NTA concentrations of 20–50 μg L^{-1} at conditions of pH 8.

8.2.4
Oxidation/Reduction Processes

Under oxidizing conditions the controlling solid may change gradually from metallic sulfides to carbonates, oxyhydroxides, oxides, or silicates, thus changing the solubility of the associated trace metals. The major process affecting the lowering of pH-values (to pH 2–3) is the exposure of pyrite (FeS$_2$) and other sulfide minerals to atmospheric oxygen and moisture, whereby the sulfidic component is oxidized to sulfate and acidity (H^{+} ions) is generated.

Field evidence for changing cadmium mobilities was reported by Holmes et al. (1974) from Corpus Christi Bay Harbor where, during the summer period when the harbor water was stagnant, cadmium precipitated as CdS at the sediment–water interface. In the winter months, however, the increased flow of oxygen-rich water into the bay resulted in a release of the precipitated metal.

In the St. Lawrence Estuary, Gendron et al. (1986) found evidence for different release mechanisms near the sediment–water interface – the profiles for cobalt resemble those for manganese and iron with increased levels downwards, suggesting a mobilization of these elements in the reducing zone and a reprecipitation at the surface of the sediment profile. On the other hand, cadmium appears to be released at the surface, probably as a result of the aerobic remobilization of organically bound cadmium.

Biological activities are typically involved in these processes; remobilization of trace metals has been explained by the removal of sulfide from pore waters via ventilation of the upper sediment layer with oxic overlying water, allowing the enrichment of dissolved cadmium that would otherwise exhibit very low concentrations due to the formation of insoluble sulfides in reduced, H_2S-containing sediments. Emerson et al. (1984) suggest a significant enhancement of metal fluxes to the bottom waters by these mechanisms. It was shown by Hines et al. (1984), using tracer experiments, that biological activity in surface sediments greatly enhances remobilization of metals by the input of oxidized water. These processes are more effective during spring and summer than during the winter months.

From enclosure experiments in Narragansett Bay, Hunt and Smith (1983) estimated that by mechanisms such as oxidation of organic and sulfidic material, the anthropogenic proportion of cadmium in marine sediments is released to the water within approximately three years. For remobilization of copper and lead, approximately 40 and 400 years, respectively, is needed, according to these extrapolations.

Metal release from tidal Elbe river sediments by a process of "oxidative remobilization" has been described by Kersten (1989) (Figure 8.2). Short (30-cm) sediment cores were taken from a site, where diurnal inundation of the fine-grained fluvial deposits take place. In the upper part of the sediment column, total particulate cadmium content was ~10 mg kg^{-1}, whereas in the deeper anoxic zone the total particulate concentration of Cd was 20 mg kg^{-1}. Sequential extractions indicate that in the anoxic zone 60–80% of the Cd was associated with the sulfidic/organic fraction. In the upper (oxic and transition) zone, the association of Cd in the carbonatic and exchangeable fractions simultaneously increase up to 40% of total Cd. This distribution suggests that the release of metals from particulate phases into the pore water and further transfer into biota is controlled by the frequent downward flux of oxygenated surface water. From the observed concentrations, it would be expected that long-term transfer of up to 50% of the Cd from the sediment subsurface would take place either into the anoxic zone located further below the sediment–water interface or released into the open water.

Pore water data from dredged material obtained at Hamburg indicate typical differences in the kinetics of proton release from organic and sulfidic sources (Table 8.1). Recent deposits are characterized by low concentrations of nitrate, cadmium and zinc; when these low-buffered sediments are oxidized during a time period of a few months to years, the concentrations of

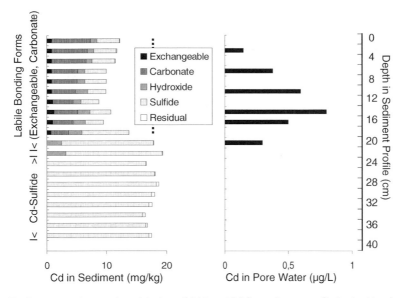

Fig. 8.2 Total concentrations and partitioning of Cd in a tidal flat sediment profile in the Heuckenlock areas near Hamburg. Sedimentation rates were determined using the ^{137}Cs-method. Cadmium pore water profile was determined at low tide (Kersten 1989).

Tab. 8.1: Mobilization of metals and nitrogen compounds from dredged material after land deposition (Maaß et al. 1985)

Element or compound	Reduced water	Oxidized water
Ammonia	125 mg L^{-1}	$< 3 \text{ mg L}^{-1}$
Iron	80 mg L^{-1}	$< 3 \text{ mg L}^{-1}$
Nitrate	$< 3 \text{ mg L}^{-1}$	120 mg L^{-1}
Zinc	$< 10 \text{ µg L}^{-1}$	5000 µg L^{-1}
Cadmium	$< 0.5 \text{ µg L}^{-1}$	80 µg L^{-1}

ammonia and iron in the pore water typically decrease, whereas those of cadmium and zinc increase (with the result that these metals are easily transferred into agricultural crops!).

8.3
Sediment Management Strategies: Remedial Options

Sediment management strategies fall into five broad categories, which are selected based upon an evaluation of site-specific risks and goals (Apitz and Power 2002):

1. No action, which is only appropriately applied if it is determined that sediments pose no risk.
2. Monitored natural recovery, which is based on the assumption that, while sediments pose some risk, it is low enough that natural processes can reduce risk over time in a reasonably safe manner.
3. In-situ containment, in which sediment contaminants are in some manner isolated from target organisms, though the sediments are left in place.
4. In-situ treatment.
5. Dredging or excavation (followed by ex-situ treatment, disposal and/or reuse).

Geochemical techniques for contaminated sediments mainly are related to categories (2) and (3), but innovative disposal technologies also apply geochemical principles for the long-term, safe storage of polluted waste materials.

Sediment remediation methods can be subdivided according to the mode of handling (e.g., in-place or excavation), or to the technologies used (containment or treatment). Important containment techniques include capping *in situ* and confined disposal. Biological processes may be applied with in-place treatment. Excavated sediments – apart from physical separation – can be treated to immobilize pollutants, most of which are metals (Table 8.2).

Remediation techniques on contaminated sediments generally are much more limited than for most other solid waste materials, except for mine wastes. The widely diverse contamination sources in larger catchment areas usually produces a mixture of pollutants, which is more difficult to treat than an industrial waste. For most sediments from maintenance dredging, there are more arguments in favor of "disposal" rather than "treatment". Mechanical separation of less strongly contaminated fractions,

however, may be a useful step prior to final storage of the residues.

8.3.1
In-Situ Remediation using Geochemical Engineering Methods

Remediation techniques are often economically unacceptable, because of the large volume of contaminated materials to be treated. In such cases the concept of "geochemical engineering" (Salomons and Förstner 1988) can provide both cost-effective and durable solutions. Geochemical engineering applies geochemical principles (such as concentration, stabilization, solidification, and other forms of long term, self-containing barriers) to determine the mobilization and biological availability of critical pollutants. In modern waste management, the fields of geochemically oriented technology include:

- the optimization of elemental distribution at high temperature processes;
- the selection of favorable milieu conditions for the deposition of large-volume wastes;
- the selection of additives for the solidification and stabilization of waste materials; and
- the development of test procedures for long-term prognoses of pollutant behavior.

In general, micro-scale methods – for example, the formation of mineral precipitates in the pore space of a sediment waste body – will be employed rather than using large-scale enclosure systems such as clay covers or wall constructions. A common feature of geochemically designed deposits, therefore, is their tendency to increase overall stability in time, due to the formation of more stable minerals and closure of pores, thereby reducing water permeation.

Tab. 8.2: Technology types for sediment remediation (Anonymous 1994)

	In place	Excavated
Containment	In situ-capping	Confined aquatic disposal/capping
	Contain/fill	Land disposal
		Beneficial use
Treatment	Bioremediation	Physical separation
	Immobilization	Chemical extraction
	Chemical treatment	Biological treatment
		Immobilization
		Thermal treatment

Recent developments in the Netherlands in "soft" (geochemical and biological) techniques on contaminated soils and sediments, both with respect to policy aspects as to technical developments have led to a stimulation of in-situ remediation options: (i) no longer do remediation actions have to be executed within a very short period of time; (ii) the result is not necessarily a "multifunctional soil"; and (iii) advantage is taken of natural processes (the self-cleaning capacity of the soil). A number of potentially relevant options for metals are summarized in Table 8.3. Phytoremediation, for example the degradation of contaminants near plant roots, may be beneficial in certain cases. As to the immobilization of contaminants by adsorption, one can think of applying clay screens, or clay layers (with or without additives). The advective dispersion of contaminants towards ground water or surface water can be reduced by capping the polluted sediment with a clay layer, with organic matter (humus) or other materials as possible additives.

It has been stressed by Joziasse and Van der Gun (2000) that for every single case, the effects of the actions (either dredging, or in-situ) on the aquatic ecosystem will have to be accounted for. In concrete cases, where a conventional approach encounters serious difficulties, an investigation dedicated to the prevailing conditions will have to provide a decisive judgment on the feasibility of an alternative (in-situ) approach.

8.3.2
Natural Attenuation on Floodplain Sediments

Unlike problems relating to conventional polluted sites, the problems in floodplains are primarily connected with the erosion and mobilization of highly contaminated soil and sediment material, and the transport and deposition of contaminated solids in downstream river and harbor sediments. The handling of such problems is a complex task which cannot be tackled by science and engineering alone. It deserves thorough consideration of legal and socioeconomic aspects including public relations. Measures taken have to be planned comprehensively and need controlling for extended periods of time.

In the framework of an international case comparison in the Spittelwasser area (Anonymous 2000), situated in the so-called Chemistry Triangle of the upper Elbe River system, the German group mainly planned investigations on the effects of plant growth and of "natural attenuation" processes of organic and inorganic contaminants in floodplain sediments and soils (Förstner et al. 2000). In practice of this concept,

Tab. 8.3: Selected options for in-situ sediment remediation (after Joziasse and Van der Gun 2000)

Remediation type	Scope (type of contaminants)	Technological concept	Technological implementation
Fixation of contaminants (sorption/immobilization)	Metals	Precipitation of metals as hydroxides or insoluble complexes	Precipitation or adsorption at plant roots (phytostabilization)
Reduction of advective dispersion towards surface waters	All contaminants	Reduction of bank erosion/wash-out	Introduction of plants
Reduction of dispersion towards ground water	All contaminants	Increased hydrological resistance	Application of a clay screen

nondestructive, "intrinsic" bonding mechanisms and their temporal development have so far found much less recognition compared to destructive processes such as biological degradation (Förstner and Gerth 2001). Yet these so-called "diagenetic" effects, which apart from chemical processes involve an enhanced mechanical consolidation of soil and sediment components by compaction, loss of water and mineral precipitations in the pore space, may induce a quite essential reduction of the reactivity of solid matrices (Table 8.4). Inclusion of these "aging" processes will provide more realistic estimation of risks and may, therefore, constitute a significant factor for saving remediation costs (Chen et al. 2000).

Regarding the chemical processes, sediments are heterogeneous at various sample, aggregate, and particle scales. Adherent or entrapped nonaqueous-phase liquids and combustion residue particulate carbon (e.g., chars, soot, and ashes) can also function as sorbents. Complex assemblages of these constituents can cause complex mass transfer phenomena, and the term "sequestration" refers to some combination of diffusion limitation, adsorption, and partitioning (Luthy et al. 1997). Some geosorbents exhibit typical nonlinear sorption behavior (Farrell and Reinhard 1994; Huang and Weber 1998).

For inorganic pollutants (mainly heavy metals and arsenic), the effect of aging predominantly consists of an enhanced retention via processes such as sorption, precipitation, coprecipitation, occlusion, and incorporation into reservoir minerals. During his investigations on the early diagenetic stages of sediments from the Rhine river, Salomons (1980) found that the proportion of cadmium which was not desorbed with sodium chloride solution (in seawater concentration), increased from 24% after 1 day to 40% after 60 days of contact time between sediment and metal solution. Sediment samples from the river barrage of Vallabreques/Rhône, which had been contaminated by artificial radionuclides from the nearby reprocessing plant, exhibit characteristic differences with respect to the extractability of geogenic and anthropogenic manganese isotopes in the reductive elution step (Förstner and Schoer 1984).

Experiments with lead and cadmium on sediment samples from the oxidized surface layer of mudflats in the South San Francisco Bay estuary, where the reaction systems were equilibrated for 24 hours at the appropriate pH for approximately 90% metal adsorption as determined by prior experiments, indicate slow release of adsorbed cadmium within a time frame of 96 hours, whereas lead was substantially non-labile

Tab. 8.4: Demobilization of pollutants in solid matrices by natural factors

Cause (Example)	Effect
Compaction	*Reduction of Matrix...*
Consolidation	Erodibility
Phytostabilization (plant roots)	Permeability
Penetration into dead-end-pores	Reactivity
Interlayer collapse of clay minerals	*Reduced Pollutant...*
Coprecipitation (high-energy sites)	Mobility
Occlusion and overcoating	Availability
Absorption/diffusion	Toxicity
"Diagenesis"	**"Natural Attenuation"**

over the 264-hour duration of the experiment (Lion et al. 1982). It was suggested that the proportion of solid organic matter constitutes the main cause for the observed irreversibility of metal sorption; this was confirmed experimentally on selected materials for copper, and – less distinctly – for nickel and cadmium (Förstner 1987). With regard to the increased fixation of zinc, nickel, cadmium and arsenic in contact with various soil constituents, among other processes long-term diffusion into the crystal lattice of goethite has been suggested by Gerth et al. (1993).

Geochemical influences on assimilation of sediment-bound metals have been evaluated by Griscom et al. (2000) in a series of experiments using suspension-feeding mussel *Mytilus edulis* and facultative deposit feeder *Macoma balthica*. Oxidized and reduced radiolabeled sediments were fed to the animals and the assimilation efficiencies (AEs) of ingested metals were determined. For oxic sediment, Cd and Co AEs in *M. edulis* decreased 3- to 4-fold with increased sediment exposure time to the metals, with smaller but significant effect also noted for Zn and Se but not Ag. Sequential extractions of the oxidized sediments showed a transfer of metals into more resistant sediment components over time, but the rate did not correlate with a decrease in metal assimilation efficiencies. The results imply that metals associated with sulfides and anoxic sediments are bioavailable, that the bioavailability of metals from sediments decreases over exposure time, that organic carbon content generally has a small effect on AEs, and that AEs of sediment-bound metals differ among species (Griscom et al. 2000).

Characterization of long-term reactivity and bioavailability of heavy metals in sediments can be performed by: (i) acid-producing potential (APP; Kersten and Förstner 1991); (ii) relationship of acid-volatile sulfide (AVS) and simultaneously extractable metals (SEM; DiToro et al. 1992); (iii) redox buffer capacities (Heron and Christensen 1995); (iv) formation of metal hydroxide surface precipitates using molecular-scale techniques (Roberts et al. 1999; Thompson et al. 1999); and (v) by microbial biosensors (Corbisier et al. 1999; Vangronsveld et al. 2000; Reid et al. 2000, in combination with an extraction procedure).

8.3.3
Subaqueous Depot and Capping

Under anoxic, strongly reducing conditions a great part of the metal content in contaminated sediments is present as practically insoluble (compared to carbonates, oxides and phosphates) sulfides. Such conditions can be provided by a permanent water cover, whereby diffusion of oxygen into the sediment is inhibited. However, it must be taken into consideration that changes in the redox-regime can be induced not only by diffusive transport of oxygen through the water-to-sediment interface, but also – and maybe more effectively – by bottom dwelling and burrowing organisms creating oxidizing microenvironments. The risk of contaminant uptake by these organisms must thus be ruled out by appropriate cap designs. Major emphasis has to be laid on the development of reactive cap additives to reduce pollutant transfer from sediment through porewater into the open water (Jacobs and Förstner 1999) and the monitoring of subaqueous depots with active barrier systems using dialysis sampler (Jacobs 2002) and diffusional gradient technique (DGT) probes (Jacobs 2003).

Cap additives have to meet a number of prerequisites such as good retention potential, chemical and physical properties suited for an underwater application, low contamination, and low cost. Some of the properties listed in Table 8.5 may be altered by appropriate treatment of the material. For example, surfaces of clays and zeolites can be modified for an enhanced sorption of organic and anionic contaminants. Fine-grained materials, such as clays or red mud, which would rather form a hydraulic barrier than a reactive, permeable one, may be granulated. However, this pretreatment may raise the capital costs. Fortunately, natural microporous materials, and in particular natural zeolites, show highly favorable chemical and physical properties with respect to their application in subaqueous capping projects along with a worldwide availability at relatively low cost.

8.4
Managing Contaminated Sediments: A Holistic Approach

The European Water Framework Directive (EU-WFD) wants to "establish a framework for the protection of inland surface waters, transitional water and groundwater" and aims at "achieving good ecological potential and good surface water chemical status at the latest 15 years from the date of entry into force of this Directive", i.e. 22 December 2000, by a combined approach using emission and pollutant standards. These consider priority pollutants from diffuse and point sources, but neglect the role of sediments as a long-term secondary sources of contamination. Such a lack of information may easily lead to unreliable risk analyses with respect to the (pretended) 'good status' (Förstner 2002).

Currently, and based on existing and upcoming regulations, there will remain an urgent need for harmonization of sediment quality targets/criteria in river catchments and those for the relocation of river sediments to the marine environment. The river catchment and the coastal zone should be treated as a continuum/single system (Gandrass and Salomons 2001). With regard to emission control at the European level, and in particular in the implementation of the EU-WFD, there are some concerns:

● The designation of water bodies as "heavily modified" results in less stringent water quality control and consequently

Tab. 8.5: Examples for potential reactive materials for active barrier systems (Jacobs and Förstner 2001)

Material	Contaminant retention	Physical/chemical suitability	Environmental accepta-bility	Availability/costs
Industrial by-products				
Fly ash	Metals	+/− (very fine-grained)	−(high equilibrium pH, potential toxicity)	+
Red mud	Metals	+/−(very fine-grained, not stable under reducing conditions)	(heavy metals)	+
Natural minerals and rocks				
Calcite	Metals, nutrients	+	+	+
Apatite	Metals	+	+	+/−
Clays (e.g., bentonite)	Metals	+/−(very fine-grained)	+	+
Zeolites (e.g., clinoptilolite)	Metals	+	+	+

in contaminated sediments. From a dredged material management point of view and for the protection of the marine environment, this qualification should be interpreted narrowly and strictly.

- The emission approach should be strengthened, the EU-WFD should not be minimized to an immission approach directive.
- At the international level, regulations concerning dredged material and contaminated sediments are part of the Oslo and Paris (OSPAR), the Helsinki and the London Convention. The OSPAR Commission is of importance in setting guidelines for the disposal of dredged material in the marine environment, which are reflected in national criteria for disposal. It is expected that OSPAR will continue in this role. The feed-back system from contaminated sediments to the reduction of emission needs to be improved.

Some recommendations for sediment assessment frameworks of environmental quality in the European Union have been addressed by Apitz and Power (2002). Keys to success which might be used to design a goal-oriented EU sediment framework are:

1. To develop guidance as a series of building blocks.
2. To assure that decision-making is transparent and somewhat standardized, but flexible enough to meet regional goals.
3. To build both natural and regional background concentrations, reference sites and site-specific bioavailability considerations.
4. Wherever possible, assure that source control is a primary requirement before other management strategies are applied.
5. In most cases, sediment guidelines should not be used as pass/fail values,

but rather as triggers for further investigations.

One point, in particular, is stressed by Apitz and Power (2002): "Consider [the] entire life-cycle of sediments and their associated contaminants with a catchment, from source to ultimate sink!"

Contamination of river sediments has so far been discussed in relation to harbor sediments (Rotterdam, Hamburg) and to the impact of pollution on coastal ecosystems. In fact, remediation and storage of contaminated dredged materials is a key issue at harbor sites. However, remediation techniques on contaminated sediments are generally much more limited than for most other solid waste materials, except for mine wastes. As a conclusion of the remediation aspects discussed above, it can be stated that the concept of reactive barriers as a general approach applies, as well as autochthonous sediment sites, as disposal sites for dredged materials (see Section 8.3.3). This results, first of all, from the economic advantages which are characteristic of passive technologies. Due to the efficiency in isolating the contaminants from the environment along with the greatly reduced or zero process costs, these technologies always represent attractive remediation alternatives, where they are technically feasible and where they conform with the legislation. However, to achieve public acceptance as to the new technology, major efforts should be undertaken in respect to the development and application of monitoring systems for long-term prognoses of both mechanical and chemical stability in the new sediment deposit.

In the upper and middle course of river systems, sediments are affected by contamination sources such as wastewater, mine water from flooded mines and atmospheric deposition. Sediments are intermittently mobilized and deposited. During floods, sediment-bound pollutants can undergo a large-scale dispersion of contaminants in flood-plains, dyke foreshores and polder areas. The complex mixtures of toxic compounds and the dimension of pollution often preclude technical measures such as chemical extraction or solidification of contaminated soil material. Instead, alternative measures have to be taken considering the different local factors such as soil, sediment and water quality, flow velocity, and the dynamics of the water level. The measures implemented should be flexible and easy to adjust to changing conditions (Förstner 2003). The Spittelwasser example (see Section 8.3.2) indicated that, unlike problems related to conventional polluted sites, the hazards here are primarily connected with the transport and deposition of contaminated solids in a catchment area, especially in downstream regions. Any problem solution strategy for such sites, therefore, has to consider both the chemical stabilization – for example, by processes of (enhanced) natural attenuation – *and* an increase in mechanical stability (reduced erodibility).

The requirements for a river basin-wide sediment concept will be even more challenging than the actual Water Framework Directive. It will include inventories of interim depots within the catchment area (underground and surficial mining residues, river-dams, lock-reservoirs), integrated studies on hydromechanical, biological and geochemistry processes, risk assessments on sedimentary biocoenoses, and last – but not least – the development of decision tools for sustainable technical measures on a river basin scale, including sediment aspects.

References

ANONYMOUS (1994) *Assessment and Remediation of Contaminated Sediments (ARCS) Program – Remediation Guidance Document.* EPA 905-R94-003. 332 S. United States Environmental Protection Agency. Great Lakes National Program Office, Chicago 1994.

ANONYMOUS (2000) *Contaminated Soil 2000 – Case Study "Comparison of Solutions for a Large Contamination Based on Different National Policies".* ConSoil 2000, 7th Intern FZK/TNO Conference on Contaminated Soil. Leipzig, 165 p.

APITZ SE and POWER EA (2002) *From risk assessment to sediment management – an international perspective.* J Soils Sediments 2:61–66.

CHEN W, KAN AT, TOMSON MB (2000) *Irreversible adsorption of chlorinated benzenes to natural sediments: implications for sediment quality criteria.* Environ Sci Technol 34:385–392.

CORBISIER P, VAN DER LELIE D, BORREMANS B, PROVOOST A, DE LORENZO V, BROWN N, LLOYD J, HOBMAN J, CSÖREGI E, JOHANNSSON G and MATTIASSON B (1999) *Whole cell- and protein-based biosensors for the detection of bioavailable heavy metals in environmental samples.* Anal Chim Acta 387:235–244.

DICKSON W (1980) *Properties of acidified waters.* In: Drablos D, Tollan A, eds. Ecological Impact of Acid Precipitation, pp. 75–83. SNSF-Project, Oslo-Aas.

DITORO DM, MAHONY JD, HANSEN DJ, SCOTT KJ, CARLSON AR and ANKLEY GT (1992) *Acid volatile sulfide predicts the acute toxicity of cadmium and nickel in sediments.* Environ Sci Technol 26:96–101.

EMERSON S, JAHNKE R and HEGGIE D (1984) *Sediment–water exchange in shallow water estuarine sediments.* J Mar Res 42:709–730.

FAGERSTRÖM T and JERNELÖV A (1972) *Aspects of the quantitative ecology of mercury.* Water Res 6:1193–1202.

FARRELL J and REINHARD M (1994) *Desorption of halogenated organics from model solids, sediments, and soil under unsaturated conditions. 1. Isotherms.* Environ Sci Technol 28:53–62.

FÖRSTNER U (1987) *Changes in metal mobilities in aquatic and terrestrial cycles.* In: Patterson JW, Passino R, eds. Metal Speciation, Separation and Recovery, pp. 3–26. Lewis Publ. Chelsea, Michigan.

Förstner U (2002) *Sediments and the European Water Framework Directive.* J Soils Sediments 2:54.

Förstner U (2003) *Geochemical techniques on contaminated sediments – river basin view.* Environ Sci Pollut Res **10**:58–68.

Förstner U and Gerth J (2001) *Natural attenuation – non-destructive processes.* In: Stegmann R, Brunner G, Calmano W, Matz G, eds. Treatment of Contaminated Soil – Fundamentals, Analysis, Applications, pp. 567–586. Springer-Verlag, Berlin-Heidelberg-New York.

Förstner U and Schoer J (1984) *Diagenesis of chemical associations of Cs-137 and other artificial radionuclides in river sediments.* Environ Technol Lett **5**:295–306.

Förstner U, Wittmann U, Gier S et al. (2000) *Case comparison Bitterfeld – German contribution.* Compiled by a temporal working group for the 7th international FZK/TNO-Conference on Contaminated Soil, Leipzig/Germany, September 18–22, 2000.

Gandrass J, Salomons W, eds. (2001) *Dredged Material in the Port of Rotterdam – Interface between Rhine Catchment Area and North Sea.* (Present and future quality of sediments in the Rhine catchment area; current and future policies and regulatory framework; substances and new criteria to watch dredged material in relation to the North Sea). POR II Project Report, GKSS Research Centre Geesthacht, 28 February 2001, 342 p.

Gendron A, Silverberg N, Sundby B and Lebel V (1986) *Early diagenesis of cadmium and cobalt in sediments of the Laurentian Trough.* Geochim Cosmochim Acta **50**:741–747.

Gerth J, Brümmer GW and Tiller KG (1993) *Retention of Ni, Zn and Cd by Si-associated goethite.* Z Pflanzenernähr Bodenk **156**:123–129.

Griscom SB, Fisher NS and Luoma SN (2000) *Geochemical influences on assimilation of sediment-bound metals in clams and mussels.* Environ Sci Technol **34**:91–99.

Heron G and Christensen TH (1995) *Impact of sediment-bound iron on redox buffering in a landfill leachate polluted aquifer (Vejen, Denmark).* Environ Sci Technol **29**:187–192.

Hesterberg D, Stigliani WM and Imeson AC (1992) *Chemical Time Bombs: Linkage to Scenarios of Socioeconomic Development.* Executive Report 20 (CTB Basic Document 2), IIASA, Laxenburg/Austria.

Huang W and Weber WJ Jr (1998) *A distributed reactivity model for sorption by soils and sediments.* 11. *Slow concentration-dependent sorption rates.* Environ Sci Technol **32**:3549–3555.

Hunt CD and Smith DL (1983) *Remobilization of metals from polluted marine sediments.* Can J Fish Aquat Sci **40**:132–142.

Jacobs PH (2002) *A new type of rechargeable dialysis pore water sampler for monitoring in-situ sediment caps.* Water Res **36**:3111–3119.

Jacobs PH (2003) *Monitoring of subaqueous depots with active barrier systems (SUBAD-ABS) for contaminated dredged material using dialysis sampler and DGT probes.* J Soils Sediments **3**:100–107.

Jacobs PH and Förstner U (1999) *The concept of sub-aqueous in-situ capping of contaminated sediments with active barrier systems (ABS) using natural and modified zeolites.* Water Res **33**:2083–2087.

Jacobs PH and Förstner U (2001) *Managing contaminated sediments. IV: Subaqueous storage and capping of dredged material.* J Soils Sediments **1**:205–212.

Joziasse J and van der Gun J (2000) *In-situ remediation of contaminated sediments: Conceivable and feasible?! In: Contaminated Soil 2000,* Vol 1, pp. 516–522. Thomas Telford, London.

Kersten M (1989) *Mechanismen und Bilanz der Schwermetallfreisetzung aus einem Süßwasserwatt der Elbe.* Dissertation Technical University of Hamburg-Harburg, 122 p.

Kersten M and Förstner U (1991) *Geochemical characterization of the potential trace metal mobility in cohesive sediment.* Geo-Marine Lett **11**:184–187.

Lion LW, Altman RS and Leckie JO (1982) *Trace metal adsorption characteristics of estuarine particulate matter: Evaluation of contribution of Fe/Mn oxide and organic surface coatings.* Environ Sci Technol **16**:660–666.

Luthy RG, Aiken GR, Brusseau ML, Cunningham SD, Gschwend PM, Pignatello JJ, Reinhard M, Traina SJ, Weber WJ Jr and Westall JC (1997) *Sequestration of hydrophobic organic contaminants by geosorbents.* Environ Sci Technol **31**:3341–3347.

Maass B, Miehlich G and Gröngröft A (1985) *Untersuchungen zur Grundwassergefährdung durch Hafenschlick-Spülfelder. II. Inhaltsstoffe in Spülfeldsedimenten und Porenwässern.* Mitt Dtsch Bodenkundl Ges **43/I**:253–258.

Reid BJ, Paton GI, Bundy JG, Jones KC and Semple KT (2000) *Determination of soil-associated organic contaminant bioavailability using a novel extraction procedure in conjunction with lux-marked microbial biosensors.* In: Contaminated

Soil 2000, Vol 2, pp. 870–871.Thomas Telford, London.

ROBERTS DR, SCHEIDEGGER AM and SPARKS DL (1999) *Kinetics of mixed Ni-Al precipitate formation on a soil clay fraction.* Environ Sci Technol **33**:3749–3754.

SALOMONS W (1980) *Adsorption processes and hydrodynamic conditions in estuaries.* Environ Technol Lett **1**:356–365.

SALOMONS W (1983) *Trace metals in the Rhine, their past and present (1920–1983) influence on aquatic and terrestrial ecosystems.* In: Proceedings International Conference Heavy Metals in the Environment, Heidelberg, September 6–9, pp. 764–771. CEP Consultants, Edinburgh.

SALOMONS W and FÖRSTNER U (1984) *Metals in the Hydrocycle.* Springer-Verlag Berlin.

SALOMONS W and FÖRSTNER U, eds. (1988) *Environmental Management of Solid Waste – Dredged Material and Mine Tailings.* Springer-Verlag, Berlin-Heidelberg-New York.

SALOMONS W and MOOK WG (1980) *Biogeochemical processes affecting metal concentrations in lake sediments (Ijsselmeer, The Netherlands).* Sci Total Environ **16**:217–229.

STIGLIANI WM (1988) *Changes in valued "capacities" of soils and sediments as indicators of non-linear and time-delayed environmental effects.* Environ Monit Assess **10**:245–307.

THOMPSON HA, PARKS GA and BROWN GE JR (1999) *Dynamic interaction of dissolution, surface adsorption and precipitation in an aging cobalt(II)-clay-water system.* Geochim Cosmochim Acta **63**:1767–1779.

VANGRONSVELD J, SPELMANS N, CLIJSTERS H, ADRIAENSENS R, CARLEER R, VAN POUCKE D, VAN DER LELIE D, MERGEAY M, CORBISIER P, BIERKENS J and DIELS L (2000) *Physico-chemical and biological evaluation of the efficacy of in situ metal inactivation in contaminated soils.* In: Contaminated Soil 2000, Vol 2, pp. 1155–1156. Thomas Telford, London.

9
Elements and Compounds in Waste Materials

Ulrich Förstner

9.1
Introduction

The widespread use of metals for different kinds of application (e.g., pigments, coatings, alloys, electronic equipment) leads to the fact that some of the utilized metals (or their compounds) end up in wastes. Metals in wastes can cause severe environmental impacts, particularly with respect to groundwater pollution.

Metal-containing waste materials include municipal solid wastes, industrial byproducts, sewage sludge, dredged material, wastes from mining and smelting operations, filter residues from waste water treatment and atmospheric emission control, ashes and slags from burning of coal and oil, and from incineration of municipal refuse and sewage sludge. All of these wastes pose most challenging problems which are not only technical in nature but also require attention to be paid with regard to social and financial aspects.

Current data on the extent of anthropogenic waste generation compared with the generation and transportation of natural solids are shown in Table 9.1. According to existing estimates, the generated quantities of municipal wastes and dredge spoils are about 1×10^9 m^3 per year, whilst sewage

Tab. 9.1: Global waste balances and comparative data (Neumann-Malkau 1991)

Household waste	$\sim 1 \times 10^9$ m^3 year^{-1}
Dredged material	$\sim 1 \times 10^9$ m^3 year^{-1}
Sewage sludge (95% H$_2$O)	$\sim 3 \times 10^9$ m^3 year^{-1}
Mining waste	$\sim 17.8 \times 10^9$ m^3 year^{-1}
(Sediment transport	$\sim 26.7 \times 10^9$ m^3 year^{-1})

sludge (with a water content of about 95%) amounts to about 3×10^9 m^3 (Baccini and Brunner 1991). The generated quantities of mining wastes at about 20×10^9 m^3, is the same magnitude as the actual erosion rate of soil and rock (Neumann-Malkau 1991).

When it comes to mining waste, the following fact should be noted: for every new car weighing 1 metric ton there are about 25 tons of waste generated. As existing resources are consumed, there is a move toward exploiting ore deposits with lower yields; that is, "per unit of metal ever larger quantities of material have to be mined and moved" (Anonymous 1983). To a significant degree, this inevitable consequence defeats the advantages associated with the more intensive use of raw materials (Schenkel and Reiche 1994). It has been estimated that the quantity of mining waste doubles every 20 to 25 years.

Using copper mining as an example (Figure 9.1), Sutter (1991) has demonstrated

Elements and their Compounds in the Environment. 2nd Edition.
Edited by E. Merian, M. Anke, M. Ihnat, M. Stoeppler
Copyright © 2004 WILEY-VCH Verlag GmbH & Co. KGaA, Weinheim
ISBN: 3-527-30459-2

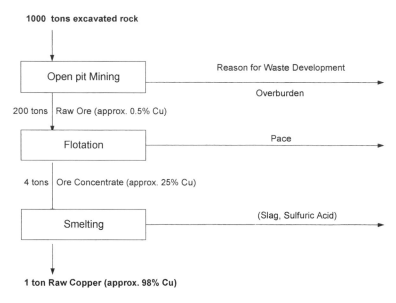

Fig. 9.1   Natural resource extraction, concentration, and residue generation. Example: raw copper (after Sutter (1991)

the relationship between raw material extraction and waste generation. In addition to the presence of host rock, sulfide copper ores contain additional impurities and auxiliary materials – for example, iron, nickel, lead, zinc, arsenic, antimony, and the rarer elements selenium, tellurium, bismuth, silver, gold, and platinum. These minerals also add to the mining waste stream and are released as residues throughout the various production stages. The latter minerals often serve as the input material for the extraction of the above metals. In the first step of strip mining, from 1000 tons of rock, sulfide copper ores with a copper content of about 0.5% are extracted and this generates about 800 tons of waste. In the next stage, almost all of the ore is processed by flotation. The end result of flotation is that the ore-forming minerals are separated from the gangue (rock) and the metal is concentrated to about 25%. The processing usually takes place near the mine. The dewatered concentrates containing a fraction of the raw ore weight are then transported for smelting to copper smelters where the sulfur is removed in the first process step (calcination). For each ton of copper in sulfide concentration, approximately 1 ton of sulfur is generated. The removal of such large quantities of sulfur can be accomplished efficiently by using recovery technologies. Consequently, the nonferrous metal smelters become significant producers of sulfur products.

9.2
Waste Composition and Impact on Adjacent Media

9.2.1
Household Waste and Industrial Waste

Waste is generally divided into the categories of solid waste and hazardous waste. The category of solid waste includes (Bilitewski et al. 1997):

- household waste,
- household-like waste from commercial and public institutions, etc.,
- bulky waste,
- street sweepings.

In other words, solid waste is generally what is collected by the municipal waste collection departments. The complex area of industrial and commercial waste has been subdivided by Bilitewski et al. (1997) as shown in Figure 9.2.

Industrial and commercial waste not only includes waste that can be disposed together with municipal waste in waste incinerators or deposited in landfills, but also waste that requires special treatment. This group of "wastes requiring special oversight" contains substances that are dangerous to human health, to the air, and water, which are explosive or combustible, or which contain transmitters of contagious diseases ("hazardous substances").

On average and in addition to other components, household waste contains about 30% organic matter, 12% paper, 9% glass, and 3% ferrous and nonferrous metals. It is estimated that the quantity of waste containing problematic substances is about 1.2 to 1.5 kg/person/year; this includes, for example, batteries, fluorescent light tubes, and paints and lacquers which contain a variety of heavy metals.

Some quantitative data regarding the contaminant concentrations in household waste fractions are available. In the paper and cardboard fraction, the quantities of heavy metals are relatively low compared to the total contamination by heavy metals in the total municipal waste stream. Newspapers, for example, contain very low concentrations of heavy metals compared to magazines, the color print of which contains heavy metals. Data from the National Household Waste Analysis conducted in Germany are listed in Table 9.2. The data reflect results that the separate collection of recyclables has on the concentrations of heavy metals and halogens in individual recyclable fractions and in the remaining waste steam. Before being mixed with the other waste, the organic fraction of household waste is as low as the fraction of

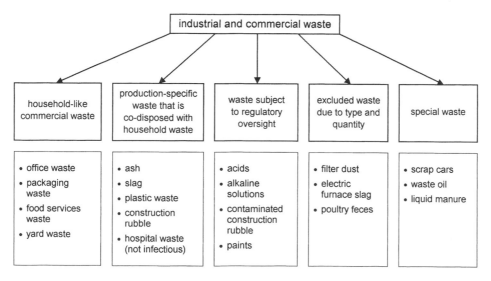

Fig. 9.2 Classification of industrial and commercial waste (Förstner 1998a after Bilitewski et al. (1997)

Tab. 9.2: Specific reduction of heavy metals from household waste by separate collection (reference value 230 kg per person per year). (From Bilitewski et al. 1997)

Material (proportion, %) collected	Share of total waste stream [%]					
	Cd	*Pb*	*Zn*	*Cu*	*Cr*	*Ni*
Paper/cardboard (15.2)	2.9	3.4	3.3	33.0	4.4	12.7
Plastics (5)	71.4	2.9	6.5	12.7	1.8	7.3
Biowaste (37)	2.9	2.7	8.6	14.9	5.7	34.7
Metals/glass (10.5)	11.4	19.3	19.1	25.5	64.2	37.3
Total (67.7)	88.6	28.3	37.5	86.1	76.1	92.0

heavy metals in food items. However, if the organic fraction is separated after it has been mixed with other waste, it contains high concentrations of heavy metals. After retroactive separation, 56% of the copper, 55% of the lead, 40% of the zinc, and 35% of the cadmium was found in the organic fraction.

9.2.2
Mining Waste

During the five decades since the Second World War, more metallic raw materials have been produced than over the entire history of mankind. While the Earth's population has almost doubled between 1959 and 1990, the advancement of the six most important base metals (aluminum, lead, copper, nickel, zinc, tin) has increased more than eight-fold. Contrary to the prognoses from the period between 1950 and the mid-1980s, no substantial bottlenecks have arisen with most ores. An important factor for the continuous availability of metallic raw materials however, is the strong increase in environmental awareness in the past three decades mainly in the industrialized nations, and to a certain extent in the developing countries (Hodges 1995).

Industrial mining and ore smelting were originally known to be among the most important causes of contamination. Since the Middle Ages, "when the mining industry in Europe began" and up until today, not much has changed with respect to deep mining and ore extraction (first comprehensively described by George Agricola in his book *De Re Metallica* published in 1556); an example is the primitive use of mercury during the current gold production in Brazil (Förstner 1998b). Today, the major problems lie with the wastes from sulfide-containing ores. Waste deposits of such nature have rested over geological times practically unchanged under overlying rocks. When they came into contact with oxygen however, acid began to develop and the acidic leachate dissolved heavy metals from the ore and the surrounding rocks.

The largest research project in the area of modeling and characterizing geochemical processes in settling plants and in the groundwater (affected by acidic leachate), was the Canadian Mine Environment Neutral Drainage (MEND) program (1988 to 1997). Based on the Canadian experience, the project 'Mitigation of the Environmental Impact of Mining Waste' was also started collaterally in Sweden by six universities together with the mining industry under financial support from the "MISTRA" governmental program.

One of the most important aims of research in this area is the development of strategies for an effective combination of different precaution and safeguard methods. These include the use of physical and hydraulic barrier systems, conceptualization of geochemical techniques, and the investigation of suitable biochemical conditions for reducing the emissions in the surroundings of partly operating mines (Paul 2003).

9.2.3
Impact on Adjacent Media

Waste disposal and mining activities are characteristic point sources for environmental pollutants in air, water, and soil (Table 9.3). Waste incineration typically releases more volatile metals such as mercury, cadmium and lead into the atmosphere; emission control, therefore, is not only crucial for smelting activities, but also the decisive prerequisite for all technologies involving higher temperature and waste materials. Landfill leachates are enriched in metals such as boron, arsenic, and cadmium; even higher concentrations of metals are found in mine effluents, particularly from acid mine tailings. With regard to the latter sources, transfer of metals into the soil involves several mechanisms such as wind erosion, river dredging, and flooding events. Apart from seasonal flooding of polder areas and flood plains, there have been catastrophic floodings in the recent past due to extreme rainfall and the failure of dams. Events such as the breaking of tailing dams in highly contaminated areas such as mining districts (e.g. Aznalcollar/Guadiamar River, Spain in 1998 and Baia Mare/Tizla River, Romania in 2000) caused considerable immediate hazards from metals (Förstner 2003a).

One characteristic source of metals in waste materials is that of the metallurgical industry (with potential target media). These include: aerosols/particulates con-

Tab. 9.3: Waste disposal and metalliferous mining and smelting as major sources of pollutant immissions into air, water and soil environments. (From Alloway and Ayres 1993.)

Waste disposal

Air:	Incineration – fumes, aerosols particulates (Cd, Hg, P, CO_2, NO_x, PCDDs, PCDFs, PAHs)
	Landfills – CH_4, VOCs
	Livestock farming wastes – CH_4, NH_3, H_2S
	Scrapyards – combustion of plastics (PAHs, PCDDs, PCDFs)
Water:	Landfill leachates, NO_3^-, NH_4^+, Cd, PCBs, microorganisms
	Effluents from water treatment – organic matter, HPO_4^{2-}, NO_3^-, NH_4^+
Soil:	Sewage sludge – NH_4^+, PAHs, PCBs, metals (Cd, Cr, Cu, Hg, Mn, Mo, Ni, Pb, V, Zn, etc.)
	Scrapheaps – Cd, Cr, Cu, Ni, Pb, Zn, Mn, V, W, PAHs, PCBs
	Bonfires, coal ash, etc. – Cu, Pb, PAHs, B, As
	Fallout from waste incinerators – Cd, PCDFs, PCBs, PAHs
	Fly tipping of industrial wastes – wide range of substances
	Landfill leachate – NO_3^-, NH_4^+, Cd, PCBs, microorganisms

Mining and smelting

Air:	SO_2, Pb, Cd, As, Hg, Ni, Tl, etc. particulates/aerosols
Water:	SO_4^{2-}, CN frothing agents, metal ions, tailings (ore minerals, e.g., PbS, ZnS, $CuFeS_2$)
Soil:	Spoil and tailings heaps – wind erosion, weathering ore particles
	Fluvially dispersed tailings – deposited on soil during flooding, river dredging, etc.
	Transported ore separates – blown from conveyance, etc. onto soil
	Ore processing – cyanides, range of metals
	Smelting – wind-blown dust, aerosols from smelter (range of metals)

taining As, Cd, Cr, Cu, Mn, Ni, Pb, Sb, Tl and Zn (air); metal ions and acid wastes from metal cleaning (water); metals in wastes, solvents, acid residues, fallout from aerosols, etc. from casting; and other pyrometallurgical processes (soil).

9.3
Waste Treatment

For bulk waste with moderately high levels of contamination, a treatment method should be used where contaminated components are pre-concentrated at the lowest possible cost. These components can then be further treated with more costly methods or they can be safely disposed of in smaller quantities. The operational principle of the Dutch research facility "TNO" is shown in Figure 9.3 (after Van Gemert et al. 1988): The "A" technologies are used on a large scale with low unit costs and relatively high flexibility with regard to changes in circumstances; the facilities can, if necessary, be made mobile or can be moved. The "B"

technologies, on the other hand, are designed for treating smaller mass streams with higher contaminant concentrations; the treatment costs per unit are higher, the technical equipment is more complex, and the demands put upon personnel are relatively high. Such facilities are generally stationary.

9.3.1
Chemical-Physical Treatment of Industrial Waste

A central aspect of pollution control technology is the treatment of solid, liquid, and gaseous waste products from industrial production. The wide variety of the hazardous wastes to be treated with chemical-physical (CP) methods ranges from relatively harmless substances to highly dangerous environmental contaminants; this is also reflected in the design of the facility.

Hazardous waste designated for CP treatment is directed toward either an inorganic or an organic treatment process; accordingly, the facility must first possess two treatment lines for wastes containing

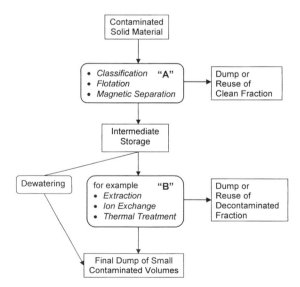

Fig. 9.3 Schematics of a treatment facility for solid waste materials (Van Gemert et al. 1988)

either organic or inorganic contaminants. Another distinction includes (Anonymous 1991):

- *detoxification methods*, which convert the environmentally harmful components of hazardous waste into more environmentally compatible components and combine or dilute it with large volumes of sludge; and
- *extraction methods*, which largely remove the undesirable substances from the primary components of the waste.

This distinction is important for waste that contains a high share of inert or useful substances (e.g., water); however, if only low concentrations of a generally safe matrix are present, then the waste as a whole must be converted – that is, either separated or burned.

In "inorganic treatment", steps such as neutralization, precipitation, flocculation, oxidation, reduction, and dewatering can be used selectively and in combination. In terms of treated quantities, there are predominantly sludge-forming reactions and these result in residues which must be disposed of. The major processes can be preceded by pretreatment steps such as homogenization, and it may also be necessary to include more thorough post-treatment, for example with activated carbon adsorption, air stripping, re-precipitation as sulfide, ion-exchange, or reverse osmosis. These facilities also have receiving and intermediate storage basins, chemical vats, reaction vessels (mono-functional, i.e., for only one treatment step, or multi-functional, for several treatment reactions), dewatering devices, post-treatment operations, wastewater containers, and flue gas scrubbers.

Standard "detoxification" processes are used for chromate-containing wastes which are generated during surface treatment and finishing in metal working and metal-plating facilities. These processes are also used for cyanide-containing wastes from case-hardening and electroplating, as well as nitrite-rich sludges, which are generated in large quantities during case-hardening (heat-treatment sludge), in machining (rust-proofing additives in the coolants) and during blackening (burnishing) of steel parts. Chromium can be precipitated as either a slightly soluble chromium sulfate (reaction with sulfur dioxide); alternately, after reduction of the chromate with sodium disulfide or iron(II) salts under acidic conditions, it is precipitated as a trivalent chromium hydroxide. Cyanide and nitrite are detoxified with oxidation reactions, for example with sodium hypochlorite.

9.3.2
Waste Incineration

The main goal of waste incineration is volume reduction. This reduction amounts to about 80% without slag recovery, and reaches 95% when slag is treated and reused. The weight reduction is approximately 60–70% by weight. Lately, efforts have been undertaken to operate these facilities as waste-to-energy power plants.

The long-term goals of research and development in the field of thermal waste treatment are as follows:

- the gaseous emissions must be reduced to an environmentally acceptable minimum;
- the solids must be treated in such a manner that they can be recovered or disposed of without adverse impacts on the environment;
- the residues, in quantities as small as feasible, contain the concentrated pollutants (primarily heavy metals) which can be recovered or disposed of underground as hazardous waste.

For every ton of municipal waste, about 250–300 kg of ash and slag is generated, 20–40 kg of filter particles and, depending on the flue gas scrubbing process employed, between 8 and 45 kg of reaction products, which is largest for dry sorption (Thomé-Kozmiensky 1989). Some of the residues are recovered, and some are landfilled. Highly contaminated filter particles, in particular, must be treated before they can be landfilled; in the future, an additional treatment step will also become necessary for ash and slag. For this reason, different methods such as flushing, sintering, or melting should be studied. Finally, wastewater is also generated in the incineration process and due to its particular composition, special treatment methods have yet to be developed.

9.3.2.1
Slags and Ashes

The quantity and individual components of the combustion products depend on the composition of the fuel, the firing system, and on the flue gas scrubbing system. Raw slag contains about:

- 3–5% unburned material,
- 7–10% ferrous and nonferrous metals (tin, copper, brass, aluminum, alloys, e.g., motors, etc.),
- 5–8% coarse material >32 mm (concrete, tile, stones, slag chunks, etc.),
- 80–83% fines <32 mm (as above, glass, stoneware, porcelain, etc.).

Fly ash is very fine; 90% of the mass has a grain size of 10 to 100 μm (Borchers et al. 1987). Chemically speaking, slag (bottom ash) and fly ash are composed of metal oxides and silicates, salts, chlorides and sulfates, and heavy metals such as zinc, lead, and cadmium. Especially the latter is highly concentrated in the fly ash (Table 9.4), and is relatively easy to extract.

9.3.2.2
Fly Ash Post-Treatment

The long-term goal of a low-emission, unmonitored disposal of residuals, a so-called "final disposal quality", can be achieved through scrubbing processes, by addition of immobilizing additives, and especially through vitrification or ceramization of flue gas scrubber residuals. When weighing these alternatives, in principle it appears advantageous that high temperature treatment leaves open the possibility for reusing these substances, for example as building materials. A survey of methods for solidifying residuals from waste incineration is provided by Faulstich and Zachäus (1992):

- *Treatment process*: separating metallic substances from the slag, and also, if necessary, sorting out unburned material and separating the residual slag for use in road construction.

- *Solidifying and scrubbing processes*: applying immobilizing additives such as hydraulic bonding agents (e.g., fly ash and cement) or clays (good adsorption capacity of clay minerals and low permeability of clays). Leaching methods in the sense of selective extraction or as pre-

Tab. 9.4: Concentration of typical heavy metals in slag and fly ash from waste incinerators. (From Baccini and Brunner 1985)

Element	Lithosphere [g kg⁻¹]	Slag [g kg⁻¹]	Slag CF	E-fly ash [g kg⁻¹]	E-fly ash CF
Zinc	0.07	4....15	140	13...39	370
Lead	0.0013	1....17	750	6...50	1200
Cadmium	0.0002	0.01...0.14	200	0.2...0.6	2000

CF: Concentration Factor compared to the element content of the lithosphere.

treatment steps preparing for subsequent treatment steps.

- Processes for *manufacturing construction materials*: producing directly usable materials from fly ash and reaction products. These processes are only used if actual reuse is taking place.
- *Low-temperature processes*: intended largely to destroy the organic contaminants in the fly ash.
- *Smelting processes*: intended to convert the residues and slag into a unleachable and usable product. The relatively small quantities of fly ash can be vitrified together with the fines in the slag.

9.3.2.3
Smelting Processes

An "inertization" of residues can be achieved by using smelting processes; several variations have be tested in pilot studies and some are already in use (Thomé-Kozmiensky 1994). The processes include: (1) plasma smelting; (2) glass vitrifying; (3) flame smelting; and (4) a smelting cyclone process. All of these processes have in common that their aluminum silicate glass products are tightly bonded to toxic elements. Depending on the process, residues such as ferrous and nonferrous metals are generated, as are stones, glass and ceramics, zinc-lead concentrate, mercury concentrate, and salts from flue gas scrubbing. Stones, glass, and ceramics are usable as filler material in dams and quarries, scrap iron can be directly marketed, and nonferrous scrap can be recycled after processing. Zinc-lead concentrates can be recovered from the waste gas of the smelting furnaces through targeted condensation and desublimation and can then be further processed in the metals industry (Means et al. 1995).

High-temperature smelting processes

The residues of thermal waste treatment are characterized by the fact that they consist of relatively environmentally harmless silicates and oxides. However, they are contaminated with heavy metals, and in part with organics, which may be harmful to the environment. Inertization, in the sense of extensively immobilizing heavy metals and destroying highly toxic chlorinated hydrocarbons, such as dioxins and furans, can be achieved through smelting processes. Similar to magmatic crystallization, it is possible to achieve a further separation of silicate, metal, and condensation products at very high smelting temperatures. The silicate phase of the RedMelt Process (Faulstich et al. 1992), where the treated slag and the added fly ash is fed into an electric arc furnace, is largely free of environmentally harmful heavy metals (see Table 9.5). During smelting, a metal product consisting of high-boiling point metals such as copper, chromium, nickel, and iron, is formed and removed from the bottom of the furnace and shipped for recycling. The condensate contains the highly volatile metals and a large share of the chlorine load. The high contents of zinc and lead thus suggest metallurgical processing in a nonferrous metal smelter.

Table 9.6 compares the residues from thermal waste treatment with respect to their long-term behavior. This assessment is usually made using leach tests; in Table 9.6, the Swiss Technical Rule limits for residuals requiring a leach strength of pH 4 were used (Anonymous 1990; see Section 9.5.3). Tests have shown that the untreated and even the scrubbed electrostatic precipitator (ESP) filter residues do not meet these limits, and even the zinc content of ESP filter residue solidified with cement is only slightly below the limit at which landfilling would be permitted at a

Tab. 9.5: Element content in fractions of the RedMelt process. (After Faulstich et al. 1992.)

Element	Input [mass-%]	Products [mass-%]		
		Silicate	*Metal*	*Condensate*
Silicon	22.0	26.2	3.4	6.4
Aluminum	5.5	6.8	0.2	0.5
Calcium	9.0	10.6	0.6	0.8
Sodium	3.9	4.2	0.08	13.7
Iron	10.1	4.5	85.0	1.4
Copper	0.3	0.03	4.4	0.3
Chromium	0.04	0.04	0.2	0.03
Nickel	0.01	0.006	0.3	0.001
Zinc	0.6	0.09	0.08	14.5
Lead	0.2	<0.01	<0.001	6.3
Cadmium	0.004	<0.00001	<0.001	0.1
Mercury	0.0001	$<$ 3E-6	<0.001	0.002
Chlorine	1.3	0.3	0.03	23.5

Tab. 9.6: Leachability of residues from waste incineration. (From Faulstich et al. 1992.)

	Lead [mg kg^{-1}]	Cadmium [mg kg^{-1}]	Zinc [mg kg^{-1}]	Copper [mg kg^{-1}]
Untreated ESP ash; WIF Oberhausen	4.2	4.5	133	1.7
Scrubbed ESP ash and wet scrubbing residue	0.77	0.94	57	0.26
TVA Limits for residue (pH = 4)	1	0.1	10	0.5
Cement-solidified ESP ash and NWR (g), BM/A = 1:2	0.14	0.08	5.1	0.05
ESP ash treated with 3R-Process	0.1	0.02	0.7	0.08
Vitrified residue from ABB-Process	<0.040	<0.030	0.11	<0.040
Silicate phase from RedMelt process	0.004	<0.001	<0.020	0.005

WIF: Waste Incineration Facility; TVA: Technical Rules-Waste (Anonymous 1990).

residual waste landfill. On the other hand, these limits can be met without a problem by more modern leaching processes (e.g., the 3R-Process) and smelting processes (e.g., ABB-Process). These already low elution values can be reduced by one magnitude with the RedMelt Process, and the leachate metals content is equivalent to the values found in natural rock such as granite. However, in this case products should be produced whose conventional manufacturing technology goes hand-in-hand with

high energy requirements – for example, mineral fibers which are also manufactured conventionally from natural rocks using smelting processes (Faulstich 1994).

9.3.2.4
Treatment of Wastewater from Waste Incineration Facilities (WIF)

Waste-specific wastewater is generated wherever water comes in contact with the combustion products during: (1) ash removal; (2) cleaning of the heat exchanger;

and (3) flue gas scrubbing (Jekel and Vater 1989).

The quantities of water from wet ash removal vary between 0.33 and 1 m³ per ton of waste. The scrubbing water serves to cool hot ashes and to seal the combustion chamber from the atmosphere. Easily soluble salts are largely leached out, but, the contaminant load of the scrubbing water remains considerable. Typical heavy metal concentrations measured in municipal WIF wastewater are listed in Table 9.7 (Reimann 1987).

Because of high pH values, some of the heavy metals have precipitated as hydroxides and can be separated mechanically. The wastewater from wet ash removal generally meets local codes without additional treatment and is thus considered nonproblematic.

In contrast to wet ash removal, small quantities of wastewater, heavily contaminated with heavy metals, are generated several times a year when heat-exchange surfaces are cleaned with high-pressure washers. The concentrations are usually clearly above local discharge limits so that the treatment with precipitation chemicals is required. The treatment can be accomplished together with the scrubbing liquids used during wet gas scrubbing.

In particular, the removal of mercury – which is contained in untreated wastewater in concentrations ranging from 3.3 to 11 mg L^{-1}, is rather insufficient and requires additional treatment. Generally, precipitation with inorganic sulfide or organic sulfur compounds, such as mercaptans (e.g., TMT-15; Reimann 1987), follows the neutralization of the wastewater and the separation of the heavy-metal hydroxides.

9.3.3
Stabilization of Wastes

9.3.3.1
Stabilizing Additives

In general, solidification/stabilization technology is considered a last approach to the management of hazardous wastes (Conner and Hoeffner 1998a,b). The aim of these techniques is a stronger fixation of contaminants to reduce the emission rate to the biosphere, and to retard exchange processes. Most of the stabilization techniques aimed at the immobilization of metal-containing wastes are based on additions of cement, water glass (alkali silicate), coal fly ash, lime or gypsum (Malone et al. 1982, Goumans et al. 1991).

Laboratory studies on the evaluation and efficiency of stabilization processes were performed by Calmano et al. (1988) on Hamburg harbor mud, using acid titration curves for limestone and cement/fly ash stabilizers. Best results are attained with calcium carbonate, since the pH-conditions are not changed significantly upon addition

Tab. 9.7: Metal concentrations in municipal Waste Incineration Facilities (WIF) wastewater (Reimann 1987; specific wastewater quantity: 0.35–0.40 m³ t^{-1} waste; sedimentation time: 2 h)

Metal	Concentration range	Mean concentration
Mercury	0.0004–0.21 mg L^{-1}	0.038 mg L^{-1}
Cadmium	< 0.01–0.66 mg L^{-1}	0.15 mg L^{-1}
Copper	0.1–1.0 mg L^{-1}	0.26 mg L^{-1}
Lead	0.20–3.2 mg L^{-1}	0.80 mg L^{-1}
Zinc	0.02–5.2 mg L^{-1}	1.8 mg L^{-1}

of $CaCO_3$. Generally, maintenance of a pH of neutrality or slightly beyond favors adsorption or precipitation of soluble metals (Gambrell et al. 1983). On the other hand, it can be expected that both low and high pH-values will have unfavorable effects on the mobility of heavy metals. In general, microscale methods (e.g., formation of mineral precipitates in the pore space of a waste body) will be employed rather than using large-scale enclosure systems such as clay covers or wall constructions (Wiles et al. 1988). An overview on various fields of environmental research and management to which mineralogical methods can be successfully applied has been given by Bambauer (1991). Before presenting the example of the stabilization of sludges from water purification, the potential use of minerals as both redox mediators and storage media will be indicated.

9.3.3.2
Storage Minerals

A new field of stabilization techniques uses the properties of certain minerals to incorporate and store critical components in their crystal lattice. For practical purposes, a special case of "mineralogical" speciation can be called a carrier mineral – for example, a crystalline compound, frequently of airborne origin, carrying a pollutant as a main or trace element (Bambauer and Pöllmann 1998). A carrier mineral may become a (storing-up) reservoir mineral, such as a mineral neoformed in a pollutant-containing milieu that is able to incorporate pollutants in its crystal structure during growth. The main properties of these mineral reservoirs after formation (e.g., in waste disposal) can be summarized as follows:
1. Highly stable or resistant and slightly soluble (low leachability) in the geochemical cycle of a given disposal and its geological environment.

2. Variable chemical composition (solid solution) to enable the incorporation of various hazardous chemicals.
3. Formation preferentially from the waste material itself (or by adding of minor admixtures).
4. Giving rise to the formation of a dense microstructure in the disposed material.

Therefore, reservoir minerals are expected to have a selective immobilization capacity for certain pollutants or trace elements. Mineral phases with special surface properties and amorphous materials may also be considered here (Table 9.8). Reservoir minerals may form in a wide range from high-temperature processes to temperatures of diagenesis.

Techniques involving reservoir minerals are mainly applied on industrial process residues. Pöllmann (1994) has demonstrated two different ways to the formation of a stabilizate, where pollutants can be stored for a long period of time. In the first case, primary reservoir minerals, which exhibit no hydraulic reaction, act as widely inert filling substances for hydraulic formations within the landfill. In the second case, a new generation of minerals, which also can incorporate pollutants from the water phase, fills the pore space. By using such condensation processes, water permeability is reduced, and consequently also the dispersion of pollutants by convection.

Experimental studies of the processes taking place with mixed residues from lignite coal incineration indicate favorable effects of incorporation of both chloride and heavy metals in newly formed minerals. Ettringite in particular, can act as a "storage mineral" for chloride and metal ions. The former may be incorporated at up to 4 kg $CaCl_2$ per m^3 of the mineral mixture. Calcium-silicate-hydrate phases may be formed in a subsequent process, and by filling further pore space these minerals can

Tab. 9.8: Examples of inorganic reservoir minerals for hazardous elements occurring in various waste materials. (From Bambauer and Pöllmann 1998.)

Compound	Hazardous element	Occurrence
I. High- to medium-temperature phases with isomorphic substitution		
Chloroellestadite $Ca_{10}Cl_2[SO_4,SiO_4]_3$ Apatite type	Zn, Cd, Pb, Sr, Ba, V, As, Se	Formation during thermal treatment of refuse incinerator flue gas purification residues
Alumosilicate glassy spheres	V, Cr, Mn, Co, Ni, Cu, Sr, Ba, etc.	Formation in fly ashes of coal-fired power plants
II. Low-temperature phases with isomorphic substitution		
Calcite $CaCO_3$	Mn, Co, Ni, Cu, Zn, Cd	Stabilized dredged sludge
Jarosite $KFe_3[(OH)_6SO_4]$	TI, As, Pb	Dumped pyrite calcines, mainly composed of hematite
Ca-monosulfo-aluminate-hydrate type compounds	Various anions (Cl^-, SO_4^{2-}); heavy metals: Cd, Cr, etc.	Stabilized residues of lignite-fired power plants and various other sludge-like waste materials
Ettringite-type: $[Ca_6Al_2(OH)_{12} \cdot 24H_2O]^{6+}$ $[(SO_4)_3 \cdot 2H_2O]^{6-}$	Various anions: heavy metals: Cr, Mn, Co, Ni, Zn, Sr, Pb, As	Stabilized residues of lignite-fired power plants and various other sludge-like waste materials

Substance	Application
III. Intracrystalline and surface sorption	
Bentonite	Additives for dump sealing materials and cements used for waste consolidation
Zeolites (natural, synthetic)	Pollutant-adsorbing additive to cements used for consolidation of industrial waste and for barrier systems
Calcium silicate hydrate	Known to adsorb chloride ions and heavy metals

significantly reduce permeability of the waste body for percolating solutions. Experimental studies of the leachability of salts and trace elements from samples of "stabilizates", with a pressure-filtration method, indicate relative high rates of release for sulfate ions, but not for zinc and cadmium in the eluate (Bambauer 1992).

9.4
Metals in Landfills

Under moderately wet climatic conditions, freshly dumped refuse has an average water content of approximately 30%. The storage capacity is much higher (450–600 L t^{-1} dry refuse = 39–46%), but

decreases with age of the landfill. Elemental composition is rather variable, and as a consequence of the inhomogeneous structure of refuse it is difficult to obtain comparable values (Ehrig 1989). Some analyses of bulk composition for selected trace metals are listed in Table 9.9.

9.4.1
Metals in "Reactor" Landfills

Subsequent to landfilling, the raw waste compounds undergo a variety of early diagenetic processes accompanying microbially mediated degradation of the organic compounds. The metabolic intermediates of organic matter decay (e.g., HCO_3^-, HPO_4^{2-}, carbohydrates and other low-

Tab. 9.9: Heavy metals (mg kg^{-1} dry mass) in municipal solid wastes. (From Ehrig 1989.)

	(1)	*(2)*	*(3)*	*(4)*	*(5)*	*(6)*
Cadmium	3.5	40–50	11	3.5	2–14	3–9
Copper	238	411–532	400	400	120–210	31–345
Lead	399	210–370	400	210	110–330	294–545
Mercury	0.6	0.3–0.4	4	1.1	1–14	–
Zinc	521	588–742	1200	1200	300–1000	310–956

References: (1) Greiner (1983); (2) Fresenius et al. (cited in Ehrig, 1989); (3) Belevi and Baccini (1987); (4) Nielsen (1978); Vogl (1978); (6) Bilitewski (1989); concentrations in wastes with 30% humidity, such wastes also contain about 76–108 mg kg^{-1} chromium and 13 mg kg^{-1} nickel

molecular organic acids) and those of the coupled inorganic reduction processes (e.g., Fe^{2+}, Mn^{2+}, S^{2-}, NH_4^+) accumulate in the interstitial water until concentrations are limited by physical convection/dispersion, by subsequent microbial utilization, or by diagenetic formation of secondary ("authigenic") minerals such as metal sulfides.

In municipal solid waste landfills, the initial conditions are characterized by the presence of oxygen and pH values between 7 and 8 (Figure 9.4). During the subsequent "acetic phase", pH values as low as 5 were

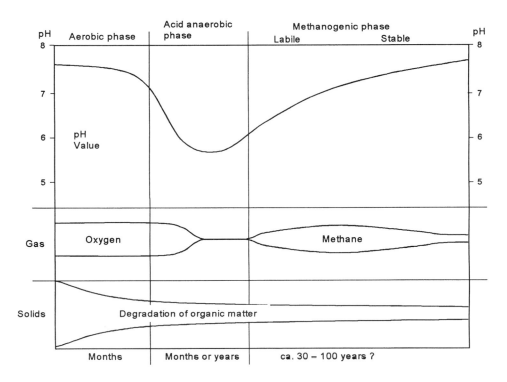

Fig. 9.4 Chemical evolution of municipal solid waste landfills ("reactor" landfill)

measured due to the formation of organic acids in an increasingly reducing milieu; concentrations of organic substances in the leachate are high. During a transition period of one to two years, the chemistry of the landfill changes from acetic to methanogenic conditions; the methanogenic phase is characterized by higher pH values and a significant drop of biochemical oxygen demand (BOD_5) values from more than 5.0 to 40.0 mg L^{-1} in the acetic phase to 20 to 500 mg L^{-1}. The long-term evolution of a "reactor landfill", subsequent to the methanogenic phase, remains an open question (Förstner et al. 1989).

Metal concentrations in leachates from municipal landfill have been analyzed since the beginning of the 1970 s. Studies performed by Quasim and Burchinal (1970), Walker (1973), Meyer (1973) and Hughes (1975) on groundwater pollution from sanitary landfill leachate and areas treated with waste compost and sewage sludge indicated that deeper fills pose fewer pollution problems than the shallower fills, which may leach the bulk of pollution in a shorter period of time, thereby exceeding the dilutional capacity of the moving groundwater. Later publications have focused on the behavior of heavy metals in the various groundwater zones (anaerobic, anoxic, aerobic) downstream from a landfill (Ehrig 1983, Nicholson et al. 1983). Differences have been found for iron, manganese, and zinc in leachates between the "acetic phase" (with high organic loadings and low pH values) and the "methanogenic phase" (with low biodegradable organics and higher pH values), whereas for other trace elements such differences have not been established (Table 9.10). The latter finding may be related to difficulties in sampling and chemical analysis since similar effects, for example, of pH on zinc mobility can be expected for other related elements such as Cd, Ni, Pb, and Cu.

A ranking of pollutants with respect to their mobility in municipal solid waste landfills was established by Christensen et al. (1989). Chloride was very mobile; sodium, ammonium, potassium, and magnesium were moderately mobile; and zinc, cadmium, iron and in most cases also manganese were only partly mobile. In particular, the heavy metals zinc and cadmium showed very restricted mobility in the anaerobic zone of even very coarse aquifer materials. Comparison of inorganic groundwater

Tab. 9.10: Concentrations of trace elements (µg L^{-1}) in leachates from municipal solid waste landfills. (From Ehrig 1989.)

Element	"Acetic Phase"		Average	Range	"Methanogenic Phase"	
	Average	Range			Average	Range
Iron	780	20–2100			15	3–280
Manganese	25	0.3–65			0.7	0.03–45
Zinc	5	0.1–120			0.6	0.03–4
Arsenic			160	5–600		
Cadmium			6	0.5–140		
Chromium			300	30–1600		
Copper			80	4–1400		
Lead			90	8–1020		
Mercury			10	0.2–50		
Nickel			200	20–2050		

constituents upstream and downstream from 33 waste disposal sites in West Germany indicated characteristic differences in pollutant mobilities (Arneth et al. 1989). High contamination factors (CF values: downstream/upstream means in the groundwater) were found for boron, ammonium, nitrate, and arsenic; the latter element may pose problems during initial phases of landfill operations (Blakey 1984). Under anaerobic conditions, soluble metals precipitate as insoluble sulfides in the landfill. Sulfide is produced by the microbiological reduction of sulfates or by the decomposition of organic compounds containing sulfur.

No empirical information is available regarding the long-term effectiveness of the liner systems until all reactions in the landfill body have ended: "In contrast to the impact of the deposited waste, the effectiveness of liner systems should be considered to be temporally limited" (Thomé-Kozmiensky 1989). Particular problems occur when leachate collection pipes are plugging during the acidic decomposition period.

The effect of released contaminants from reactor landfills on groundwater quality after the failure of a base liner has been calculated by Baccini et al. (1992). The calculation was made for a fictitious model region called "Metaland" where, for 50 years, one million people "managed" a groundwater

aquifer with a water volume of $2 \times 10^9 \, m^3$ and encompassing an area of 2500 km² (Table 9.11).

During this very long time period, processes may occur under the impact of changing hydrochemical conditions that lead to a new phase of contaminant mobilization. It is easy to visualize how a process may be initiated when oxidized precipitation infiltrates a post-methanogenic landfill body and a front of increased metal concentrations moves toward the groundwater over a long period of time as a consequence of dissolution and precipitation reactions (Förstner et al. 1987). This possibility of metal mobilization from "reactor landfills" should always be considered when low-organic hazardous waste is deposited, especially if it contain sulfides or easily soluble components.

Detailed reviews on the behavior of metals under different redox conditions are given by Christensen et al. (1994, 2001).

9.4.2
Metals and Final Storage Quality

The final storage approach is one way to develop and control landfills on a conceptual basis. It has been defined by the Swiss Federal Government in 1986 (Anonymous, 1986) and received wider attention in the book *Landfills – Reactor and Final Storage*

Tab. 9.11: Estimate of the increase in annual concentrations in the groundwater, after landfill liner failure. (From Baccini et al. 1992.)

Mean concentration	C_{org} [mg L⁻¹]	Cl	Zn [µg L⁻¹]	Cd
In leachate after 50 years	600	500	600	2
In uncontaminated groundwater	0.5	3	5	0.02
Annual increase in groundwater	0.24	0.2	0.24	0.0008
Annual increase [%]	50	7	5	4

Quantity of municipal waste in the landfill: 40000 kg per inhabitant, specific leachate influx (for aquifer size $2 \times 10^9 \, m^3$) 0.02 L kg⁻¹ municipal waste per year.

(Baccini 1989): "Landfills with solids of final storage quality need no further treatment of emissions into air and water".

Solid residues with final storage quality should have properties very similar to the Earth's crust (natural sediments, rocks, ores, soil). This can be achieved in several ways, for example by assortment or thermal, chemical and biological treatment. In most cases, this standard is not attained by simple incineration of municipal solid waste – that is, by only the reduction of organic fractions. There is, in particular, the problem of easily soluble minerals such as sodium chloride. Future efforts should be aimed at optimizing the incineration process in a sense that critical components are concentrated in the filter ash and in the washing sludge, whereas the quality of the bottom ash is improved in such a way that deposition is facilitated and even reuse of this material is possible due to either the low concentrations or chemically inert bonding forms of metals.

How reactive are incineration products of municipal solid waste? Studies conducted by Baccini and colleagues (1993) have shown that, under long-term considerations, even products such as incineration bottom ash are not really inert. After the release of chloride and sulfate, two developments can be distinguished. First, part of the calcium content, which was formerly present as hydroxide, is now carbonatized. Iron and aluminum are transformed to oxides. In the long term, the carbonate buffer in the slag will be emptied, and this will correlate with relatively low pH-values. Compared to the former conditions, elevated trace metals concentrations in the leachate can be expected.

The long-term release of calcium, chloride, sulfate and heavy metals from bottom ash and slag has been estimated by the Swiss EKESA-Project (Anonymous 1992) and Kersten et al. (1995) (Table 9.12):

- Phase 1 is characterized by very high concentrations of easily soluble components such as chloride and sulfate; the slag is not yet carbonatized; portlandite and gypsum may influence leachate composition to some extent.

- Phase 2 follows after emptying the contents of chloride and sulfate in the slag. Part of the calcium content, which was formerly present as hydroxide, is now carbonatized. It can be expected that cadmium and lead are present as carbonates, since these compounds are less soluble than the respective hydroxides. Zinc concentrations are widely unchanged. Chromium contents, however, will depend on sorption processes on iron oxide (Kersten et al. 1998) or barium sulfate (Johnson et al. 1999).

Tab. 9.12: Estimation of long-term evolution of leachates from municipal solid waste incineration residues. (From Anonymous 1992.)

	Phase 1	Phase 2	Phase 3
pH	≥ 8.3	7.3–8.3	< 5–6
Calcium [mg L^{-1}]	520	16 –60	< 16
Chloride [mg L^{-1}]	100–5000	< 100	< 100
Sulfate [mg L^{-1}]	100–000	< 100	< 100
Chromium [µg L^{-1}]	13	Much higher?	Diluted?
Zinc [µg L^{-1}]	4	Unchanged	High
Duration	Tens of years	Hundreds of years	Thousands of years

● Phase 3 exhibits a far-reaching emptying of the carbonate buffer in the slag. This is correlated with low pH-values, buffered by the still-present oxides and silicates. Compared to phase 2, elevated trace metal concentrations can be expected.

Temporal development of alkalinities in several examples from Swiss MSWI incineration landfills (Baccini et al. 1993) indicate the expected sequence of depletion of individual buffer substances. In the young deposit there are still calcium hydroxide and other effective bases, but these components are depleted in the older deposits. In the medium-age deposits (~10 years old), there is a typical reduction of the calcium silicate and bicarbonate, whereas calcium carbonate is not changed to any great extent.

Calculated and measured dependencies of metal solubility from pH-variations indicate (Kersten 1996; Johnson et al. 1998) that at between pH 11 and pH 7, concentrations of Pb and Cd remain below 10^{-6} mol L^{-1}; at pH 7, remobilization of Cd begins, whereas Pb is significantly mobilized only below pH 5. At pH 4, approximately 25% of Cu, 50% of Pb, 45% of Cd, and 70% of the total Zn-inventory has been mobilized. Therefore, the long-term potential for metal mobilization by pH decrease can be considered as significant. It has been argued that the major reason for the reduced mobilization of the metals compared with the calculated solubility for carbonate and hydroxide species might be found in the chemical incorporation of the metals in hydrated cement phases such as calcium silicate hydrate (CSH).

9.4.3
Geochemical Engineering Concept for Landfills

"Landfills can be seen as an accumulation of mobilizable contaminants; the pragmatist can only be concerned with delaying the risk of spreading the contaminants into surface- and groundwater and to spatially limit their geographic expansion." (Schenkel 1987). The "multi-barrier concept" (Stief 1986) means that the waste in a landfill can be deposited "safely" only if several functioning barriers are in place that work independently of each other:

● Barrier 1 "Geology": site selection after careful consideration of hydrogeological and geotechnical aspects.
● Barrier 2 "Liner": constructing a liner system that effectively encapsulates the landfill consisting of a base liner, side wall liners, and surface cover.
● Barrier 3 "Disposal": optimally functioning systems for collecting, treating, and disposing of leachate and landfill gas.
● Barrier 4 "Operation": operating the landfill on the basis of the known "state-of-the-art" technologies and knowing all forms of emission reduction.
● Barrier 5 "Monitoring".
● Barrier 6 "Long-Term monitoring and Care".

In the history of the barrier development (Ryser 1989), initially only the site location was considered; in the next phase, landfill liners were added as surface covers and base liners to facilitate leachate collection (during the 1970s). Since the early 1980s, attempts have been made to control reactions inside the landfill, for example by waste compaction and gas collection. In the early the 1990s, it became evident that in order to effectively minimize landfill emissions, the prior separation of contaminants was necessary. The culmination of this development will be reached, when the inert- and residual landfill will become the norm.

With this new strategic reorientation, and since the German Technical guidance for waste dumping (TASi) was set in 1993, all landfills are to be planned, established and operated following the state of the art in such a way to create independent effective barriers that help in preventing the release and propagation of pollutants. The TASi (Anonymous 1993), aims at converting the key-process of integrating thermally treated waste-deposits and binding these primary pollutants to a waste matrix or an "Internal barrier" rather than an external one. Through monitoring the materials and reactions, and through improved testing procedures, it will be possible in the future to depict the security of the long-term behavior of pollutants (Förstner 2003b).

9.5
Prognostic Tools for Metal Release from Wastes

9.5.1
Factors Influencing Release of Metals from Solid Waste Materials

With regard to the potential release of metals from solid waste materials, changes in pH and redox conditions are of prime importance. It can be expected that changes from reducing to oxidizing conditions, which involve transformations of sulfides and a shift to more acid conditions, increase the mobility of typical "B-" or "chalcophilic" elements, such as Hg, Zn, Pb, Cu, and Cd. On the other hand, the mobility is character-istically lowered for Mn and Fe under oxidizing conditions. Elements able to form anionic species, such as S, As, Se, Cr, and Mo are solubilized, for example, from fly ash sluicing/ponding systems at neutral to alkaline pH conditions (Dreesen et al. 1977, Turner et al. 1982) (Table 9.13)

The major process affecting the lowering of pH-values (down to pH 2–3) is the exposure of pyrite (FeS_2) and of other sulfide minerals to atmospheric oxygen and moisture, whereby the sulfidic component is oxidized to sulfate and the acidity (H^+-ions) is generated. Bacterial action can assist the oxidation of Fe^{2+}(aq) in the presence of dissolved oxygen.

Acidity is perhaps the most serious long-term threat from metal-bearing wastes. Water seeping from mine refuse has been passing increased metal concentrations into receiving waters for decades. The threat is especially great in waters with little buffer capacity – that is, in carbonate-poor areas, where dissolved-metal pollution can be spread over great distances. The acidity production can develop many years after disposal, when the neutralizing or buffering capacity in a pyrite-containing waste is exceeded. High concentration factors have been determined in inland waters affected by acidic mine effluents.

Primary emissions of high metal concentration occur from waste rocks and tailings, while secondary effects on groundwater take place from the ponds. An important and

Tab. 9.13: Trace element concentrations in ash-pond effluent water relative to cooling lake intake and outlet. (From Dreesen et al. 1977.)

Concentration ratio	Lake intake	Lake outlet
> 50	B, F, Mo, Se, V	Se
10–50	As	B
2–10	–	As, Cr, F, Mo, V
< 2	Cd, Cu, Zn	Cd, Cu, Zn

long-term source of metals are the sediments reworked from the floodplain, mainly by repeated oxidation and reduction processes. The acid-producing potential not only is related to the oxidation of sulfides, but oxidation of organic matter must also be considered. It has been shown by Swift (1977) that the contribution of protons from organic-N and organic-S in a sample containing approximately 5% organic carbon is equivalent to the acid-producing potential of 1% FeS_2. Studies of the long-term evolution and diagenesis in a sewage sludge landfill and similar natural sediments (peat, organic soils) by Lichtensteiger et al. (1988) suggest that the transformation of organic material will last for geological time scales (10^3 to 10^7 years).

9.5.2
Test Procedures

Several types of test procedure have been developed for the prediction of potentially adverse effects from waste materials on adjacent media. The main interest of waste studies is focused on the removal of contaminants in the solution phase. Consequently, elution (extraction) tests are central. Typical objective of leaching tests applied to waste materials are listed in Table 9.14.

9.5.2.1
Development and Harmonization of Leaching Test Procedures

From a practical point of view, leaching test procedures can be classified into static, dynamic and diffusion tests, and there are special concepts related to the long-term prognosis of metal release from waste materials:

- *Static tests*, which are classified by the type of contact into shaking and stationary tests, establish an equilibrium between the solid and the solution.
- *Dynamic tests* provide data on the maximum leachable volume of contaminants by multiple or continuous replacement of the leaching solution and allows for estimation of the leaching time frame.
- *Diffusion tests* on a sample are primarily used in the analysis of solidification prod-

Tab. 9.14: Objectives of leaching tests applied to waste materials. (From Chandler et al. 1995.)

Objective	Description
Environmental impact assessment	Estimate potential impact of waste disposal or utilization on the environment
Quality control in waste treatment	Verify the efficiency of a treatment process using a simple pass/fail criterion
Waste classification	Compare wastes against performance criteria for classification, e.g. as hazardous or nonhazardous
Identification of leachable constituents	Determine which constituents of a waste are subject to dissolution upon contact with a liquid
Evaluation of process modifications	Determine of modifications to a waste-generating process result in less leachable waste
Design of leachate treatment systems	Obtain a typical leachate to use for treatability experiments
Field concentration estimates	Express leaching over time (e.g., to be used as a source in transport modeling)
Parameter quantification for modeling	Quantify partition coefficients and kinetic parameters for us in transport modeling

ucts, because their elution is primarily controlled by diffusion when the integrity of the sample is not compromised.

The most common shaking test is the 1986 US EPA TCLP-test ("toxicity characterizing leaching procedure"), where size-reduced waste material is extracted with an acidic leaching liquid, the pH of which is determined by the alkalinity of the waste sample (Francis et al. 1989). The SOSUV-test (named after the Studiegroep Outwikkeling Standard Uitloogentesten Verbrandingsresiduen (Van der Sloot et al. 1984), was developed in the Netherlands and combines a column test with a five-stage dynamic vessel test (cascade test). The leaching liquid is water acidified with nitric acid to a pH of 4, which is primarily meant to simulate the effects of acid precipitation. In Switzerland, a CO_2 test is used for the evaluation of pretreated waste; this is a dynamic bottle test with two leaching steps where CO_2-saturated water (pH 4–4.5) is used as the extractant (Tobler 1990).

Among the diffusion tests, the very thoroughly evaluated ANS 16.1 (American Nuclear Society (Cote and Isabel 1984) leach test should be mentioned. In the Dutch Tube Procedure (Van der Sloot et al. 1987), the material to be studied is spiked with radionuclides (heavy metals or with organic contaminants marked with ^{14}C) and is introduced together with unspiked material in a polyethylene diffusion tube.

The long-term behavior of heavy metals in waste material can be simulated in a circulation system by enhancing the mobilizing factors (pH, redox, ion strength, organic complex formers) under realistic conditions (reverse current, continuous flow, suspension slurry) (Schoer and Förstner 1987). The more recent methodological development, which is primarily focused on evaluating residuals from thermal treatment, begins by titrating the suspended solids.

One example is the pH-stat method, which was developed by Obermann and Cremer (1992) and commissioned by the Office of Water and Waste of the State of Northrhine-Westphalia, Germany.

There are methodological and strategic questions with respect to the further development of test procedures – for example, whether the evaluation can be somehow standardized in order to simplify administrative decisions and actions. It has been shown that a complex task such as the evaluation of long-term inertization requires stepwise implementation, and where limits should be established only when if underlying processes have been understood.

A discussion of available test procedures for short- and long-term prognosis has been presented exemplarily by Förstner (1997):

1. Composition of pore water is a highly sensitive indicator for reactions between chemicals on solid substrates and the aqueous phase which contacts them; while the direct recovery and analysis of water-borne constituents can be seen as a major advantage of this approach, there are several difficulties, particularly arising from the sampling and sample preparation.

2. The advantage of short-term elutriate tests is that especially important parameters can be directly observed and particularly unfavorable conditions can be simulated; interpretation may become difficult for systems undergoing redox variations, which is usually the case both in aquatic sediments and land-disposed dredged materials.

3. The undisputed advantage of sequential extraction procedures lies in the fact, that rearrangement of specific solid phases can be evaluated prior to the actual remobilization of certain proportions of an element into the dissolved

phase; when proceeding further in the extraction sequences, usually a reduction in prognostic reliability takes place.

4. Simultaneous application of standard sequential leaching techniques can be used for geochemical characterization of anoxic, sulfide-bearing sediments in relation to the potential mobility of critical trace metals (Kersten and Förstner 1991).

5. With the pH-stat-technique (Obermann and Cremer 1992), a combined determination of metal releases and buffer capacity can be made at different time intervals; the sum curve of acid consumption provides information on the potential changes of the matrix composition during acidification.

Regarding procedures for characterization of waste, harmonization on various objectives and leaching/extraction test methods is undertaken by European Standardization Network CEN, Technical Committee (TC) 292, and a distinction is made between three types of leaching tests which will be applied in different circumstances (Van der Sloot et al. 1997, Van der Sloot 1998, 2002):

● *Basic characterization tests* which will generate knowledge on the materials investigated.

● *Compliance tests* to check previously extensively described waste against limit values.

● *On-site verification tests* to provide a quick tool at the site entrance to check that waste matches earlier descriptions.

A compliance leaching test for granular waste material has been published (prEN 12457) which describes three different procedures: shake test at liquid/solid (L/S) ratio = 2, L/S = 10; and a two-step shake test at L/S = 2 and 10. The leachant used is demineralized water. It is likely that the standard will be validated. The driving force behind this standard was the draft European Directive on the landfill of waste.

Actually basic characterization tests are developed as standards combining percolation test and pH static leach test.

9.5.3
Lysimeter Test on MSWI Residues

With regard to the long-term behavior of municipal solid waste incineration residues (MSWI), there are typical kinetic effects especially acid buffering by silicates and the formation of secondary phases such as iron oxides or clay minerals, by which the metal release pattern is typically influenced by time. This is the reason why several laboratory lysimeter tests have been conducted, and these have been run with water for one year or acidified water (pH-controlled) for some months (Hirschmann and Förstner 1998). The leachates of the lysimeters with pH-control show a near-constant pH of 11 and an increasing redox potential from slightly reduced to oxidized conditions. The soluble salts and organic compounds are released very quickly with increasing L/S ratio. The calcium leaching at high pH-values above 10 is limited by the solubility of gypsum.

The leachates of the pH-controlled lysimeters reached pH 4 after 5 or 6 months, and showed three buffering plateaus (Figure 9.5). A small plateau is apparent at pH 9.5 and a stronger one at between pH 7 and 5.5. Below pH 5, the pH again decreases very slowly. The release patterns of calcium and carbonate clearly show that dissolution of calcium carbonate is mainly responsible for the buffering zone between pH 7 and 5.5. Other calcium phases such as calcium silicate hydrates buffer the added acid at pH 9.5. Aluminum and iron oxides/hydroxides in addition to silicates are responsible for buffering at pH-values below 5.

The heavy metal release pattern of the pH-controlled lysimeters are similar to each

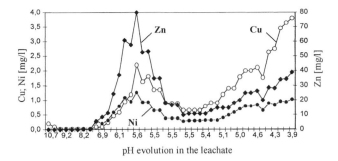

Fig. 9.5 Concentrations of Cu, Ni and Zn related on pH in the leachate of municipal solid waste inciner-ation residues (Hirschmann and Förstner 1998)

other (see Figure 9.5). Zinc, nickel and copper are released firstly between pH 7 and 5.5 in the carbonate buffering zone (characteristic for zinc), and a second peak occurs at pH values below 5 (characteristic for copper).

These experimental test procedures, together with calculations of the acid-pro-ducing potential (APP) and acid-neutraliz-ing capacity (ANC), are characteristic tools for the medium- and long-term prediction of metal release from all types of waste materials. Application of such prognostic tool will become even more important, when the rate of waste material utilization/recycling is enhanced (in Germany up to 100% by the year 2020), and will intensify the contact of the secondary products with the soil environment.

In accordance with the activities of CEN/TC 292 "characterization of waste" (Anony-mous 1994a, 1999) and the International Ash Working Group (IAWG; Anonymous 1994b), the following sequence of test proce-dures were recommended by Hirschmann (1999, 2003) for MSWI residues:

● Evaluation of gas formation, reactions in contact with water, and – at higher organic matter contents – degradability of organic substances.

● Measurement of pH- and redox evolution during aeration, and determination of

acid- neutralizing capacity and acid-buf-fering behavior by titration.

● Determination of principal leaching mechanisms (elution with water at increasing L/S ratio, pH-dependent release [e.g., stepwise pH adjustment] and metal availability).

9.6
Material Management: Recycling of Metals

9.6.1
Avoidance as the Primary Task of Pollution Control Technology

Processes for avoidance, reuse, and disposal should be viewed as part of the material economy. Figure 9.6 enhances this by show-ing a model of the total economic material flow, including waste management. The return of certain waste components into the value-added chain is made possible by the technically available measures of avoid-ance/recycling. The value-added sequence displayed in Figure 9.6 (raw and basic mate-rial extraction, initial product, half-finished product, and final product manufacturing) generally consists of a greater or lesser number of processes. In turn, these proc-esses encompass additional processes or production steps. Each process is associated process: (1) roasting; (2) multi-level sulfuric acid leaching; (3) neutralization and iron precipitation as jarosite; (4) leaching by cementation of Cu, Ni, Co and Cd with Zinc powder; and (5) zinc electrolysis. From an environmental protection point of view, the traditionally used jarosite procedure shows environmental incompatibility in two regards (v. Röpenack 1991, Gock et al. 1996):

- particles and sulfur dioxide emissions in the roasting stage; and
- pollution of the jarosites by zinc, lead, cadmium, arsenic, thallium and others.

Ruhr-Zinc GmbH (the first zinc smelter in Europe) defined a concept, which eliminated these two problems (Veltman and Weitz 1982). It used sulfuric acid, oxidizing pressure-leaching of zinc sulfide, with which Fe(III) ions take over a catalytic role by constant oxidation with oxygen. The resulting elementary sulfur can be produced in either liquid or solid form by further pressure-leaching of the residues. Pyrite (FeS) can also be used for meeting the sulfuric

acid demand in the procedure (pyrite usually appears as a secondary mineral).

With this Hematite process, the step of jarosite precipitation is avoided (v. Röpenack 1982). After the reduction of Fe(III) by zinc blend concentrate and after the separation of the solute noble metals, ferrous iron is oxidized to ferric iron which then precipitates in the form of hematite. The intermediate noble metal yields from the Jarosite and Hematite processes are compared in Figure 9.7. The quality of the hematite produced permits its utilization in the cement industry as a coloring material. For use in the steel industry, the zinc content of hematite must be reduced from 1% to 0.1%.

Apart from these fundamental improvements via procedural change, a number of further measures are specified in Table 9.15, with which a sustainable reduction of the emissions during ore smelting can be achieved (Elgersma et al. 1995). For air pollution abatement from heavy metals, the use of highly effective woven filters as a replacement of electrostatic dust samplers

Tab. 9.15: Operational improvements for decreasing heavy metal emissions during lead-zinc lodging and the related advantages. (From Elgersma et al. 1995.)

Emission	Improvements	Economic use
Dust in storage section	Spraying with water, minimization of material employment	Less material employment, less interest rates
SO_2 from the sulfuric acid plant	Double contact acid plant, Ammonia-separator, High-speed catalyst	Valuable byproduct (acid), Ammonium sulfate as byproduct
Hg in exhaust	Hg-separation plant	Valuable byproduct
Cd/Co/Cu-waste (cement)	Optimized reproduction	High-content valuable material (Cd/Co/Cu)
Zinc in gypsum residues	Zinc byproduction	Extended zinc sales
Heavy metals in wastewater	Biosulfide formation	Improved use of the useable material
Furnace cinder	Higher temperature	Commercial utilization as building material

is highlighted; the removal of mercury (e.g., by organic sulfur binding) improves the value of the produced acid as an important byproduct. A particularly interesting issue is the solid residue from the incineration process. With an economically and technically acceptable increase of the temperature, the heavy metal contents in the cinders can be reduced as long as they remain usable as building material (Elgersma et al. 1995).

9.6.3
Recycling of Industrial Waste

Residues rich in metals accumulate in various forms in industries and businesses. Because of the presence of heavy metals, the wastewater streams from plating, pickling, and etching operations are especially problematic (Martinetz 1980). The disposal of sludges from these operations – overwhelmingly hydroxides – is also problematic. Among other things, approximately 35 tons of zinc, 5 tons of copper and nickel as well as 15 tons of chromium are contained in 1000 tons of metal hydroxide sludge. Recovery of these metals can be accomplished with electrolysis or precipitation processes (Winkler and Worch 1986). In electrolysis, the hydroxide sludges are dissolved in sulfuric acid and processed in simple electrolysis cells with plate-like vertically and horizontally arranged electrodes. The metals are then separated according to their respective precipitation voltages. This process is particularly effective if the sludges are present in a pure form. In the sulfide precipitation process, the heavy metals are separately isolated by varying the pH. Large-scale copper recovery with subsequent smelting is already being carried out using hydrogen sulfide, sodium, or ammonium sulfide.

Methods for the hydrometallic processing of metallic wastes have been developed, for example, at the Montana College of Mineral Science and Technology (Ball et al. 1987). After dissolving the metal oxides, the iron is precipitated and removed as jarosite (for problems, see Section 9.6.2), followed successively by copper and zinc (liquid extraction with various complexing agents, stripping with sulfuric acid and subsequent precipitation crystallization), chromium (oxidation and precipitation as lead chromate) and nickel (crystallization as sulfate). The estimated costs are around $200 ton^{-1}, which is only slightly above the costs for disposal.

Many separation techniques that have gained great importance in pollution control technology were derived from conventional mechanical or thermal processes. A variety of processes and their use in phase separation are listed in Table 9.16.

Basic mechanical operations play a central role in the recovery of metal-rich consumer goods, which proceed on the levels of "logistics", "disassembly", "pretreatment", "separation and sorting", "fines treatment", "material recycling of pure scrap", and "residual recovery". The process step "disintegration by size-reduction" is of critical importance (Pötzschke 1991); because of the composite design and miniaturization of many components, it is possible to generate a high share of so-called "fines" (particle size < 15 to 20 mm), which can only be further recovered by using extensive refining processes.

The scrap automobile is a typical example of the treatment, recycling, and recovery of so-called secondary mixed scrap (Sattler 1988). The unwanted automobile is collected by the scrap dealer, partially disassembled by a recycler (e.g., lead battery, aluminum rims, radiator, variety of parts), and is then reduced in size by a shredder. Steel and nonferrous metals are sent to secondary smelting plants, while nonmetallic components are further processed. The lead battery

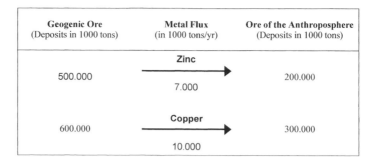

Geogenic Ore (Deposits in 1000 tons)	Metal Flux (in 1000 tons/yr)	Ore of the Anthroposphere (Deposits in 1000 tons)
	Zinc	
500.000	7.000	200.000
	Copper	
600.000	10.000	300.000

Fig. 9.8   Rough estimation of the global deposits for zinc and copper (Lichtensteiger 2001). The arrows are annual fluxes according to the global production quantities. Geogenic deposits are estimated predominantly according to their terrestrial ore occurrence that is worthy of exploitation

If zinc was deposited as zinc ore, then a regional reserve for an approximately 10-year demand would have been formed. Global geogenic and anthropogenic deposits of zinc and copper are compared in Figure 9.8 (Lichtensteiger 2001). Landfills are referred to as anthropogenic deposits; the major part of the anthropogenic zinc and copper deposits are found in existing buildings.

Based on these examples, a hypothesis can be formulated that the modern urban infrastructure represents a stock of raw material which, in the past decades and centuries, was enriched with materials in such a manner that in the future the mining industry can be reduced and partly replaced by the anthropogenic exploitation of this stock. In the future, the exploitation of raw materials from the Earth's crust should be diminished and replaced by the exploitation from the anthroposphere (buildings, networks, capital goods and consumer goods).

Acknowledgements

The author thanks Imad Kordab for help during preparation of the sections on geochemical techniques, and Dr. Günther Hirschmann for his conceptual and experimental ideas during their combined studies on the long-term behavior of municipal solid waste incineration bottom ash.

References

ALLOWAY BJ and AYRES DC (1993) *Chemical Principles of Environmental Pollution.* Blackie Academic & Professional, London-Glasgow-New York.

ANONYMOUS (1983) *Umwelt – Weltweit.* Report of United Nations Environmental Programme (UNEP) 1972–1982. Erich Schmidt Verlag, Berlin.

ANONYMOUS (1986) *Leitbild für die schweizerische Abfallwirtschaft.* Schriftenreihe Umweltschutz Nr. 51. Eidgen. Kommission für Abfallwirtschaft, Bundesamt für Umweltschutz, Bern.

ANONYMOUS (1989) *Umweltforschung und Umwelttechnologie.* Programme 1989 bis 1994. Der Bundesminister für Forschung und Technologie, Bonn.

ANONYMOUS (1990) *Technische Verordnung über Abfälle (TVA), Entwurf einer Richtlinie zur Durchführung des Eluat-Testes für Inertstoffe und endlagerfähige Reststoffe.* Schweizerischer Bundesrat. Directive from 10.12.1990.

ANONYMOUS (1991) *Abfallwirtschaft.* Special Expert Report. September 1990. Der Rat von Sachverständigen für Umweltfragen. 718 p. Metzler-Poeschel Verlag, Stuttgart.

ANONYMOUS (1992) *Emissionsabschätzung für Kehrichtschlacke (Projekt EKESA).* MBT Umwelttechnik Zürich/EAWAG, Abteilung Abfallwirtschaft und Stoffhaushalt, Dübendorf.

ANONYMOUS (1993) *TA Siedlungsabfall.* Dritte Allgemeine Verwaltungsvorschrift zum Abfallgesetz vom 14. Mai 1993. Bundesanzeiger, Jahrgang 45, Nr. 99a, Berlin.

ANONYMOUS (1994a) *Comité Européen de Normalisation (CEN) Technical Committee 292 Characterization of Waste in Europe.* State of the art report for CEN TC 292.STB/94/28, Brussels.

ANONYMOUS (1994b) *An international perspective on characterization and management of residues from municipal solid waste incineration.* International Ash Working Group (IAWAG) Summary Report, London.

ANONYMOUS (1994c) *Die Industriegesellschaft gestalten – Perspektiven für einen nachhaltigen Umgang mit Stoff- und Materialströmen.* Bericht der Enquete-Kommission "Schutz des Menschen und der Umwelt – Bewertungskriterien und Perspektiven für umweltverträgliche Stoffkreisläufe in der Industriegesellschaft" des 12. Deutschen Bundestages. Economica, Bonn.

ANONYMOUS (1999) *Comité Européen de Normalisation (CEN) Technical Committee 292/Working Group 6. Basic Characterization Test for Leaching Behaviour: pH Dependence Test, Draft Version 5.* CEN Central Secretariat, rue de Strassart 36, B-1050 Brussels.

ARNETH JD, MILDE G, KERNDORFF H and SCHLEYER R (1989) *Waste deposits influences on ground water quality as a tool for waste type and site selection for final storage quality.* In: Baccini P, ed. The Landfill – Reactor and Final Storage. Lecture Notes in Earth Sciences 20:399–416. Springer, Berlin-Heidelberg-New York.

BACCINI P, ed. *The Landfill – Reactor and Final storage.* Lecture Notes in Earth Sciences No 20, 438 p. Springer, Berlin Heidelberg New York.

BACCINI P and BADER H-P (1996) *Regionaler Stoffhaushalt – Erfassung, Bewertung und Steuerung.* 420 p. Spektrum, Heidelberg.

BACCINI P and BRUNNER PH (1985) *Behandlung und Endlagerung von Reststoffen aus Kehrichtverbrennungsanlagen.* Gas-Wasser-Abwasser 65:403–409.

BACCINI P and BRUNNER PH (1991) *Metabolism of the Anthroposphere.* Springer, Berlin-Heidelberg-New York.

BACCINI P, BELEVI H and LICHTENSTEIGER T (1992) *Die Deponie in einer ökologisch orientierten Volkswirtschaft.* Gaia 1:34–49.

BACCINI P, BADER H-P, BELEVI H, FERRARI S, GAMPER B, JOHNSON A, KERSTEN M, LICHTEN-

STEIGER T and ZELTNER C (1993) *Deponierung fester Rückstände aus der Abfallwirtschaft – Endlagerqualität am Beispiel Müllschlacke.* vdf Hochschulverlag, Zürich.

BALL RO, VERRET GP, BUCKINGHAM PL and MAHFOOD S (1987) *Economic feasibility of a state-wide hydrometallurgical recovery facility.* In: Patterson JW, Passino R, eds. Metals Speciation, Separation, and Recovery. Pp. 689–708. Lewis Publishers, Chelsea, Michigan.

BAMBAUER HU (1991) *The application of mineralogy to environmental management. An overview.* Proc ICAM 91, Vol. 1. Pretoria, South Africa.

BAMBAUER HU (1992) *Mineralogische Schadstoffimmobilisierung in Deponaten – Beispiel: Rückstände aus Braunkohlenkraftwerken.* BWK Umwelt-Spezial März 1992:S29–S34. Düsseldorf.

BAMBAUER HU and PÖLLMANN H (1998) *Concepts and methods for applications of mineralogy to environmental management.* In: Marfunin AS, ed. Mineral matter in space, mantle, ocean floor, biosphere, environmental management, and jewelry. Advanced Mineralogy, Vol. 3, pp. 279–292. Springer, Berlin-Heidelberg-New York.

BELEVI H and BACCINI P (1987) *Water and element fluxes from sanitary landfills.* In: Process Technology and Environmental Impact of Sanitary Landfills, Cagliari/Sardinia, October 19–23, 1987.

BILITEWSKI B (1989) *Opportunities for fuel from municipal solid waste (in German).* Entsorgungs-Praxis 3:74–78.

BILITEWSKI B, HÄRDTLE G and MAREK K (1997) *Waste Management.* Springer, Berlin-Heidelberg-New York.

BLAKEY NC (1984) *Behavior of arsenical wastes co-disposed with domestic solid wastes.* J Water Pollut Control Fed 56:69–75.

BORCHERS H-W, FAULSTICH M and THOMÉ-KOZMIENSKY KJ (1987) *Maßnahmen zur Schadstoffreduzierung bei der Abfallverbrennung.* In: Thomé-Kozmiensky KJ, ed. Müllverbrennung und Umwelt 2. Pp. 1–150. EF-Verlag für Energie- und Umwelttechnik, Berlin.

BRUNNER P (1990) *RESUB – Der Regionale Stoffhaushalt im Unteren Bünztal. Die Entwicklung einer Methodik zur Erfassung des regionalen Stoffhaushaltes.* Eidgen. Anstalt für Wasserversorgung, Abwasserreinigung und Gewässerschutz (EAWAG). Dübendorf/Schweiz.

CALMANO W (1988) *Stabilization of dredged mud.* In: Salomons W, Förstner U, eds. Environmental management of solid waste – dredged material and mine tailings, pp. 80–98. Springer, Berlin-Heidelberg-New York.

CHANDLER AJ, EIGHMY TT, HARTLEN J, HJELMAR O, KOSSON DS, SAWELL SE, VAN DER SLOOT HA and VEHLOW J (1995) *An International Perspective on Characterization and Management of Residues from Municipal Solid Waste Incineration.* Summary Report prepared by the International Ash Working Group (IAWG), London.

CHRISTENSEN TH, KJELDSEN P, LYNGKILDE J and TJELL JC (1987) *Behavior of leachate pollutants in groundwater.* In: Process Technology and Environmental Impact of Sanitary Landfills, Cagliari/Sardinia, October 19–23, 1987, Paper No. XXXVIII.

CHRISTENSEN TH, KJELDSEN P, ALBRECHTSEN H-J, HERON G, NIELSEN PH, BJERG PL and HOLM PE (1994) *Attenuation of pollutants in landfill leachate polluted aquifers.* Crit Rev Environ Sci Technol **24**:119–202.

CHRISTENSEN TH, KJELDSEN P, BJERG PL, JENSEN DL, CHRISTENSEN JB, BAUN A, ALBRECHTSEN H-J and HERON G (2001) *Biogeochemistry of landfill leachate plumes.* Appl Geochem **15**:659–718.

CONNER JR and HOEFFNER SL (1998a) *The history of stabilization/solidification technology.* Crit Rev. Environ Sci Technol **28**:325–396.

CONNER JR and HOEFFNER SL (1988b) *A critical review of stabilization/solidification technology.* Crit Rev Environ Sci Technol **28**:397–462.

COTE PL and ISABEL D (1984) *Application of a dynamic leaching test to solidified hazardous wastes.* In: Jackson D, Rohlik A and Conway P, eds. Hazardous and Industrial Waste Management and Testing, ASTM STP 851, pp. 48–60. Washington DC.

DREESEN DR, GLADNEY ES, OWENS JW, PERKIN BL, WIENKE CL and WANGEN LE (1977) *Comparison of levels of trace elements extracted from fly ash and levels found in effluent waters from a coal-fired power plant.* Environ Sci Technol **11**:1017–1019.

EHRIG HJ (1983) *Quality and quantity of sanitary landfill leachate.* Waste Manage Res **1**:53–68.

EHRIG HJ (1989) *Water and element balances of landfills.* In: Baccini P, ed. The Landfill – Reactor and Final Storage. Lecture Notes in Earth Sciences 20:83–116. Springer, Berlin-Heidelberg-New York.

ELGERSMA F, SCHINKEL JN and WEIJNEN MPC (1995) *Improving environmental performance of a primary lead and zinc smelter.* In: Salomons W, Förstner U and Mader P, eds. Heavy Metals – Problems and Solutions. Pp. 193–207. Springer, Berlin-Heidelberg-New York.

FAULSTICH M (1994) *Übersicht über die derzeit verfügbaren thermischen Inertisierungsverfahren und Entwicklungstendenzen.* In: Wilderer P and Schindler U, eds. Inertisierung durch thermische Abfallbehandlung. 17. Mülltechnisches Seminar. Reports from Wassergüte- und Abfallwirtschaft Technische Universität München No. 119. Pp. 45–69, München.

FAULSTICH M and ZACHÄUS D (1992) *Verfahren zur Behandlung von Rückständen aus der Müllverbrennung.* In: Faulstich M, ed. Rückstände aus der Müllverbrennung. Pp. 1–159. EF-Verlag für Energie- und Umwelttechnik, Berlin.

FAULSTICH M, FREUDENBERG A, KÖCHER P and KLEY G (1992) *RedMelt-Verfahren zur Wertstoffgewinnung aus Rückständen der Abfallverbrennung.* In: Faulstich M, ed. Rückstände aus der Müllverbrennung, pp. 703–727. EF-Verlag für Energie- und Umwelttechnik, Berlin.

FÖRSTNER U (1998a) *Integrated Pollution Control.* Springer, Berlin-Heidelberg-New York.

FÖRSTNER U (1998b) *The global problem of the impact of the production of energy, metals, materials, chemicals, and radionuclides in the modern industrial society on air, water, and soil pollution.* In: Bambauer U, ed. Advanced Mineralogy, Vol. III. Mineral Matter in Space, Mantle, Ocean Floor, Biosphere, Environmental Management, Jewelry (Series Editor: A. S. Marfunin). Chapter 5.1, pp. 268–278. Springer, Berlin-Heidelberg-New York.

FÖRSTNER U (2003a) *Geochemical techniques on contaminated sediments – river basin view.* Environ Sci Pollut Res **10**:58–68.

FÖRSTNER U (2003b) *Technische Geochemie – Konzepte und Praxis.* In: Förstner U and Grathwohl P, eds. Ingenieurgeochemie. Pp. 1–150. Springer, Berlin-Heidelberg-New York.

FÖRSTNER U (1987) *Demobilisierung von Schwermetallen in Schlämmen und festen Abfallstoffen.* In: Straub H, Hösel G and Schenkel W, eds. Handbuch Müll- und Abfallbeseitigung. Nr. 4515, 20 S. Erich Schmidt, Berlin.

FÖRSTNER U and HIRSCHMANN G (1997) *Langfristiges Deponieverhalten von Müllverbrennungsschlacken.* BMBF-Verbundvorhaben Deponiekörper, Teilvorhaben 1. Förderkennzeichen 1460799A. Umweltbundesamt Fachgebiet III 3.6. Projektträger Abfallwirtschaft und Altlastensanierung (PT AWAS) im Auftrag des Bundesministeriums für Bildung, Wissenschaft, Forschung und Technologie (BMBF) Bonn. 202 p. Berlin.

FÖRSTNER U, CALMANO W and KIENZ W (1991) *Assessment of long-term metal mobility in heat-processing wastes.* Water Air Soil Pollut **57/ 58**: 319–328.

FÖRSTNER U, KERSTEN M and WIENBERG R (1989) *Geochemical processes in landfills.* In: Baccini P, ed. The landfill – Reactor and Final Storage. Lecture Notes in Earth Sciences 20: 39–81. Springer, Berlin-Heidelberg-New York.

FRANCIS CW, MASKARINEC MP and LEE DW (1989) *Physical and chemical methods for the characterization of hazardous waste.* In: Baccini P.(Ed.): The Landfill – Reactor and Final Storage. Pp. 371–399. Springer-Verlag, Berlin.

FRIEGE H (1998) *Stoffstrommanagement für Cadmium.* In: Friege H, Engelhardt C and Henseling KO, eds. Das Management von Stoffströmen. Pp. 82–85. Springer, Berlin-Heidelberg-New York.

GAMBRELL RP, REDDY CN and KHALID RA (1983) *Characterization of trace and toxic materials in sediments of a lake being restored.* J Water Pollut Control Fed **55**: 1271–1279.

GOCK E, KÄHLER J and VOGT V (1996) *Produktionsintegrierter Umweltschutz bei der Aufbereitung und Aufarbeitung von Rohstoffen.* In: Brauer H, ed. Handbuch des Umweltschutzes und der Umweltschutztechnik. Band 2: Produktions- und produktintegrierter Umweltschutz. Pp. 79–237. Springer, Berlin-Heidelberg-New York.

GOUMANS JJJM, VAN DER SLOOT HA and AALBERS ThG, eds. (1991) *Waste Materials in Construction.* Studies in Environmental Science 48. Elsevier, Amsterdam.

GREINER B (1983) *Chemisch-physikalische Analyse von Hausmüll.* Umweltbundesamt Berichte 7/83. Erich Schmidt, Berlin.

HIRSCHMANN G (1999) *Langzeitverhalten von Schlacken aus der thermischen Behandlung von Siedlungsabfällen.* Dissertation an der Technischen Universität Hamburg-Harburg. Fortschr.-Ber. VDI Reihe 15 (Umwelttechnik) Nr. 220. 266 p. VDI, Düsseldorf.

HIRSCHMANN G (2003) *Langzeitverhalten von Deponien.* In: Förstner U and Grathwohl P, eds. Ingenieurgeochemie. Pp. 273–297. Springer, Berlin-Heidelberg-New York.

HIRSCHMANN G and FÖRSTNER U (1998) *Long term metal release from deposits of municipal solid waste incineration (MSWI) bottom ash.* In: Contaminated Soil '98, pp. 851–853. Thomas Telford, London.

HODGES CA (1995) *Mineral resources, environmental issues and land use.* Science **268**: 1305–1312.

HUGHES JL (1975) *Evaluation of Ground-Water Degradation Resulting from Waste Disposal near Barstow, California.* US Geol Surv Prof Paper 878, 33 p.

JEKEL M and VATER C (1989) *Umweltverträgliche Behandlung der Abwässer aus Hausmüllverbrennungsanlagen.* Abfallwirtsch J **1**: 9–13.

JOHNSON CA, KAEPPELI M, BRANDENBERGER S, ULRICH A and BAUMANN W (1999) *Hydrological and geochemical factors affecting leachate composition in municipal solid waste incinerator bottom ash, Part II: The geochemistry of leachate from landfill Lostorf, Switzerland.* J Contam Hydrol **40**: 239–259.

JOHNSON CA, RICHNER GA, VITVAR T, SCHITTLI N and EBERHARD M (1998) *Hydrological and geochemical factors affecting leachate composition in municipal solid waste incinerator bottom ash, Part I: The hydrology of landfill Lostorf, Switzerland.* J Contam Hydrol **33**: 361–376.

KERSTEN M (1996) *Emissionspotential einer Schlackenmonodeponie. Schwermetalle im Sickerwasser von Müllverbrennungsschlacken – ein langfristiges Umweltgefährdungspotential.* Geowiss **14**: 180–185.

KERSTEN M and FÖRSTNER U (1991) *Geochemical characterization of potential trace metal mobility in cohesive sediment.* Geo-Mar Letts **11**: 184–187.

KERSTEN M, JOHNSON CA and MOOR CH (1995) *Emissionspotential einer MV-Schlackendeponie für Schwermetalle.* Müll u. Abfall **11**: 748–758.

KERSTEN M, SCHULZ-DOBRICK B, LICHTENSTEIGER T and JOHNSON A (1998) *Speciation of Cr in leachates of a MSWI bottom ash landfill.* Environ Sci Technol **32**: 1398–1403.

KUMMERT R and STUMM W (1988) *Gewässer als Ökosysteme.* Verlag der Fachverein, Zürich.

LICHTENSTEIGER T (2001) *Die petrologische Evaluation als Ansatz zu erhöhter Effizienz im Umgang mit Rohstoffen.* In: Huch M, Matschullat J and Wycisk P, eds. Im Einklang mit der Erde – Geowissenschaften für die Zukunft, pp. 193–208. Springer, Berlin-Heidelberg-New York.

LICHTENSTEIGER T, BRUNNER PH and LANGMEIER M (1988) *EAWAG-Projekt Nr. 30–681, Klärschlamm in Deponien.* COST-681-Research Project. Dübendorf/Schweiz.

MALONE PG, JONES LW and LARSON RJ (1982) *Guide to the Disposal of Chemically Stabilized and Solidified Waste.* SW-872. Office of Water and Waste Management. U.S. Environmental Protection Agency, Washington DC.

10
Elements and Compounds on Abandoned Industrial Sites

Ulrich Förstner and Joachim Gerth

10.1
Introduction

Abandoned contamination sites include old garbage dumps and industrial production residues, contaminants from industrial facilities, areas in the vicinity of smoke stacks and discharge pipes, the concomitant contaminations and consequences of two world wars, military installations of the past and present, leaking wastewater lines, and buildings that were constructed with materials that have adverse effects on human health.

From a regulatory perspective, abandoned landfills are "abandoned and inactive waste disposal sites, regardless of the point in time at which they were rendered inactive, illegal waste disposal sites that existed before the enactment of the respective waste laws (so-called 'illegal dumps')" and "other abandoned/inactive dumps or fills"; whereas abandoned contamination sites are "sites of inactive installations that handled environmentally hazardous substances" (i.e., these are primarily old industrial and commercial facilities).

On such industrial sites, production residues were often superficially buried, or production input, intermediate, and end-products were stored without any protective measures (former gas utilities, insecticide plants). Other subsurface contamination was caused by leaking pipelines used for chemicals, petroleum products, etc., or leaking above ground storage tanks (ASTs), for example, abandoned refineries and airports. Oil and gasoline leakage from underground storage tanks (LUSTs) has caused considerable soil contamination (e.g., gas stations). Soil contamination was also caused, for example, in the port of Hamburg, Germany, by the effects of war, where organic chemicals and petroleum products from destroyed above-ground tanks and/or operating facilities seeped into the subsurface (Förstner 1998). An overview on typical industrial sites and their relevant metal contaminants is provided in Table 10.1 (Anonymous 1990, 1995).

10.2
Treatment of Contaminated Industrial Sites

What should be done once soil contamination on abandoned industrial sites has been discovered? The following are some available alternatives (Thomé-Kozmiensky 1989):

1. *Leaving* the contaminated soil in place and limiting the use of the site.

Elements and their Compounds in the Environment. 2nd Edition.
Edited by E. Merian, M. Anke, M. Ihnat, M. Stoeppler
Copyright © 2004 WILEY-VCH Verlag GmbH & Co. KGaA, Weinheim
ISBN: 3-527-30459-2

Tab. 10.1: Suspected abandoned contamination sites and possible relevant substances (Anonymous 1990, 1995)

Batteries, accumulators	antimony, arsenic, lead, cadmium, chromium, fluoride, copper, nickel, mercury, acids/bases, selenium, zinc
Basic inorganic chemicals	antimony, arsenic, beryllium, lead, cadmium, chromium, copper, nickel, mercury, acids/bases, selenium, thallium, vanadium, zinc
Fertilizers	arsenic, cadmium, copper, acids/bases, thallium
Plastics	lead, cadmium, chromium, acids/bases, selenium, zinc
Paints and coatings	antimony, arsenic, lead, cadmium, chromium, copper, mercury, acids/bases, selenium, thallium, zinc
Plant protection, herbicides, pesticides, etc.	arsenic, lead, chromium, copper, mercury, selenium, thallium, zinc
Ammunition and explosives	antimony, arsenic, lead, chromium, copper, mercury, acids/bases
Coal mines, gas works, coking plant	arsenic, (asbestos), lead, chromium, acids/bases
Crude oil processing/ petroleum storage (incl. waste oil)	arsenic, chromium, copper, lead, nickel, acids/bases, selenium, tetraethyl lead, vanadium, zinc
Iron and steel production	antimony, arsenic, beryllium, cadmium, chromium, copper, lead, mercury, nickel, acids/bases, selenium, thallium, vanadium, zinc
Nonferrous refinery	antimony, arsenic, beryllium, cadmium, chromium, copper, lead, mercury, nickel, acids/bases, zinc
Nonferrous metallurgical plant	acids/bases, antimony, arsenic, beryllium, cadmium, chromium, copper, lead, mercury, nickel, selenium, thallium, vanadium, zinc
Surface treatment and hardening of metals	antimony, arsenic, lead, cadmium, chromium, copper, nickel, mercury, acids/bases, selenium, zinc
Metal foundries	antimony, arsenic, lead, cadmium, chromium, fluorides, copper, nickel, mercury, selenium, zinc
Working, treating, and processing of wood	arsenic, acids/bases, chromium, copper, nickel, mercury, zinc
Paper, cardboard, and textiles	antimony, acids/bases, chromium, copper, lead, mercury, thallium, zinc
Processing of rubber, plastics, and asbestos	antimony, arsenic, cadmium, chromium, copper, lead, mercury, zinc
Manufacturing and processing of leather	arsenic, chromium
Edible oils and fats	acids/bases, nickel
Junk yards, salvage yards	cadmium, chromium, lead, zinc
Airports	lead alkyls, bromine compounds
Gas stations	leaded alkyls

2. *Capping* or *encapsulating* the soil in place with impermeable material and applying a layer of clean topsoil.
3. *Excavating* the contaminated soil and disposing of it at a solid or hazardous waste landfill.
4. *Remediating* the contaminated soil in-situ, "on-site", i.e., at the site, or off-site, at a facility located elsewhere.

The choice of the remediation method is also of importance for the later use of the site or of the soil. The *intensity* of handling – that is, changing the soil – increases from "biological treatment", to "soil flushing" to "thermal soil treatment". It is possible for some biological and flushing methods to drastically change the original chemical soil properties by adding *chemicals* and *nutrients* and by enhancing the growth of microorganisms; the physical soil properties, however, remain the same and the disturbed soil biology can *regenerate* within a few years and will again adapt to site conditions. Subsequent *land use restrictions* can be the consequence of possible groundwater contamination from applying nitrates and from the release of nitrogen from the endogenous decomposition of microorganisms (Slenders et al. 1997; Stegmann et al. 2001).

In the following sections, two groups of treatment techniques are described, both of which were applied to metal contamination in abandoned industrial sites. It should be noted, however, that compared to the treatment of organic contaminants (which is mainly by biodegradation), the results of these procedures may be limited, either due their low efficiency for extracting critical metals pollutants (e.g., during chemical treatment) or due to negative side effects, such as increase in volume and consumption of costly additives (e.g., during solidification and stabilization).

10.2.1
Solidification/Stabilization

The US Environmental Protection Agency (EPA; Anonymous 1986) defines these two terms as waste treatment methods that have the following objectives:

- To improve manageability and physical properties, e.g., by the sorption of free liquids.
- To reduce the surface area of waste that can be caused by contaminant migration and/or loss.
- To limit the solubility of hazardous waste components, e.g., by pH adjustment, or by sorption processes.

Solidification describes a process where a bonding agent is mixed with the waste material to create a mechanically solid product.

The associated testing methods usually originate in soil mechanics and soils testing (strength, permeability, temperature and moisture resistance, etc.) Together with its technical implementation, this term describes primarily a method of waste treatment.

Stabilization describes the goal of solidification as it relates to harmful components, which is, to convert the waste material into a chemically more stable form and to limit solubility of its hazardous constituents. The degree of stabilization is determined by leach tests, and studies of sorption, diffusion, and volatilization. At best, stabilization results in immobilization through solidification: the migration of contaminants from the wastes' surface area is prevented or at least minimized.

Inertization describes the mechanisms that cause stabilization or immobilization. The type of inertization is determined by special physical (e.g., electron-optical or X-ray) or chemical methods (e.g., for heavy metals with valence-specific sequential extractions).

Some solidification methods serve only to improve transportability and storability. To do that, liquid and pasty wastes are converted so that any seepage of liquids is prevented and above-ground storage is therefore allowed. In other cases, the goal is material recovery (recycling), especially for large-volume waste materials such as dredge spoils and power plant fly ash.

The processes and techniques of stabilization/solidification (S/S) have matured into an accepted part of environmental technology (Means et al. 1995; Conner and Hoeffner 1998a). There are different generic and proprietary S/S processes that can be conveniently categorized as follows:

- Chemical processes: cement-based, pozzolan-based, lime-based, phosphate-based, additive-intensive.

- Physical processes: macroencapsulation/containerization, nonchemical microencapsulation.
- Thermal processes: thermoplastic polymer encapsulation, vitrification.

For metals, the primary factors affecting immobilization are pH control, chemical speciation, and redox potential control. For organics, immobilization of constituents can be broken down into two primary classifications: (1) reactions that destroy or alter organic compounds; and (2) physical processes such as adsorption and encapsulation (Conner and Hoeffner 1998b).

S/S processes develop a wide variety of strength and durability values, depending on many factors which include waste type, water content, reagent type, reagent addition ratio (mix ratio), curing time, and temperature. Contrary to the opinion held by many, rock-hard solids are not always desirable. In landfill operations, a friable, compactable material is usually preferable, and low permeability, whilst desirable from a leaching point of view, may make operation of a landfill difficult in wet weather (Conner and Hoeffner 1998b).

The compatibility of typical metal waste components with solidification agents is discussed by Wiedemann (1982), Wiles (1987) and Conner and Hoeffner (1998a,b) (Table 10.2).

10.2.1.1
Cement

Cement (Portland cement) consists of a mixture of oxidized calcium, aluminum, and silicium compounds which are induced into a reactive condition in a cement kiln and, when mixed with water, form hydrated aluminates and silicates, become solidified and water-resistant. Absorbent materials such as diatomaceous earth and powdered clay can be added to absorb certain liquids. The immobilization of multi-valent metal ions in the form of slightly soluble hydroxides or alkaline carbonates is promoted by a high pH. This is the main area of use for these comparatively expensive additives whose advantages include a variety of applications and proven methodologies. Furthermore, it is advantageous that sludge dewatering is not necessary: for example, the treatment of flotation sludge in its original condition with cement leads to rapid settling and solidification of particles; the excess water is clear and can be drawn off. The disadvantages include the presence of components in the waste, such as sulfate and organic substances, and possibly sodium,

Tab. 10.2: Solidification/stabilization processes and agents for metal-rich wastes (after Anonymous 1990)

Material	Additives	Binding mechanism	Potential applications
Cement (Portland cement)	Water, perhaps bentonite	Stabilization, sometimes fixation	Subsurface injection, watery sludges
Liquid glass (sodium silicate); pulverized silicate	Not known	Compaction	Soil compaction (sub-base of contaminated soil)
Lime	Hydrophobing agent	Dispersion	Oil sludges, leachate
Brown Coal Power Plant Ash	Scrubber residues, desulfurization wastewaters	Solidification, stabilization	Brown coal combustion residues and desulfurization residues
Glass (electrical energy)	Electrolytical salts	Vitrification	"In-situ" wastes and soils

manganese, lead, and zinc, which compromise the solidity of the products.

10.2.1.2
Glass

Two reactions can convert alkali silicate glass into a solid mass with which contaminated sludges can be bound: (1) by adding acid to form silica gel, whereby the evaporation of water does not occur (alternative to vaporization!); and (2) reaction with multi-valent metal ions (e.g., calcium chloride) while forming aqueous metal silica gel, where heavy metals are precipitated and are mechanically bonded into the gel structure. The CHEM-FIX-Process is primarily used in the United States for inorganic contaminants; the Belgian SOLIROC-Process contains additives to make it usable for the solidification of organic wastes.

10.2.1.3
The pozzolanic effect

This describes the specific curing behavior of fly ash, cement dusts, and certain steel works byproducts, is based on the reaction of silicate and aluminous materials with quick lime. Here too, as with the above-mentioned additives, a higher pH causes the precipitation of metal hydroxides and carbonates. The British SEALOSAFE-Process uses fly-ash plus Portland cement, or alkali silicate glass and Fe/Al hydroxides to solidify a broad spectrum of wastes. In the POZ-O-TEC-Process, the wastes from flue gas scrubbers are solidified together with grate ash and fly-ash. The pozzolanic processes have the advantage of excellent long-term stability; however, the products solidify rather slowly and are susceptible to acids.

10.2.1.4
Lime

Lime in the form of calcium oxide (quicklime) or calcium hydroxide has long been used in the chemical stabilization of soil; indeed, some reactions represent a "pozzolanic effect". Quicklime that has been treated with reaction inhibitors (hydrophobic) is used especially for the solidification of oily sludges and contaminated soils (Bölsing 1986).

When evaluating these processes, the question of bonding stability becomes paramount. Whilst with inorganic contaminants such as heavy metals, the addition of a few additives permits a completely different form of contaminant fixation, during the solidification of organic components, a bonding change in critical contaminants is often not intended. Overall, it should be concluded that little practical experience is available of the long-term behavior of most S/S products. Should tests reveal that the immobilizing effect of a bonding agent is insufficient, then the large blocks of solidified material would, in effect, have to be leveled using mining techniques, or would need to be encapsulated (Anonymous 1995).

10.2.2
Washing and Electrochemical Methods

All remediation methods which remove, convert, or destroy contaminants in a matrix, i.e., soil or groundwater, except thermal and biological methods, are defined as chemical–physical methods. It is possible to pursue a variety of strategies with chemical–physical methods (Offutt et al. 1988):

● The generation of small quantities of concentrated contaminants through conversion and separation.

● The generation of relatively large quantities of diluted contaminant streams from which concentrated contaminants have to be separated before disposal.

Typical examples of separation methods include washing and extraction processes used for relatively highly concentrated liq-

The US EPA (1999) considers monitoring a key element of this remediation approach. It is therefore termed "Monitored Natural Attenuation" (MNA) and is based on the understanding and quantitatively documenting naturally occurring processes at a contaminated site that protect humans and ecological receptors from unacceptable risks of exposure to hazardous contaminants. MNA is a "knowledge-based" remedy and, instead of imposing active controls, as in engineered remedies, scientific and engineering knowledge is used to understand and document naturally occurring processes. A thorough engineering analysis informs the understanding, monitoring, predicting, and documenting of the natural processes (US EPA 2001). It can only be applied to those sites at which the contaminants are controlled by processes that destroy or strongly immobilize the contaminants.

Consideration of MNA as a remedy or remedy component is the result of a complex process including evaluation of MNA along with other remedial approaches and technologies. In particular, source control measures should be evaluated for all sites and implemented at most sites where practicable. These measures include the removal, treatment or containment of sources (US EPA 1999). If human action is involved to accelerate the attenuation processes the term "Enhanced Natural Attenuation" (ENA) is used.

10.3.2
Natural Attenuation Mechanisms

For most organic pollutants the key mechanism of NA is mass reduction by microbial decomposition. Most inorganic pollutants cannot be reduced in mass and are mainly attenuated by sorption processes. For this group of contaminants, NA aims at a retention of pollutants by sorption and immobilization reactions, dispersion, dilution and in part also by volatilization. The latter two mechanisms are not accepted as a basis for an NA concept (Track and Michels 1999). In the case of inorganic pollutants which are also found in nonpolluted systems as a natural background, the dilution eventually leads to "natural" concentration levels. This could favor a "do-nothing" option, although a certain degree of dilution must be accepted because the average fraction of sorbed contaminant that cannot be remobilized increases with dilution. Eventually, the extent to which dilution can be accepted as a mechanism of NA is limited by target values at well-defined distances from the source.

Sorption processes are very effective and include adsorption/desorption (reversible binding at the solid–water interface), absorption (diffusion of pollutants into the solid matrix), precipitation and coprecipitation (incorporation into a freshly formed solid), and occlusion (sequestration of adsorbed pollutants during mineral growth). The most important factors for retention processes are pollutant concentration, the composition of the solid matrix, solution composition (e.g., complexing agents) and E/pH conditions (Brady and Borns 1997).

The degree of reversibility can depend on the amount of time that the pollutant has been in contact with the solid. Sorption onto iron hydroxides, organic matter and metal carbonate minerals is often observed to be irreversible over time spans exceeding years (Brady et al. 1999). Immobilization reactions are indicated by slow reaction kinetics as found, for example, for the binding of nickel, zinc, and cadmium to different soil constituents which (Gerth 1985). When in contact with iron oxides, these elements are immobilized by matrix diffusion (Brümmer et al. 1988; Gerth et al. 1993).

Ros Vincent and Duursma (1976) extracted sediment samples which had been equilibrated with different metals for seven months to find that zinc, cesium, and cobalt had become immobilized in the mineral structure, while strontium and cadmium remained in exchangeable positions. Rhine sediments were shown to retain increasing proportions of cadmium at prolonged contact time (Salomons 1980). High-affinity reactions occur with organic matter which was found to immobilize copper (Förstner 1987) and lead (Lion et al. 1982). Clay minerals showed a time-dependent sorption of heavy metals with an increasing, irreversibly bound fraction (Helios-Rybicka and Förstner 1986; Gerth 1985).

In sediments, the immobilization of heavy metals often occurs by sulfide formation. Pollutants can become remobilized by sediment oxidation after lowering of the water table or by erosion (Förstner 1995). The solution composition of the interstitial water phase is the most sensitive indicator of pollutant reactions with surfaces (Förstner et al. 1999). A comprehensive description of the reactions of inorganics in anoxic sediments was provided by Song and Müller (1999).

Pollutant release is determined by the type of binding, which can be reversible or irreversible. Reversible binding is in equilibrium with the dissolved species and is reduced with decreasing solution concentration. It is also affected by other changes in solution composition. Irreversibly bound pollutants are not in equilibrium with dissolved species. Normally, a clear distinction between irreversibly and reversibly bound forms is not possible, and in most cases the binding strength lies between these two extremes (Brady and Borns 1997). Characterization becomes even more complicated if the integrity of the matrix depends on the E/pH conditions, as is the case with iron oxides and sulfides. After depletion of oxygen by flooding and at appropriate levels of organic matter even "irreversibly" bound pollutants can be released. "Irreversibility" in the binding of pollutants can only be relied on in the context of stable geochemical conditions.

10.3.3
Natural Attenuation Concepts

10.3.3.1
Pollutant Degradation Concept

The original concept of NA refers to a source/plume scenario of degradable pollutants in the saturated zone and the control of pollutant dissemination in such a way that the extension of the plume is stabilized or even reduced by mass reduction. An evaluation of different contaminated sites revealed that the extension of plumes is limited depending on the properties of the particular organic pollutant (Schiedek et al. 1997). Accordingly, NA processes can be used as an alternative to conventional remediation strategies to control migrating pollutants. Without removal of the source, NA is a long-term remedial action. In combination with source control, the NA principle can be assumed to operate within reasonable time frames. In such case it is, as in Germany, common practice to imply that NA processes successfully degrade cut-off plumes without control and maintenance (Teutsch et al. 2000). It is necessary, however, to use NA on a legal and administrative basis and well-defined technical guidelines which are being developed.

The steps and criteria necessary for the implementation of NA include (after Teutsch et al. 2000):

1. Site investigation: evaluate existing data and collect additional information on the type, the amount, and distribution

of pollutants present (extension of the plume), the hydrogeologic and geochemical data, the site-specific biological processes and retention capacity of the aquifer.

2. Auxiliary information: evaluation of experience gained from other cases, literature data on plume lengths (not site-specific), sorption behavior and degradability of pollutants, possible toxic metabolites, general information on suitability of site conditions.

3. Modeling of pollutant transport considering the site-specific processes and parameters together with additional information (see point 2) to verify site data; plausibility check of the model using existing data; quantify uncertainties of predictive analyses regarding the site-specific, spatiotemporal development of the plume.

4. Acceptance considerations: the comprehensible and quantitative NA concept has to meet regulatory requirements, for decision-making authorities may have to take into account further criteria such as future use, property aspects, possibility of combining NA with other techniques.

5. Site-specific monitoring concept: monitoring has to gauge effectiveness and demonstrate that NA is occurring as expected; contingency measures have to be provided for such case that NA fails to perform as predicted and performance criteria cannot be met.

The plume control concept is used in connection with organic pollutants such as fuels and fuel additives and chlorinated solvents.

10.3.3.2
Immobilization Concept

Most inorganic and nondegradable organic pollutants can only be attenuated by nondestructive processes such as sorption and immobilization. In many cases the source of pollutants are contaminated soils, with contaminant transfer from the surface soil to the groundwater occurring by vertical leachate migration. The leachate fluids mix with the lateral groundwater flow stream and spread horizontally from the point of entry in the downgradient direction of groundwater flow. On their way through the unsaturated zone mobile pollutants come into contact with reactive surfaces and are retarded by adsorption and, in part, also by immobilization.

According to the German Soil Protection Act (BBodSchG 1999), thresholds of insignificance for pollutant concentrations must be met in the mixing zone of soil leachate and groundwater. The regulation precludes dilution by mixing of leachate concentrations beyond regulatory thresholds with the groundwater. Otherwise, active measures must be taken. This concept was designed for precautionary purposes when recycling contaminated materials, for example for construction and also in the after-care of old sites. It sets regulatory margins for NA in the unsaturated zone of areas contaminated on a large scale in the unsaturated zone, as may be typical for old industrial sites.

The immobilization concept is not restricted to the unsaturated zone. Inorganics such as arsenic and mobile heavy metals are found in the groundwater and can also become immobilized under saturated conditions.

The NA pathways of inorganics and data necessary to characterize the process are summarized in Table 10.3. However, the processes that affect speciation, fate, and transport of arsenic and metals are not sufficiently understood. According to US EPA (2001), there is need to:

- further elucidate attenuation mechanisms governing the immobilization of arsenic and other metals;

Tab. 10.3: Natural attenuation pathways and data needs for metals and other inorganics (after Brady et al. 1999)

Chemical	Natural attenuation pathways	Data needs
Pb	Sorption to iron hydroxides, organic matter, carbonate minerals, formation of insoluble sulfides	Iron hydroxide availability; pH, alkalinity, and Ca^{2+} levels to answer if calcium carbonate is stable. E_H, and if E_H is low, sulfide levels. Organic carbon content
Cr(VI) as CrO_4^{2-}	Reduction by organic matter, sorption to iron hydroxides, formation of $BaCrO_4$	E_H, electron donor levels, pH (reduction rates are faster at low pH).
As(III or V)	Sorption to iron hydroxides, formation of sulfides	E_H, and, if E_H is low, sulfide levels
Zn	Sorption to iron hydroxides, carbonate minerals, formation of sulfides	Iron hydroxide availability; pH, alkalinity, and Ca^{2+} levels to answer if calcium carbonate is stable. E_H, and, if E_H is low, sulfide levels.
Cd	Sorption to iron hydroxides, carbonate minerals, formation of insoluble sulfides	Iron hydroxide availability; pH, alkalinity, and Ca^{2+} levels to answer if calcium carbonate is stable. E_H, and, if E_H is low, sulfide levels.
Ba	Sorption to iron hydroxides, formation of insoluble sulfate minerals	Sulfate levels
Ni	Sorption to iron hydroxides, carbonate minerals	Iron hydroxide availability; pH, alkalinity, and Ca^{2+} levels to answer if calcium carbonate is stable. E_H, and, if E_H is low, sulfide levels.
Hg	Formation of insoluble sulfides	E_H, and, if E_H is low, sulfide levels
U(VI)	Sorption to iron hydroxides, precipitation of insoluble minerals, reduction to insoluble valence states	Iron hydroxide availability; pH, availability of reducing compound
Pu(V and VI)	Sorption to iron hydroxides, formation of insoluble hydroxides	Iron hydroxide availability; pH, availability of reducing compound
Sr	Sorption to carbonate minerals, formation of insoluble sulfates	Iron hydroxide availability; pH, and Ca^{2+} levels to answer if calcium carbonate is stable.
Am	Sorption to carbonate minerals	Iron hydroxide availability; pH, and Ca^{2+} levels to answer if calcium carbonate is stable.
Cs	Sorption to clay interlayers	Clay content, cation exchange capacity.
I	Sorption to sulfides, organic matter	Metal sulfide mineral content
Tc(VII) as TcO_4^-	Possible reductive sorption to reduced minerals (e.g., magnetite), forms insoluble reduced oxides and sulfides	E_H, and, if E_H is low, sulfide levels.
Co^{2+}	Sorption to iron hydroxides, organic matter, and carbonate minerals	Iron hydroxide availability; pH, and Ca^{2+} levels to answer if calcium carbonate is stable.

- evaluate the effect changes in geochemical conditions have on the re-mobilization of once immobilized contaminants;
- understand the fate and behavior of metals in co-mingled organic/inorganic environments;
- develop guidelines for obtaining field and analytical data needed for an NA remedy, demonstrating sufficient understanding

of performance of the immobilization processes, using models reconciling laboratory and field data, and incorporating uncertainty analysis.

So far, there is no technique to predict rates of the nonbiodegradation NA processes like irreversible sorption or solid formation (Brady and Borns 1997). There still is a lack of appropriate tools and parameters

Tab. 10.4: Processes, level of understanding and likelihood of success of natural attenuation (after Mac-Donald 2000)

Chemical class	Dominant attenuation process	Current level of understanding	Likelihood of success
Organic			
Hydrocarbons			
BTEX	Biotransformation	High	High
Gasoline, fuel oil	Biotransformation	Moderate	Moderate
Nonvolatile aliphatic compounds	Biotransformation, immobilization	Moderate	Low
Polycyclic aromatic hydrocarbons	Biotransformation, immobilization	Moderate	Low
Creosote	Biotransformation, immobilization	Moderate	Low
Oxygenated hydrocarbons			
Low molecular-weight alcohols, ketones, esters	Biotransformation	High	High
MTBE	Biotransformation	Moderate	Low
Halogenated aliphatics			
Tetrachloroethylene (TCE), carbon tetrachloride	Biotransformation	Moderate	Low
TCA	Biotransformation, abiotic transformation	Moderate	Low
Methylene chloride	Biotransformation	High	High
Vinyl chloride	Biotransformation	Moderate	Low
Dichloroethylene	Biotransformation	Moderate	Low
Halogenated aromatics			
Highly chlorinated			
PCBs, tetrachlordibenzofuran, pentachlorophenol, multichlorinated benzenes	Biotransformation, immobilization	Moderate	Low
Less chlorinated			
PCBs, dioxins,	Biotransformation	Moderate	Low
Monochlorbenzene	Biotransformation	Moderate	Moderate
Nitroaromatics			
TNT, RDX (Hexogen)	Biotransformation, immobilization, abiotic transformation	Moderate	Low
Inorganic			
Metals			
Ni	Immobilization	Moderate	Moderate
Cu, Zn	Immobilization	Moderate	Moderate
Cd	Immobilization	Moderate	Low
Pb	Immobilization	Moderate	Moderate
Cr	Biotransformation, immobilization	Moderate	Low to moderate
Hg	Biotransformation, immobilization	Moderate	Low
Nonmetals			
As	Biotransformation, immobilization	Moderate	Low

Tab. 10.4: (Continued)

Chemical class	Dominant attenuation process	Current level of understanding	Likelihood of success
Se	Biotransformation, immobilization	Moderate	Low
Oxyanions			
Nitrate	Biotransformation	High	Moderate
Perchlorate	Biotransformation	Moderate	Low
Radionuclides			
^{60}Co	Immobilization	Moderate	Moderate
^{137}Cs	Immobilization	Moderate	Moderate
^{3}H	Decay	High	Moderate
^{90}Sr	Immobilization	High	Moderate
^{99}Tc	Biotransformation, immobilization	Low	Low
^{238,239,240}Pu	Immobilization	Moderate	Low
^{235,238}U	Biotransformation, immobilization	Moderate	Low

BTEX = benzene, toluene, ethylbenzene, xylene; MTBE = methyl *tert*-butyl ether; TCE = trichloroethylene; TCA = trichloroethane; PCBs = polychlorinated biphenyls.

for determining the site-specific and contaminant-specific NA potential. For inorganics in particular, an assessment must consider the effect of future use on previously immobilized contaminants (Brady et al. 1999).

10.3.4
Potential of Application

Natural attenuation was re-assessed by the US National Research Council (NRC) and is considered as an established remedy with a high likelihood of success (at more than 75% of contaminated sites) for only a few types of organic contaminants such as BTEX, low-molecular-weight alcohols, ketones and esters, and methylene chloride. For inorganics such as metals, metalloids, oxyanions and radionuclides, the likelihood of success is rated as moderate or low (at more than 50% or at less than 25% of contaminated sites, respectively; Table 10.4). It is assumed that any given site will have the

correct conditions for natural attenuation of the particular contaminant. The ratings are based on field evidence and the current understanding of the attenuation processes (MacDonald 2000). Apparently, NA as a remedy for inorganics, in particular, is given moderate potential only.

The NRC's view is that natural attenuation should be selected only when the mechanisms responsible for destroying or immobilizing the contaminant are scientifically recognized, documented to be working now at the site, and sustained for as long as the contamination source is present (MacDonald 2000).

Natural attenuation processes are always site-specific, so every site needs to be evaluated individually. The accepted mechanisms for inorganics, including radionuclides, are immobilization and/or biotransformation. For tritium, decay is acceptable. In most of the cases listed in Table 10.2, the attenuation process is only moderately understood. Although recognized as being important,

WIEDEMEIER TH, RIFAI HS, NEWELL CJ and WILSON JT (1999) *Natural attenuation of fuels and chlorinated solvents in the subsurface.* John Wiley & Sons, New York.

WILES CC (1987) *A review of solidification/stabilization technology.* J Hazard Mater **14**:5–21.

WILICHOWSKI M (2001) *Remediation of soils by washing processes – an historical overview.* In: Stegmann R, Brunner G, Calmano W, Matz G, eds. Treatment of contaminated soil – fundamentals, analysis, applications, pp. 417–433. Springer-Verlag, Berlin-Heidelberg-New York.

11
Elements and Their Compounds in Indoor Environments

Pat E. Rasmussen

11.1
Introduction

Prior to the mid-1980s, there was a relative scarcity of information on metals and metalloids in indoor environments, compared to the large literature on volatile synthetic organic chemicals, simple inorganic compounds and moieties, radon, and asbestos (Fishbein, 1991; Fishbein, 1989). By the early 1990s, growing evidence of a positive association between Pb concentrations in common household dust and blood Pb levels in children indicated that risks associated with childhood Pb exposure were not limited to smelter towns, minesites, and other industrial hotspots, but were present in many urban residential environments (Charney et al., 1980; Duggan and Inskip, 1985; Rabinowitz et al., 1985; Bornschein et al., 1985; Laxen et al., 1987; Davies et al., 1990; Centers for Disease Control (CDC) 1991; Health Canada, 1994). Observations of young children in the United Kingdom showed that at least 97% of their total daily Pb intake is from ingestion of house dust, food, and water, and only a small proportion ($< 3\%$) is through inhalation (Davies et al., 1990). Further studies in the U.K. found that 50% of the daily Pb intake of 2-year-old urban children occurs by ingestion of house dust through normal, hand-to-mouth activities (Thornton et al., 1994). Such findings triggered a large research effort into residential exposures to Pb, advantages and disadvantages of various indoor sampling approaches, and abatement technologies (see Lanphear et al., 1998; Adgate et al., 1998; Sutton et al., 1995; Mushak, 1998 and references cited therein), and application of stable Pb isotope analysis for identification of sources and exposure pathways (Gulson et al., 1994; Gulson et al., 1995; Rabinowitz, 1995; Maddaloni et al., 1998; Manton et al., 2000).

Recognition that Pb-contaminated house dust is a major source of Pb exposure for young children inspired research into other inorganic constituents of house dust, and by the early 2000s multi-element indoor data became available for many towns and cities in the U.K. and continental Europe, North America, Asia, Australia, and New Zealand. This research has revealed another widespread phenomenon: concentrations of many key metals and metalloids, including Pb, Hg, As, Cd, Cu, Zn, and Sb, are commonly elevated in indoor dust compared to exterior dust and soil in ordinary urban environments (e.g., Thornton et al., 1985; Culbard et al., 1988; Fergusson and Kim, 1991; Kim and Fergusson, 1993;

Elements and their Compounds in the Environment. 2nd Edition.
Edited by E. Merian, M. Anke, M. Ihnat, M. Stoeppler
Copyright © 2004 WILEY-VCH Verlag GmbH & Co. KGaA, Weinheim
ISBN: 3-527-30459-2

Gulson et al., 1995; Kim et al., 1998; Rasmussen et al., 2001). The precise causes for elemental enrichment in household dust are in most cases unknown, and there is little evidence as to the toxicological impact (with the exception of Pb), but the widespread nature of the observation points to the importance of understanding indoor accumulation processes and indoor sources.

This chapter outlines current approaches for characterizing elements in settled dust and indoor air, factors influencing the geochemistry of household dust, potential indoor and outdoor sources, and the need for reliable bioavailability data to improve assessments of residential exposures to metals. The emphasis of this chapter is on studies that examine a variety of metals and metalloids in indoor environments. For information on individual elements, the reader is referred to the elemental chapters in Part III, and for information on other indoor sources and pathways (e.g., food and drinking water) and potential effects on human health, the appropriate chapters in Part I and II should be considered.

11.2
Composition of Indoor Dust

The composition of house dust varies widely, depending on construction material and architectural design, proximity to industrial and vehicular pollution sources, mode of heating and cooking, temperature and humidity, and variations in air exchange and particle infiltration rates in different climates and geographic regions. Even in houses of comparable age and construction in the same community, the elemental profile of dust in an individual house will depend on the lifestyle and activities of the inhabitants, their occupations, hobbies and

crafts, smoking habits, preferences for paint, furnishings, and other consumer products, and housekeeping practices.

House dust contains carpet fibers, chips of paint and metal-plated objects, human and animal hairs, with particles clinging to many of the hairs, and soil tracked in from outside (Harrison, 1979). House dust also contains molds, pollen, allergens, bacteria, viruses, arthropods, skin particles, ash, soot, cooking and heating residues, and building components (Lioy et al., 1993). Fergusson and Kim, 1991) reported the following components (by weight): 32–50% limestone/soil, 29–46% cement materials, 2–14% metal particles, 6–11% coal/fly ash, and 0–2% salt. Recently, Mølhave et al. (2000) analyzed the physical, biological and chemical composition of airborne particles and vacuum dust sampled from seven Danish office buildings. The origin and content of 179 airborne dust particles trapped on 0.22-μm membrane filters were determined by optical polarization microscopy, scanning electron microscopy (SEM), and energy dispersive X-ray fluorescence (ED-XRF). Results indicated that 53% of the airborne particles were of inorganic or mineral origin (including quartz, calcite, feldspar, gypsum, salt), 18% were of organic origin (including organic macromolecules and microorganisms), 4% were of human origin (hair and skin fragments), 3% were organic fibers (wood, paper, textile), 2% were metal particles (including Fe, Cu, Ni), and the remainder were unidentified.

Measurement of the size distribution indicated that 10 wt.% of the airborne dust particles was less than 2.5 μm and 50 wt.% was greater than 10 μm. Gravimetric analysis indicated that water content was 2.5 $\pm$ 1%, and that organic matter content (estimated by percent weight loss-on-ignition) ranged from 58 to 70% and averaged 66%. Vacuum dust samples (sieved using ASTM mesh 8) collected in the same study (Møl-have et al., 2000) contained about 2% water, about 32–33% organic matter, and had a specific density of 1.0 $g\,cm^{-3}$. Minerals in vacuum dust samples identified by SEM included apatite, biotite, calcite, feldspar, gypsum, ilmenite, halite, hematite, horn-blende, quartz, limonite, and pyrite. Most of the mineral content was judged to origi-nate from the outdoor environment, includ-ing gypsum and halite derived from road salt.

11.2.1
Element Speciation

Information on element speciation in house dust is limited, as most current analytical methods are designed to determine total or near-total element concentrations. In air-borne particulate matter, metals occur both as salts and complexed to inorganic and organic components, including humic-like substances (Ghio et al., 1996). Using induc-tively-coupled plasma (ICP) optical emis-sion spectroscopy, (Ghio et al., 1996) found a correlation between acid-soluble concen-trations of first-row transition metals and the humic-like substance content of parti-cles trapped on PM_{10} filters, and concluded that humic-like substances in air pollution particles act as organic metal chelators. These findings are relevant to the indoor environment, as organic matter is a signifi-cant component of house dust. Fergusson and Kim (1991) reported 40–50% organic matter content in residential house dust (estimated by percent weight loss on igni-tion), similar to the office environments studied by Mølhave et al. (2000) and higher than the range of 3 to 20% typically found in street dust. Mølhave et al. (2000) observed that airborne dust has a higher organic con-tent than settled dust collected in the same indoor environment, likely caused by the low resuspension rate of higher density inor-ganic particles.

Sequential extraction procedures devel-oped by Tessier et al. (1979) were applied to house dust and street dust by Fergusson and Kim (1991). Results indicated that most of the Pb (65–85%), Zn (70–95%), and Cd (60–80%) was associated with the carbonate phase and the amorphous iron/manganese hydrous oxide phases, but it was noted that the distribution of metals amongst different matrix components varied with proximity to different industrial sources. In another study, Zn was associated with the calcium-rich matrix, whereas Cd and Pb were associated with the silicon-rich matrix (Johnson et al., 1982).

11.2.2
Particle Size

Particle size is an important influence on the metal concentration of house dust, and is considered a critical sampling parameter in most study designs (see review by Paus-tenbach et al. (1997). Higher elemental con-centrations are often (but not always) associ-ated with smaller particle size fractions, which have a greater surface area-to-mass ratio than coarse fractions. Gulson et al. (1995) reported that Pb concentrations are two- to nine-fold higher in fine fractions compared to bulk (unsieved) fractions. Using ED-XRF, Lisiewicz et al. (2000) deter-mined concentrations of Cr, Ni, Cu, Zn, Pb, and Br in three size fractions ($< 32\ \mu m$, $32–63\ \mu m$, and $63–125\ \mu m$) of household vacuum samples collected in Warsaw, the capital city of Poland. Results showed that elemental concentrations of Cr, Ni, Cu, and Pb increased by about 5% to 30% for each decrease in size fraction (Figure 11.1), while Br and Zn did not appear to vary systematically with particle size. Particle size is also a relevant parame-

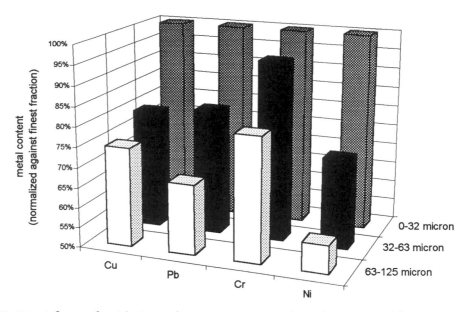

Fig. 11.1   Influence of particle size on element concentrations in house dust, interpreted from geometric mean data for 23 vacuum samples collected in Warsaw, Poland by Lisiewicz et al. (2000). Concentrations in coarser size fractions (32–63 µm and 63–125 µm) are expressed as percentages of concentrations in fine size fraction (<32 µg, where Cr=100 µg g^{-1}; Ni=58 µg g^{-1}; Cu=162 µg g^{-1}; and Pb=194 µg g^{-1}).

ter in exposure assessments. Smaller size fractions (< 100 µm) tend to adhere to a child's hand and are thus more likely to be ingested; once ingested, these tend to have a higher bioavailability than larger particles (Barltrop and Meek, 1979; Duggan and Inskip, 1985; Davies et al., 1990; Mushak, 1998; Gulson et al., 1995; Paustenbach et al., 1997 and references cited therein). With respect to the inhalation pathway, particle size is a key parameter: the smaller the aerodynamic diameter, the more likely that a particle will be re-suspended and available for inhalation during cleaning activities.

11.3
Measurement of Elements in Indoor Environments

The characterization of elements and their compounds in the indoor environment is a newly emerging field, and requires researchers to develop methods that depart from standard occupational health and safety protocols. A variety of techniques are used currently to characterize metals and metalloids associated with airborne particulate matter and settled dust in indoor environments. Consequently, differences in sample collection procedures, extraction methods, instrumental detection limits, reported size fractions, and units of measure amongst different studies often result in data sets that are not directly comparable.

11.3.1
Vacuum Methods

Most indoor studies report metal concentrations in vacuum dust samples using mass concentration units (µg g^{-1}; Table 11.1). In studies aimed at estimating residential exposures, settled dust samples typically are col-

Tab. 11.1: Element concentrations ($\mu g\ g^{-1}$) in common household dusts of cities and towns worldwide (not including mining towns or other industrial/geochemical hotspots). See text for description of analytical methods

	Particle/sieve size	Pb range (geo. mean)	Cd range (geo. mean)	Cu range (geo. mean)	Zn range (geo. mean)
London, England[1]					
n = 683	<1 mm	5–36 900 (1007)	<1–336 (7.7)	9–5300 (208)	81–114 800 (1324)
England, Scotland & Wales[1]					
n ~3957	<1 mm	13–34 500 (507)	<1–8040 (6.8)	3–48 800 (204)	128–70 100 (1060)
Germany[2]					
n ~ 3900	nr	<0.1–37 000 (5.9)	<0.05–220 (0.86)	<33–12 540 (80)	<212–30 600 (475)
Warsaw, Poland[3]					
n = 23	63–125 μm	64–318 (131)	nr	48–336 (121)	534–4080 (1250)
Christchurch, New Zealand[4]					
n = 120	Not sieved	101–3510 (573)	0.557–21.0 (4.2)	54–1010 (165)	871–205 000 (10 400)
Louisiana, USA[5]					
n = 39	nr	11.9–38 604.3 (nr)	0.1–620.1 (nr)	22–1976 (nr)	nr
Ottawa, Canada[6]					
n = 48	100–250 μm	50–3226 (233)	1.12–34.94 (4.42)	59–601 (171)	239–1840 (628)
State of Bharain[7]					
n = 8	30 mesh	120–600 (nr)	nr	nr	15–114 (nr)
Hong Kong, China[8]					
n = 151	<250 μm	0.1–1415 (157*)	0.2–2341 (4.3*)	46–32 611 (311*)	72–12 940 (1409*)
Taejon area, Korea[9]					
n = 37	<180 μm	19–491 (80)	nr	26 –1360 (128)	95–2500 (320)

*median; nr, not reported. [1] Thornton et al. (1985) and Culbard et al. (1988); [2] Seifert et al. (2000); [3] Lisiewicz et al. (2000); [4] Kim and Fergusson (1993); [5] Lemus et al. (1996); [6] Rasmussen et al. (2001); [7] Akhter and Madany (1993); [8] Tong and Lam (2000); [9] Kim et al. (1998).

lected from floors and carpets in common living areas using vacuum samplers, and are dried, sieved, homogenized, and analyzed according to protocols originally developed for soils. In other types of studies, such as those aimed at obtaining historical archives of household dust, samples may be collected from less accessible places like attics and other building cavities (Cizdziel and Hodge, 2000; Lead Group Inc., 1999). Vacuum samplers vary from simple house-

hold vacuum cleaners to specialized units that have been custom-designed to improve sampling consistency and minimize loss of fine particles (see comparisons by U.S. EPA, 1995).

Table 11.1 provides examples of recent multi-element vacuum dust surveys in urban environments around the world, not including industrial or geochemical hotspots. The data in Table 11.1 are summarized using geometric means or medians (where available), as indoor geochemical data tend to be log-normally distributed. This is not an exhaustive inventory, and many studies provide data for a variety of other elements. Moreover, summaries such as that in Table 11.1 cannot adequately convey the spatial and temporal heterogeneity of indoor geochemical data, and it must be emphasized that concentrations of Pb and other elements in house dust demonstrate a high degree of variability within an individual house, amongst houses within the same community, and amongst communities within the same geographic region (Laxen et al., 1988; Thornton et al., 1994; Sutton et al., 1995; Adgate et al., 1998; Rasmussen et al., 2001).

Analytical schemes represented in Table 11.1 include total element methods such as ED-XRF (Lisiewicz et al., 2000) or total digestion using a mixture of strong acids followed by ICP-MS (Rasmussen et al., 2001) or atomic absorption spectroscopy (Thornton et al., 1985; Culbard et al., 1988; Tong and Lam, 2000). Akhter and Madany, 1993) and Kim et al., 1998) used a near-total digestion with aqua regia followed by atomic absorption spectroscopy (AAS), while Kim and Fergusson (1993), Seifert et al. (2000), and Lemus et al. (1996) digested samples using nitric acid alone followed by various optical spectroscopic methods. Different particle size fractions were analyzed in the various surveys (see Table 11.1). Comparisons between studies should be made with caution, as these differences in analytical approaches may affect absolute values for certain metals.

Summary data (Table 11.1) suggest that worldwide geometric mean concentrations of Pb in house dust generally vary between 80 and 1007 $\mu g\ g^{-1}$, while Cd concentrations fall within a tighter range (4.2 to 7.7 $\mu g\ g^{-1}$), as do Cu concentrations (121 to 311 $\mu g\ g^{-1}$). An exception is the German survey by Sei-

fert et al., 2000) in which geometric means of Pb (5.9 $\mu g\ g^{-1}$), Cd (0.86 $\mu g\ g^{-1}$), and Cu (80 $\mu g\ g^{-1}$) are at least an order of magnitude lower than in the other surveys (Table 11.1). The cause of this is unknown. Geometric mean concentrations of Zn generally fall within the range 320 to 1409 $\mu g\ g^{-1}$, with the exception of the New Zealand survey (Kim and Fergusson, 1993), in which Zn is at least an order of magnitude higher (Table 11.1). Kim and Fergusson (1993) attributed the source of Zn in New Zealand homes to galvanized roofing material and attrition of carpet backings.

11.3.2
Dust and Metal Deposition Techniques

Dust and metal deposition rates are measured by setting out a collection vessel, such as a beaker or Petri dish, on an exposed surface inside the home to collect dust over a specified period of time, yielding values of mass per unit area per unit time (e.g., mg m^{-2} day^{-1} and $\mu g\ m^{-2}$ day^{-1}; Table 11.2). Gulson et al. (1995) evaluated the advantages and disadvantages of this method, and concluded that it is a simple and informative technique that deserves more attention than it has received. Dust and metal deposition data from two large surveys of children's homes in Germany are compared in Table 11.2. Seifert et al. (2000) sampled 600 households representative of the whole of Germany, with the purpose of obtaining exposure information for the general population, while Meyer et al. (1999) sampled 454 households in Hettstedt, a city in eastern Germany having a long history of mining and smelting and other industrial sources of metals (Table 11.2). In both studies, a sampling cup was placed in each home for a period of one year, after which As, Cd, and Pb concentrations were determined using nitric acid digestion followed by electrothermal AAS (Meyer et al., 1999; Seifert et al., 2000). Table 11.2 shows that geometric mean and median Pb and As deposition rates are three-fold higher in the smelter town, while Cd deposition rates are 1.5-fold higher, compared to the general population.

11.3.3
Surface Wipe Techniques

Common wipe methods use pre-moistened paper towels or commercial towelettes to wipe dust from smooth surfaces such as floors and windowsills (see variations in U.S. EPA, 1995), yielding a value for mass of metal per unit area (e.g., $\mu g\ m^{-2}$). This approach helps to define the highly variable parameter of housecleaning practices, and is considered by many to provide the best reflection of potential human exposure to both Pb (Davies et al., 1990; Lanphear et al., 1995) and Cr (Lioy et al., 1992).

The choice of dust sampling method depends on the purpose of the study. A number of publications compare various wipe and vacuum methods for Pb sampling (e.g., U.S. EPA, 1995; Sutton et al., 1995; Laxen et al., 1988). Wipe methods have the advantage of being simple and inexpensive compared to vacuum techniques. However, vacuum methods have some distinct advantages over common wipe methods: vacuum

methods can collect dust from carpets whereas wipe methods cannot; vacuum methods permit the reporting of element concentrations as both area concentration units ($\mu g\ m^{-2}$) and mass concentration units ($\mu g\ g^{-1}$); and vacuum methods permit sieving of the dust sample into relevant size fractions. Although the surface wipe method is commonly termed "Pb loading," elapsed time of dust accumulation is not included as a measurement parameter (in contrast to dust deposition methods; Table 11.2). In a study of surface wipe sampling in the workplace, Caplan (1993) reported that ratios between surface and air concentrations vary over a million-fold range, with most of the data showing a thousand-fold range. Caplan (1993) concluded that, although wipe samples are a quick and easy indicator of the need for better housekeeping in Health Physics occupational settings, there is no quantitative relationship between wipe samples and concurrent airborne concentrations of the same contaminant.

11.3.4
Indoor Air Sampling

Metal concentrations in indoor air are reported in units of $\mu g\ m^{-3}$ or $ng\ m^{-3}$ (Table 11.3), and are generally orders of

Tab. 11.2: Measurement of dust and metal deposition rates in children's homes in Germany, from two studies using the same sampling and analytical techniques (described in text)

	Dust [mg m⁻² day⁻¹]	Pb [μg m⁻² day⁻¹]	Cd [μg m⁻² day⁻¹]	As [μg m⁻² day⁻¹]
Representative German population (600 households with children aged 6–14 years)[1]				
Geometric mean	5.4	0.39	0.016	0.008
Median	5.8	0.38	0.016	0.008
Smelter town in eastern Germany (454 households with children aged 4–15 years)[2]				
Geometric mean	8.9	1.14	0.024	0.023
Median	9.2	1.12	0.022	0.021

[1] Seifert et al. (2000); [2] Meyer et al. (1999)

Tab. 11.3: Airborne elemental concentrations (in units of µg m^{-3} or ng m^{-3} as shown), measured inside homes and public places, compared with outdoor concentrations where available. See text for description of analytical methods

Element	Units	Air concentration	Air concentration
Residential monitoring (UK)[1]		Indoor range (Geo. mean) n = 607	Outdoor range (Geo. mean) n = 605
Pb	µg m^{-3}	0.08–0.88 (0.26)	0.12–1.53 (0.43)
Residential Monitoring (The Netherlands)[2]		Indoor range (Geo. mean) n = 101	Outdoor range (Geo. mean) n = 30
Pb	µg m^{-3}	0.13–0.74 (0.26)	0.28–0.52 (0.41)
Residential Monitoring (USA)[3]		Indoor range (median) n = 48	Outdoor (median) n = 4
Pb	µg m^{-3}	0.0015–0.088 (0.011)	(0.032)
Al	µg m^{-3}	ND – 0.739 (0.0992)	(0.44)
Ca	µg m^{-3}	ND – 2.71(0.481)	(4.1)
Fe	µg m^{-3}	ND – 1.46 (0.263)	(1.6)
Mg	µg m^{-3}	0.0071–0.746 (0.137)	(1.38)
Na	µg m^{-3}	ND – 2.13 (0.302)	(0.57)
Zn	µg m^{-3}	ND – 0.17 (0.045)	(0.106)
Cu	µg m^{-3}	ND– 0.044 (0.0009)	(0.045)
Mn	µg m^{-3}	ND – 0.06 (0.012)	(0.111)
K	µg m^{-3}	ND – 0.564 (0.184)	(0.253)
Ni	µg m^{-3}	ND – 0.008 (0.003)	(0.0034)
Se	µg m^{-3}	ND – 0.007 (0.0003)	(0.002)
Residential monitoring (USA)[4]		Indoor range (median) n = 50 houses at 6 sites	Outdoor range (median) n = 50 houses at 6 sites
Cr	ng m^{-3}	1–17 (5)	4–7 (6)
Indoor public places (USA)[5]		Smoking places n = 8 range (average)	Non–smoking places n = 1
As	ng m^{-3}	<0.1–1.0 (0.4)	<0.13
Br	ng m^{-3}	4–39 (18)	<2.4
Cd	ng m^{-3}	4.0–37.9 (14.4)	<1.8
Cl	µg m^{-3}	0.5–10.1 (5.0)	0.07
K	µg m^{-3}	<0.1–8.5 (4.6)	<0.6
Na	µg m^{-3}	0.5–1.9 (1.4)	0.1
Sb	ng m^{-3}	0.5–1.9 (1.4)	<0.03
Zn	µg m^{-3}	0.02–0.30 (0.14)	<0.03

[1]Davies et al. (1990); [2] Diemel et al. (1981); [3] Van Winkle and Scheff (2001); [4] Lioy et al. (1992); [5] Landsberger and Wu (1995). ND, not detected.

magnitude lower than indoor occupational guidelines. Thus, for the reliable measurement of airborne metals in homes and offices, very low detection limits are required for both gravimetric analysis and elemental analysis, compared to the requirements of standard occupational health and safety protocols.

Air particle samples are collected inside homes using personal or stationary air sam-

plers; usually with low flow rates to minimize noise and disruption while the residents are carrying out their normal daily
routines. Low-volume samplers are typically
operated below 10 L min^{-1}, in contrast to
dichotomous samplers (15–20 L min^{-1}),
and high-volume samplers (1000 L min^{-1}
or more). The disadvantage of low-volume
samplers is that under typical North American indoor conditions ($<$ 100 µg m^{-3} particulate matter), they yield very small particle
masses ($<$ 1 mg per filter). For example,
Wu and Feng (2000) found that a MINI-
VOL^{TM} device operated at between 4 and
5 L min^{-1} for 24 h yielded particle masses
ranging from 0.015 to 0.274 mg, and a
PEM^{TM} operated at between 9 and 10 L
min^{-1} for 24 h yielded particle masses ranging from 0.34 to 0.54 mg. For gravimetric
analysis of these samples, Wu and Feng
(2000) achieved an instrumental detection
limit of 5 µg per filter based on laboratory
blanks, and a method detection limit of
20 µg per filter based on field blanks. In contrast, high-volume samplers operated under
similar conditions yield particle masses
ranging from 3 to 210 mg per filter (n =
36 samples) for sampling periods of 12 to
24 h (Biran et al., 1996).

For the above reasons, indoor air filters
often yield particle masses that are orders of
magnitude below the mass requirements
for quantitative ED-XRF analysis – the
multi-element detection method traditionally
used for air analysis. Researchers are currently seeking alternative multi-element analytical techniques with adequate sensitivity
for indoor air samples. Indoor and outdoor
air concentration data, from a variety of studies which employed different sampling and
analytical approaches, are summarized in
Table 11.3. Van Winkle and Scheff (2001)
used high-volume samplers equipped with
37-mm quartz fiber filters and an air flow
rate of 25 L min^{-1} to monitor 10 homes in

southeast Chicago (Table 11.3). A total of 48
24-h air filter samples (collected over a one-
year period) were analyzed using electrothermal AAS for Pb, As, and Se, and using ICP-
atomic emission spectroscopy (AES) for the
remainder of the elements. Despite the relatively high flow rate, more than 75% of the
filters were below detection limit for As,
Cr, Cd, V, Sn, Co, and Mo and all were
below the detection limit for Sb. For the elements shown in Table 11.3, 25–75% of the
filters were above the detection limit for K,
Ni and Se, and $>$ 75% were above the detection limit for the remainder of the elements.

Landsberger and Wu (1995) used instrumental neutron activation analysis to determine metal concentrations in particulate
matter generated from cigarette smoking
in a variety of public indoor environments
in Champaign-Urbana, Illinois, U.S.A
(Table 11.3). Samples were collected using
personal samplers with air flow rates up to
5 L min^{-1}, equipped with 47-mm Teflon filters, in eight restaurants and clubs where
tobacco smoking is permitted and in one
tobacco-free setting. Very low detection
limits (0.03 ng per filter for Sb and Zn;
0.13 ng per filter for As; 2 ng per filter for
Cd) were achieved using epithermal neutron
irradiation and a Compton suppression
gamma-ray detection system.

In a study of residential exposure to Cr-
laden wastes in Hudson County, New
Jersey, Lioy et al., 1992) collected 3-day and
4-day composite samples using low-volume
air samplers equipped with 37-mm Teflon
filters. Total Cr in airborne particles was
determined using XRF with a detection
limit of 2 µg g^{-1} (shown in Table 11.3), and
extractable Cr (not shown) was determined
using nitric acid/sulfuric acid digestion
and ICP-AES. The mean ratio of extractable
Cr to total Cr was 0.3.

Measurements of airborne Pb concentrations in Birmingham, England by Davies

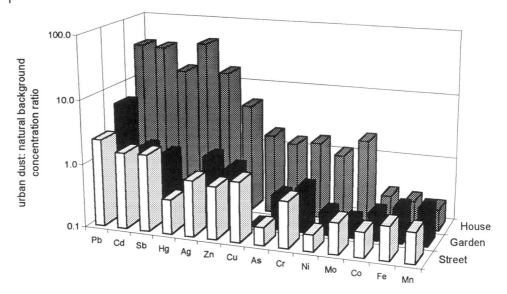

Fig. 11.2 Variation in multi-element profiles of street dust, garden soil, and house dust, sampled from 48 houses in the city of Ottawa, Canada. Total element concentrations in the 100–250 μm size fraction are normalized against natural background concentrations, using the 95th%ile for all media (modified from Rasmussen et al., 2001).

to accumulate in the indoor environment (Paustenbach et al., 1997). Alternatively, in cases where indoor dust samples are not sieved, but exterior soil samples are sieved, I/O ratios may be underestimated. Thus, comparisons should be based on a consistent size fraction, to account for potential differences in particle size distribution in indoor versus outdoor environments. I/O ratios within studies are not likely to be influenced by the analytical method where the efficiency is independent of the matrix (e.g., XRF). However, if a weak acid digestion is used, a bias may exist for elements that are encapsulated in a refractory matrix which is more prevalent in one environment than another (i.e., indoors versus outdoors). Note that the comparisons in Figure 11.2 are based on an aggressive digestion method and the same particle size range (100–250 μm) for all three media (Rasmussen et al., 2001).

11.5
Sources of Elements in the Indoor Environment

Where elemental concentrations are higher in indoor dust than in soil and exterior dust, it may be expected that indoor sources, including construction materials, paint, furnishings and other consumer products, are an important influence. Relative contributions of indoor sources of metals are difficult to quantify. Harrison (1979) reported that abrasion of household objects causes deposition of metal particles in house dust, but observed that the majority of these particles are not readily recognizable by microscopy. Elements such as Al, Ba, Si, and Ti have been proposed as "conservative tracers" to estimate the relative contribution of soil to house dust (Trowbridge and Burmaster, 1997; Calabrese and Stanek, 1992), but these estimates are based on the assumption that there are no indoor sources of these elements. In fact, there may be numerous indoor sources of these elements (e.g., talc, spray deodorant, paint), and such methods are limited by their inability to distinguish elements derived from soil versus common household products. Even in the case of Pb, which is the most thoroughly studied metal in the indoor environment, the sources of variability are not fully understood. Sutton et al., 1995) concluded that only 13% of the variability in dust lead levels in 933 California households could be explained by the environmental factors measured in the survey (interior and exterior paint, soil, Pb-related occupation, and age of home). Gulson et al. (1995) used a combination of SEM and stable Pb isotope analysis to apportion sources of Pb in a range of urban and mining environments in Australia, and found that total Pb concentration measurements of bulk dust samples are meaningless for source apportionment.

11.5.1
House Characteristics

Many studies report that the age of the home is a chief predictor of Pb concentration in household dust, largely due to the prevalence of Pb-based paint in older homes (Sutton et al., 1995; Rasmussen et al., 2001; Adgate et al., 1998; Davies et al., 1990; Thornton et al., 1994; Meyer et al., 1999). Older paints had concentrations of up to 50% Pb, while later paints had lower Pb concentrations but increased concentrations of Ti, Zn and Ba (U.S. EPA, 1986). Thus, indoor Pb levels tend to increase while houses are being renovated, particularly if the renovation involves electric sanding or burning with a blow lamp (Laxen et al., 1988; Davies et al., 1990). Galvanized iron roofing material, used commonly in Australia and New Zealand, con-tributes to elevated indoor metal concentrations (Kim and Fergusson, 1993). Interior dust deposition and Pb and Cd deposition rates are strongly influenced by other aspects of the construction design, including ventilation and insulation systems, and state of maintenance or repair (Meyer et al., 1999). Humidity, which promotes particle coagulation and deposition in indoor environments, is an important factor controlling dustiness. Meyer et al. (1999) found that dust loading rates were 1.73 times higher in damp houses than in dry houses, and that dampness was associated with higher Pb, Cd, and As levels in house dust.

11.5.2
Mode of Cooking and Heating

In a study of ten U.S. homes, Van Winkle and Scheff (2001) found that cooking activity was the dominant source of indoor element emissions. Elevated air concentrations of Al, Ca, Fe, Mg, and Zn were associated with the cooking of fried foods. Several U.S. studies reviewed by Wallace (1996) identified cooking as an important indoor source of particles, with contributions ranging from about 10 to 20 $\mu g\ m^{-3}$. Half of the world's households, located mainly in rural areas of Africa, Asia, and South America, cook their daily meals using wood, crop residues, or animal dung in simple stoves made of rock or clay (Fishbein, 1991). Biomass combustion in Himalayan residences of Nepal generates airborne particle concentrations in the range of 3 to 42 $mg\ m^{-3}$ (total suspended particles), with Ag, Cd, Cu, Pb, and Zn enriched more than 10-fold above their average crustal abundances (Fishbein, 1991 citing data from Davidson et al., 1986).

Previously, Fishbein (1991) provided data showing that charcoal, wood, and biomass combustion can significantly enhance indoor metal concentrations, particularly where these fuels are burned to heat enclosed, poorly ventilated living areas. Heating with coal is associated with elevated As loading rates, compared to central heating (Meyer et al., 1999). Metal-bearing smoke particles can enter living spaces from improperly installed, maintained, or operated stoves and fireplaces, or by penetration from outdoor sources such as chimneys of neighboring homes (Fishbein, 1991). In certain parts of Berlin, 40–50% of houses are still heated with coal, and Moriske et al. (1996) found that these homes are characterized by higher dust and metal deposition rates than centrally heated homes – with the exception of one malfunctioning central heating system which caused higher indoor air pollution than any other home in the study.

11.5.3
Activities of Residents

Metal loading rates tend to increase with the number of people who occupy a house, especially if some members of the household are occupationally exposed, or pursue hobbies involving metals. Metal-laden dust regularly carried home on employees' clothes and skin ultimately leads to metal enrichment in the home environment (Niosh 1995; Sutton et al., 1995; Fergusson and Kim, 1991, and references cited therein). In a German smelter town, Meyer et al. (1999) found a significant relationship between occupational exposure of either parent and elevated Pb and Cd levels in house dust, consistent with previous findings. Home hobbyists often use products containing the same metal compounds found in the workplace. Examples include ceramic glazes and fluxes, paint pigments, toners used in photographic processing, came strips of Pb and solder used in stained

glass, and inks used in screen printing (Paul, 1993).

Smoking is an important indoor source of fine and coarse particles, with estimated increases of 25 to 45 µg m^{-3} PM$_{2.5}$ in homes with smokers (Wallace, 1996). Previously, Fishbein (1991) summarized concentration data for a range of metals, metalloids, and radioactive elements in cigarette tobacco, mainstream smoke, and sidestream smoke. Ligocki et al. (1995) estimated that indoor air metal concentrations in homes with smokers exceed those in homes without smokers by an average increment of 1.3 ng m^{-3} for Cd, 0.18 ng m^{-3} for As, and 21 pg m^{-3} for Cr, consistent with other studies (e.g., Lioy et al., 1992; Leaderer et al., 1994; Landsberger and Wu, 1995). However, some surveys revealed no correlation between metal levels and smoking behavior (e.g., Meyer et al., 1999; Kim et al., 1998), possibly due to the interplay of other factors.

11.5.4
Consumer Products

Elevated concentrations of airborne Hg in homes and schools are generally attributed to Hg spilled from thermometers, gas meters, and other consumer products. More localized occurrences are associated with ethnic practices such as sprinkling liquid Hg under beds, or wearing amulets filled with liquid Hg (Forman et al., 2000). It has been estimated that up to 10% of U.S. households may have levels of airborne Hg above the U.S. EPA reference concentration (300 ng m^{-3}) due to historic accidents with Hg-containing devices (Carpi and Chen, 2001). In one New York City residence, where indoor Hg concentrations (523 ± 6 ng m^{-3}) exceeded the U.S. EPA reference concentration, breakage of a Hg thermometer had occurred within the 6-month period prior to monitoring (Carpi and Chen, 2001). It is likely that house dust also plays an important role in the indoor Hg cycle, based on a Canadian survey of 50 houses which showed geometric mean Hg concentrations to be 28-fold higher in indoor dust than in exterior soils (Rasmussen et al., 2001).

Attrition of rubber carpet backing contributed significantly to elevated concentrations of Cd in household dust sampled in an urban setting in Illinois, U.S. by Solomon and Hartford (1976), although this source of Cd was absent from British homes sampled by Harrison (1979). In some areas, carpet backing may be a source of Zn, in addition to other household sources of Zn such as rubber, paints, and fillers used in linoleum (Fergusson and Kim, 1991).

Nriagu and Kim (2000) measured the amount of Pb and Zn released from 14 different brands of candles with metal wicks sold in Michigan, U.S. Emissions of Pb from burning candles ranged from 0.5 to 66 µg h^{-1} and emissions of Zn ranged from 1.2 to 124 µg h^{-1}. Van Winkle and Scheff (2001) estimated element emission factors (also in µg h^{-1}) for a variety of consumer products found in urban homes in southeast Chicago. Emission factors were considered significant (at $P < 0.05$) for copper from spot-remover, closet storage of chemicals, and glue storage; As and Se from mothball storage; Co and Mg from washer/drier use; and Mg from hairspray storage (Van Winkle and Scheff, 2001).

Aluminum trihydrate, magnesium hydroxide, calcium and zinc molybdates, antimony pentoxide, and zinc borate are examples of inorganic compounds used as flame retardants in the manufacture of household furniture, upholstery, wall coverings, draperies, and carpets (National Research Council (NRC), 2000). Antimony trioxide is sometimes used in combination with zinc borate or brominated flame retardants, and is also used in paper, adhesives, plastics and as a paint pigment. It is likely that antimony (Sb) compounds and other flame-retardant chemicals accumulate in house dust, as studies have reported elevated Sb levels in house dust in relation to its natural crustal abundance (see Chapter 1). In four U.K. towns, the median house dust Sb concentration was 13 µg g^{-1}, with maxima exceeding 100 µg g^{-1} (Thompson and Thornton 1997), while in Ottawa, Canada the median house dust Sb concentration was 5 µg g^{-1}, with a maximum of 57 µg g^{-1} (Rasmussen et al., 2001).

Personal care products that are potential indoor sources of elements include home remedies, underarm antiperspirants, and cosmetics. Zinc oxide and titanium oxide are used in many commercial sunscreen products, and zinc oxide (as calamine) has long been recognized for its healing properties (Lansdown and Taylor 1997). Aerosolized aluminum chlorohydrate, a common antiperspirant formulation, is one of many potential indoor sources of Al (Fishbein, 1991). Sainio et al., 2000) determined Ni, Pb, Co, As, and Cr concentrations in commercial eyeshadows, and found that 75% of the colors contained more than 5 µg g^{-1} of at least one of the elements. Lead-containing remedies and cosmetics used by some Middle Eastern and Asian communities are listed by ATSDR (2000). Two Mexican folk remedies which contain Pb are azarcon and greta, which are used to treat "empacho," which is a colic-like illness (ATSDR, 2000).

11.6
Bioavailability of Metals in the Indoor Environment

There are many data gaps and areas of controversy associated with quantifying exposures to metals in house dust (see Mushak, 1998; Paustenbach et al., 1997). A major source of uncertainty is the estimation of gastrointestinal absorption (Diamond et al., 1998). Complex interactions between ingested particles and gut physiological processes can either increase or decrease metal bioavailability (see Mielke and Heneghan, 1991). Absorption of metals in humans is influenced by a variety of factors including the exposure pathway, the chemical form of the metal, the matrix composition, the age of the host, contents of the gastrointestinal tract, temporal pattern of meal consumption, diet, and nutritional status (Diamond et al., 1998; Fishbein, 1991). Recent reviews indicate that no single *in vitro* test has emerged as the acceptable choice for estimating metal absorption in either the gastrointestinal tract (Diamond et al., 1998; Mushak, 1998; Canady et al., 1997; Ruby et al., 1999) or the respiratory system (Ansoborlo et al., 1999). Appropriate animal models, combined with detailed knowledge of the sample mineralogy and chemical speciation, are generally recommended for reliable estimates of metal bioavailability in a given matrix (Mushak, 1998).

One of the major factors influencing biological availability and toxic action of particle-bound metal compounds is their solubility under the pH and ionic strength conditions of human body fluids. Mullins and Norman (1994) measured metals in wind-blown dust from mine wastes as a function of particle size, which determines where in the respiratory tract the particle-bound metals will eventually settle, and as a function of solubility in different body fluids. They found that compounds of Cd and Mn in the < 10-μm fraction were highly soluble in simulated stomach fluid (40 to 91%) and simulated lung fluid (27 to 100%). Compounds of As were fairly soluble in stomach fluid (17 to 37%) but not in lung fluid (0.1 to

0.4%), while solubilities of Pb compounds varied widely amongst waste sites (from 0.3 to 25% in lung fluid and from 4 to 36% in stomach fluid).

In the field of toy safety, guidelines have been developed for the determination of the migration of certain metals and metalloid elements into simulated stomach acid (EN-71, 1995). In addition to toys and other articles used by children, the EN-71 guidelines apply to childcare products that come into frequent contact with children and give rise to ingestion (Bowin, 1999). The elements covered by the EN-71 definition are: As, Sb, Ba, Cd, Cr, Pb, Hg, and Se. The method applies to parts that are small enough to be swallowed, and to particles that may become detached and ingested if the article is bitten, and thus migration into 0.07 M HCl (for 2 h at pH 1.5 and 37 °C) is determined. The guidelines do not apply to articles that are sucked rather than bitten, as migration of certain metals into saliva is slower than migration into the HCl solution specified by EN-71 (Bowin, 1999). While the EN-71 method does not account for the complexities of the human gastrointestinal tract, it does provide a simple and reproducible test for screening the large numbers of unknown samples encountered in the monitoring of consumer products.

A few studies have explored the potential of the toy safety test for screening household dust samples in areas where elevated metal levels are suspected, such as inner urban settings and mining towns. Duggan and Williams (1977) used the standard 0.07 M HCl extraction protocol to determine available Pb in street dust compared to total Pb extracted with aqua regia (HCl/HNO_3). Later, Harrison (1979) applied the same approach for the determination of Pb, Cr, Co, Cu, Cd, Ni, and Zn in household and exterior urban dusts. Both studies yielded total Pb:available Pb ratios in the range of 1.3 to 2.1. Gulson et al. (1994) used a similar approach (0.1 M HCl) to determine available Pb in house dust, soil, and weathered ore material in the Broken Hill mining community in Australia, where in 1992 about 85% of children displayed blood Pb levels above 10 µg dL^{-1}. Gulson et al. (1994) found that a high proportion of Pb in the ore-derived dust is soluble and available for uptake, comparable with the estimates for urban dust (Harrison, 1979; Duggan and Williams, 1977).

In general, metal speciation in house dust, and the range of proportions of various metal compounds across different types of house dust, is an unknown. The above studies have shown that the solubility of Pb in ore-derived dust varies widely amongst different mine settings (Mullins and Norman, 1994; Gulson et al. (1994). For the investigation of metals in dust samples of unknown composition, Mullins and Norman (1994) and many others have recommended that duplicate samples be obtained, one for total metal content and the other for dissolution in acidic simulated stomach fluid. Significant differences in metal content between the two samples should then be taken into account to assess the potential hazard for systemic toxicity (Mullins and Norman, 1994).

References

ADGATE JL, WILLIS RD, BUCKLEY TJ, CHOW JC, WATSON JG, RHOADS GG and LIOY PJ (1998) *Chemical mass balance source apportionment of lead in house dust.* Environ Sci Technol. 32:108–114.

AKHTER SM and MADANY IM (1993) *Heavy metals in street and house dust in Bahrain.* Water Air Soil Poll 66:111–119.

ANSOBORLO E, HENGE-NAPOLI MH, CHAZEL V, GIBERT R and GUILMETTE RA (1999) *Review and critical analysis of available in vitro dissolution tests.* Health Physics 77:638–645.

ATSDR (2000) Agency for Toxic Substances and Disease Registry Case Studies in Environmental Medicine: *Lead Toxicity.* http://www.atsdr.cdc.gov/HEC/CSEM/lead/exposure_pathways.html

BARLTROP D and MEEK F (1979) *Effect of particle size on lead absorption from the gut.* Arch Environ Health 34:280–285.

BIRAN R, TANG Y, BROOK JR, VINCENT R and KEELER GJ (1996) *Aqueous extraction of airborne particulate matter collected on hi-vol Teflon filters.* Int J Environ Anal Chem 63:315–322.

BORNSCHEIN RL, SUCCOP PA, DIETRICH KN, CLARK CS, QUE HEE S and HAMMOND PB (1985) Environ Research 38:108–118.

BOWIN J (1999) *Child safety: risk assessment and design solutions.* Child Safety, 1st Edn. Standardiseringsgruppen STG och SIS Förlag AB.

CALABRESE EJ and STANEK EJ (1992) *What proportion of household dust is derived from outdoor soil?* J Soil Contam 1:253–263.

CANADY RA, HANLEY JE and SUSTEN AS (1997) *ATSDR science panel on the bioavailability of mercury in soils: lessons learned.* Risk Analysis 17:527–532.

CAPLAN KJ (1993) *The significance of wipe samples.* Am Indust Hyg Assoc J 53:70–75.

CARPI A and CHEN YF (2001) *Gaseous elemental mercury as an indoor air pollutant.* Environ Sci Technol 35:4170–4173.

CDC (Centers for Disease Control) (1991) *Preventing Lead Poisoning in Young Children*: A Statement by the Centers for Disease Control. U.S. Department of Health and Human Services. Atlanta, GA, October 1991.

CHARNEY E, SAYRE J and COULTER M. (1980) *Increased lead absorption in inner city children: where does the lead come from?* Pediatrics 65:226–231.

CIZDZIEL JV and HODGE VF (2000) *Attics as archives for house infiltrating pollutants: trace elements and pesticides in attic dust and soil from southern Nevada and Utah.* Microchem J 64:85–92.

CULBARD E, MOORCROFT S, WATT J and THORNTON I (1983) *A nationwide reconnaissance survey of metals in urban dusts and soils.* Minerals and the Environment 5:82–84.

CULBARD EB, THORNTON I, WATT J, WHEATLEY M, MOORCROFT S and THOMPSON M (1988) *Metal contamination in British urban dusts and soils.* J Environ Qual 17:226–234.

DAVIDSON CI, LIN SW, OSBORN JF, MANDEY MR, RASMUSSEN RA and KHALLI MAK (1986) *Indoor and outdoor air pollution in the Himalaya.* Environ Sci Technol **20**:561–567.

DAVIES DJA, THORNTON I, WATT JM, CULBARD EB, HARVEY PG, DELVES HT, SHERLOCK JC, SMART GA, THOMAS JFA and QUINN MJ (1990) *Lead intake and blood lead in two-year-old U.K. urban children.* Sci Total Environ **90**:13–29.

DIAMOND GL, GOODRUM PE, FELTER SP and RUOFF WL (1998) *Gastrointestinal absorption of metals* [corrected and republished article originally printed in Drug Chem Toxicol 1997 Nov; **20**:345–368]. Drug Chem Toxicol **21**:223–251.

DIEMEL JAL, BRUNEKREEF B, BOLEIJ JSM, BIERSTEKER K and VEENSTRA SJ (1981) *The Arnhem Lead Study: indoor pollution and indoor outdoor relationships.* Environ Research **25**:449–456.

DUGGAN MJ and WILLIAMS S (1977) *Lead in dust in city streets.* Sci Total Environ **7**:91–97.

DUGGAN MJ and INSKIP MJ (1985) *Childhood exposure to lead in surface dust and soil: a community health problem.* Public Health Rev **13**:1–54.

EN-71 (1995) *Safety of Toys – Part 3: Specification for migration of certain elements.* British Standard EN 71 –3:1995, Brussels: European Committee for Standardization.

FERGUSSON JE and KIM ND (1991) *Trace elements in street and house dusts: sources and speciation.* Sci Total Environ **100**:125–150.

FISHBEIN L. (1989) *Metals in the indoor environment.* Toxicol Environ Chem **22**:1–7.

FISHBEIN L (1991) *Indoor Environments: The Role of Metals.* In: Merian E, ed. Metals and Their Compounds in the Environment, pp. 287–309. VCH Verlagsges; Weinheim-New York-Basel-Cambridge.

FORMAN J, MOLINE J, CERNICHIARI E, SAYEGH S, TORRES JC, LANDRIGAN MM, HUDSON J, ADEL HN and LANDRIGAN PJ (2000) *A cluster of pediatric metallic mercury exposure cases treated with meso-2, 3-dimercaptosuccinic acid (DMSA).* Environ Health Perspect **108**:575–577.

GHIO AJ, STONEHUERNER J, PRITCHARD RJ, PIANTADOSI CA, QUIGLEY DR, DREHER KL and COSTA DL (1996) *Humic-like substances in air pollution particulates correlate with concentrations of transition metals and oxidant generation.* Inhal Toxicol **8**:479–494.

GULSON BL, DAVIS JJ, MIXON KJ, DORSCH MJ and BAWDEN-SMITH J (1995) *Sources of lead in soil and dust and the use of dust fallout as a sampling medium.* Sci Total Environ **166**:245–262.

GULSON BL, DAVIS JJ, MIZON KJ, KORSCH MJ, LAW AJ and CSIRO Exploration and Mining (1994) *Lead bioavailability in the environment of children:*

blood lead levels in children can be elevated in a mining community. Arch Environ Health **48**:326–331.

HARRISON RM (1979) *Toxic metals in street and household dusts.* Sci Total Environ **11**:89–97.

HEALTH CANADA (1994) *Update of evidence for low-level effects of lead and blood lead intervention levels and strategies.* Final Report of the Working Group. Environmental Health Directorate, September 1994. Ottawa, Ontario, Canada.

HILTS SR, PAN UW, WHITE ER and YATES CL (1995) *Trail Lead Program: Exposure Pathways Investigations.* Final Report, Trail, British Columbia, Canada: Trail Community Lead Task Force.

JOHNSON DL, FORTMANN R and THORNTON I (1982) *Individual particle characterization of heavy metal rich household dusts.* Trace Subst Environ Health **16**:116–123.

KIM K-W, MYUNG J-H, AHN JS and CHON H-T (1998) *Heavy metal contamination in dusts and stream sediments in the Taejon area, Korea.* J Geochem Explor **64**:409–419.

KIM N and FERGUSSON J (1993) *Concentrations and sources of cadmium, copper, lead, and zinc in house dust in Christchurch, New Zealand.* Sci Total Environ **138**:1–21.

LANDSBERGER S and WU D (1995) *The impact of heavy metals from environmental tobacco smoke on indoor air quality as determined by Compton suppression neutron activation analysis.* Sci Total Environ **173/174**:323–337.

LANPHEAR BP, MATTE TD, ROGERS J, CLICKNER RP, DIETZ B, BORNSCHEIN RL, SUCCOP P, MAHAFFEY KR, DIXON S, GALKE W, RABINOWITZ M, FARFEL M, ROHDE C, SCHWARTZ J, ASHLEY P and JACOBS DE (1998) *The contribution of lead contaminated hose dust and residential soil to children's blood lead levels.* Environ Res (Section A) **79**:51–68.

LANPHEAR BP, EMOND M, JACOBS DE, WEITZMAN M, TANNER M, WINTER NL, YAKIR B and EBERLY S (1995) *A side-by-side comparison of dust collection methods for sampling lead-contaminated house dust.* Environ Res **68**:114–123.

LANSDOWN ABG and TAYLOR A (1997) *Zinc and titanium oxides: Promising UV-absorbers but what influence do they have on the intact skin?* Intern J Cosmetic Sci **19**:167–172.

LAXEN DPH, LINDSAY F, RAAB GM, HUNTER R, FELL GS and FULTON M (1988), *The variability of lead in dusts within the homes of young children.* Environ Geochem Health **10**:3–9.

LAXEN DPH, RAAB GM and FULTON M. (1987) *Children's blood lead and exposure to lead in household dust and water – a basis for an environ-*

mental standard for lead in dust. Sci Total Environ **66**:235–244.

LEAD Group Inc (1999) *Research into Ceiling Dust.* Lead Action News **7**:1.

LEADERER BP, KOUTRAKIS P and BRIGGS SL (1994) *The mass concentration and elemental composition of indoor aerosols in Suffolk and Onandaga counties, New York.* Indoor Air **7**:23–34.

LEMUS R, ABDELGHANI AA, AKERS TG and HORNER WE (1996) *Health risks from exposure to metals in household dusts.* Rev Environ Health **11**:179–189.

LI XD, POON CS and LIU PS (2001) *Heavy metal contamination of urban soils an street dusts in Hong Kong.* Appl Geochem **16**:1316–1368.

LIGOCKI MP, STIEFER PS, ROSENBAUM AS, ATKINSON RD and AXELRAD D (1995) *Cumulative exposures to air toxics: indoor sources.* In: Proceedings, 88th Annual Meeting – Air Waste Management Association, pp. 1–16.

LIOY PJ, FREEMAN NCG, WAINMAN T, STERN AHBR, HOWELL T and SHUPACK SI (1992) *Microenvironmental analysis of residential exposure to chromium laden waste in and around New Jersey homes.* Risk Analysis **12**:287–300.

LIOY PJ, WAINMAN T and WEISEL C (1993) *A wipe sampler for the quantitative measurement of dust on smooth surfaces: laboratory performance studies.* J Exp Anal Environ Epidemiol **3**:315–320.

LISIEWICZ M, HEIMBURGER R and GOLIMOWSKI J (2000) *Granulometry and the content of toxic and potentially toxic elements in vacuum-cleaner collected, indoor dusts of the city of Warsaw.* Sci Total Environ **263**:69–78.

MADDALONI M, LOLACONO N, MANTON W, BLUM C, DREXLER J and GRAZIANO J (1998) *Bioavailability of soilborne lead in adults by stable isotope dilution.* Environ Health Perspect **106**:1589–1594.

MANTON WI, ANGLE CR., STANEK KL, REESE YR and KUEHNEMANN TJ (2000) *Acquisition and retention of lead by young children.* Environ Res **82**:60–80.

MEYER I, HEINRICH J and LIPPOLD U (1999) *Factors affecting lead, cadmium, and arsenic levels in house dust in a smelter town in Eastern Germany.* Environ Res **81**:A32–A44.

MIELKE HW and HENEGHAN JB (1991) *Selected chemical and physical properties of soils and gut physiological processes that influence lead bioavailability.* Chemical Speciation & Bioavailability **3**:129–134.

MORISKE H-J, DREWS M, EBERT G, MENK G, SCHELLER C, SCHÖNDUBE M and KONIECZNY L (1996) *Indoor air pollution by different heating sys-*

tems: coal burning, open fireplace and central heating. Toxicol Lett **88**:349–354.

MULLINS MJP and NORMAN JB (1994) *Solubility of metals in windblown dust from mine waste dump sites.* Appl Occup Environ Hygiene **9**:218–223.

MUSHAK P (1998) *Uses and limits of empirical data in measuring and modeling human lead exposure.* Environ Health Perspect **106**:1467–1484.

MØLHAVE L, SCHNEIDER T, KJÆRGAARD SK, LARSEN L, NORN S and JØRGENSEN O (2000) *House dust in seven Danish offices.* Atmos Environ **34**:4767–4779.

NRC (National Research Council) *(2000) Toxicological Risks of Selected Flame-Retardant Chemicals,* Washington, D. C.: National Academy Press.

NRIAGU JO and KIM M-J (2000) *Emissions of lead and zinc from candles with metal-core wicks.* Sci Total Environ **250**:37–41.

PAUL, M. (1993) *Common household exposures.* In: Occupational and Environmental Reproductive Hazards: A Guide for Clinicians. Baltimore, Williams and Wilkins. pp. 361–378.

PAUSTENBACH DJ, FINLEY BL and LONG TF (1997) *The critical role of house dust in understanding the hazards posed by contaminated soils.* Int J Toxicol **16**:339–362.

RABINOWITZ MB (1995) *Stable isotopes of lead for source identification.* Clin Toxicol **33**:649–655.

RABINOWITZ MB, WATERNAUX C, BELLINGER DC, LEVITON A and NEEDLEMAN HL (1985) *Environmental correlates of infant blood lead levels in Boston.* Environ Res **38**:96–107.

RASMUSSEN PE, SUBRAMANIAN KS and JESSIMAN BJ (2001) *A multi-element profile of house dust in relation to exterior dust and soils in the city of Ottawa, Canada.* Sci Total Environ **267**:125–140.

RUBY MV SCHOOF R, BRATTIN W, GOLDADE M, POST G, HARNOIS M, MOSBY DE, CASTEEL SW, BERTI W, CARPENTER M, EDWARDS D, CRAGIN D and CHAPPELL W (1999) *Advances in evaluating the oral bioavailability of inorganics in soil for use in human health risk assessment: critical review.* Environ Sci Technol **33**:3697–3705.

SAINIO EL, JOLANKI R, HAKALA E and KANERVA L (2000) *Metals and arsenic in eye shadows.* Contact Dermatitis **42**:5–10.

SEIFERT B, BECKER K, HELM D, KRAUSE C, SCHULZ C and SEIWERT M (2000) *The German environmental survey 1990/1992 (Ger ES II): reference concentrations of selected environmental pollutants in blood, urine, hair, house dust drinking water and indoor air.* J Exposure Analysis Environ Epidemiol **10**:552–565.

Solomon RL and Hartford JW (1976) *Lead and cadmium in dusts and soils in a small urban community.* Environ Sci Technol **10**:773–777.

Sutton PM, Athanasoulis M, Flessel P, Giurgius G, Haan M, Schlag R and Goldman LR (1995) *Lead levels in the household environment of children in three high-risk communities in Florida.* Environ Res **8**:45–57.

Tessier A, Campbell PCG and Bisson M (1979) *Sequential extraction procedure for the speciation of particulate trace metals.* Anal Chem **51**:844–951.

Thompson M and Thornton I. (1997) *Antimony in the domestic environment and SIDS.* Environ Technol **18**:117–119.

Thornton I, Culbard E, Moorcroft S, Watt J, Wheatley M and Thompson M (1985) *Metals in urban dusts and soils.* Environ Technol Lett **6**:137–144.

Thornton I, Watt JM, Davies DJA, Hunt A, Cotter-Howells J and Johnson DL (1994) *Lead contamination of U.K. dusts and soils and implications for childhood exposure: an overview of the work of the Environmental Geochemistry Research Group, Imperial College, London, England 1981–1992.* Environ Geochem Health **16**:113–122.

Tong STY and Lam KC (2000) *Home sweet home? A case study of household dust contamination in Hong Kong.* Sci Total Environ **256**:115–123.

Trowbridge PR and Burmaster DE (1997) *A parametric distribution for the fraction of outdoor soil in indoor dust.* J Soil Contamination **6**:161–168.

U.S. EPA (1986) *Air Quality Criteria for Lead.* Environmental Criteria and Assessment Office, Office of Research and Development: Research Triangle Park, NC; U.S. Environmental Protection Agency PA 600/8-83-028a-d.

U.S. EPA (1995) *Sampling House Dust for Lead: Basic Concepts and Literature Review.* Final Report EPA 747-R-95-007. U.S. Environmental Protection Agency, Washington, DC, 20460.

Van Winkle MR and Scheff PA (2001) *Volatile organic compounds, polycyclic aromatic hydrocarbons, and elements in the air of ten urban homes.* Indoor Air **11**:49–64.

Wallace L (1996) *Indoor particles: a review.* J Air Waste Manage Assoc **46**:98–126.

Wu S and Feng X (2000) *Appendix B: A collocation study to evaluate samplers used for collecting airborne particulate matter.* In: The Alberta Oil Sands Community Exposure and Health Effects Assessment Program: Methods Report, B-1 to B-20, Alberta, Canada: Alberta Department of Health and Wellness.

12

From the Biological System of the Elements to Biomonitoring

Bernd Markert, Stefan Fraenzle and Annette Fomin

12.1
Introduction

Due to industrialization, the increasing consumption of resources has led to a redistribution of matter to an extent that, during the past 200 years, has grossly changed the living conditions of most biological species. In earlier times, these changes were very slow, in line with the long time-scales of evolution. Such changes have great influence on both living and inanimate beings, such that there appears to be a form of predictable, natural direction in evolutionary development.

Resources are used in a conservative and cautious manner, with the main aim being the greatest possible efficiency, irrespective of whether the extraction of a component or the derivation of (chemical) energy is involved. Among the most recently developed organisms, mammals which live on land became partially independent of the water cycle and therefore gained an evolutionary advantage. This partial independence from the life-elixir of water in both space and time allowed the rapid evolution of mental ability in man as a species, and this species underwent a highly complex evolution in order to reach the present levels. This change in turn brought about

yet another competitive advantage that, in the past, was only achieved by very few microorganisms, for example plague bacteria and smallpox viruses. However, since these organisms in time came to be controlled by healthcare systems and medicinal measures developed by man during the past decades, their advantages in this respect were lost.

Usually, such a strategy is not considered before use, and today is often subject to feedback in one form or another. The genomes bearing the "genetic files" are no longer beyond manipulation; consequently, it is possible that civilized mankind might be subjected to extreme hazards should microbial epidemics be either catalyzed or supported by genetic engineering. Examples of these include bovine spongiform encephalopathy, AIDS, and many other epidemics that have been the subject of much less discussion in the media.

Until now, the ranges (minimum and maximum values) of natural stressors have remained essentially constant along evolutionary time-scales of millions of years. Accordingly, species have been able to adapt to changing environmental conditions, though there was also a natural "background" stress situation which is necessary to maintain biological organization at all

Elements and their Compounds in the Environment. 2nd Edition.
Edited by E. Merian, M. Anke, M. Ihnat, M. Stoeppler
Copyright © 2004 WILEY-VCH Verlag GmbH & Co. KGaA, Weinheim
ISBN: 3-527-30459-2

levels. An important property of all living systems is their ability to respond and react to stressors; in contrast, species and entire ecosystems could evolve no further without stressors. Stress, therefore, could be regarded as the "motor" of evolution (Oehlmann and Markert 1999).

Changes in the consumption of matter that have occurred during the past few centuries have been larger than ever before. Novel substances (e.g., xenobiotics and radionuclides) which did not exist previously have subsequently been released into the environment by man. In addition, other potentially hazardous compounds that were already present in the environment (albeit in much smaller quantities than today; e.g., heavy metals, naturally occurring radionuclides) have also been released. Such novel stressors usually bring about multiple effects, adding to those from natural stressors or combining with one another in a manner that may surpass the tolerance of most living beings. The latter in turn are limited by individual or evolutionary adaptation (Oehlmann and Markert 1999). This book is concerned with metals rather than organic contaminants of any types, and all metals (or all elements) must be considered from a double perspective. In order to deal with the complexities and dynamics of living beings, analytical data must be linked to the biological functions or roles of the elements. Thus, this chapter deals with the "Biological System of the Elements" (BSE) to provide a comprehensive approach towards bioindication of metal-based components in organisms.

12.2
From the Biological System of the Elements towards Biomonitoring

Working groups whose interests are focused towards either nutrition, physiology, or ecotoxicology mainly consider the essential (positive, increasing vitality), indifferent (undetectable), and toxic actions of any single chemical element. However, a given element will often have ambivalent or multivalent functions due to different actions imposed by the same substance, let alone antagonisms (A/Se; Hg/Se). What actually happens is dependent upon the different concentration levels, on unlike types of chemical speciation, and will also differ among different target organisms (Markert and Fraenzle 2000). Neither essentiality nor chronic nor acute toxicities of chemical elements can be inferred from positions and classifications in the Periodic System of Elements developed by Mendeleyev and Meyer as early as 1869. Of course, the rules of physical chemistry apply to molecular aspects of biological processes, but there are deviations from straightforward expectations. These are primarily due to the adaptation of all types of organic life to aqueous milieus. For reasons of different hydration, sodium ions in water are larger than potassium ions; because size relationships are the reverse of that with unsolvated crystalline ions, the transport rate of Na^+ through biological membranes is less than that for K^+. While this can be readily understood by established chemical knowledge, the situation worsens considerably when an attempt is made to explain more complex physiological processes using small amounts of information based on chemistry and physics. The periodic system of the elements relies on the number of outer-sphere electrons in an atom. However, in order to interpret and understand functions in biology, another system must be developed which takes into account data from both animal and plant physiologies, respectively. These include:

- data on all chemical elements and their corresponding interelementary correlation coefficients;
- accumulation mechanisms and preferences of certain groups of organisms for specific elements or groups of the latter;
- stoichiometric network analyses to determine mechanisms of action that depend on speciation; and
- any other system components, including results and research topics from biogeochemistry.

Although biological diversity is brought about by genetics and ecosystems-related adaptation, the question remains as to whether biological diversity is accompanied by some extent of chemical diversity on the ecosystems level (Markert and Fraenzle 2000).

For this purpose, *experimental* results on interelementary correlations, mechanisms of take-up and biological functions must first be gathered (Figure 12.1) (Markert 1994a), and then corroborated by *theoretical* aspects from stoichiometric network analysis (Fraenzle and Markert 2000a,b). The latter method permits the prediction as to whether – given its properties – a chemical element might be essential at all and, if so,

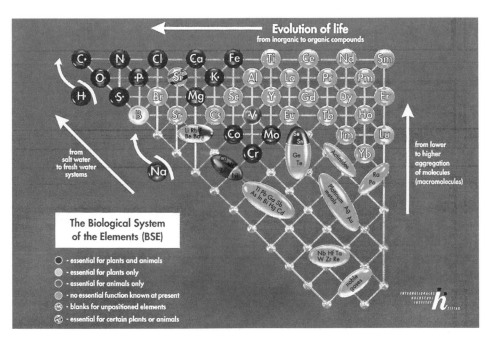

Fig. 12.1 The Biological System of the Elements compiled from data on correlation analysis, physiological function of the individual elements in the living organism, evolutive development out of the inorganic environment, and with respect to their uptake form by the plant organism as a neutral molecule or charged ion. The elements H and Na exercise such diverse functions in biological systems that they are not conclusively fixed. The ringed elements can at present only be summarized as groups of elements with a similar physiological function since there either is a lack of correlation data or else these data are too imprecise (from Markert 1994a)

under which (biochemical) conditions this might occur. Conditions to be considered in this respect include redox potentials and chemical binding modes, including co-ordinated and organic ones, with the latter being produced, for example, by biomethylation.

For purposes of bioindication and biomonitoring, clearly both a highly specific approach relating to each single chemical species of an element and a comprehensive treatment of general features are required. The latter is included in a Biological System of the Elements.

12.3
Definitions

It seemed clear from the start that bioindication and biomonitoring are promising (and also possibly low-cost) methods to observe the impact of external factors on ecosystems and their development over a long period, or to differentiate between an unpolluted site and a polluted site (Markert et al. 2003a). The overwhelming enthusiasm shown in developing these methods has resulted in a problem that is still unsolved: the definitions of bioindication and biomonitoring respectively – and therefore the expectations associated with these methods – have never led to a common approach by the international scientific community, so that different definitions (and expectations!) now exist simultaneously. A fine overview of the various definitions is given by Wittig (1993). As a first starting point for the difficult use of bioindication methods, the following references may be helpful: Altenburger and Schmitt (2003), Arndt (1992), Bargagli (1998), Breulmann et al. (1997), Breulmann et al. (1998), Carreras et al. (1998), Djingova and Kuleff (2000), Farago (1994), Figueiredo et al. (2001), Fraenzle (1993), Fraenzle and Markert (2002), Freitas et al.

(1999), Garty (1998), Genßler et al. (2001), Herpin et al. (1997, 2001), Klumpp et al. (2000), Kostka-Rick et al. (2001), Lieth (1998), Markert (1993), Markert et al. (2003b), Siewers and Herpin (1998), Siewers et al. (2000), Vtorova et al. (2001), Vutchkov (2001) and Wolterbeek et al. (1995).

In the following section, some definitions are provided which have been developed and used by the present authors over the past 20 years (Markert et al. 1999), and it is felt that these differentiate clearly between bioindication and biomonitoring using the qualitative/quantitative approach to chemical substances in the environment. This makes bioindicators directly comparable to instrumental measuring systems. From that viewpoint, it is possible to distinguish clearly between *active* and *passive* bioindication (biomonitoring). Especially where the bioindication of metals is concerned, the literature often makes a distinction between "accumulation indicators" and "effect indicators" in respect of the reaction of the indicator/monitor to changes in environmental conditions. Here, we should bear in mind that this differentiation does not imply a pair of opposites; it merely reflects two aspects of analysis. As the accumulation of a substance by an organism already constitutes a reaction to exposure to this substance which – at least in the case of high accumulation factors – is measurably reflected in at least one of the parameters used in defining the term "effect indicator/monitor" (e.g., morphological changes at the cellular level; formation of metal-containing intracellular granules in many invertebrates after metal accumulation), we should discuss whether it is worthwhile distinguishing between accumulation and effect indicators, or whether both terms fall under the more general expression "reaction indicator". Also, it is often not until a substance has been accumulated in organisms that intercellular or

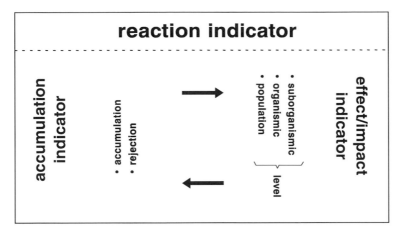

Fig. 12.2   Illustration of the terms reaction, accumulation and effect/impact indicator.

intracellular concentrations are attained which produce effects that are then analyzed in the context of effect and impact monitoring (Figure 12.2).

From these preliminaries we come to the following definitions (see Markert et al. 1997, 1999):

- A *bioindicator* is an organism (or part of an organism or a community of organisms) that contains information on the *quality* of the environment (or of some part of the environment).
- A *biomonitor*, on the other hand, is an organism (or a part of an organism or a community of organisms) that contains information on the *quantitative* aspects of the quality of the environment. A biomonitor always is also a bioindicator, but a bioindicator does not necessarily meet the requirements for a biomonitor.
- We speak of *active bioindication* (biomonitoring) when bioindicators (biomonitors) created in laboratories are exposed in a standardized form in the field for a defined period of time. At the end of this exposure time, the reactions provoked are recorded or the xenobiotics taken up by the organism are analyzed.

In the case of *passive* biomonitoring, organisms already occurring naturally in the ecosystem are examined for their reactions. This classification of organisms (or communities of these) is according to their "origin".

A classification of organisms (or communities of these) according to their "mode of action" (Figure 12.2) is as follows:

- *Accumulation* indicators/monitors are organisms that accumulate one or more elements and/or compounds from their environment. *Effect* or *impact indicators/ monitors* are organisms that demonstrate specific or unspecific effects in response to exposure to a certain element or compound or a number of substances. Such effects may include changes in their morphological, histological or cellular structure, their metabolic-biochemical processes, their behavior or their population structure. In general, the term "reaction indicator" also includes accumulation indicators/monitors and effect or impact indicators/monitors as described above.

When studying accumulation processes it would seem useful to distinguish between the paths by which organisms take up elements/compounds. Various mechanisms contribute to overall accumulation (*bioaccumulation*), depending on the species-related interactions between the indicators/monitors and their biotic and abiotic environment.

- *Biomagnification* is the term used for absorption of the substances from nutrients via the epithelia of the intestines. It is therefore limited to heterotrophic organisms and is the most significant contamination pathway for many land animals, except in the case of metal-(oid)s that form highly volatile compounds (e.g., Hg, As) and are taken up through the respiratory organs (e.g., tracheae, lungs).

- *Bioconcentration* means the direct uptake of the substances concerned from the surrounding media – that is, the physical environment – through tissues or organs (including the respiratory organs). Besides plants (which can only take up substances in this way, mainly through roots or leaves), bioconcentration plays a major role in aquatic animals. The same may also apply to soil invertebrates with a low degree of solarization when they come into contact with the water in the soil.

Besides the classic floristic, faunal and biocoenotic investigations that primarily record rather unspecific reactions to pollutant exposure at higher organizational levels of the biological system, various newer methods have been introduced as instruments of bioindication. Most of these are biomarkers and biosensors.

- *Biomarkers* are measurable biological parameters at the suborganismic (genetic, enzymatic, physiological, morphological) level in which structural or functional changes indicate environmental influences in general, and the action of pollutants in particular in qualitative and

sometimes also in quantitative terms. Examples include:

1. enzyme or substrate induction of cytochrome P-450 and other Phase I enzymes by various halogenated hydrocarbons;
2. the incidence of forms of industrial melanism as markers for air pollution;
3. tanning of the human skin caused by UV radiation;
4. changes in the morphological, histological or ultrastructure of organisms or monitor organs (e.g., liver, thymus, testicles) following exposure to pollutants.

- *Biosensors* are measuring devices that produce a signal in proportion to the concentration of a defined group of substances through a suitable combination of a selective biological system (e.g., enzyme, antibody, membrane, organelle, cell or tissue) and a physical transmission device (e.g., potentiometric or amperometric electrode, optical or optoelectronic receiver). Examples include:

1. Toxiguard bacterial toximeter;
2. EuCyano bacterial electrode.

- *Biotest* (*bioassay*): a routine toxicological-pharmacological procedure for testing the effects of agents (environmental chemicals, pharmaceuticals) on organisms, usually in the laboratory but occasionally in the field, under standardized conditions (with respect to biotic or abiotic factors). In the broader sense, this definition covers cell and tissue cultures when used for testing purposes, enzyme tests and tests using microorganisms, plants and animals in the form of single-species or multi-species procedures in model ecological systems (e.g., microcosms and mesocosms). In the narrower sense, the term only covers single-species and model system tests, while the other procedures may be called suborganismic tests. Bioassays use certain biomarkers or – less often – specific biosensors and can be used in bioindication or biomonitoring.

With regard to genetic and non-genetic adaptation of organisms and communities to environmental stress, differentiation must be made between the terms tolerance, resistance, and sensitivity:

- *Tolerance* (Oehlmann and Markert 1997) is the desired resistance of an organism or community to unfavorable abiotic factors (climate, radiation, pollutants) or biotic factors (parasites, pathogens), where adaptive physiological changes (e.g., enzyme induction, immune response) can be observed.

- *Resistance*, unlike tolerance, is a genetically derived ability to withstand stress (Oehlmann and Markert 1997).This means that all tolerant organisms are resistant but not all resistant organisms are tolerant. However, in ecotoxicology the dividing line between tolerance and resistance is not always so clear. For example, the phenomenon of pollution-induced community tolerance (PICT) is described as the phenomenon of community shifts towards more tolerant communities when contaminants are present. It can occur as a result of genetic or physiological adaptation within species or populations, or through the replacement of sensitive organisms by more resistant organisms (Blanck et al. 1988, Rutgers et al. 1998).

- *Sensitivity* of an organism or a community means its susceptibility to biotic or abiotic change. Sensitivity is low if the tolerance or resistance to an environmental stressor is high, and sensitivity is high if the tolerance or resistance is low.

12.4
Comparison of Instrumental Measurement: Bioindicators/biomonitors and Harmonization/Quality Control

The strong similarity in terms between instrumental chemical analysis (qualitative and quantitative measurements) and the field of bioindicators (as a qualitative approach to pollution control) and biomonitors (as a quantitative approach) makes it worthwhile to compare the two techniques. The discussion follows the lines of Markert et al. (2003a).

12.4.1
Instruments and Bioindicators

The more technical details of instrumental analysis are shown in Figure 12.3 (see also Vol. III, Part V, Chapters 2 and 3), which details typical procedures for measuring chemical substances, enzyme activities or other ecosystem-relevant parameters using either atomic absorption spectrometry (AAS) or photometry. In many spectrometric methods, a specific wavelength is used to obtain a signal by analyzing a sample placed in a cuvette (photometer); examples include flame and graphite furnace atomic absorption spectrometry (FAAS, GFAAS), inductively coupled plasma (ICP), optical emission spectrometry (OES) or mass spectrometry (MS), supported by photo-multipliers, amplifiers, and other equipment and finally evaluated by detector systems. The main sources of error are the sampling procedure (up to 1000%) and sample preparation (up to 300%). A detailed discussion of typical errors in orders of magnitude is given by Markert (1996).

The direct comparison with a biological measuring device (bioindicator) in Figure 12.3 shows that the whole process of instrumental measurement is very often

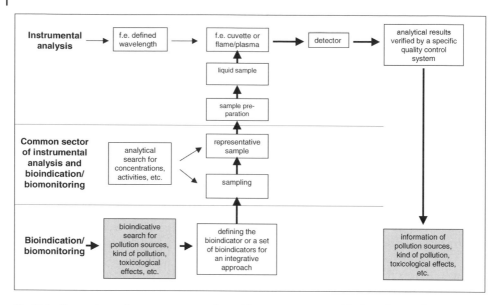

Fig. 12.3 Comparison of measurements performed by spectrometers and by bioindicators/biomonitors. In practice, instrumental measurements are often an integral part of bioindication (from Markert et al. 2003a). A full instrumental flow chart for instrumental chemical analysis of environmental samples can be found in Markert (1996).

integrated into the procedure of bioindication, at least when samples must be analyzed for chemical compounds. This means that laboratory investigations on bioindicators depend heavily on instrumental measuring equipment in order to obtain additional information from the bioindicator. Hence, when the question "bioindication or direct instrumental measurement?" is asked, it seems that this relationship was not fully understood. The practical laboratory problems encountered in biomonitoring are often the same as in chemical analysis. Take, for example, paradigm 1 of the sampling process: "The samples collected must be representative for the scientific question under review" (Markert 1996). The representative collection of samples for monitoring or/and instrumental measurement must be made with the greatest care. This prerequisite is mentioned and

explained in numerous excellent articles and textbooks, and so is not discussed here in detail (e.g., Keith 1988, Markert 1994b, Rasemann and Markert 1998, Wagner 1992, Klein and Paulus 1995).

12.4.2
Precision and Accuracy

In addition to the similar need for highest representative quality of the sample to be analyzed or to be used as a bioindicator, most general rules and prerequisites of quality control in chemical analysis must be taken into account in biomonitoring activities. During the past 20 years, a strict differentiation between the terms "precision" (reproducibility) and "accuracy" (the "true" value) has been established in chemical analytical research. The practical application of this differentiation makes it possible to determine the "true" or real content of a substance "X" in a sample "Y". The purpose of determining the precision of the data by repeatedly measuring the analytical signal is to track down and eliminate errors which might be generated, for example, by insufficient long-term stability of the measuring device (device-specific misadjustment). If the analytical procedures are not too complex, the precision will range from 1 to 5%, and for most analytical problems this can be considered sufficiently exact. However, the mere fact that a signal is readily reproducible does not permit any statement about its accuracy. Even highly precise data can diverge greatly from the "true" (e.g., element) content of a sample. Correct analytical results can only be obtained if the entire analytical process is subjected to targeted quality control, where every result is checked for its precision and accuracy. Basically, two methods are now used to check the accuracy of analytical results:

1. The use of certified reference materials (commercially available samples with a certified content of the compound to be measured and a matrix similar to the original samples to be measured in the laboratory).
2. The use of independent analytical procedures.

With bioindicators we can, of course, carry out repeated sampling to obtain an idea of how "stable" the bioindicator under investigation is with respect to site and time variations. A more difficult problem is that of accuracy during the sampling procedure, for which we have at present no "certified reference system" as a calibrator for accuracy in representative sampling. As a rule, "polluted" and "unpolluted" systems will be compared, but there is no way to be sure of working accurately. The only possible strategy here is that of "independent methods", when different research groups have the task of working in the same area with

the same indicators, so that the data – when obtained independently – can be compared. This is a very expensive method that can only be used in very special bioindication proposals where method development is of general concern, for example in for European Union (EU) or United States (US) directives.

12.4.3
Calibration

In general, considerable problems exist with bioindicators themselves, but these do not usually arise with instrumental measurement techniques: the calibration of the biological system as such (Figure 12.4). The limits within which organisms can indicate exposure become especially obvious in attempts to quantify environmental qualities, for example in biomonitoring in the stricter sense of the term (Markert et al. 1997). Although the number of potential bioindicators is growing virtually by the hour, it is difficult to find organisms (in nature) that meet the criteria of an active or passive biomonitor. For instance, the analysis of individual accumulation indicators for body burdens of certain substances does not necessarily permit conclusions to be made about concentrations in the environment. Many plants and animals display high accumulation factors for certain substances at low environmental concentrations, but the accumulation factors decrease sharply at higher environmental levels. The result is more or less a plateau curve for environmental concentration/body burdens (Figure 12.4). On the other hand, many organisms succeed in keeping their uptake of toxic substances very low over a wide range of concentrations in the environment (Markert et al. 1997). Not until acutely toxic levels in the environment are exceeded do the regulatory mechanisms break down,

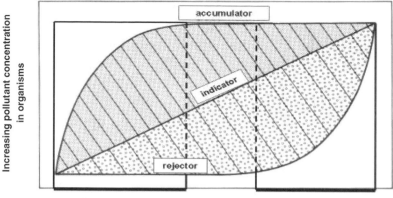

Fig. 12.4 Correlation between the environmental concentration of the pollutant to be monitored and the concentration in the organism. Linear ranges for calibration are very limited for both accumulators and rejectors (from Markert et al. 1997)

then resulting in a high degree of accumulation (Figure 12.4). Exceptions are, of course, substances that are not taken up actively but enter the body by way of diffusion processes – doubtlessly rare in the case of inorganic metal compounds.

This often means that the relationship between the bioindicator/biomonitor and its environment in respect of the concentration of the compound to be accumulated is not linear, but logarithmic. Even when linearity of the logarithmic function is achieved by mathematical conversion, the linear relationship between the two measurements is restricted to a small range. However, organisms can only provide unequivocal information about their environment if a linear relationship exists which is comparable to the calibration line of measuring instruments.

Compared to spectrometric instrumental analytical methods – for example, where the linear calibration range normally covers several orders of magnitude – a linear range for bioindicators is more difficult to achieve since living organisms are constantly changing their "hardware" by bio-

logical processes. Standardization of bioindicators therefore seems unrealistic at the moment, which means that harmonization between users of the same indicators is of specific and real concern for the future.

An interpretable or even linear relationship between burden and biological/biochemical signal only is to be expected when there is a high degree of genetic – and thus metabolic – homogeneity in the test organisms. Obviously most abundant organisms display substantial genetic diversity, notable exceptions among vertebrates (both very abundant and homogeneous) being Syrian (golden) hamster and man. These are thus best suited for biomonitoring.

12.4.4
Harmonization

Just as interlaboratory tests have for years enabled different laboratories to use real samples to optimize the quality of their own analyses in the field of analytical chemistry, greater attention must be given to harmonizing the use of the same indicators in different places for the "calibration" of bioindicators. This is not so much true of work carried out in the laboratory, since bioassays as tests for chemicals, for example, are highly standardized and thus reproducible; it applies chiefly to all aspects of the use of bioindicators in the field. First of all, more cooperative planning in program design seems absolutely necessary in order to compare results from individual working groups. On a regional and national level this is relatively easy to achieve, but on a global and intercontinental level the geographic distances between the research groups sometimes cause problems. For example, the International Atomic Energy Agency (IAEA 2001) tries to carry out biomonitoring of elements in different continents, and the high cost of personal meetings for an exchange of views must be taken into account. Training and crash courses over a defined period of time (e.g., weeks) seem to be the first and best step towards harmonizing scientific and (sometimes) cultural differences. And this should not be underestimated in a globalizing world: bioindication in its different facets and on its different scientific levels can be performed by practically anybody, so that especially cross-border projects, have a tremendous intercultural impact. We should beware of over-optimism, but "bioindication may be seen as a gateway to intercultural understanding and as a catalyst for peaceful international cooperation". Questions to be answered during this exchange of information might include how to relate observations of the same phenomena made by different techniques, such as remote sensing and on-site information (Roots 1996, Smodis 2003). Scaling problems in space and in time are partly a matter of program design. Program design includes choice of measurements, sensors and recording methods and finally questions of information delivery and information technologies. Good examples of "questions in mind before starting the job" can be found in numerous national and international sampling campaigns for environmental observation and in literature dealing specifically with these harmonization steps (e.g., Schroeder et al. 1996, Bosch and Pinborg 2003, Lazorchak et al. 2003, Matthiessen 2003, Parris 2003).

12.4.5
An Example of Effect Biomonitoring

In the following section, an example of effect monitoring that fulfils the requirements of Sections 12.4.2 to 12.4.4, where precision, calibration, and harmonization will be discussed in the use of snails to monitor bisphenol A (BPA). The more detailed experimental features of this investigation may be found elsewhere (Schulte-Oehlmann et al. 2001).

Within a Federal Environmental Agency research project to develop a biological test for hormone-mimetic compounds using the freshwater snail *Marisa cornuarietis*, the effects of the suspected xenoestrogenic substance BPA, not only on freshwater but also on marine prosobranch snails, were investigated (Schulte-Oehlmann et al. 2001).

For the laboratory experiments the ramshorn snail *M. cornuarietis* and the ovoviviparous snail *Potamopyrgus antipodarum* were considered as freshwater species, and two marine prosobranchs – the netted whelk *Nassarius reticulatus* and the dog whelk *Nucella lapillus* – were additionally employed. *N. reticulatus*, as a typical sediment-living species, was exposed via artificial sediments, while the three other prosobranchs were exposed via water.

The test series with *M. cornuarietis* covered a nominal concentration range between 1 and 100 µg BPA L^{-1} in a 5-month experiment with adult snails and a complete life cycle test for 12 months. Additionally, a third test in the nominal range between 0.05 and 1 µg BPA L^{-1} (measured: 0.0079 to 0.404 µg L^{-1}) was performed with adult-snails for 6 months. In these experiments, BPA induced a complex syndrome of physiological and morphological alterations in female *Marisa* referred to as the induction of "super females". Affected specimens were characterized by the formation of additional female organs, an enlargement of the accessory pallial sex glands, gross malformations of the pallial oviduct section resulting in an increased female mortality, and a massive stimulation of oocyte and spawning mass production. For these tests, a Lowest Observed Effect Concentration (LOEC) of 48.3 ng L^{-1}, a No Observed Effect Concentration (NOEC) of 7.9 ng L^{-1} and an Effective Concentration 10 (EC_{10}) of 13.9 ng L^{-1} were calculated. Super females occurred also in the BPA exposure experiment with the other snail species, but comparable oviduct malformations as in *Marisa* were not found, probably due to species differences in the gross anatomical structure of the pallial oviduct.

During the 9-week test with *P. antipodarum* in the nominal concentration range between 1 and 100 µg L^{-1}, BPA induced an enhancement of embryo production even in the sexual repose phase of the reproductive cycle. A characteristic inverted U-type concentration response relationship was found.

N. reticulatus was exposed via BPA-spiked artificial sediments (nominal concentration range 10 to 1000 µg kg^{-1} dry wt.) for 3 months. BPA exhibited a significant and concentration-dependent uterotrophic effect which could be detected not only by an enlargement but also by a weight increase of the accessory pallial gland complex in the pallial oviduct section.

Adult *N. lapillus* were tested for 3 months in the laboratory in a nominal concentration range between 1 and 100 µg L^{-1} BPA. Super females in the dog whelk were also characterized by enlarged accessory pallial sex glands and an enhancement of egg production, but the test compound also affected the males in this species. A lower percentage of exposed specimens had ripe sperm stored in their vesicula seminalis, and male *Nucella* exhibited a reduced length of penis and prostate gland when compared to controls.

Because statistically significant effects were observed already at the lowest nominal test concentration (1 µg L^{-1}), it can be assumed that even lower concentrations may have a negative impact on the snails. The results show that prosobranch snails are affected by BPA at lower concentrations compared to other systematic taxa in the animal kingdom. Consequently the results of these experiments should be considered for the current EU risk assessment for BPA in order to achieve a sufficient protection of wildlife in aquatic ecosystems.

12.5
Integrative Biomonitoring

The following information reflects only a very small part of the overall existing and proposed strategies and concepts for bioindication. Much greater detail on specific programs are provided, for example, by the Environmental Protection Agency (EPA, US), the OECD, and the EEA. Further international and national organizations [the International Standards Organization (ISO), CH], the European Union (EU, Belgium), especially in its section on "Measurement and Testing" [the former Bureau Community of Reference (BCR, Belgium)], Deutsches Institut für Normung (DIN, FRG) and others have elaborated various

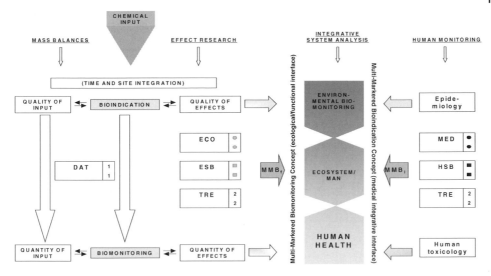

Fig. 12.5 The Multi-Markered Bioindication Concept (MMBC) with its functional and integrated windows of prophylactic health care (from Markert et al. 2003a). Explanations in the text. DAT (for data), ECO (for ecology), ESB (for environmental specimen banking), TRE (for trend), MED (for medicine) and HSB (for human specimen banking) designate individual toolboxes and their test sets.

programs for environmental control, observation and protection which are available on request via literature search or (more effective) via the internet.

The future development and coordination of bioindication methods should follow a two-leveled (A and B) parallel line:

- Level A optimizes the development and harmonization of existing and new indicators to make them suitable for practical use in risk management.
- Level B (which is discussed in Markert et al. 2003a, and in Wünschmann et al. 2001, 2002), represents a strongly integrated approach with environmental and health indicators to fill the gap between environmental biomonitoring and human health aspects.

One concept of an integrated approach to bioindication based on forward-looking strategies is described below.

12.5.1
The Multi-Markered Bioindicator Concept (MMBC)

The dilemma of bioindication lies in the fact that conclusions about the "overall condition" of an ecosystem have to be drawn from observations of a few representative indicator species. So, because of the demands made on bioindication, we must ensure that the use of bioindicators is not carried *ad absurdum,* for its own sake, as a result of the extreme complexity of systems in conjunction with a high level of dynamic development. In future, simplifications (i.e., the reduction of a great diversity of species to a few representative bioindicators) should be carried out in a less isolated manner.

Besides increasing the specificity of bioindicators, it is essential to place more emphasis on examining their functional interactions and interdependence. A summary of the so-called Multi-Markered Bioindication Concept (MMBC) is provided in Figure 12.5. The sole objective of this concept is to relate toxicological effects detected in a test system to a potential hazard to human health. The aim is to combine ecotoxicological data with data from environmental medicine by means of a toolbox model and the integrated use of various instrumental and bioindicative methods. As Figure 12.5 shows, possible mass balances for a particular pollutant are initially established by means of bioindicators and instrumental measurement methods; subsequently, they are traced back qualitatively and quantitatively to their probable sources, using the toolboxes ECO, ESB, and TRE to facilitate the method (left side of Figure 12.5). At the same time, data on human toxicology from the fields of both environmental and ecological medicine are compiled using toolboxes MED, HSB, and TRE (right side of Figure 12.5). The MMB Concept is an attempt to combine data from human toxicology and ecotoxicology via "windows" in the context of an integrated system analysis in order to permit health care of a prophylactic and predictable nature. Intelligent calculation methods are required to take both functional (MMB_f) and integrated (MMB_i) aspects into account. Some of these methods have yet to be developed by basic research, since there is too little knowledge of certain functional and integrated connections.

12.5.2
Environmental Specimen Banks

The purpose of environmental specimen banks is to acquire samples capable of providing ecotoxicological information and to store them without change over long periods to permit retrospective analysis and evaluation of pollution of the environment with substances that could not be analyzed, or did not seem relevant, at the time the samples were taken.

In Europe (Germany) and the USA, the concept of Environmental Specimen Banking in support of monitoring was proposed and discussed at several international meetings (e.g., Berlin et al. 1979, Luepke 1979, Lewis et al. 1984). Close cooperation between Germany and the USA led subsequently to the construction of banking facilities and the current performance of banking projects in both countries and some similar activities around the world (Stoeppler et al. 1982, Wagner 1992, Stoeppler and Zeisler 1993, Emons 1997). Individual aspects and background mainly for the German approach is given in some detail in Kettrup (2003). The general tasks and objectives of environmental sample banks may be outlined as follows (Klein 1999):

- to determine the concentrations of substances that had not been identified as pollutants at the time the samples were stored, or which could not be analyzed with sufficient accuracy (retrospective monitoring);
- to check the success or failure of current and future prohibitions and restrictions in the environmental sector;
- regular monitoring of the concentrations of pollutants already identified by systematic characterization of the samples before archiving;
- prediction of trends in local, regional and global pollution;
- description of standardized sampling methods;
- documentation of the conditions under which the sample material is stored as a requirement for obtaining comparable results.

The German Specimen Bank strategy which will be mainly discussed here as a particular example also assumes that pollution at a

Tab. 12.1: Sample species collected in the German Environmental Specimen Bank. (From Klein 1999.)

Sample species	Target compartment
Spruce (*Picea abies*) / pine (*Pinus sylvestris*)	Annual shoots
Red beech (*Fagus sylvatica*) / Lombardy Poplar (*Populus nigra "Italica"*)	Leaves
Domestic pigeon (*Columba livia* f. *domestica*)	Eggs
Roe deer (*Capreolus capreolus*)	Liver (kidneys)
Earthworm (*Lumbricus terrestris/Aporrectodea longa*)	Worm body without gut contents
Zebra mussel (*Dreissena polymorpha*)	Soft parts
Bream (*Abramis brama*)	Muscle tissue and liver
Brown algae (*Fucus vesiculosus*)	Thallus
Edible mussel (*Mytilus edulis*)	Soft parts
Blenny (*Zoarces viviparus*)	Muscle tissue and liver
Herring gull (*Latus argentatus*)	Eggs
Lugworm (*Arenicola marina*)	Worm body without gut contents

particular location cannot be demonstrated by one bioindicator alone because of the different degrees of exposure of the organisms in an ecosystem to pollutants and their different genetic predeterminants (Klein 1999). Only a set of suitable bioindicators is capable of reflecting the pollutants present in the ecosystem.

Table 12.1 shows the bioindicators available at the German Federal Environmental Specimen Bank. The criteria for choice of the sample species are discussed in detail in Klein and Paulus (1995). The expected functional connections between ecosystems are shown in Figure 12.6.

A problem posed by the environmental samples, which are carefully stored and

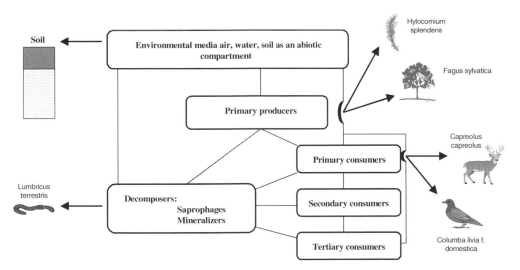

Fig. 12.6   Selected sets of sample species at the ecosystem level for the German Federal Environmental Sample Bank (derived from Klein 1999).

refrigerated under liquid nitrogen, is the rather high operating cost of the facility and the required high experience of the scientists involved in ESB activities. There is at present also a certain lack of flexibility in taking in or handing out a bioindicator organism that has been analyzed previously and over a period of years. The highly specific sampling guidelines often make it diffi-cult to carry out comparisons with "normal" sampling protocols. These problems could be solved by integrating the results from an Environmental Specimen Bank with other bioindication studies. In the MMBC this is shown by integrating the toolboxes ECO and MED with ESB and HSB in Figure 12.5.

Tab. 12.2: Types of environmental monitoring networks used in ecological observation in Germany. (From Wagner 1992.)

Types of monitoring network	Objectives	Characteristics of the network	Methods, examples
1 Permanent measuring stations/permanent observation sites, including ecosystem approaches	Reference and back-ground data; time lines; integrated pollution and effect surveys; basis for comparisons for environ-mental quality standards	Strictly according to regional statistics, avoid-ing local sources of interference; selected measuring points or sites to be observed	Widest possible range of methods as a reference basis, e.g. "Integrated Monitoring" – Baden Württemberg, also UBA monitoring network, ecosystem research + UPB*, DWD*
2 Monitoring networks for individual states	Overview of regional sta-tistics; background data	Coordinate-based, wide-meshed networks (10– max. 50 km, avoiding local sources of interfer-ence)	Preferably passive bio-monitoring, e.g. Bavarian moss and spruce moni-toring network, Saarland poplar/spruce network
3 Regional monitoring networks	Screening (identification and delimitation of pol-luted areas or zones); integrated effects of complex or unknown types of pollution	Usually regular, relatively close-meshed measuring networks (approx. 1– 10 km) limited in size (e.g., rural district, county, "polluted area")	Active and passive bio-monitoring, effect cadaster in polluted and "clean air" regions, with-out reference to specific emitters
4 Emitter-related moni-toring networks	Determining extent of spread of pollution and the pollutant effects of an emitter	Usually close-meshed, often radial or linear net-works or transects ($<$ 1–10 km between measuring points)	Primarily active or exper-imental methods geared to specific emitters or pollutants
5 Environmental impact analyses	Determining degrees of existing pollution and maximum tolerated burden before planned measures take effect (preservation of evi-dence)	As 4	As 4. Possibly additional unspecific methods + UPB as preservation of ecotoxicological evidence

* UPB $\hat{=}$ ESB; DWD $\hat{=}$ German Metrological Service.

12.5.3
Time- and Site Integration

The chief objective of biomonitoring is to permit statements about pollution and changes in biodiversity on various spatial and temporal scales. The site dependency of bioindicators/biomonitors is often affected by different biotopes which are characterized by different population structures and climatic, soil, and food conditions. The latter can be delimited fairly easily by sampling the bioindicator from various locations at the same time. For this purpose, Wagner (1992) developed a system (Table 12.2) for fitting the sampling network to the quality of pollution control to be expected from the selected bioindicators (biomonitors) in use.

Compared with parameters resulting from the site, however, the behavior of the bioindicator (biomonitor) along the time axis is much more difficult to determine. Especially in temperate climates, the great variation of seasonal effects causes variations of the pollutant concentration in one and the same bioindicator organism. For example, the seasonal fall in most of the heavy metal concentrations in spring (northern hemisphere) can be explained by the dilution effect of the first biomass of the year (Markert and Weckert 1993). In particular, a comparison of data obtained by different working groups using the same bioindicator must be carefully checked with site-dependent and especially time-dependent parameters.

In conclusion, there is very much interest on integrated monitoring which will require an interdisciplinary design and also the formation of research groups in future surveys. This would permit rapid and flexible adjustment of the working groups to the particular frame of reference and enable a quick exchange of information between the individual disciplines.

Acknowledgment

The authors thank Dipl.-Ing. Simone Wuenschmann (IHI Zittau) for her extraordinary help in the coordination of difficult parts of the manuscript during its preparation.

References

ALTENBURGER R and SCHMITT M (2003) *Predicting toxic effects of contaminants in ecosystems using single species investigations.* In: Markert BA, Breure AM, Zechmeister HG, eds. Bioindicators and Biomonitors, pp. 153–198. Elsevier, Amsterdam.

ARNDT U (1992) *Key reactions in forest disease used as effects criteria for biomonitoring.* In: McKenzie DH, Hyatt DE and McDonald VJ, eds. Ecological indicators. Proc. Int. Symp. Fort Lauderdale USA, pp. 829–840. Elsevier, London.

BARGAGLI R, ed. (1998) *Trace Elements in Terrestrial Plants – an Ecophysiological Approach to Biomonitoring and Biorecovery.* Springer-Verlag, Berlin, Heidelberg.

BERLIN A, WOLFF AH and HASEGAWA Y, eds. (1979) *The Use of Biological Specimens for the Assessment of Human Exposure to Environmental Pollutants.* Proc. of the International Workshop at Luxembourg, 18–22 April 1977. Martinus Nijhoff Publishers, The Hague, Boston, London.

BLANCK H, WAENGBERG SA and MOLANDER S (1988) *Pollution-Induced Community Tolerance. A new ecotoxicological tool.* In: Cairns JJ and Pratt JR, eds. Functional testing of aquatic biota for estimating hazards of chemicals. ASTM STP 988, pp. 219–230. American Society for Testing and Materials, Philadelphia.

BOSCH P and PINBORG U (2003) *Bioindicators and the indicator approach of the European Environment Agency.* In: Markert BA, Breure AM and Zechmeister HG, eds. Bioindicators and Biomonitors, pp. 903–916. Elsevier, Amsterdam.

BREULMANN G, MARKERT B, LEFFLER U, WECKERT V, HERPIN U, LIETH H, OGINO K, ASHTON PS, LA FRANKIE JV, HUA SANG LEE and NINOMIYA I

SCHROEDER W, FRAENZLE O, KEUNE H and MANDY P, eds. (1996) *Global Monitoring of Terrestrial Ecosystems*. Ernst & Sohn Verlag, Berlin.

SCHULTE-OEHLMANN U, TILLMANN M, CASEY D, DUFT M, MARKERT B and OEHLMANN J (2001) *Östrogenartige Wirkungen von Bisphenol A auf Vorderkiemenschnecken (Mollusca: Gastropoda: Prosobranchia)*. UWSF-Z Umweltchem Ökotox **13**: 319–333.

SIEWERS U and HERPIN U (1998) *Schwermetalleinträge in Deutschland. Moos-Monitoring 1995/96*. Geol. Jb. Sonderheft SD **2**: 1–200.

SIEWERS U, HERPIN U and STRASSBURG S (2000) *Schwermetalleinträge in Deutschland. Moos-Monitoring 1995/96. Teil 2*. Geol. Jb. Sonderheft SD **3**: 1–121.

SMODIŠ B (2003) *IAEA approaches to assessment of chemical elements in atmosphere*. In: Markert BA, Breure AM and Zechmeister H, eds. Bioindicators and Biomonitors, pp. 875–902. Elsevier, Amsterdam.

STOEPPLER M, DUERBECK HW and NUERNBERG HW (1982) *Environmental specimen banking, a challenge in trace analysis*. Talanta **29**: 963–972.

STOEPPLER M and ZEISLER R, eds. (1993) *Biological Environmental Specimen Banking (BESB)*. First International Symposium on Biological Environmental Specimen Banking, Vienna, Austria, 22–25 September 1991. Sci Total Environ, Special Issue, Vols. 139–140.

VTOROVA V, KHOLOPOVA L, MARKERT B and LEFFLER U (2001) *Multi-Elemental Composition of Tropical Plants and Bioindication of the Environmental Status*. In: Biogeochemistry and Geochemical Ecology. Selected Presentations of the 2nd Russian School of Thought 'Geochemical Ecology and the Biogeochemical Study of Taxons of the Biosphere', January 25–29, 1999, Moscow, pp. 177–189.

VUTCHKOV M (2001) *Biomonitoring of air pollution in Jamaica through trace-element analysis of epiphytic plants using nuclear and related analytical techniques*. In: Co-ordinated research project on validation and application of plants as biomonitors of trace element atmospheric pollution, analyzed by nuclear and related techniques. IAEA, NAHRES-63, Vienna.

WAGNER G (1992): *Einsatzstrategien und Meßnetze für die Bioindikation im Umweltmonitoring*. Ecoinforma, Bayreuth, pp. 1–8.

WITTIG R (1993) *General aspects of biomonitoring heavy metals by plants*. In: Markert B, ed. (1993) Plants as biomonitors – Indicators for heavy metals in the terrestrial environment, pp. 3–27. VCH-Publisher, Weinheim.

WOLTERBEEK HT, KUIK P, VERBURG TG, HERPIN U, MARKERT B and THOENI L (1995) *Moss interspecies comparisons in trace element concentrations*. Environ Monit Assess **35**: 263–286.

WUENSCHMANN S, OEHLMANN J, DELAKOWITZ B and MARKERT B (2001) *Untersuchungen zur Eignung wildlebender Wanderratten (Rattus norvegicus) als Indikatoren der Schwermetallbelastung, Teil 1*. UWSF-Z Umweltchem Ökotex **13 (5)**: 259–265.

WUENSCHMANN S, OEHLMANN J, DELAKOWITZ B and MARKERT B (2002) *Untersuchungen zur Eignung wildlebender Wanderratten (Rattus norvegicus) als Indikatoren der Schwermetallbelastung, Teil 2*. UWSF-Z Umweltchem Ökotex **14 (2)**: 96–103.

Part II
Effects of Elements in the Food Chain and on Human Health

1
Essential and Toxic Effects of Elements on Microorganisms

Dietrich H. Nies

1.1
Introduction

1.1.1
Bioelements

Six nonmetallic major bioelements or macroelements are needed to form the multitude of macromolecules in a living cell (C, O, H, N, P, S), and four additional metals to neutralize the predominantly negative charges of these macromolecules (K, Mg, Ca, Na). While the elements from carbon to calcium are undisputed major bioelements, some authors may remove sodium (Na) from this list and insert iron (Fe) instead. However, iron – like most trace elements – is a transition element, whereas sodium is an element of a major group of the periodic system like the other three metallic major bioelements.

The remaining chemical elements are either trace elements (also minor bioelements or microelements), solely toxic bioelements, or elements without biological importance. The definition of trace element follows that of Arnon and Stout (1939), as discussed in more detail by G. Schilling in Part II, Chapter 2 of this book. Before the biological importance of the naturally occurring elements of the periodic system are detailed, the rules that define the impact of a given chemical element on living cells will be defined. This will also eliminate elements without biological importance from future consideration in this chapter.

1.1.2
Parameters Defining the Biological Impact of a Chemical Element

1.1.2.1
Rule Number 1 (Availability rule)
There are three simple, almost trivial rules that define whether a chemical element is a major bioelement, a trace element, a purely toxic element, or is without biological importance. The first rule is that the element must be available in a cell for biochemical reactions. This rule appears trivial, but it is in fact the consequence of a variety of physical, chemical, and biological aspects. First, the element must have been produced in the nuclear fire of a star in the past. The probability of a given element for being produced declines hyperexponentially with increasing atomic number (Schaifers 1984). Thus, elements with high atomic numbers are usually rare in the universe, while elements with low atomic numbers are frequently occurring. Exceptions are Li, Be, and B that are used again during ele-

Elements and their Compounds in the Environment. 2nd Edition.
Edited by E. Merian, M. Anke, M. Ihnat, M. Stoeppler
Copyright © 2004 WILEY-VCH Verlag GmbH & Co. KGaA, Weinheim
ISBN: 3-527-30459-2

mental synthesis in a working star, and iron and its neighbors in the periodic systems that are preferentially produced during certain supernova events (Schaifers 1984).

Second, an element must be soluble in water to be available for life. The concentration of the chemical elements in sea water may give us the result of both effects, elemental synthesis in stars and release from the Earth's crust into water (Table 1.1). Table 1.1 also provides information (enrichment E) if a given element has been enriched (or to the contrary) on its way from ancient stars to our sea water and from the Earth's crust to sea water. The term is expressed as the decadic logarithm; thus, for example, $\lg(E) = 2$ means a 100-fold enrichment, and $\lg(E) = -2$ a 100-fold impoverishment. Since the abundance of an element in the universe is compared to that of hydrogen, $\lg(E) = 0$ for H in Table 1.1.

Third, the element must be taken up into the cell and yield a sufficiently high cytoplasmic concentration. This requires the presence of uptake systems for an element and a sufficient concentration of the element in the environment. Finally, an element must be chemically active to be of any biological importance. This excludes the rare or noble gases from any biological importance because they are chemically inactive. Helium – the second most abundant element in the universe – is strongly impoverished in sea water compared to the universe and reaches only nanomolar concentrations, probably due to diffusion losses during formation of the Earth.

Therefore, some elements are rare in the universe and nearly not available to life on earth (At, Fr, Ir, Nb, Os, Pa, Pd, Po, Pt, Ra, Re, Rh, Rn, Ru, Ta, Tc, Th). These elements may be toxic when applied in artificially high concentrations. The toxicity of Pd, Pt, and Ru was even used for the construction

of anti-cancer compounds such as cisplatin, the latter being imported into cells by copper-uptake systems (Ishida et al. 2002). Soluble rhodium compounds were shown to be genotoxic, most likely by oxidative damage induction (Migliore et al. 2002). However, toxic concentrations of these elements probably do not occur under natural conditions.

The actinides (Th, Pa, U) are the three naturally occurring elements with the highest atomic masses. None of these three has a stable isotope, which makes Th and Pa rare elements in the Earth's crust and sea water. In contrast, the uranium concentration in sea water is 12.6 nM, present as U(VI) in an oxyanionic soluble form. This oxyanion may serve as an electron acceptor for anaerobic respiration in bacteria (Lovley et al. 1991, Payne et al. 2002, Holmes et al. 2002). The resulting U(IV) is insoluble and can be precipitated by microbial cells (Yong and Macaskie 1998), which in turn opens up a biotechnological route for the bioremediation of uranium-contaminated environments.

Other elements may occur in the Earth's crust, but their concentration in (sea) water is negligible due to a very low solubility. Especially trivalent and tetravalent cations form insoluble hydroxides at neutral pH-values, which diminishes the bioavailability of many elements, especially of groups IIIa (Ga, In), IVa (Ge, Sn), Va (Bi), IIIb (Sc, Y, La, Ac), IVb (Hf, Ti, Zr) and the lanthanides. Lanthanum was not accumulated into the cytoplasm of bacterial cells (Bayer and Bayer 1991), and this may also be true for the other elements of these groups.

On the other hand, the "availability rule" highlights which elements are used by living cells: eight of the nine major components of sea water are also major bioelements. These elements are present in molar (H and O as components of water)

Tab. 1.1: Elements of life

No^a	Element	Biol. func.^{b}	Form^c	Chem. Cons.^{d}	OSW^e	Lg(E)^f, Crust	Lg(E)^g, Stars
1	Hydrogen	Bio	H(I)	Sol	107 M	1.89	0.00
2	Oxygen	Bio	O(−II)	Sol	54 M	0.27	2.90
3	Chlorine	Trace	Cl(−I)	Sol	536 mM	2.16	2.70
4	Sodium	Bio	Na(I)	Sol	457 mM	−0.35	3.33
5	Magnesium	Bio	Mg(II)	Sol	55.5 mM	−1.24	1.21
6	Sulfur	Bio	S(VI)	Oxy	2.7.6 mM	0.53	1.21
7	Calcium	Bio	Ca(II)	Sol	9.98 mM	−2.02	1.67
8	Potassium	Bio	K(I)	Sol	9.72 mM	−1.74	1.46
9	Carbon	Bio	C(IV)	Sol	2.33 mM	−0.85	−1.26
10	Bromine	Trace	Br(−I)	Sol	813 µM	1.41	
11	Boron	Trace	B(III)	Sol	425 µM	−0.34	3.60
12	Silicon	Trace	Si(IV)	Ins	107 µM	−4.97	−1.60
13	Strontium		Sr(II)	Sol	92.4 µM	−1.67	
14	Fluorine	Trace	F(−I)	Sol	68.4 µM	−2.68	−2.59
15	Nitrogen	Bio	N(−III)	Sol	35.7 µM	−1.60	−2.38
16	Lithium		Li(I)	Sol	25.9 µM	−2.05	2.38
17	Argon		Ar	Nob	15.0 µM	−0.77	−1.65
18	Phosphorus	Bio	P(V)	Oxy	2.26 µM	−4.18	−3.08
19	Rubidium	Trace	Rb(I)	Sol	1.40 µM	−2.88	
20	Iodine	Trace	I(−I)	Sol	473 nM	−0.92	
21	Aluminum		Al(III)	Ins	371 nM	−6.92	−2.86
22	Barium		Ba(II)	Sol	218 nM	−4.15	
23	Iron	Trace	Fe(III)	Ins	179 nM	−6.75	−4.28
24	Indium		In(III)	Ins	< 174 nM	−0.70	
25	Zinc	Trace	Zn(II)	Sol	153 nM	−3.85	
26	Molybdenum	Trace	Mo(VI)	Oxy	104 nM	−2.18	
27	Nickel	Trace	Ni(II)	Sol	92.0 nM	−4.14	−3.37
28	Copper	Trace	Cu(II)	Sol	47.2 nM	−4.26	
29	Arsenic	Toxic	As(V)	Oxy	40.0 nM	−2.78	
30	Vanadium	Trace	V(V)	Oxy	39.3 nM	−4.83	−4.34
31	Manganese	Trace	Mn(II)	Sol	36.4 nM	−5.68	−4.77
32	Krypton		Kr	Nob	29.8 nM	1.40	
33	Tin		Sn(IV)	Ins	25.3 nM	−2.82	
34	Titanium		Ti(IV)	Ins	20.9 nM	−6.76	−4.41
35	Uranium	Toxic	U(VI)	Oxy	12.6 nM	−2.95	
36	Neon		Ne	Nob	6.94 nM	−1.55	−6.19
37	Cobalt	Trace	Co(II)	Sol	4.58 nM	−4.97	−5.47
38	Cesium		Cs(I)	Sol	3.76 nM	−3.30	
39	Yttrium		Y(III)	Ins	3.37 nM	−7.00	
40	Cerium		Ce(III)	Ins	2.85 nM	−5.18	
41	Silver	Toxic	Ag(I)	Sol	2.78 nM	−2.37	
42	Antimony	Toxic	Sb(V)	Oxy	2.71 nM	−2.78	
43	Helium		He	Nob	1.72 nM	−3.06	−9.59
44	Selenium	Trace	Se(IV)	Ins	1.14 nM	−2.74	
45	Cadmium	(Toxic)	Cd(II)	Sol	979 pM	−3.26	
46	Germanium		Ge(IV)	Ins	964 pM	−4.89	
47	Chromium	(Toxic)	Cr(VI)	Oxy	962 pM	−6.30	−6.15
48	Tungsten	Trace	W(VI)	Oxy	544 pM	−4.18	

group, and are major bioelements. Both are required to form bridges between negatively charged components of biological macro-molecules, the smaller Mg(II) within the cytoplasm, the bigger Ca(II) within membranes.

1.2.1.3
Group IIIA (**B**, **Al**, Ga, In, **Tl**)

Boron is the only nonmetal of this group and has, as borate, a much higher bioavailability than the four metals (425 µM; Table 1.1). Although aluminum is the third-most abundant element in the Earth's crust (after oxygen and silicon; Weast 1984), its bioavailability is very low due to insoluble hydroxide complexes. Solubility increases only at low pH-values (Macdonald and Martin 1988). Al(III) becomes toxic under these conditions. Aluminum-stress is answered by plants by excretion of aluminum-complexing compounds as malate or citrate (Kataoka et al. 2002, Pineros et al. 2002, Tesfaye et al. 2001, Ma et al. 2001b, Yang et al. 2000, Ma, 2000), silicon (Desouky et al. 2002), or the synthesis of efflux systems (Sasaki et al. 2002). Since cells do usually not import trivalent cations [exceptions are Fe(III) as siderophores and Cr(III) that originates after reduction of chromate, see below], Al(III) may only be imported instead of Fe(III) in the center of siderophore compounds, making it unlikely that aluminum will be identified as a trace element.

Bioavailability of thallium, mainly a monovalent cation, is low. A thallium tolerance has been described (Sensfuss et al. 1986), and this is based on the mutation of a potassium uptake system, though a bacterium faces very unlikely toxic thallium concentrations.

This leaves boron as the only element of this group that is bioavailable. Boron was recently identified as a trace element, and is known to be essential for plant life (Brown et al. 2002b). The element was found to be part of a chemical signal that is exchanged as part of bacterial cell-to-cell communication (Coulthurst et al. 2002, Chen et al. 2002). This is the first "hard" biochemical evidence for the function of boron as trace element, and it should be simply a question of time before the function of boron in higher organisms is uncovered.

1.2.1.4
Group IVA (**C**, **Si**, Ge, Sn, **Pb**)

Carbon as a major bioelement makes life possible. In its most oxidized state, CO_2, carbon is diffusible, but also soluble as carbonate in water. Carbon is able to form covalent bonds with itself and other nonmetals. It can also build hydrophilic as hydrophobic molecules – features that are not matched by any other element. Carbon is formed in stars during one possible route of hydrogen-burning to helium, and is finally formed in high amounts in old stars when they burn their helium up to carbon. Thus, in addition to water, carbon is the most important prerequisite of life, and the two elements are similarly present in high amounts in the universe – which makes our universe very life-friendly.

Silicon – although being element number two in the Earth's crust – has a limited solubility, and concentrations reach only 107 µM in sea water and comparable ecosystems (see Table 1.1). Silicon is used as a building material in diatoms (Grachev et al. 2002) and also to sequester metal cations at toxic concentrations (Rogalla and Romheld 2002, Liang and Ding 2002, Iwasaki et al. 2002, Desouky et al. 2002, Neumann and zur Nieden 2001). Plants require silicon and also accumulate the element (Ma et al. 2001a).

Lead forms the divalent heavy metal cation Pb(II), which is a toxic-only cation (Godwin 2001). Its affinity for thiol compounds is much too high to allow this element to become a trace element (Nies 2003). Pb(II) is imported into cells by the uptake systems for other divalent metal cations (Bannon et al. 2002), and exported by ATP-hydrolyzing efflux systems that belong to the protein family of the soft-metal-transporting P-type ATPases (Rensing et al. 1999).

1.2.1.5
Group VA (**N**, **P**, **As**, **Sb**, Bi)

Nitrogen is the only nonmetallic major bioelement that is able to form positively charged chemical groups, while phosphorus is able to form high-energy-storing acid anhydrides. This makes both major bioelements, although their bioavailability is low compared to the others. In the case of arsenic, in the $+5$ oxidation state the oxyanion arsenate (AsO_4^{3-}), is structurally very similar to phosphate (PO_4^{3-}). Therefore, arsenate is taken up by the phosphate uptake systems and interferes with the biological function of phosphate, and this is the basis of its toxicity (Nies 1999). In bacteria, the detoxification of arsenate requires reduction to arsenite (Ji and Silver 1992, Ji et al. 1994, Liu and Rosen 1997), followed by export via arsenite-efflux pumps that are also capable of detoxifying antimonite. This interaction – the use of arsenate by bacteria for anaerobic respiration and interaction of eukaryotic cells with arsenate – has recently been reviewed in detail (Mukhopadhyay et al. 2002). Antimonite is accumulated by aquaglycerol facilitators into bacterial (Sanders et al. 1997) and mammalian (Liu et al. 2002) cells and detoxified by arsenate resistance determinants.

1.2.1.6
Group VIA (**O**, **S**, **Se**, **Te**, Po)

Oxygen and sulfur are major bioelements, while polonium is almost unavailable to living organisms. Selenium, with a nanomolar bioavailability, is the only known trace element of this chemical group (Andreesen and Ljungdahl 1973, Fu et al. 2002). Selenium has a much lower pKa value than its neighbor sulfur, and this results in selenol groups that are deprotonated under physiological pH-values in contrast to the mainly protonated thiol groups. Selenium is imported into cells as selenate, activated as selenophosphate, and incorporated into a serine-tRNA leading to a seleno-cysteine-tRNA. During translation, specific mRNA structures and proteins are required for the incorporation of this amino acid into the growing polypeptide chain (Selmer and Su 2002). This makes selenium a unique trace element that is not only required as essential element for most organisms, but is the only trace element which is incorporated into proteins during translation.

Selenate may also be used as electron acceptor for anaerobic respiration by certain bacteria (Rathgeber et al. 2002), which are mostly also able to reduce tellurite (Di Tomaso et al. 2002). Tellurium has a very low bioavailability. Tellurite resistance has long been known (Summer and Jacoby 1977, Taylor et al. 1987, Kormutakova et al. 2000, Turner et al. 2001), but tellurium is most likely not a trace element, unlike selenium.

1.2.1.7
Group VIIA (**F**, **Cl**, **Br**, **I**, At)

The halogenides are mostly monovalent anions under physiological conditions, and are commonly used by many organisms for the formation of halometabolites. Whereas bromometabolites are mainly found in marine environments, chlorometabolites are predominantly produced by terrestrial organisms. Iodo- and fluorocompounds are synthesized only in rare cases (van Pee 2001). Iodine has long been known to be a trace element for humans, and is needed for the formation of the hormone thyroxine (Aquaron et al. 2002).

Chlorine has recently been described as a signal molecule in gene regulation of a moderately halophilic bacterium (Roeßler and Müller 2002), and some bacteria have – under specific conditions – chloride-specific transport channels (Iyer et al. 2002). Most importantly, Cl is integral part of the water-splitting apparatus of photosystem II in cyanobacteria and chloroplasts (Yachandra et al. 1993).

1.2.2
Transition Metals

1.2.2.1
Group VB (**V**, Nb, Ta)
Due to its electronic configuration of $3d^3$ $4s^2$, vanadium is mostly present in the $+5$ oxidation state that forms the oxyanion vanadate (VO_4^{3-}). Due to its structural similarity to phosphate PO_4^{3-}, VO_4^{3-} is taken up by phosphate uptake systems and treated like phosphate by the cell (Mahanty et al. 1991). However, since vanadate does not form stable molecules as does phosphate (Ivancsits et al. 2002), this leads to vanadate toxicity. This ability is also used for in-vitro and in-vivo vanadate-inhibition experiments (Rensing et al. 1997).

Vanadate can probably be reduced to a less toxic form in the cytoplasm (Capella et al. 2002), so that oxidation states other than $+5$ [e.g., V(IV) or V(III)] may also be present (Michibata et al. 2002, Nagaoka et al. 2002). As it can be reduced under biological conditions, vanadate also serves as electron acceptor for anaerobic respirations (Yurkova

and Lyalikova 1990, Lyalikova and Yurkova 1992). As its toxicity (by interfering with cellular phosphate metabolism) reduces the usefulness of vanadium, its function as a trace element is limited to a few known cases (Rehder 1992), though it is an essential trace element for humans (Nagaoka et al. 2002). Nitrogenase, the enzyme that reduces molecular nitrogen to ammonium ion, is usually a molybdenum-containing protein (see Section 1.2.2.2); however, the bacterium *Azotobacter chroococcum* is able the express a vanadium-containing enzyme under molybdenum starvation (Thiel 1996).

Besides other occurrences of vanadium where the underlying biochemical mechanism is not understood (Rehder 1992, Mohammad et al. 2002a,b, Semiz et al. 2002, Semiz and McNeill 2002), vanadate-dependent non-heme oxidases are involved in the halogenation of organic compounds (see Section 1.2.1.7; Ohshiro et al. 2002, Sarmah et al. 2002, Tanaka et al. 2002, Ohsawa et al. 2001). Due to its high availability and its unique chemical features, more functions for vanadium as trace element may be uncovered in the future.

1.2.2.2
Group VIB (**Cr**, **Mo**, **W**)
This group of transition metals is exceptional in many respects. First, molybdenum and tungsten (synonym wolfram) are present in the Earth's crust and sea water in much higher amounts than their high atomic numbers would predict. Second, the trace elements of this group are mainly Mo and W of the second and third transition period, while Cr of the first period is a trace element only in rare cases. This makes W the trace element with the highest known atomic mass.

All three metals are able to form oxyanions, with the oxidation states $+6$, chromate, molybdate and tungstate. Chromate is accumulated by sulfate-uptake systems (Nies and Silver 1989) and interferes with sulfate metabolism (Peitzsch et al. 1998, Juhnke et al. 2002), as do vanadate and arsenate with phosphate metabolism. Moreover, chromate reduction to Cr(III) produces radicals, which makes this metal very toxic. Chromate is therefore toxic, allergenic ("mason's allergy") and carcinogenic (Costa 1997). This should diminish the usefulness of chromium as a trace element; however, the chromium cation binds in humans to a low-molecular mass peptide at a ratio of four Cr per peptide, and the resulting complex is able specifically to activate the insulin receptor (Davis and Vincent 1997a,b). This makes chromium an essential trace element in humans. As chromium is one of the few trivalent cations which appears in the cytoplasm, more trace element functions for this element may be awaiting scientific exploration. Chromate is detoxified in bacteria by efflux in addition to reduction (Nies 2003).

As judged from our ecological model compound sea water, the bioavailability of molybdenum (104 nM) is higher than that of chromium (962 pM) (see Table 1.1). While chromium is insoluble as Cr(III) in the Earth's crust, the reduction of molybdate is not as easy as chromate reduction, which leads to a factor of 10000 when the release of chromium and molybdenum from the Earth's crust into sea water is compared. Together with its low toxicity (Nies 1999), this makes molybdate the prime choice for biochemical reactions requiring oxyanion catalysis (Williams and da Silva 2002).

Molybdate is imported into the cells by highly specific uptake systems (ABC family). The imported molybdate is than used for the formation of a specific molybdenum-containing cofactor that is the Mo-containing element of most Mo-dependent enzymes such as nitrate reductase (Menen-dez et al. 1997, Anderson et al. 1992, Hochheimer et al. 1998, Schindelin et al. 1996, Romão et al. 1995). In contrast, molybdenum in nitrogenase enzymes is not bound to a pterin, but rather to a histidine residue, to isocitrate and sulfur (Bolin et al. 1993, Chan et al. 1993).

The availability of tungsten is half of that of chromate, and hence much lower than that of molybdenum. However, in evolutionary terms, during the early days of life and when conditions were still very anaerobic and reduced, there may have been more tungsten available to the early living forms than molybdenum (Williams and da Silva 2002). Thus, both anaerobic bacteria of the present day and evolutionarily aged organisms such as archaea use tungsten instead of molybdenum for oxyanion catalysis (Kletzin 1997, Hochheimer et al. 1998, Andreesen and Ljungdahl 1973, Raaijmakers et al. 2002). Uptake of tungstate is by an (ABC) uptake system similar to that used for molybdate (Makdessi et al. 2001).

1.2.2.3
Group VIIB (**Mn**, Tc, Re)
Manganese is one of the most prominent trace elements. In a $3d^5 4s^2$ electronic configuration, it can assume all oxidation states between $+2$ and $+7$. This makes the element an "electron buffer" for biochemical reactions. Most prominent is the water-splitting ability of the manganese-containing photosystem II of cyanobacteria and chloroplasts. This system is responsible for nearly all the oxygen in the atmosphere of Earth and the oxidized state of the Earth's crust. In this membrane-bound protein complex, four manganese atoms are bound to histidine residues close to a tyrosine radical residue, which may be required to remove protons from the substrate water molecules (Brudvig 1987, Noguchi et al. 1997, Tang et al. 1994, Gilchrist et al. 1995,

Hoganson and Babcock 1997). Manganese alters between Mn(III) and Mn(IV), accepting as a tetranuclear complex (that contains also calcium and chlorine) four electrons from two water molecules, leaving molecular oxygen (Brudvig 1987, Yachandra et al. 1993, Abramowicz and Dismukes 1984, Ahrling et al. 1997). Driven by light, the redox potential of these electrons is ultimately lowered to negative values that allow the reduction of NAD^+ or $NADP^+$, which are used to assimilate carbon dioxide in an autotrophic life process. By using the "solvent of life", water, as redox-donor for autotrophic carbon fixation, cyanobacteria were able to settle in all environments of the Earth, including the interior of eukaryotic cells as chloroplasts, and to produce nearly all of the biomass present on our globe. This seems only possible with the help of manganese, the "electron buffer", which makes this function of Mn much more important than other Mn-dependent reactions, for example in superoxide dismutases or sporulation in certain Gram-positive bacteria (Chou and Tan 1990, Gerlach et al. 1998, Guan et al. 1998, Meier et al. 1994a,b, 1995, Polack et al. 1996, Francis and Tebo 2002), although the water-splitting complex of photosystem II may have originated from a Mn-containing superoxide dismutase (Stallings et al. 1985, Guan et al. 1998).

Mn(II) is taken up into bacteria by a multitude of different uptake systems. Toxicity of Mn(II) to bacteria is generally low compared to other transition metals (Nies 1999). Manganese is toxic to plants and the central nervous system of mammals (Fageria et al. 2002, Ingersoll et al. 1995). It may be detoxified in plants by sequestration to silicon (Iwasaki et al. 2002, Rogalla and Romheld 2002), while efflux-mediated manganese resistance in bacteria has not been described. It may, however, be used as an electron acceptor in anaerobic respiration processes and as an electron donor for chemolithoautotrophic bacteria (Langenhoff et al. 1997, Francis and Tebo 2002).

1.2.2.4
Group VIIIB1 (**Fe**, Ru, Os)

Iron is counted as the only major bioelement of the transition elements, or as the most important trace element. This importance is based on a unique feature of iron: depending on the species and distance of the ligands of this metal in complex compounds, high- or low-spin complexes with a broad variety of redox potentials can be formed, namely iron- sulfur clusters and heme compounds. Therefore, iron forms the most electron-transferring prosthetic groups in the cell, enabling respiration and other important redox reactions.

In aerobic ecosystems, iron is mostly present as the insoluble Fe(III). Cells must synthesize, excrete and re-import iron-specific chelators, siderophores, for a sufficient supply with iron under these conditions (Braun et al. 1998). In contrast, under anaerobic conditions iron is mainly present as Fe(II). Similar to Mn(II), Fe(II) can be imported into the cell by a variety of uptake systems.

1.2.2.5
Group VIIIB2 (**Co**, Rh, Ir)

Cobalt is the trace element with the lowest concentration in sea water (see Table 1.1), indicating a general low bioavailability of this metal. The metal has seven electrons in the respective incompletely filled d-orbitals, allowing formation of complex compounds. Cobalt occurs mainly as divalent and soluble cation Co(II).

The low concentration of Co(II) is sea water of ~5 nM is just sufficient to allow micromolar cytoplasmic concentrations to be formed by the action of chemiosmotic uptake systems. Like other divalent heavy metal cations, Co(II) is important in (CorA-like) magnesium uptake systems (Gibson et al. 1991, Nelson and Kennedy 1971, Ross 1995, Kucharski et al. 2000), in other (NRAMP) transport systems (Picard et al. 2000) and may also be complexed by citrate (Krom et al. 2000). Uptake of Co(II) (present at e.g., 5 nM) by CorA or NRAMP systems must compete with the uptake of bio-metals such as Mg(II), which are present at much higher concentrations [e.g., 55 mM in the case of Mg(II) and sea water]. Since the affinities of the transport systems for both cations are similar in the case of magnesium uptake systems (Nies and Silver 1989), the uptake of Co(II) may not be very efficient under natural conditions. Cobalt as a trace element is mostly present in the heme-compound cobalamin (vitamin B_{12}), which is the prosthetic group in enzymes involved in C–C or C–H rearrangements (Nies 1999). Since cobalamin is a stable substance, B_{12} may be the actual cobalt-containing compound exchanged between organisms, making uptake of Co(II) superfluous under most conditions (Nies 1999).

Interestingly, if organisms contain non-B_{12} cobalt enzymes such as nitrilases (Kobayashi and Shimizu 1998), an additional slow, chemisosmotically driven uptake system (NiCoT protein family) is co-expressed with the enzyme (Komeda et al. 1997). ATP-hydrolyzing uptake systems for cobalt (e.g., ABC-transport systems) are not known. This indicates that cobalt for B_{12}-enzymes may indeed be imported as cobalamin, Co(II) for other enzymes by NiCoT transport systems and that Co(II)-import by other systems may not be important in the natural environment of the cells. Co(II) is of medium toxicity and is detoxified by efflux systems (CDF protein family, RND-driven CBA-export systems) in bacteria and yeasts (Nies 2003).

While Fe(II) can be oxidized to Fe(III) under physiological conditions, this reaction is very difficult in the case of Co(II). However, Co(III) can be used under anaerobic conditions by specialized bacteria that use halogenated compounds as electron acceptors – a process called reductive dehalogenation. Cobalt is bound to a corinoid cofactor in the enzyme required for this reaction (Neumann et al. 2002).

1.2.2.6
Group VIIIB3 (**Ni**, Pd, Pt)

This group of heavy metals is, in theory, still able to form octagonal complex compounds. However, in contrast to the heavy metals on the left-hand side of the periodic system (e.g., Co, Fe, Mn), there is no difference between the high-spin and the low-spin state in these complexes. If Ni(II) forms an octagonal complex, the three non-binding orbitals are completely filled with three electron pairs, while the two anti-binding orbitals each contain a single electron. This gives the bond towards two ligands a radical-like character. Nickel is therefore used in highly specialized enzymes which deal mainly with small molecules that are being formed, bound or split: for example, molecular hydrogen in hydrogenases, carbon monoxide in the bifunctional carbon monoxide deydrogenase/acetyl-Coenzyme A (CoA) synthase in anaerobic bacteria, urea in urease, methane using cofactor F430, and superoxide radicals in superoxide dismutase (Goubeaud et al. 1997, Thauer et al. 1980, 1983, Thauer and Bonacker 1994, Mobley et al., 1995, Hausinger, 1987, Lee et al. 2002a). Urease is especially important in the pathogenicity of *Helicobacter pylori*, a Gram-negative bacterium which causes gastritis and peptic ulcer disease in humans, because urease is needed to survive in this acidic environment by the production of ammonia (for a review, see Nies 1999).

The nickel-containing reactive center is unique in the bifunctional CO dehydrogenase/acetyl-CoA synthase. This enzyme reduces carbon dioxide to carbon monoxide at a nickel/four iron/five sulfur cluster that is composed of a nickel/three iron/four sulfur cubus bridged to a mononuclear iron containing site (Dobbek et al. 2001, Drennan et al. 2001). The resulting CO migrates through a 13.8 nm-long channel within the protein complex to the second site. The acetyl-CoA is assembled there at a cuban iron-sulfur cluster bridged to a binuclear copper-nickel site (Doukov et al. 2002).

These complicated nickel-containing sites are assembled by helper proteins. For the assembly of nickel-containing hydrogenases, seven maturation enzymes plus ATP, GTP and carbamoyl phosphate are required (Blokesch et al. 2002). Assembly of the corresponding sites in CO dehydrogenase and urease requires similar factors and mechanisms (Maier et al. 1993, Cheesman et al. 1989, Fu et al. 1995, Mulrooney and Hausinger 1990, Gollin et al. 1992, Park et al. 1994, Kerby et al. 1997, Rey et al. 1994, Watt and Ludden 1998, Jeon et al. 2001), all of which involve GTPases. The uptake of nickel by cells is managed by the (CorA) magnesium uptake system, but also by highly specific but slow (NiCoT) transporters such as HoxN and by ATP-hydrolyzing (ABC) uptake systems (Degen and Eitinger 2002, de Pina et al. 1995).

Some of these helper proteins such as UreE and HybB bind the nickel cations meant for incorporation into nickel-containing enzymes (Lee et al. 1993, Park et al. 1994, Lee et al., 2002b, Song et al. 2001, Remaut et al. 2001). These proteins act as metal chaperones, which discriminate between "anabolic" nickel that serves as a trace element and toxic nickel. Ni(II) binds to polyphosphate, like many other divalent cations (Gonzalez and Jensen 1998), but

specifically to histidine in many organisms (Krämer et al. 1996, Joho et al. 1992, 1995). In *E. coli*, the toxic action of Ni(II) seems to be based on interference with histidine and Fe(II) metabolism (D. H. Nies, unpublished results). It is possible that this toxicity has limited the usefulness of nickel as a trace element, allowing its use predominantly in anaerobic bacteria. Nonetheless, if bacteria are confronted with high nickel concentrations in the environment or, like *H. pylori* are nickel-dependent, they contain nickel efflux systems (Nies 1999, 2003).

1.2.2.7
Group IB (**Cu, Ag, Au**)

Copper, when present in aerobic ecosystems as Cu(II), is reduced by most microbial cells to Cu(I), which is the predominant form in the cytoplasm. This easy one-electron redox reaction makes copper very useful as cofactor for the reaction with radicalic compounds such as molecular oxygen. The last enzyme complex of the respiratory chain in mitochondria and in many fully aerobic bacteria is that of cytochrome c oxidase. This contains two copper centers: one for uptake of electrons from cytochrome c (Cu A center); the other (Cu B center) for the reduction of molecular oxygen to water using these electrons. This reaction generates a proton-motive force that may be used for the synthesis of ATP (Iwata et al. 1995, Michel et al. 1998, Ostermeier and Michel 1997).

The main advantage of copper – its radicalic character – also brings about danger. Copper is able easily to form hydroperoxide radicals (Rodriguez Montelongo et al. 1993), interact with the cell membrane (Suwalsky et al. 1998), and also bind to thiol compounds (Nies 2003). The use of copper is therefore strictly controlled by the cells. Similar to nickel, "anabolic" copper is bound to copper chaperones (O'Halloran and Culotta

2000), which keep this metal away from mischief. Surplus copper is exported by P-type ATPases of the soft-metal group that have been found in all kingdoms of life (Rensing et al. 1999, Mandal et al. 2002, Riggle and Kumamoto 2000). Additionally, efflux systems (CBA-type) are involved in copper-detoxification in Gram-negative bacteria (Nies 2003). It is possible that these efflux complexes are able to detoxify Cu(I) directly from the periplasm of these bacteria.

Silver is a highly toxic-only metal, ranking in toxicity second only to mercury (Nies 1999). The toxicity of silver is based on its extremely high affinity to thiol compounds. Bacteria are able to detoxify silver using efflux systems of the P-type (Rensing et al. 1999, Mandal et al. 2002, Riggle and Kumamoto 2000) and/or CBA-type (Gupta et al. 1999, Franke et al. 2001) that are identical or similar to the Cu(I)-detoxifying transporters (Nies 2003). Gold is a noble metal and therefore mostly present in the metallic form. Au(III) comes next to Ag(I) when toxicity to bacterial cells is counted (Nies 1999), but the bioavailability of gold is only 2.5% that of silver (see Table 1.1), making toxic gold concentrations a rare event. Gold may be reductively precipitated by microbial cells (Kashefi et al. 2001). Due to their toxicity, silver and gold will probably be never identified as trace elements.

1.2.2.8
Group IIB (**Zn**, **Cd**, **Hg**)

Zinc as trace element is as important as iron, but the biochemical function of zinc is opposite to that of iron: while iron is the most important redox-active transition metal, zinc is the most important redox-inactive one. Zn(II) is used as Lewis acid and to tether domains of macromolecules into a distinct and concise structure. This ability of zinc results from the completely filled 3d-orbitals of the zinc atoms.

The apparent Zn(II) concentration in bacterial cells is probably higher than 200 µM and up to 1 or 2 mM (Nies 2003). In *E. coli*, 200 000 Zn(II) per cell are needed only for eight of the 48 known zinc-containing enzymes, for example the RNA polymerase that binds 10 000 Zn(II) (Outten and O'Halloran 2001). Surplus Zn(II) cations were thought to be bound to cellular compounds, but "free" zinc also seems to exist in bacterial cells (Nies 2003).

Zinc is also toxic (Nies 1999). The high bioavailability of zinc makes it very probable for microbes to encounter toxic zinc concentration (Nies 2000). Therefore, zinc-detoxification systems which rely on efflux (P-, CBA- and CDF-type export systems) are frequently found in bacteria (Nies 2003). Zn(II) is imported into microbial cells by many systems (Nies 1999, Hantke 2001).

Cadmium, again with a much higher affinity to thiol compounds, is a toxic-only element. Only one case of Cd(II) as a trace element has been reported: in the absence of Zn(II), Cd(II) was used by some diatoms in the active site of carbonic anhydrase (Lane and Morel 2000). Cd(II) enters the cell by (CorA- and NRAMP-like) uptake systems (Nies 1999), binds to thiol compounds (thereby exerting toxicity), and is then re-exported by efflux systems (P-, CBA, or CDF type) (Nies 2003).

Mercury showed the highest toxicity of all metals examined (Nies 1999). The Hg(II) cation diffuses into the cells as a hydroxo-complex and adheres to thiols. For efficient detoxification, Hg(II) is actively imported into bacterial cells and reduced to the metallic form, which then diffuses out of the cell and its environment. Due to their toxicity, mercury is not a trace element, and cadmium is only one under rare conditions (Brown et al. 2002a).

References

Due to lack of space, only exemplary references were cited. A full-length version of this chapter containing all references is available under http://bionomie.mikrobiologie.uni-halle.de

ABRAMOWICZ DA and DISMUKES GC (1984) *Manganese proteins isolated from spinach thylakoid membranes and their role in O2 evolution. II. A binuclear manganese-containing 34 kilodalton protein, a probable component of the water dehydrogenase enzyme.* Biochim Biophys Acta **765**:318–328.

AHRLING KA, PETERSON S and STYRING S (1997) *An oscillating manganese electron paramagnetic resonance signal from the S0 state of the oxygen evolving complex in photosystem II.* Biochemistry, **36**:13148–13152.

ALTENDORF K and EPSTEIN W (1996) *The Kdp-ATPase of Escherichia coli.* In: Dalbey RE, ed. Advances in Cell and Molecular Biology of Membranes and Organelles, Vol. 5, pp. 401–418. JAI Press, Inc., Greenwich, London.

ANDERSON GL, WILLIAMS J and HILLE R (1992) *The purification and characterization of arsenite oxidase from Alcaligenes faecalis, a molybdenum-containing hydroxylase.* J Biol Chem **267**:23674–23682.

ANDREESEN JR and LJUNGDAHL LG (1973) *Formate dehydrogenase of Clostridium thermoaceticum: incorporation of selenium-75, and the effects of selenite, molybdate and tungstate on the enzyme.* J Bacteriol **116**:867–873.

AQUARON R, DELANGE F, MARCHAL P, LOGNONE V and NINANE L (2002) *Bioavailability of seaweed iodine in human beings.* Cell Mol Biol **48**:563–569.

ARNON DI and STOUT PR (1939) *The essentiality of certain elements in minute quantity for plants with special reference to copper.* Plant Physiol **14**:371–375.

BANNON DI, PORTNOY ME, OLIVI L, LEES PSJ, CULOTTA VC and BRESSLER JP (2002) *Uptake of lead and iron by divalent metal transporter 1 in yeast and mammalian cells.* Biochem Biophys Res Commun **295**:978–984.

BAYER ME and BAYER MH (1991) *Lanthanide accumulation in the periplasmic space of Escherichia coli B.* J. Bacteriol **173**:141–149.

BLOKESCH M, PASCHOS A, THEODORATOU E, BAUER A, HUBE M, HUTH S and BÖCK A (2002) *Metal insertion into NiFe-hydrogenases.* Biochem Soc Trans **30**:674–680.

BOLIN JT, CAMPOBASSO N, MUCHMORE SW, MORGAN TV and MORTENSON LE (1993) *The structure and environment of the metal clusters in the nitrogenase MoFe protein from Clostridium pasteurianum.* In: Stiefel EI, Coucouvanis D and Newton WE, eds. Molybdenum enzymes, cofactors and model systems, pp. 186–195. American Chemical Society, Washington, D. C.

BRAUN V, HANTKE K and KOSTER W (1998) *Bacterial iron transport: mechanisms, genetics, and regulation.* Met Ions Biol Syst **35**:67–145.

BROWN NL, SHIH YC, LEANG C, GLENDINNING KJ, HOBMAN JL and WILSON JR (2002a) *Mercury transport and resistance.* Biochem Soc Trans **30**:715–718.

BROWN PH, BELLALOUI N, WIMMER MA, BASSIL ES, RUIZ J, HU H, PFEFFER H, DANNEL F and ROMHELD V (2002b) *Boron in plant biology.* Plant Biol **4**:205–223.

BRUDVIG GW (1987) *The tetranuclear manganese complex of photosystem II.* J Bioenerg Biomembr **19**:91–104.

CAPELLA LS, GEFE MR, SILVA EF, AFFONSO-MITIDIERI O, LOPES AG, RUMJANEK VM and CAPELLA MAM (2002) *Mechanisms of vanadate-induced cellular toxicity: role of cellular glutathione and NADPH.* Arch Biochem Biophys **406**:65–72.

CHAN MK, KIM J and REES DC (1993) *The nitrogenase FeMo-cofactor and P-cluster pair: 2.2 Å resolution structures.* Science **260**:792–794.

CHEESMAN MR, ANKEL FUCHS D, THAUER RK and THOMPSON AJ (1989) *The magnetic properties of the nickel cofactor F430 in the enzyme methylcoenzyme M reductase of Methanobacterium thermoautotrophicum.* Biochem J **260**:613–616.

CHEN X, SCHAUDER S, POTIER N, VAN DORSSELAER A, PELCZER I, BASSLER BL and HUGHSON FM (2002) *Structural identification of a bacterial quorum-sensing signal containing boron.* Nature **415**:545–549.

CHOU FI and TAN ST (1990) *Manganese(II) induces cell division and increases in superoxide dismutase and catalase activities in an aging deinococcal culture.* J Bacteriol **172**:2029–2035.

COSTA M (1997) *Toxicity and carcinogenicity of Cr(VI) in animal models and humans.* Crit Rev Toxicol **27**:431–442.

COULTHURST SJ, WHITEHEAD NA, WELCH M and SALMOND GPC (2002) *Can boron get bacteria talking?* Trends Biochem Sci **27**:217–219.

DAVIS CM and VINCENT JB (1997a) *Chromium oligopeptide activates insulin receptor tyrosine kinase activity.* Biochemistry **36**:4382–4385.

DAVIS CM and VINCENT JB (1997b) *Isolation and characterization of a biologically active chromium oligopeptide from bovine liver.* Arch Biochem Biophys **339**: 335–343.

DE PINA K, NAVARRO C, McWALTER L, BOXER DH, PRICE NC, KELLY SM, MANDRAND BERTHELOT MA and WU LF (1995) *Purification and characterization of the periplasmic nickel-binding protein NikA of Escherichia coli K12.* Eur J Biochem **227**: 857–865.

DEGEN O and EITINGER T (2002) *Substrate specificity of nickel/cobalt permeases: Insights from mutants altered in transmembrane domains I and II.* J Bacteriol **184**: 3569–3577.

DESOUKY M., JUGDAOHSINGH R, McCROHAN CR, WHITE KN and POWELL JJ (2002) *Aluminum-dependent regulation of intracellular silicon in the aquatic invertebrate Lymnaea stagnalis.* Proc Natl Acad Sci USA **99**: 3394–3399.

DI TOMASO G, FEDI S, CARNEVALI M, MANEGATTI M, TADDEI C and ZANNONI D (2002) *The membrane-bound respiratory chain of Pseudomonas pseudoalcaligenes KF707 cells grown in the presence or absence of potassium tellurite.* Microbiology **148**: 1699–1708.

DOBBEK H, SVETLITCHNYI V, GREMER L, HUBER R and MEYER O (2001) *Crystal structure of a carbon monoxide dehydrogenase reveals a [Ni-4Fe-5S] cluster.* Science **293**: 1281–1285.

DOUKOV TI, IVERSON TM, SERAVALLI J, RAGSDALE SW and DRENNAN CL (2002) *A Ni-Fe-Cu center in a bifunctional carbon monoxide dehydrogenase/Acetyl-CoA synthase.* Science **298**: 567–572.

DRENNAN CL, HEOJY, SINTCHAK MD, SCHREITER E and LUDDEN PW (2001) *Life on carbon monoxide: X-ray structure of Rhodospirillum rubrum Ni-Fe-S carbon monoxide dehydrogenase.* Proc Natl Acad Sci USA **98**: 11973–11978.

FAGERIA NK, BALIGAR VC and CLARK RB (2002) *Micronutrients in crop production In: Advances in Agronomy,* Vol. 77, pp. 185–268. Academic Press, Inc., San Diego.

FRANCIS CA and TEBO BM (2002) *Enzymatic manganese(II) oxidation by metabolically dormant spores of diverse Bacillus species.* Appl Environ Microbiol **68**: 874–880.

FRANKE S, GRASS G and NIES DH (2001) *The product of the ybdE gene of the Escherichia coli chromosome is involved in detoxification of silver ions.* Microbiology **147**: 965–972.

FU CL, OLSON JW and MAIER RJ (1995) *HypB protein of Bradyrhizobium japonicum is a metal-binding GTPase capable of binding 18 divalent nickel ions per dimer.* Proc Natl Acad Sci USA **92**: 2333–2337.

FU L-H, WANG X-F, EYAL Y, SHE Y-M, DONALD LJ, STANDING KG and BEN-HAYYIM G (2002) *A selenoprotein in the plant kingdom: mass spectrometry conforms that opal codon (UGA) encodes selenocysteine in Chlamydomonas reingardtii glutathione peroxidase.* J Biol Chem **277**: 25983–25991.

GERLACH D, REICHARDT W and VETTERMANN S (1998) *Extracellular superoxide dismutase from Streptococcus pyogenes type 12 strain is manganese-dependent.* FEMS Microbiol Lett **160**: 217–224.

GIBSON MM, BAGGA DA, MILLER CG and MAGUIRE ME (1991) *Magnesium transport in Salmonella typhimurium: the influence of new mutations conferring Co²⁺ resistance on the CorA Mg²⁺ transport system.* Mol Microbiol **5**: 2753–2762.

GILCHRIST ML, JR, BALL JA, RANDALL DW and BRITT RD (1995) *Proximity of the manganese cluster of photosystem II to the redox-active tyrosine YZ.* Proc Natl Acad Sci USA **92**: 9545–9549.

GODWIN HA (2001) *The biological chemistry of lead.* Curr Opin Chem Biol **5**: 223–227.

GOLLIN DJ, MORTENSON LE and ROBSON RL (1992) *Carboxyl-terminal processing may be essential for production of active NiFe hydrogenase in Azotobacter vinelandii.* FEBS Lett **309**: 371–375.

GONZALEZ H and JENSEN TE (1998) *Nickel sequestering by polyphosphate bodies in Staphylococcus aureus.* Microbios **93**: 179–185.

GOUBEAUD M, SCHREINER G and THAUER RK (1997) *Purified methyl-coenzyme-M reductase is activated when the enzyme-bound coenzyme F430 is reduced to the nickel(I) oxidation state by titanium(III) citrate.* Eur J Biochem **243**: 110–114.

GRACHEV MA, DENIKINA NN, BELIKOV SI, LIKHOSHVAI EV, USOL'TSEVA MV, TIKHONOVA IV, ADEL'SHIN RV, KLER SA and SCHERBAKOVA TA (2002) *Elements of the active center of silicon transporters in diatoms.* Mol Biol **36**: 534–536.

GUAN Y, HICKEY MJ, BORGSTAHL GE, HALLEWELL RA, LEPOCK JR, O'CONNOR D, HSIEH Y, NICK HS, SILVERMAN DN and TAINER JA (1998) *Crystal structure of Y34F mutant human mitochondrial manganese superoxide dismutase and the functional role of tyrosine 34.* Biochemistry **37**: 4722–4730.

GUPTA A, MATSUI K, LO JF and SILVER S (1999) *Molecular basis for resistance to silver in Salmonella.* Nature Med **5**: 183–188.

HANTKE K (2001) *Bacterial zinc transporters and regulators.* Biometals **14**: 239–249.

HAUSINGER RP (1987) *Nickel utilization by microorganisms.* Microbiol Rev **51**: 22–42.

Hochheimer A, Hedderich R and Thauer RK (1998) *The formylmethanofuran dehydrogenase isoenzymes in Methanobacterium wolfei and Methanobacterium thermoautotrophicum: induction of the molybdenum isoenzyme by molybdate and constitutive synthesis of the tungsten isoenzyme.* Arch Microbiol 170:389–393.

Hoganson CW and Babcock GT (1997) *A metalloradical mechanism for the generation of oxygen from water in photosynthesis.* Science 277:1953–1956.

Holmes DE, Finneran KT, O'Neil RA and Lovley DR (2002) *Enrichment of members of the family Geobacteraceae associated with stimulation of dissimilatory metal reduction in uranium-contaminated aquifer sediments.* Appl Environ Microbiol 68:2300–2306.

Ingersoll RT, Montgomery EB, Jr and Aposhian HV (1995) *Central nervous system toxicity of manganese. I. Inhibition of spontaneous motor activity in rats after intrathecal administration of manganese chloride.* Fundam Appl Toxicol 27:106–113.

Ishida S, Lee J, Thiele DJ and Herskowitz I (2002) *Uptake of the anticancer drug cisplatin mediated by the copper transporter Ctr1 in yeast and mammals.* Proc Natl Acad Sci USA 99:14298–14302.

Ivancsits S, Pilger A, Diem E, Schaffer A and Rudiger HW (2002) *Vanadate induces DNA strand breaks in cultured human fibroblasts at doses relevant to occupational exposure.* Mutat Res Genet Toxicol Environ Mutagen 519:25–35.

Ivshina IB, Peshkur TA and Korobov VP (2002) *Efficient uptake of cesium ions by Rhodococcus cells.* Microbiology 71:357–361.

Iwasaki K, Maier P, Fecht M and Horst WJ (2002) *Leaf apoplastic silicon enhances manganese tolerance of cowpea (Vigna unguiculata).* J Plant Physiol 159:167–173.

Iwata S, Ostermeier C, Ludwig B and Michel H (1995) *Structure at 2.8 Å resolution of cytochrome c oxidase from Paracoccus denitrificans.* Nature 376:660–669.

Iyer R, Iverson TM, Accardi A and Miller C (2002) *A biological role for prokaryotic chloride channels.* Nature 419:715–718.

Jeon WB, Cheng JJ and Ludden PW (2001) *Purification and characterization of membrane-associated CooC protein and its functional role in the insertion of nickel into carbon monoxide dehydrogenase from Rhodospirillum rubrum.* J Biol Chem 276:38602–38609.

Ji G and Silver S (1992) *Reduction of arsenate to arsenite by the ArsC protein of the arsenic resistance operon of Staphylococcus aureus plasmid pI258.* Proc Natl Acad Sci USA 89:9474–9478.

Ji GY, Garber EAE, Armes LG, Chen CM, Fuchs JA and Silver S (1994) *Arsenate reductase of Staphylococcus aureus plasmid pI258.* Biochemistry 33:7294–7299.

Joho M, Inouhe M, Tohoyama H and Murayama T (1995) *Nickel resistance mechanisms in yeasts and other fungi.* J Ind Microbiol 14:164–168.

Joho M, Ishikawa Y, Kunikane M, Inouhe M, Tohoyama H and Murayama T (1992) *The subcellular distribution of nickel in Ni-sensitive and Ni-resistant strains of Saccharomyces cerevisiae.* Microbios 71:149–159.

Juhnke S, Peitzsch N, Hübener N, Grosse C and Nies DH (2002) *New genes involved in chromate resistance in Ralstonia metallidurans strain CH34.* Arch Microbiol 179:15–25.

Kashefi K, Tor JM, Nevin KP and Lovley DR (2001) *Reductive precipitation of gold by dissimilatory Fe(III)-reducing bacteria and archaea.* Appl Env Microbiol 67:3275–3279.

Kataoka T, Stekelenburg A, Nakanishi TM, Delhaize E. and Ryan PR (2002) *Several lanthanides activate malate efflux from roots of aluminium-tolerant wheat.* Plant Cell Environ 25:453–460.

Kerby RL, Ludden PW and Roberts GP (1997) *In vivo nickel insertion into the carbon monoxide dehydrogenase of Rhodospirillum rubrum: molecular and physiological characterization of cooCTJ.* J Bacteriol 179:2259–2266.

Kletzin A (1997) *Tungsten-containing aldehyde ferredoxine oxidoreductases.* In: Winkelmann G and Carrano CJ, eds. Transition metals in microbial metabolism, pp. 357–390. OPA, Amsterdam, The Netherlands.

Kobayashi M and Shimizu S (1998) *Metalloenzyme nitrile hydratase: structure, regulation, and application to biotechnology.* Nature Biotechnol 16:733–736.

Komeda H, Kobayashi M and Shimizu S (1997) *A novel transporter involved in cobalt uptake.* Proc Natl Acad Sci USA 94:36–41.

Kormutakova R, Klucar L and Turna J (2000) *DNA sequence analysis of the tellurite-resistance determinant from clinical strain of Escherichia coli and identification of essential genes.* Biometals 13:135–139.

Krämer U, Cotterhowells JD, Charnock JM, Baker AJM and Smith JAC (1996) *Free histidine* as a metal chelator in plants that accumulate nickel. Nature 379:635–638.

KROM BP, WARNER JB, KONINGS WN and LOLKEMA JS (2000) *Complementary metal ion specificity of the metal-citrate transporters CitM and CitH of Bacillus subtilis.* J Bacteriol 182:6374–6381.

KUCHARSKI LM, LUBBE WJ and MAGUIRE ME (2000) *Cation hexaammines are selective and potent inhibitors of the CorA magnesium transport system.* J Biol Chem 275:16767–16773.

LANE TW and MOREL FMM (2000) *A biological function for cadmium in marine diatoms.* Proc Natl Acad Sci USA 97:4627–4631.

LANGENHOFF AAM, BRONWERS-CEILER DL, ENGELBERTING JHL, QUIST JJ, WOLKENFELT JPN, ZEHNDER AJB and SCHRAA G (1997) *Microbial reduction of manganese coupled to toluene oxidation.* FEMS Microbiol Ecol 22:119–127.

LEE JW, ROE JH and KANG SO (2002a) *Nickel-containing superoxide dismutase.* In: Superoxide Dismutase, Vol. 349, pp. 90–101. Academic Press, Inc., San Diego.

LEE MH, PANKRATZ HS, WANG S, SCOTT RA, FINNEGAN MG, JOHNSON MK, IPPOLITO JA, CHRISTIANSON DW and HAUSINGER RP (1993) *Purification and characterization of Klebsiella aerogenes UreE protein: a nickel-binding protein that functions in urease metallocenter assembly.* Protein Sci 2:1042–1052.

LEE YH, WON HS, LEE MH and LEE BJ (2002b) *Effects of salt and nickel ion on the conformational stability of Bacillus pasteurii UreE.* FEBS Lett 522:135–140.

LIANG YC and DING RX (2002) *Influence of silicon on microdistribution of mineral ions in roots of salt-stressed barley as associated with salt tolerance in plants.* Sci China Ser C-Life Sci 45:298–308.

LIU J and ROSEN BP (1997) *Ligand interactions of the ArsC arsenate reductase.* J Biol Chem 272:21084–21089.

LIU Z, SHEN J, CARBREY J, MUKHOPADHYAY R, AGRE P and ROSEN BP (2002) *Arsenite transport by mammalian aquaglyceroporins AQP7 and APQ9.* Proc Natl Acad Sci USA 99:6053–6058.

LOVLEY DR, PHILLIPS EJP, GORBYYA and LAND AER (1991) *Microbial reduction of uranium.* Nature 350:413–416.

LYALIKOVA NN and YURKOVA NA (1992) *Role of microorganisms in vanadium concentration and dispersion.* Geomicrobiol J 10:15–26.

MA JF (2000) *Role of organic acids in detoxification of aluminum in higher plants.* Plant Cell Physiol 41:383–390.

MA JF, GOTO S, TAMAI K and ICHII M (2001a) *Role of root hairs and lateral roots in silicon uptake by rice.* Plant Physiol 127:1773–1780.

MA JF, RYAN PR and DELHAIZE E (2001b) *Aluminium tolerance in plants and the complexing role of organic acids.* Trends Plant Sci 6:273–278.

MACDONALD TL and MARTIN RB (1988) *Aluminium ion in biological systems.* Trends Biochem Sci 13:13–15.

MAHANTY SK, KHAWARE R, ANSARI S, GUPTA P and PRASAD R (1991) *Vanadate-resistant mutants of Candida albicans show alterations in phosphate uptake.* FEMS Microbiol Lett 68:163–166.

MAIER T, JACOBI A, SAUTER M and BÖCK A (1993) *The product of the hypB gene, which is required for nickel incorporation into hydrogenases, is a novel guanine nucleotide-binding protein.* J Bacteriol 175:630–635.

MAKDESSI K, ANDREESEN JR and PICH A (2001) *Tungstate uptake by a highly specific ABC transporter in Eubacterium acidaminophilum.* J Biol Chem 276:24557–24564.

MANDAL AK, CHEUNG WD and ARGUELLO JM (2002) *Characterization of a thermophilic P-type Ag^+/Cu^+-ATPase from the extremophile Archaeoglobus fulgidus.* J Biol Chem 277:7201–7208.

MEIER B, MICHEL C, SARAN M, HUTTERMANN J, PARAK F and ROTILIO G (1995) *Kinetic and spectroscopic studies on a superoxide dismutase from Propionibacterium shermanii that is active with iron or manganese: pH-dependence.* Biochem J 310:945–950.

MEIER B, SEHN AP, MICHEL C and SARAN M (1994a) *Reactions of hydrogen peroxide with superoxide dismutase from Propionibacterium shermanii – an enzyme which is equally active with iron or manganese – are independent of the prosthetic metal.* Arch Biochem Biophys 313:296–303.

MEIER B, SEHN AP, SCHININA ME and BARRA D (1994b) *In vivo incorporation of copper into the iron-exchangeable and manganese-exchangeable superoxide dismutase from Propionibacterium shermanii. Amino acid sequence and identity of the protein moieties.* Eur J Biochem 219:463–468.

MENENDEZ C, OTTO A, IGLOI G, NICK P, BRANDSCH R, SCHUBACH B, BOTTCHER B and BRANDSCH R (1997) *Molybdate-uptake genes and molybdopterin-biosynthesis genes on a bacterial plasmid-characterization of MoeA as a filament-forming protein with adenosine triphosphatase activity.* Eur J Biochem 250:524–531.

MICHEL H, BEHR J, HARRENGA A and KANNT A (1998) *Cytochrome c oxidase: structure and spec-*

troscopy. Annu Rev Biophys Biomolec Struct **27**: 329 – 356.

MICHIBATA H, UYAMA T, UEKI T and KANAMORI K (2002) *Vanadocytes, cells hold the key to resolving the highly selective accumulation and reduction of vanadium in ascidians.* Microsc Res Tech **56**: 421 – 434.

MIGLIORE L, FRENZILLI G, NESTI C, FORTANER S and SABBIONI E (2002) *Cytogenetic and oxidative damage induced in human lymphocytes by platinum, rhodium and palladium compounds.* Mutagenesis **17**: 411 – 417.

MOBLEY HL, ISLAND MD and HAUSINGER RP (1995) *Molecular biology of microbial ureases.* Microbiol Rev **59**: 451 – 480.

MOHAMMAD A, SHARMA V and McNEILL JH (2002a) *Vanadium increases GLUT4 in diabetic rat skeletal muscle.* Mol Cell Biochem **233**: 139 – 143.

MOHAMMAD A, WANG J and McNEILL JH (2002b) *Bis(maltolato)oxovanadium(IV) inhibits the activity of PTP1B in Zucker rat skeletal muscle in vivo.* Mol Cell Biochem **229**: 125 – 128.

MUKHOPADHYAY R, ROSEN BP, PUNG LT and SILVER S (2002) *Microbial arsenic: from geocycles to genes and enzymes.* FEMS Microbiol Rev **26**: 311 – 325.

MULROONEY SB and HAUSINGER RP (1990) *Sequence of the Klebsiella aerogenes urease genes and evidence for accessory proteins facilitating nickel incorporation.* J Bacteriol **172**: 5837 – 5843.

NAGAOKA MH, YAMAZAKI T and MAITANI T (2002) *Binding patterns of vanadium ions with different valence states to human serum transferrin studied by HPLC/high-resolution ICP-MS.* Biochem Biophys Res Commun **296**: 1207 – 1214.

NELSON DL and KENNEDY EP (1971) *Magnesium transport in Escherichia coli: Inhibition by cobaltous ion.* J Biol Chem **246**: 3042 – 3049.

NEUMANN A, SEIBERT A, TRESCHER T, REINHARDT S, WOHLFARTH G and DIEKERT G (2002) *Tetrachloroethene reductive dehalogenase of Dehalospirillum multivorans: substrate specificity of the native enzyme and its corrinoid cofactor.* Arch Microbiol **177**: 420 – 426.

NEUMANN D and ZUR NIEDEN U (2001) *Silicon and heavy metal tolerance of higher plants.* Phytochemistry **56**: 685 – 692.

NIES DH (1999) *Microbial heavy metal resistance.* Appl Microbiol Biotechnol **51**: 730 – 750.

NIES DH (2000) *Heavy metal resistant bacteria as extremophiles: molecular physiology and biotechnological use of Ralstonia spec. CH34.* Extremophiles **4**: 77 – 82.

NIES DH (2003) *Molecular physiology of efflux-mediated heavy metal resistance in prokaryotes.* FEMS Microbiol Rev **27**: 313 – 339.

NIES DH and SILVER S (1989) *Metal ion uptake by a plasmid-free metal-sensitive Alcaligenes eutrophus strain.* J Bacteriol **171**: 4073 – 4075.

NOGUCHI T, INOUE Y and TANG X S (1997) *Structural coupling between the oxygen-evolving Mn cluster and a tyrosine residue in photosystem II as revealed by Fourier transform infrared spectroscopy.* Biochemistry **36**: 14705 – 14711.

O'HALLORAN TV and CULOTTA VC (2000) *Metallochaperones, an intracellular shuttle service for metal ions.* Journal of Biological Chemistry **275**: 25057 – 25060.

OHSAWA N, OGATA Y, OKADA N and ITOH N (2001) *Physiological function of bromoperoxidase in the red marine alga, Corallina pilulifera: production of bromoform as an allelochemical and the simultaneous elimination of hydrogen peroxide.* Phytochemistry **58**: 683 – 692.

OHSHIRO T, HEMRIKA W, AIBARA T, WEVER R and IZUMI Y (2002) *Expression of the vanadium-dependent bromoperoxidase gene from a marine macro-alga Corallina pilulifera in Saccharomyces cerevisiae and characterization of the recombinant enzyme.* Phytochemistry **60**: 595 – 601.

OSTERMEIER C and MICHEL H (1997) *Cytochrome c oxidase – the key enzyme of aerobic respiration.* In: Winkelmann G and Carrano CJ, eds. Transition metals in microbial metabolism, pp. 311 – 328. Harwood Academic Publishers, Amsterdam.

OUTTEN CE and O'HALLORAN TV (2001) *Femtomolar sensitivity of metalloregulatory proteins controlling zinc homeostasis.* Science **292**: 2488 – 2492.

PARK IS, CARR MB and HAUSINGER RP (1994) *In-vitro activation of urease apoprotein and role of UreA as a chaperone required for nickel metallocenter assembly.* Proc Natl Acad Sci USA **91**: 3233 – 3237.

PATEL S, YENUSH L, RODRIGUEZ PL, SERRANO R and BLUNDELL TL (2002) *Crystal structure of an enzyme displaying both inositol-polyphosphate-1-phosphatase and 3'-phosphoadenosine-5'-phosphate phosphatase activities: a novel target of lithium therapy.* J Mol Biol **315**: 677 – 685.

PAYNE RB, GENTRY DA, RAPP-GILES BJ, CASALOT L and WALL JD (2002) *Uranium reduction by Desulfovibrio desulfuricans strain G20 and a cytochrome c3 mutant.* Appl Environ Microbiol **68**: 3129 – 3132.

PEITZSCH N, EBERZ G and NIES DH (1998) *Alcaligenes eutrophus as a bacterial chromate sensor.* Appl Environ Microbiol **64**: 453 – 458.

PICARD V, GOVONI G, JABADO N and GROS P (2000) *Nramp 2 (DCT1/DMT1) expressed at the plasma membrane transports iron and other divalent cations* into a calcein-accessible cytoplasmic pool. J Biol Chem 275: 35738–35745.

PINEROS MA, MAGALHAES JV, ALVES VMC and KOCHIAN LV (2002) *The physiology and biophysics of an aluminum tolerance mechanism based on root citrate exudation in maize.* Plant Physiol 129: 1194–1206.

POLACK B, DACHEUX D, DELIC ATTREE I, TOUSSAINT B and VIGNAIS PM (1996) *Role of manganese superoxide dismutase in a mucoid isolate of Pseudomonas aeruginosa: adaptation to oxidative stress.* Infect Immun 64: 2216–2219.

RAAIJMAKERS H, MACIEIRA S, DIAS JM, TEIXEIRA S, BURSAKOV S, HUBER R, MOURA JJG, MOURA I and ROMAO MJ (2002) *Gene sequence and the 1.8 angstrom crystal structure of the tungsten-containing formate dehydrogenase from Desulfovibrio gigas.* Structure 10: 1261–1272.

RATHGEBER C, YURKOVA N, STACKEBRANDT E, BEATTY JT and YURKOV V (2002) *Isolation of tellurite- and selenite-resistant bacteria from hydrothermal vents of the juan de fuca ridge in the pacific ocean.* Appl Environ Microbiol 68: 4613–4622.

REHDER D (1992) *Structure and function of vanadium compounds in living organisms.* Biometals 5: 3–12.

REMAUT H, SAFAROV N, CIURLI S and VAN BEEUMEN J (2001) *Structural basis for Ni^{2+} transport and assembly of the urease active site by the metallochaperone UreE from Bacillus pasteurii.* J Biol Chem 276: 49365–49370.

RENSING C, GHOSH M and ROSEN BP (1999) *Families of soft-metal-ion-transporting ATPases.* J Bacteriol 181: 5891–5897.

RENSING C, MITRA B and ROSEN BP (1997) *The zntA gene of Escherichia coli encodes a Zn(II)-translocating P-type ATPase.* Proc Natl Acad Sci USA 94: 14326–14331.

REY L, IMPERIAL J, PALACIOS JM and RUIZARGUESO T (1994) *Purification of Rhizobium leguminosarum HypB, a nickel-binding protein required for hydrogenase synthesis.* J Bacteriol 176: 6066–6073.

RIGGLE PJ and KUMAMOTO CA (2000) *Role of a Candida albicans P1-type ATPase in resistance to copper and silver ion toxicity.* J Bacteriol 182: 4899–4905.

RODRIGUEZ MONTELONGO L, DE LA CRUZ RODRIGUEZ LC, FARIAS RN and MASSA EM (1993) *Membrane-associated redox cycling of copper mediates hydroperoxide toxicity in Escherichia coli.* Biochim Biophys Acta 1144: 77–84.

ROESSLER M and MÜLLER V (2002) *Chloride, a new environmental signal molecule involved in gene regulation in a moderately halophilic bacterium, Halobacillus halophilus.* J Bacteriol 184: 6207–6215.

ROGALLA H and ROMHELD V (2002) *Role of leaf apoplast in silicon-mediated manganese tolerance of Cucumis sativus L.* Plant Cell Environ 25: 549–555.

ROMÃO MJ, ARCHER M, MOURA I., MOURA JJG, LEGALL J, ENGH R, SCHNEIDER M, HOF P and HUBER R (1995) *Crystal structure of the xanthine-oxidase-related aldehyde oxidoreductase from D. gigas.* Science 270: 1170–1176.

ROSS IS (1995) *Reduced uptake of nickel by a nickel resistant strain of Candida utilis.* Microbios 83: 261–270.

RYVES WJ, DAJANI R, PEARL L and HARWOOD AJ (2002) *Glycogen synthase kinase-3 inhibition by lithium and beryllium suggests the presence of two magnesium binding sites.* Biochem Biophys Res Commun 290: 967–972.

SANDERS OI, RENSING C, KURODA M, MITRA B and ROSEN BP (1997) *Antimonite is accumulated by the glycerol facilitator GlpF in Escherichia coli.* J Bacteriol 179: 3365–3367.

SARMAH S, HAZARIKA P, ISLAM NS, RAO AVS and RAMASARMA T (2002) *Peroxo-bridged divanadate as selective bromide oxidant in bromoperoxidation.* Mol Cell Biochem 236: 95–105.

SASAKI T, EZAKI B and MATSUMOTO H (2002) *A gene encoding multidrug resistance (MDR)-like protein is induced by aluminum and inhibitors of calcium flux in wheat.* Plant Cell Physiol 43: 177–185.

SCHAIFERS KT (1984) *Meyers Handbuch Weltall,* Bibliographisches Institut, Mannheim, Wien, Zürich.

SCHINDELIN H, KISKER C, HILTON J, RAJAGOPALAN KV and REESE DC (1996) *Crystal structure of DMSO reductase: redox-linked changes in molybdopterin coordination.* Science 272: 1615–1621.

SELMER M and SU XD (2002) *Crystal structure of an mRNA-binding fragment of Moorella thermoacetica elongation factor SelB.* EMBO J 21: 4145–4153.

SEMIZ S and MCNEILL JH (2002) *Oral treatment with vanadium of Zucker fatty rats activates muscle glycogen synthesis and insulin-stimulated protein phosphatase-1 activity.* Mol Cell Biochem 236: 123–131.

SEMIZ S, ORVIG C and MCNEILL JH (2002) *Effects of diabetes, vanadium, and insulin on glycogen synthase activation in Wistar rats.* Mol Cell Biochem 231: 23–35.

SENSFUSS C, REH M and SCHLEGEL HG (1986) *No correlation exists between the conjugative transfer of the autotrophic character and that of plasmids in*

Tab. 2.1: Classification of elements in higher plants, animals, and humans. (From Kieffer 1991, Marschner 1995, Anke et al. 2000, and further different sources.)

Higher plants	Animals and human bodies
1. Essential macroelements (= macronutrient elements): Ca, K, Mg, N, P, S (and C, H, O)	1. Essential macroelements: Ca, Cl, K, Mg, Na, P, S
2. Trace elements – with deficiency symptoms (= micronutrient elements): B, Cl, Cu, Fe, Mn, Mo, Ni, Zn – beneficial elements: Co, Na, (Se), Si, (Al)	2. Trace elements – with deficiency symptoms: (Cr ?) Co, Cu, Fe, I, Se, Zn – without deficiency symptoms: Mn, Mo, Ni
3. Ultratrace (bulky) elements (without known function and mostly with low toxicity threshold): As, Ba, Br, Cd, Cr, F, Hg, I, Li, Nb, Pb, Rb, Sn, Sr, Ti, Tl, U, V, W, Zr,...	3. Ultratrace elements (without known function and with low toxicity threshold): Al, As, B, Ba, Br, Cd, F, Hg, Li, Nb, Pb, Rb, Si, Sn, Sr, Ti, Tl, U, V, W, Zr,...

examples for various assignments. Such differences are caused by physiological features of the organism groups under consideration, and can be based on differences in uptake, accumulation, release, and/or metabolism of the compounds containing the elements. Additionally, however, the differences are caused by various modes of subdividing the classes.

In plant nutritional science it is usual to regard also N, C, H, and O as essential elements, and to characterize the entire group as **macronutrient elements** (see Table 2.2). N, Ca, K, Mg, P, and S represent in this context the subgroup mineral macronutrient elements. Additionally, in plants the degree of essentiality (Table 2.1) is a more precisely considered criterion than in animals and humans (see below). This facilitates consideration of the major metabolic differences between many species.

Among higher plants, the class of **trace elements** contains 13 members. The class comprises both elements that are essential for all higher plants (micronutrient elements) and also the "beneficial" group. The ions or compounds of the latter subclass either stimulate plant growth without being essential, or they are essential only for certain plants, or they act only under specific conditions. Se and Al are placed in brackets because direct stimulation is doubtful, but cannot be excluded (Santosh et al. 1999). Thus, the beneficial effects of Se compounds on *Astragalus* plants in nutrient solution cultures were based on preventing the accumulation of toxic phosphate levels in leaves (for details, see Läuchli 1993). Growth-stimulating Al concentrations in nutrient solutions (nominally given, possible precipitations, e.g., as phosphate not considered) varied between < 1 and 5 mg kg^{-1} in sugar beet, maize, rice, and some legumes, but were higher in tea plants (Bollard 1983, Marschner 1995). According to Asher (1991), these effects may be of secondary nature because Al^{3+} affects both the uptake and toxicity of other mineral nutrient elements such as P, Cu, and Zn (for details, see Section 2.3). Finally, Cl, Na, and Si are assigned to the trace elements despite their relatively high concentrations in plant materials (see Table 2.2), mainly because only tiny amounts are needed to induce their special effects. Therefore, in plants "content" and "effectiveness" are not synonymous.

Tab. 2.2: Element contents of some higher plants. (Epitomized from Schilling et al. 2000.)

Assignment in plant nutritional science	Element	Concentration in dry matter (%)[a]			
		Oats (Avena sativa L.)		Potato (Solanum tub. L.)	
		Grain	Straw	Shoots	Tubers
Macronutrient elements					
– Volatile during dry ashing	C	In mean of all higher plants 40–50			
	H	In mean of all higher plants 5–7			
	O	In mean of all higher plants 42–48			
	N	1.5–2.5	0.4–0.6	0.2–2.3	0.35–1.2
– Non volatile during dry ashing	Ca	0.22	0.9	3.0	0.07
	K	0.5–0.7	1.16–1.4	0.3–4.2	0.5–1.9
	Mg	0.12	0.06–0.1	0.6	0.02–0.08
	P	0.35–0.43	0.12–0.13	0.14	0.06–0.28
	S[b]	0.2	0.17	0.33	0.3
Trace elements					
– Micronutrient elements	B	2×10^{-4}	4×10^{-4}	4×10^{-3}	–
	Cl[b]	0.14	1.2	0.26–3.2	0.13
	Cu	6×10^{-4}	8×10^{-4}	9×10^{-4}	–
	Fe	1×10^{-2}	2×10^{-2}	0.18	3×10^{-2}
	Mn	6×10^{-3}	6×10^{-2}	9×10^{-3}	1.3×10^{-3}
	Mo	0.2×10^{-4}	0.3×10^{-4}	4.8×10^{-5}	4.3×10^{-5}
	Ni	In the investigated plants 3×10^{-5}–6×10^{-4}			
	Zn	4×10^{-3}	4×10^{-3}	7×10^{-3}	–
– Beneficial elements	Co	3×10^{-6}	9×10^{-6}	6.6×10^{-5}	5×10^{-6}
	Na		0.13[c]	0.1–1.1	8×10^{-2}
	Si		0.8–1.1	0.55	0.04
Ultratrace (bulky) elements	V	3×10^{-6}	8×10^{-6}	1.3×10^{-4}	7×10^{-6}

[a] Often given as μg g^{-1} (= ppm); 1 μg g^{-1} = 1×10^{-4} %. [b] During dry ashing partially volatile. [c] Shoots in time of jointing.

The class of **ultratrace (bulky) elements** comprises components with no known function. Higher concentrations are in all cases toxic, unless a special tolerance is in existence. It is possible that more elements will be added to this class when the sensitivity of the analytical methods used is improved. On the other hand, the concept should not be excluded that tiny quantities of certain bulky elements are essential because today it is impossible to establish a culture medium or test diet that is absolutely free of such elements. The same situation is applicable to the ambient air.

2.2
Functions of Nutrient Elements in Higher Plants

2.2.1
Principles of Substance Formation and Role of Nutrient Elements

The higher plant represents a spatially divided system (organs, tissues, cells, cell compartments) that forms distinct groups of organic compounds in each part. The fundamental reaction is the photosynthetic splitting of water in shoots. The derived hydrogen is transferred to ferredoxin, followed by reduction of the CO_2 which has been taken up from the ambient air (for details, see Bowyer and Leegood 1997). Initially, glyceraldehyde-3-phosphate is formed, and this is converted into D-fructose. After transformation to sucrose and partial transport into other plant organs, various high molecular-weight compounds are produced (Figure 2.1). These complicated reactions consist of a series of linked enzymatic processes in which additional substances, e.g. nutrient elements, are incorporated into the C-skeletons. A portion of the formed compounds enlarges the basis for the substance production (system growth). This comprises the enzymatic system of the plant, together with its proteins, lipids, and genetic information carriers (DNA, RNA). The other portion is deposited as final products which are not involved in further production of substances (e.g., storage proteins, cellulose and lignin in cell walls, storage starch, many alkaloids). The increase in biomass which results from the production of such compounds is named "product growth". The peculiarity of the plant consists, in simple terms, of a capability to renew, to enlarge, and to change the entire enzyme apparatus in a continual manner and in interaction with environmental factors such as temperature, light, water, and nutrient elements. This occurs by changing the pattern of gene activity (differential gene activation), leading to ontogenesis. In this connection, phytohormones play a role in

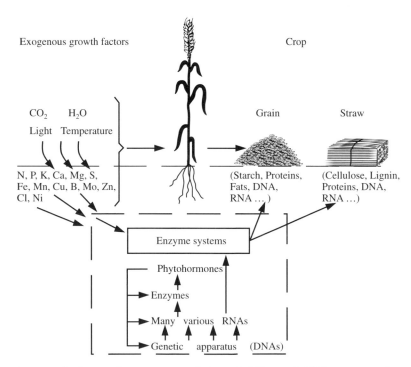

Fig. 2.1 Diagram of substance formation in cereal plants (Schilling et al. 2000). Exogenous factors induce differential gene (DNA) activation that catalyzes (via messenger RNA) the formation of proteins with enzymatic character. Some enzymes help to synthesize phytohormones which induce in other organs the typical enzyme pattern of these plant's parts.

the transfer of chemical information from organ to organ. Hence, ontogenesis is under the control of many environmental factors within the genetically fixed limits.

In this chapter, attention is paid to the role of the **nutrient elements** listed in Table 2.1 – which are occasionally also referred to as "nutrients". However, nutrients are in reality those forms of elements which are taken up and involved in metabolism, for example K^+, Ca^{2+}, Mg^{2+}, Mn^{2+}, NH_4^+, $CO(NH_2)_2$, NO_3^-, and $H_2PO_4^-$. Many elements are absorbed in more than one form, but because the element concerned – and not its ligands – is related to the nutritive value, it is referred to only occasionally as a nutrient in the literature. This also holds true for the element's oxides (e.g., P_2O_5, K_2O) which are, in practice, not present in plants. Such a nomenclature is incorrect and so is not used in this chapter.

The following criteria are commonly accepted as typical characteristics for essential nutrient elements (Arnon and Stout 1939):

- The plant cannot complete its life cycle in absence of the element.
- The element must not be replaceable by another element.
- The element must be directly involved in plant metabolism.

The essential elements of Table 2.1 meet these demands. In all cases they are components of the metabolic system in cell or of important final products; for example, cellulose for the upright standing of the plant. The function as constituents of such compounds is clear for C, H, and O. These three elements are together components of nearly all organic compounds in the plant [only hydrocarbons (e.g., carotins) are free of O], and therefore they build up the plant's shape. A similarly clear situation holds true for N and P, both of which are constituents of the information carriers DNA and RNA. N is a component of their purine and pyrimidine bases, while phosphoric acid esters of D-ribose or 2-deoxy-D-ribose form the backbone of their nucleotide sequences. Moreover, P plays a very important role in energy metabolism, the key compounds being nucleotide phosphates (e.g., adenosine triphosphate, ATP) (see Scheme 2.1) and the homologous molecules

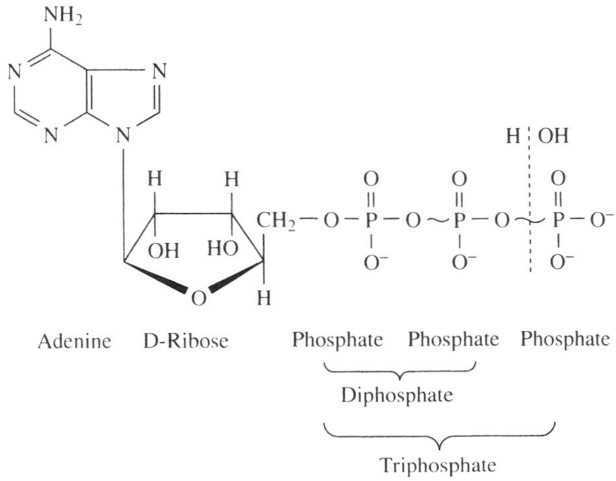

| Adenine | D-Ribose | Phosphate | Phosphate | Phosphate |

Diphosphate

Triphosphate

Scheme 2.1 ATP and ADP

uridine triphosphate (UTP) and guanosine triphosphate (GTP).

The bonds between the phosphate residues are energy-rich, which means that their hydrolysis delivers free enthalpy, namely $\Delta G° = -35 \text{ kJ mol}^{-1}$ for the final phosphate. The hydrolytic splitting of this molecule by ATPases leads to the production of adenosine diphosphate (ADP) and inorganic phosphate (P_{in}) (see Scheme 2.1). Due to the exergonic character of this hydrolysis, the phosphate may also be transferred directly to alcoholic OH groups (e.g., in sugars), to carboxyl groups, or to guanidino groups. These molecules are phosphorylated in this way. The phosphorylation is an endergonic reaction, and so the liberated energy ($-\Delta G°_1$) of ATP hydrolysis is used as a driving force for this second chemical reaction ($+\Delta G°_2$). This means that an energetic coupling exists between the first and second reaction, and the esters of phosphoric acid so formed are energy-rich. It is possible to link these with other organic compounds under splitting off the phosphate residue (designated as Ⓟ). In this way, higher molecular-weight products are produced in step-wise fashion (e.g., starch, see Scheme 2.2) via several phosphorylations (for details, see Schilling et al. 2000).

It is of interest that the substrate for the ATPases appears to be a MgATP complex rather than free ATP (Rea and Sanders 1987, Rea 1999). In addition, MgATP can be utilized very well for the transfer of energy-rich phosphoryl groups (Balke and Hodges 1975). K^+ also plays a role in such processes (Mansour et al. 1998). Therefore, the simultaneous involvement of a variety of nutrient elements in such reactions is verified. ATP that underwent decomposition for the phosphorylations is later rebuilt from ADP + P_{in}. In plants, the energy for this process is obtained in chloroplasts directly from sunlight (the process of photophosphorylation), whilst in cells and compart-

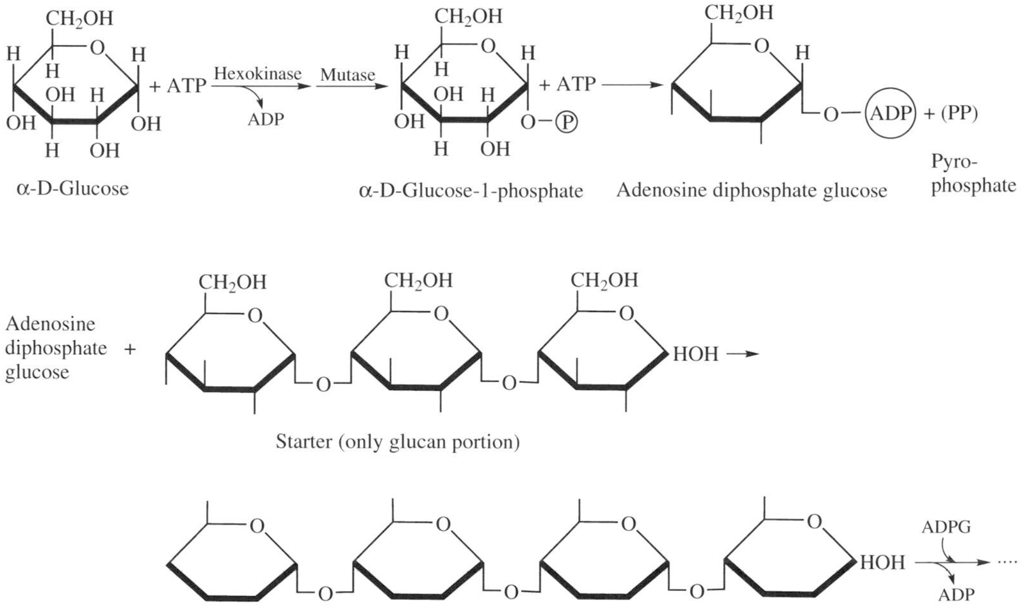

Scheme 2.2 Starch biosynthesis

ments that are free of chlorophyll the energy is provided by respiration. As the substrate for the latter process is provided by photosynthesis, sunlight drives (either directly or indirectly) all energy-consuming processes in higher plants, and P is a key element in this context.

The energetic coupling process outlined is also necessary for the biosynthesis of proteins, many of which are enzymes or parts thereof. Besides C, H, and O, proteins also contain the macronutrient elements N and S. The 20 amino acids that are linked in protein molecules by virtue of peptide bonds also carry side chains with functional groups, and these point outwards from the axis. Some of these groups contain S (see Scheme 2.3).

The free SH-groups of the incorporated cysteine have structural importance for the protein molecule. In nature, the polypeptide chains possess the shape of a right-handed helix or a β-pleated sheet structure. There are about 3.6 amino acids per turn in an α-

helix, and this so-called "secondary" structure is stabilized by H-bridges between neighboring threads. The helices, in turn, are tangled up into knots, thereby forming the "tertiary" structure in which various segments are held together by (among other possibilities) −S−S− bridges. The bridges are formed when SH-groups of cysteine, either in the same polypeptide chain or among different polypeptides of a multisubunit protein, are in opposition and become oxidized, with the release of H.

In higher plants the SH-groups are also essential for other reasons, including the tripeptide, glutathione (see Scheme 2.4). This compound is one component of the protection system against oxidative attacks of trace gases in the atmosphere (O_3, peroxyacylnitrate, NO_2). The SH-groups are able to prevent such oxidations by reducing the oxidants.

In case of diminishing glutathione content as a result of Cu excess in plant (De Vos et al. 1992) or by the oxidation of SH-

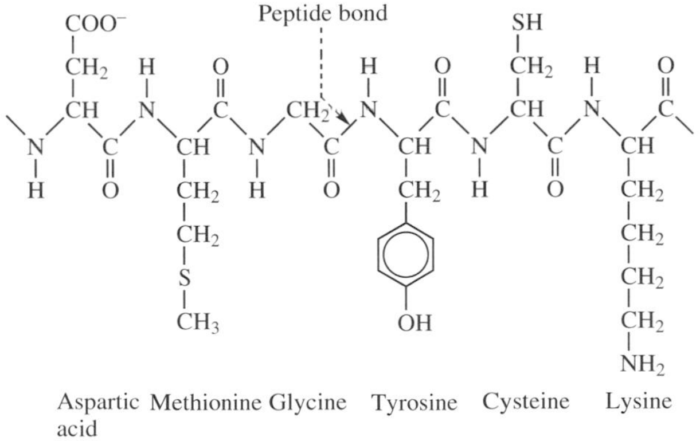

Aspartic acid | Methionine | Glycine | Tyrosine | Cysteine | Lysine

Scheme 2.3 Part of a peptide chain in protein

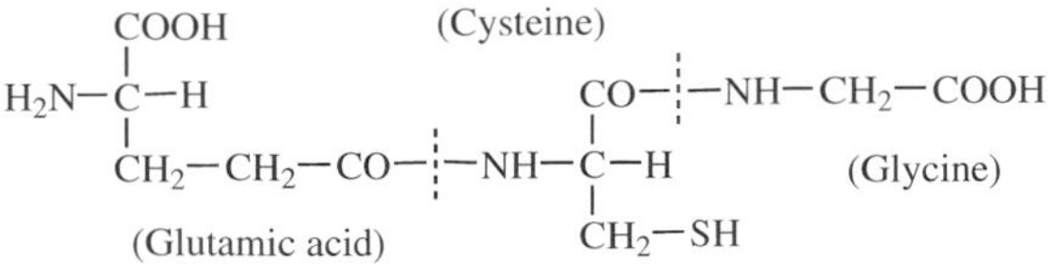

Scheme 2.4 Glutathione

groups to $-SO_3H$, this protective capability is lost. Consequently, radicals formed from the trace gases may attack the lipids of the cell membranes (see below), causing decomposition of these components (Elstner und Hippeli 1995). The process begins with withdrawal of one H atom from the hydrocarbon chain of a multiple-unsaturated fatty acid. This reaction induces binding of O_2 from the air at the vacant position and generating a hydroperoxide group. By this way, the fatty acid molecule becomes unstable and breaks sooner or later. Fragments are a small hydrocarbon molecule (e.g., ethane) and an aldehyde which is not capable to accomplish the function of the fatty acid in the lipid. This, in turn, leads to destruction of the membranes and possible bleaching of the pigments because the unprotected pigment molecules can be oxidized now rapidly. Cations of mineral nutrient elements such as Mg^{2+}, Ca^{2+}, and K^+ are liberated by this process and are leached away by rain. This effect plays a role in so-called "forest decay", and also explains the temporary alleviation of symptoms by the application of Mg-containing K fertilizers and by simultaneous liming. This shows that macronutrient elements are not only responsible for the growth and development of plants but can also prevent damage caused by environmental factors.

Many proteins function as enzymes and are, therefore, the most important biocatalysts found in organisms. Of course, the enzymes often contain additional components which act as cofactors (firmly linked as prosthetic groups or reversibly bound as coenzymes). The cofactors are not proteins; rather, they are macro- or micronutrient elements, or small molecules which contain such elements, and are essential for correct functioning of the enzyme. In addition there are effectors which either increase or decrease the activity of a specific enzyme, but are not essential. Effectors are also often macro- or micronutrient elements, or can be derived from them. Examples of these enzyme components are listed in Table 2.3. Clearly, enzymes with equal activity may have different amino acid compositions in different species, and so it is not possible to transfer the details of one object to all other situations.

Nevertheless, from Table 2.3 it may be deduced that all elements designated as macro- or micronutrient elements in Table 2.1 play a role in this context (the exception is B). Therefore, their essentiality is evident alone from this standpoint. In several cases ions are capable of replacing one another, this being due to equality in electrical charge and similar diameter. Indeed, some characteristics of a catalyzed reaction are affected by such exchanges, an example being the Michaelis constant (K_m).

It appears evident therefore that the main function of most nutrient elements is as a constituent of the enzymes required to build up the organic matter within plants. Many such reactions run side by side in cells and tissues, and this is made possible by the presence of biomembranes that allow the build-up and decomposition of compounds, without mixing the components. Biomembranes subdivide cells into reaction spaces (e.g., nucleus, plastids, mitochondria, ribosomes, vacuoles, cytosol), and they permit well-ordered substance exchange between the compartments. Such processes are also responsible for ion uptake by root cells from the soil solution. Despite certain differences, all biomembranes have a similar chemical structure, the basic components being double lamellae of P-containing lipids (Figure 2.2) such as phosphatidylserine and glycolipids (Table 2.4). Proteins are movably incorporated into these double lamellae (see

Tab. 2.3: Instances of non-protein components of enzymes (from different sources)

Function	Substances and abbreviations, for effectors complete enzymes	Carrier for ... or reaction[a]	Important mineral nutrient elements
Cofactors: Coenzymes or prosthetic groups a) Free of metal	Nicotinamide adenine dinucleotide phosphate NADP$^+$	Hydrogen	N, P
	Flavin mononucleotide	Hydrogen	N, P
	Adenosine 5'-triphosphate ATP/ Adenosine 5'-diphosphate ADP	Phosphate	N, P
	Pyridoxal 5'-phosphate	Amino group	N, P
	Uridine 5'-diphosphate (during sucrose and cellulose formation)	Monosaccharides	N, P
	Biotin (fatty acid formation)	CO_2	N, S
	Coenzyme A HS-CoA (e.g., fatty acid formation)	Acetyl group	N, P, S
b) Metalliferous	Chlorophyll (Mg porphyrin)	Electrons	N, Mg
	Hemin (Fe protoporphyrin)	Electrons	N, Fe
	Cytochrome oxidase (complete enzyme)	Electrons	N, Fe, Cu, (Zn, Mg ?)
	Heme (in cytochrome c)	Electrons	N, Fe
c) Metals (in many cases linked with the protein by metal-S-clusters)	Fe in ferredoxin	Electrons	N, Fe, S
	Cu in plastocyanin	Electrons	N, Cu
	Mo and Fe in nitrate reductase	Electrons	N, Fe, Mo
	Cu in phenoloxidases	Oxidation	N, Cu
	Zn in carbonic anhydrase	$H^+ + HCO_3^- \rightleftarrows H_2O + CO_2$	N, Zn
	Ni in urease	Urea hydrolysis	N, Ni
	Mn in photosystem II (water splitting)	Electrons	N, Mn, Ca, Cl
Effectors: (in single cases cofactor function not to exclude)	Aldehyde dehydrogenase (from yeast, in higher plants similar)	$CH_3CHO \rightarrow$ acetic acid	K$^+$, NH$_4^+$ (30% of K$^+$effectivity), Na$^+$(4% of K$^+$ effectiv.)
	Aldolase (from chloroplasts of various plants)	Fructose-1,6-bi-phosphate $\rightleftarrows$ 2 triose phosphate	K$^+$, Zn^{2+} (presumably bound)
	ATPase (from pea roots)	$ATP + H_2O \rightarrow ADP + Pin$	Mg^{2+}(partially replaceable by Ca^{2+}, Mn^{2+}), K$^+$, Na$^+$
	Different organisms (DNA, RNA)	Activation of genetic messages	Mg^{2+}, Zn^{2+}, Mn^{2+}
	α-Amylase (from barley a.o.)	Starch hydrolysis	Ca^{2+}
	Asparagine synthetase (from lupins)	Glutamine + aspartate $\rightarrow$ asparagine + glutamate	Cl$^-$
	Isocitric acid dehydrogenase (from oats coleoptiles)	Isocitric acid + NADP$^+ \rightarrow$ Oxalosuccinic acid + NADPH + H$^+$	Mg^{2+} (K$_m$ = 4.5×10^{-4} M), Mn^{2+} (K$_m$ = 1×10^{-5} M)

[a]In some instances various functions occur. This is not considered here.

Tab. 2.4: Some components of biomembranes

Chemical structure	Second alcohol component	Name
1 $CH_2O-\overset{\overset{O}{\|\|}}{C}-R_1$ 2 $HC-O-\overset{\overset{O}{\|\|}}{C}-R_2$ 3 $CH_2O-\overset{\overset{O^{\ominus}}{\|}}{\underset{\|}{P}}-\boxed{OH \quad H}O-CH_2-\underset{\underset{NH_2}{\|}}{CH}-COOH$ O	Serine	Phosphatidyl-serines
$\text{HO}\diagdown\!\!\!\overset{CH_2OH}{\underset{OH}{\bigcirc}}\!\!\!\overset{O}{\diagup}\,O\dashline{H\text{-}\text{-}C_3}$ OH	D-Galactose	Monoglactosyl-glycerol

R_1, $R_2 =$ Fatty acid residues; $C_3 =$ Third C atom of glycerol of which the HO-group is linked with D-galactose by glycosidic linkage.

Figure 2.2), and can carry side chains which point outward from the membrane. In many cases the proteins are enzymes. Among these enzymes the intrinsic membrane proton ATPases deserve special attention as they pump H^+ outward from the membrane and OH^- inwards (Briskin 1986), according to the scheme:

$$ATP + nH_2O \rightarrow ADP + P_{in}$$
$$+ n - 1\,(H^+) + n - 1\,(OH^-)$$

In this way, a proton gradient arises which drives the ion uptake into the cell by means of the following mechanism. The ion to be taken up is loaded outside with H^+ and is thereby drawn inwardly by the OH^-. In the case of anions, this is only possible if more protons are transported than negative charges. The cation uptake does not suffer from such problems, but there is sometimes a symport together with H^+ from outside to inside. So-called transporters assist in this process. These are incorpo-

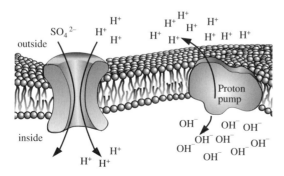

Fig. 2.2 Model of the proposed mechanism of SO_4^{2-} transport throug a plant membrane by a high-affinity SO_4^{2-}-H^+ symporter (Smith 1999, enlarged). Transport is energized by a large integral membrane proton ATPase that pumps H^+ to the outside of the membrane, and OH^- to the inside.

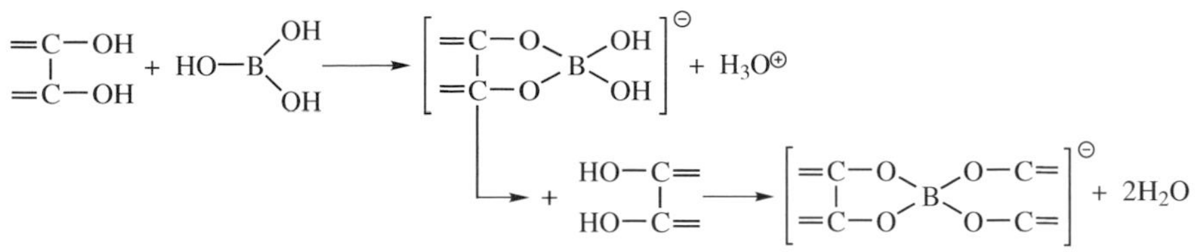

$$\begin{array}{l}=\!\!C\!-\!OH \\ | \\ =\!\!C\!-\!OH\end{array} + HO\!-\!B\!\!\begin{array}{c}OH \\ \\ OH\end{array} \longrightarrow \left[\begin{array}{l}=\!\!C\!-\!O \\ | \\ =\!\!C\!-\!O\end{array}\!\!B\!\!\begin{array}{c}OH \\ \\ OH\end{array}\right]^{\ominus} + H_3O^{\oplus}$$

$$\longrightarrow + \begin{array}{l}HO\!-\!C\!= \\ | \\ HO\!-\!C\!=\end{array} \longrightarrow \left[\begin{array}{l}=\!\!C\!-\!O \\ | \\ =\!\!C\!-\!O\end{array}\!\!B\!\!\begin{array}{l}O\!-\!C\!= \\ | \\ O\!-\!C\!=\end{array}\right]^{\ominus} + 2H_2O$$

Scheme 2.5   Formation of mono- and diesters of boric acid with cis-diols

rated into the membrane (Figure 2.2). According to recent findings (Hawkesford and Smith 1997), these are relatively large polypeptides that range in mass from ~53 to 75 kDa. This corresponds to molecules of between 500 and 600 amino acids in length (Smith 1999). They ensure a selective ion transport through the membrane by means of their ion specificity. Current research to identify such transporters is ongoing with the use of molecular biological methods (Heiss et al. 1999, Figueira et al. 2002). However, the molecular mechanisms of this catalyzed ion transport (e.g., closeable channels for K^+ and for Ca^{2+} as well as other possibilities) are in most cases unclear, as separation of the transporters from the lipid membrane induces a loss in natural tertiary structure and destroys their ability to function.

The membrane structure described must be maintained if the metabolism in cells is to function normally, and for this a variety of macro- and microelements is employed:

● The element **B** forms stable mono- and diesters with cis-diols such as sugars and sugar alcohols. Thereby, unexchangeable complexes originate within cell walls (cellulose), middle lamellae (rhamnogalacturonan II; Match and Kobayashi 1998), and as plasma membrane constituents such as glycoproteins or glycolipids (see Scheme 2.5). These complexes stabilize the structure (Cakmak and Roemheld 1997), and this may be especially important at the inter-

face between the cell wall and plasma membrane. The effect on H^+ pump achievement (ion uptake!) demonstrated in membrane vesicles from B-sufficient and B-deficient sunflower roots may be explained in this way (Ferrol et al. 1993). Other functions of B are independent of this, and concern especially the binding of 6-phosphogluconic acid; this leads to a preferred conversion of D-glucose from photosynthesis into cellulose, hemicelluloses, pectins, and lignins. This explains, together with the functions of B in DNA and RNA metabolism, the serious damage seen at the apical growing point as a consequence of B deficiency (literature see Goldbach et al. 2002).

● Zn^{2+} also stabilizes membranes (Mohamed et al. 2000), and Zn deficiency causes a considerable increase in plasma membrane permeability, especially in roots (Welch et al. 1982); this is indicated by leakage of small-molecule solutes such as sugars, amino acids, and K^+. The phospholipid content also decreases (Cakmak and Marschner 1988), perhaps due to inadequate stability of the protein structure in membranes. As Zn^{2+} is bound to the SH-groups of cysteine, to histidine and to glutamate or aspartate residues, its absence destabilizes such structures. In this connection, more toxic oxygen radicals are generated ($2O_2 + NADPH \rightarrow NADP^+ + 2O_2^- + H^+$) because Zn is lacking to interfere with membrane-bound $NADPH/H^+$ oxidase (Cakmak 2000).

- Ca^{2+} is responsible for membrane stability, and Ca-deficient plants of cucumber show leakage of low molecular-weight solutes via the roots (Matsumoto 1988). Similar processes are seen in the tissues of tomato fruits (Van Goor 1966) and in potato tubers (Coria et al. 1998), and are due mainly to the disintegration of membrane structures (Hecht-Buchholz 1979, Saure 2001). Ca^{2+} stabilizes these membrane structures, presumably by bridging dissociated phosphate and carboxylate groups of the phospholipids. Similar reactions are possible between proteins at the membrane surfaces, and in this way Ca^{2+} counteracts the negative effects of H^+ on plasma membrane integrity and their unfavorable consequences for proton pumping. Ca^{2+} can easily be displaced from its binding sites by other metal cations (including Na^+); therefore, a high Ca^{2+} concentration is required in saline soils. In addition to these membrane-stabilizing effects, Ca^{2+} is involved in processes such as cell division and extension, energy metabolism (Ca^{2+}-ATPases), and the formation and activation of enzymes (e.g., as a cofactor for photosynthetic O_2 evolution; Matysik et al. 2000). Moreover, the structure of cell walls and middle lamellae between adjoining cells is stabilized by Ca^{2+} binding to $RCOO^-$ groups of polygalacturonic acids (pectins) in interaction with B (Match and Kobayashi 1998).

In summarizing these results, it becomes clear that the mineral nutrient elements of Table 2.1 are not only components of the metabolic system and of the final products, but are also responsible for the cooperation of all these substances during the formation of the plant's shape.

2.2.2
Special Effects of Beneficial Elements

Co, Na, and Si belong to the beneficial elements that are not necessary for growth in all higher plants.

- **Cobalt** seems to be essential only for N_2-fixing symbioses. This is current for legumes in the same manner as for nodules of nonleguminous plants (e.g., alder). When Co is lacking, initially all proliferation of rhizobia is inhibited in *Lupinus angustifolius* L., followed by deficiency of leghemoglobin (Riley and Dilworth 1985a, 1985b). The latter effect is clearly based on a lack of 5′-deoxycobalamin (coenzyme B_{12}, discovered in nodules in 1963) in which about 12% of the total nodule Co is bound. The compound is formed by four partially hydrogenated pyrrole rings, with Co as the central atom. Its structure resembles that of chlorophyllid. In *Rhizobia* and closely related N_2-fixing microbes, three enzymes are presumably cobalamin-dependent: methionine synthase (Watson et al. 2001); ribonucleotide reductase; and methylmalonyl-coenzyme A mutase which are needed for leghemoglobin formation (Dilworth and Bisseling 1984). Leghemoglobin controls the O_2 supply of the N_2-fixing enzyme nitrogenase. On the basis of results with *L. angustifolius* L., it is possible that the growing root nodules support plant growth not only by N_2 fixation but also by other processes such as cytokinin (phytohormone) production (Gladstones et al. 1977). At present further growth-enhancing functions of Co in higher plants are not known.

- **Sodium** is essential for distinct C_4 plants. Their special form of photosynthesis (Figure 2.3) is characterized by primary incorporation of atmospheric CO_2 into

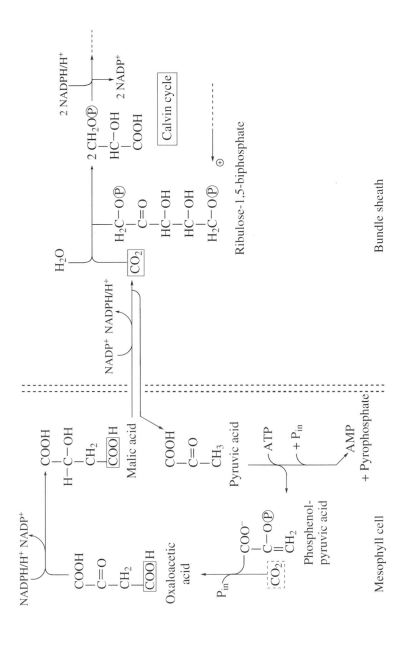

Fig. 2.3 CO_2 incorporation of C_4 plants during photosynthesis (summarized from various sources)

phosphoenolpyruvic acid. This occurs in the chloroplasts of mesophyll cells. The formed oxaloacetic acid (C_4 body, therefore C_4 plants) is reduced to malic acid which is transported into the so-called bundle sheath. This cell circle envelopes the vessel system of leaf. There, the compound is decarboxylated, with formation of pyruvic acid and CO_2. While the first-mentioned compound is retranslocated and converted into phosphenolpyruvic acid for the next cycle, CO_2 enters into the Calvin cycle where it is incorporated into ribulose-1,5-biphosphate (C_5 body), reduced by H from the photosynthetic water splitting and converted into two molecules of glyceraldehyde-3-phosphate and further monosaccharides. These are used for cell growth and for re-formation of ribulose-1,5-biphosphate, which then is introduced into the next cycle. The Calvin cycle is present in all green plants; indeed, the special feature of C_4 plants is that of "preliminary carboxylation", with the formation of malic acid. This process facilitates CO_2 incorporation into ribulose-1,5-biphosphate by enriching this substrate. Within the preliminary carboxylation, Na^+ seems to play an important role in distinct C_4 species. It enhances pyruvate uptake into isolated mesophyll chloroplasts of hog millet (*Panicum miliaceum* L.), with a stoichiometry of about $1:1$ (Ohnishi et al. 1990). This suggests a $Na^+/$ pyruvate cotransport through the envelope into the chloroplasts, and also explains the need for Na^+ in these plants. In other species, Na^+ is not essential but beneficial, mainly because Na^+ can partly replace K^+. Both ion types resemble one another (radius in hydrated state of Na^+ 0.36 nm, of K^+ 0.33 nm), and they are the sole form of these elements in higher plants. Of course, Na^+ is not capable of replacing

K^+ completely, and the percentage replacement is dependent on both species and cultivars (Subbarao et al. 1999). Apparently, this situation is due to the variety of Na^+ transport systems seen in different plants, and by prevailing of replaceable K^+ functions in leaves. In species such as common bean (*Phaseolus vulgaris* L.), Na^+ is not able to replace K^+ because there is an effective exclusion mechanism for Na^+ transport from root to shoot. The natrophile sugar beet (*Beta vulgaris* ssp. *altissima*), however, behaves quite differently. Here, Na^+ is transported to shoots easily (Marschner et al. 1981), and a partial substitution of potassium in the substrate results often in an increased dry matter yield. The reason for this might be that the replaceable functions comprise mostly cell expansion and water balance of leaves. Na^+ surpasses K^+ in the generation of turgor, as its water envelope is larger and it accumulates preferentially in vacuoles. Therefore, it is more effective in this respect than K^+, and the percentage of replacement can be higher in mature leaves than in growing ones (Lindhauer et al. 1990). Moreover, Na^+ seems to improve the water balance of plants by inducing good stomata regulation (Hampe and Marschner 1982).

- **Silicon** is essential for unicellular diatoms. Nevertheless its absence impairs growth and development of some other plants also. Thus, the stress tolerance of rice (*Oryza sativa* L.) is decreased by Si deficiency because the synthesis and function of cell walls are impaired (Agarie et al. 1998); likewise, the yield of sugar beet decreases under such conditions (Anderson 1991). This beneficial effect of Si can be explained on the basis of some chemical characteristics of the element. Si forms only at high dilution ($< 2 \times 10^{-3}$ mol L^{-1}) and at pH 2–3 orthosilicic acid [Si $(OH)_4$] which is stable to some degree. However, this compound is altered under other conditions. Thus, water can be separated off intramolecularly, leading to the formation of pyrosilicic acid [$(HO)_3$ Si–O–Si $(OH)_3$]. Further removal of water generates polysilicic acids. The smaller molecules of these compounds are taken up by plant roots, together with water. After xylem transport to the different organs, Si is deposited mostly as $SiO_2 \cdot nH_2O$ (opal) at sites where the transpiration stream is ending. Some depositions in cell walls of xylem have also been found that prevent compression of these vessels when the transpiration rates are very high (Raven 1983). The other depositions stabilize the plant structure by improving leaf erectness, decreasing susceptibility to lodging in cereals, diminishing cuticular transpiration, and inducing resistance of tissues against fungal attack. It may be possible to determine the degree of polymerization of such deposits by using fluorescence spectra of a substituted oxazole (Shimizu et al. 2001). Some metabolic effects of Si compounds have also been identified. For example, both the content and metabolism of polyphenols are affected in xylem cell walls (Parry and Kelso 1975), and this may be important in lignin biosynthesis. In addition, silicic acid seems to form esters with HO-groups in xylem, and these stabilize the structures (e.g., effects on rice, see above). Finally, Si alleviates the toxicity of Mn by inducing a more homogenous distribution of the micronutrient element in leaves. As a consequence, the brown speckles in mature leaves (containing MnO_2 and oxidized polyphenols; Wissemeier and Horst 1992) are toned down, though the mechanism of this effect is not clear.

2.2.3
Visual Symptoms as Consequences of Nutritional Disorders

The pathways of metabolism affected by mineral nutrient elements are illustrated diagrammatically in Figure 2.4. Substance formation begins with CO_2 assimilation, which in turn leads to glyceraldehyde-3-phosphate as a first coordination point in metabolism. From here: (i) all other carbohydrates are formed by condensation reactions; and (ii) acetyl CoA originates via glycolysis. The last-mentioned compound is regarded as a second crossing-point in metabolism, because from here pathways lead to fatty acids and lipids, to terpenes (e.g., carotenoids) and into the citric acid cycle. 2-Oxoglutarate, an intermediate of the cycle, is the starting point for the biosynthesis of amino acids and proteins. Figure 2.4 shows that the plant derives from these materials further compounds which have totally different chemical characteristics; for example, purines and pyrimidines for DNA and RNA, alkaloids, and indole-3-acetic acid (IAA) as a representative of the 5–6 phytohormone groups.

The mineral nutrient elements take part in many processes. Interestingly enough, most elements are involved simultaneously in different reactions of metabolism. Thus, P, K, Mg, Ca, and B are important for the formation of nucleic acids, production of nucleotide phosphates, energy metabolism, and stabilization of membrane structures. Zn is a component of many different enzyme processes, and even Mn and Fe are involved in various reactions. Therefore, no clear assignment of individual elements to distinct areas of metabolism is possible. Indeed, results obtained with different vege-

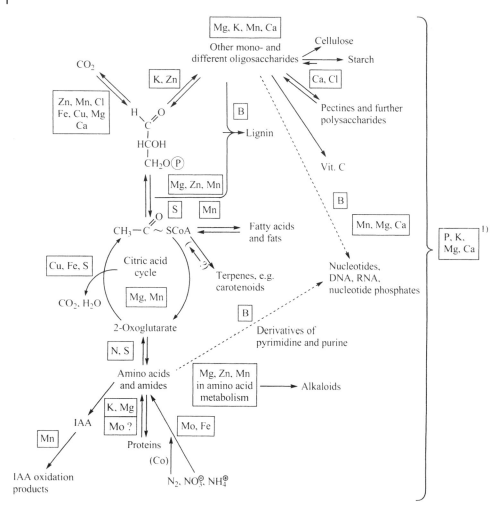

Fig. 2.4 The most important steps of metabolism in which mineral nutrient elements are involved. The specific functions of elements in the individual reactions (e.g., constituent of an enzyme) are not considered (summarized from various sources). [1] Participation especially in energy transfer.

tation experiments have shown that to supply plants with adequate amounts of P, K, and Mg generally enhances the formation of highly polymerized compounds. The role of the above-mentioned elements in energy metabolism may be decisive in this respect. When the N supply is low in such cases mostly polysaccharides and fats (in oil plants) are formed from the low molecular C-skeletons. In the case of sufficient N pro-

visioning, however, this element must be incorporated under the formation of proteins, and this occurs at the expense of polysaccharides and fats. Only the contents of cellulose and lignin remain almost unchanged (Polz 1965). When other elements are lacking for protein synthesis, low molecular N compounds are accumulated (especially amides and amines) because amino acids can not be linked to polypepti-

Tab. 2.5: Classification ranges for nutritional states with essential mineral elements

Designation	Absolute deficiency	Latent deficiency	Sufficient supply	High supply	Toxic range
Characteristics	Deficiency symptoms, low crop yields	Not any symptoms but addition of the essential element increases growth and crop yields	Not any symptoms, max-imum crop yields, addition of the essential element does not increase crop yields	Luxury con-sumption	Depression of growth and crop yields, toxicity symptoms
Content of the element in shoots	**low**				→ **high**

des and proteins. In the case of K deficiency, putrescine [$H_2N-(CH_2)_4\,NH_2$] is formed as a degradation product of the amino acid L-arginine (Tachimoto et al. 1992), and this effect may be used as an indicator of K deficiency. In summarizing these results, a deliberate influencing of the composition and quality of crops is possible by using these relationships (Finck 1991, Schilling et al. 2000). The most important measure of inducing such effects is a differentiated nitrogen fertilization.

Deficiency of a nutrient element leads to the situation that its functions in metabolism can not be accomplished, and in this way growth is confined. If the plant appears externally normal, it exhibits latent deficiency (Table 2.5), but the quality of the crop may deteriorate under such conditions. Absolute deficiency arises by the aggravation of a lack of that element and is characterized by definitive symptoms. In the reverse situation, when more of an element is supplied than is necessary for growth, the plant conducts luxury consumption without changes of growth and ontogenesis. Further increase of uptake causes toxicity, which is generally characterized by growth depression and the occurrence of visual symptoms. Among the essential elements, B and Mn are most problematic as their levels of suffi-

ciency and toxicity are close together. Among the bulky elements, the heavy metals are most toxic (Bergmann 1993).

The assignment of visual phenomena to the deficiency or toxicity of an element is difficult because different factors often induce the same process, and so the symptoms may be similar, or even virtually identical. Thus, yellowing of a cereal population in spring may be caused not only by N or Mg deficiency but also by stagnant moisture (CO_2 excess in the root space). All three factors restrict protein synthesis in leaves, and this leads to a decreased formation and destruction of chloroplasts. Membrane injuries as a result of contact with air pollutants generate similar effects. Therefore, a correct diagnosis demands the simultaneous registration of various phenomena, and in particular the distribution of symptoms among the plant parts requires attention. Thus, deficiency of Mn, Fe, Cu, Ca, B, and S appears first within the younger organs because the ions or compounds of these elements are lacking there. The reason for this is their almost exclusive transport by the transpiration stream, followed by deposition as "transpiration residues" in old leaves, where mobilization is barely possible. A different situation exists for N, P, K, and Mg, as their transportable forms are translocated in

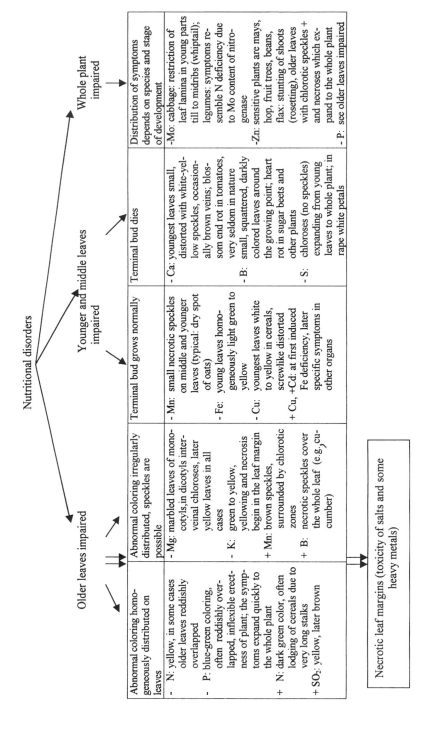

Fig. 2.5 Assignment of visual symptoms to nutritional disorders (−= deficiency, += surplus. The data demonstrate the principles of diagnosis, but are not sufficient for identification of damage. The symptoms vary, and are species-dependent (but not all elements and effects are included).

the phloem such that they always move to growing plant parts and may even be re-translocated there from old leaves. Therefore, deficiency symptoms appear first on old leaves. These and other criteria must be considered when a diagnostic system is developed. (For a specified description of this procedure, see Bergmann 1993, Schilling et al. 2000.)

The first step is to describe the damage which is present. This includes observations of population and of its roots. For example, soil acidification will liberate Al^{3+}, and this leads to short, thick roots with black, chapped apices. Following this initial evaluation, the symptoms must be assigned to one of the following groups of phenomena: changes in color (yellow, red, brown, black?); wilting; necroses; rots; habit anomalies; and feeding damage. The next step is to distinguish between biotic and abiotic causes of damage. In the first-mentioned case, the presence of harmful organisms (viruses, fungi, bacteria, animals) must be proven; in the second case, distinction must be made between anomalies of environment (temperature, wind, precipitations), emissions (position to emitters), faulty farming (pH, application of pesticides and growth regulators), and nutritional disorders. If the results suggest that there may be a nutritional disorder, the symptoms are assigned to one typical group of phenomena (Figure 2.5).

A differential diagnosis is then necessary. This involves comparing the symptoms with reproductions in a pictorial atlas (e.g., Wallace 1961, Bergmann 1993). It must always be taken into account that one and the same deficiency may manifest differently in various plant species. For example, in monocotyls Mg deficiency generates small chlorophyll accumulations between the vascular bundles of older leaves. Thereby, the organ has a marbled appearance before homogeneous yellowing begins. In dicotyls, however, the equivalent leaves show interveinal chloroses without marbling. Consequently, exact diagnoses are often difficult, since several overlapping causes can exist. Therefore, in cases of doubt only the results of additional chemical analyses of plant material are suitable to clarify the cause. Deficiency is reflected by low contents of the regarded element, whereas too-high concentrations refer to toxicity. Tabulated data (e.g., Bergmann and Neubert 1976) serve as measures for valuation, and these are designated specifically for species and different organs of plants. It must also be taken into account that the deficiency of an element may be evoked by an excess of another. Such interactions are existing in the case of Cu^{2+}/Fe^{2+}, Mn^{2+}, Zn^{2+}; K^+/Mg^{2+}, Ca^{2+}, Na^+; Ca^{2+}/Mg^{2+}, Al^{3+}, Zn^{2+}, Mn^{2+}, Sr^{2+}; SO_4^{2-}/MoO_4^{2-}; NO_3^-/Cl^-; Mn^{2+}/Mg^{2+}, Fe^{2+}, Zn^{2+}; and Fe^{2+}/Zn^{2+}, Ni^{2+}, Co^{2+}. The deficient ion may be excluded either during its uptake or later in metabolism.

2.3
Mechanisms of Toxicity and Tolerance in Higher Plants

Each element can act as a poison – either when its content is too high, or if it causes a diminution of the efficacy of another. The following elements have very low toxicity thresholds: As, Al, B, Cd, Co, Cr, Cu, F,

$$\text{Adenosine-O-}\begin{bmatrix} O & O \\ \| & \| \\ P \sim O\text{-Se} \\ | & \| \\ OH & O \end{bmatrix}\text{-OH} \rightarrow \text{H-Se-CH}_2-\text{CH(NH}_2)\text{-COOH} \rightarrow \text{CH}_3\text{-Se-CH}_2-\text{CH}_2-\text{CH(NH}_2)\text{-COOH}$$

APSe   Selenocysteine   Selenomethionine

Mn, Ni, Pb, Tl. The most important causes of toxic effects include: (i) replacement of an essential element in enzymes or intermediates without fulfilling the functions of the original constituent; (ii) chemical modification of essential cell components; (iii) changes in membrane permeability; and (iv) displacement of essential ions.

The replacement of S by Se in S-containing amino acids is an example of the substitution of one essential element by another (Eustic et al. 1981, Brown and Shrift 1982). After its uptake, selenate is activated by ATP under the formation of adenosine phosphoselenate (APSe). Following some conversions inclusive reduction, Se appears as selenocysteine and later in selenomethionine.

The selenoamino acids are incorporated into proteins. These do either not or at least much less function as enzymes than the original S-containing proteins. This means that Se cannot accomplish the function of S in this case.

A second possible cause of toxicity – the chemical modification of proteins or of metabolites by heavy metals – is of major importance. The thiol (HS−) groups of proteins and polypeptides are the most sensitive sites for binding Cd, Hg, Co, Ni, Pb, and As. The metals replace H in mind of a thiolate bond, or they form chelates. In this way, the protein structure is altered considerably and no disulfide bridges can be linked. Moreover, those enzymes are blocked in which the thiol group of L-cysteine participates directly in the reaction (e.g., glyceraldehyde-3-phosphate dehydrogenase). Stable complexes of Cd, Hg, Pb, and Zn with glutathione also exist; these lower the level of antioxidants in cells (Beyersmann 1991). Hence, the above-mentioned elements impair all redox processes in metabolism (Gallego et al. 1999 for Cd), and this leads to membrane destruction.

Changes of membrane permeability as third possibility for toxic effects may be caused by the enhancement of oxidative processes, though specific effects also exist. Thus, Pb^{2+} disorganizes Ca^{2+} functions when it replaces this cation (Beyersmann 1991), while Zn^{2+} deficiency (replacement by other M^{2+} ions) increases membrane permeability for phosphate and raises the P content of cotton plants, possibly to toxic levels (Cakmak and Marschner 1986). Vanadate ions inhibit transfer of the terminal ATP-orthophosphate to membrane-bound ATPases (Briskin 1986) in the first step of ATP hydrolysis, and consequently ion uptake is disorganized. NH_3 depolarizes membranes by binding H^+ for the formation of NH_4^+ with uncoupling of photophosphorylation (Krogmann et al. 1959). All things considered, membrane effects may have serious consequences for metabolism.

Finally, displacement of essential ions by others often overlaps the effects outlined above. The extent to which the activity of some enzymes is altered by the exchange of ions as effectors is shown in Table 2.3. (Further details are available in the 1st edition of this book; see also Beyersmann 1991). Moreover, such events play a special role in ion uptake. Thus, a decrease in pH diminishes loading of the root cell walls with Mg^{2+}, Ca^{2+}, Zn^{2+}, and Mn^{2+} because many sorption sites are occupied by H^+ and Al^{3+} (soil cultivation). This confines the uptake of the nutrient element ions. It is clear that Al^{3+} plays a special role in this connection (Marschner 1991). Moreover, Al^{3+} can inhibit Ca^{2+} uptake also by blocking Ca^{2+} channels in the plasma membrane (Huang et al. 1992) and Mg^{2+} absorption by occupying binding sites of transporters (Rengel and Robinson 1989). Because K^+ uptake is not decreased in such cases (K^+ channels remain open), ion balance in cells is disturbed with consequences for

Ca^{2+} binding to pectins and all other Ca-dependent processes. Additionally, Al phosphates can be precipitated in the vascular system, and this inhibits water and substance transports between plant organs; the result is seen as plant wilting.

In many higher plants mechanisms exist which facilitate adaptation to such unfavorable conditions. This can be achieved by avoidance of the stress factor, by tolerance, or by both strategies. The relative importance of the various mechanisms depends on plant species and ecotypes, as well as on the special element. The following seven possibilities are most important:

1. *Release of organic acids* for complexing ions outside the root apoplast plays a role in Al^{3+}-tolerant cultivars of maize and wheat. The excreted (or exuded) malic and/or citric acid incorporates Al^{3+} into negatively charged stable anion complexes which are harmless for roots and which are not taken up (Jones et al. 1996).

2. *Binding to cell walls* is important for cations of Cd, Zn, Fe, and Cu. These are sorbed at negatively charged sites of polygalacturonic acids, or precipitated for instance as Zn silicate (Neumann and Zur Nieden 2001). Copper can also be bound to glycoproteins or proteins (Van Cutsem and Gillet 1982).

3. *Restricted influx* through the plasma membrane is the most important mechanism for excluding distinct ions from the protoplasm. The ion specificity of transporters is operating within the limits outlined above.

4. *Active efflux pumps* are existing for example for Na^+ in maize (Schubert and Läuchli 1990) and for Ca^{2+} in wheat roots (Olbe and Sommarin 1991). They prevent the development of too-high concentrations in root cells.

5. *Compartmentation in vacuoles* plays an important role for different ions. It is essential for turgor regulation and maintenance of low cytosolic concentrations of Na^+ (Garbarino and Du Pont 1989) and Ca^{2+} (Chanson 1991).

6. *Complexing and chelate forming* of toxic ions are possible to prevent their direct contact with sensitive enzymes. For Al^{3+}, the complexes with organic acids are important within the plant (Ma et al. 1997, Wenzel et al. 2002). In other instances, proteins or phytochelatins (polypeptides consisting of repetitive glutamylcysteine units) are formed which bind the toxic ion. While the role of an additional synthesized protein in Al-tolerant wheat genotypes merits further consideration (Taylor et al. 1997), the detoxification of Cd^{2+} (Tukendorf and Rauser 1990) by phytochelatins is evident.

7. *Volatilization of toxic compounds* may on occasion play a role. For instance, dimethylselenide ($CH_3-Se-CH_3$) can be formed by degradation of selenomethionine. The rates of volatilization vary between the species considerably. With a supply of 10 µM in nutrient solution, sugar beet plants volatilized <15 µg Se m^{-2} leaf area, but rice and cabbage released more than 200 µg Se m^{-2} per day (Terry et al. 1992).

The mechanisms outlined co-operate in many cases. Thereby, distinct species are adapted to special sites (for details, see the 1st edition of this book).

2.4
Requirement of Mineral Nutrient Elements and its Estimation for Crops

The quantitative requirement of nutrient elements corresponds to both the yield of a crop and the content in the plant. The yield, in turn, depends on the available nutrient element in culture substrate. The ratio between the supply of a distinct nutrient element in available form and the yield is described by the "law of yields", the best-known form of which is the Mitscherlich equation (1st approximation; Mitscherlich 1909, 1956):

$$dy/dx = c_0(A - y);$$

resolved to y under using decadic logarithms:

$$y = A(1 - 10^{-cx}).$$

This means that the increase of yield (dy) per minimal addition of the regarded nutrient element (dx) is proportional to the yield difference (A−y) to the maximum yield (A). The maximum yield A is defined as the yield which can not be exceeded by increasing the doses of the regarded nutrient element without to alter amount and ratio of the other growth factors. $c = c_0$ 0.4343 is the effectiveness factor for the observed nutrient element; this is very high in the case of micronutrient elements, but low for N. The equation describes an exponential function for yield formation in dependence on the supply with each nutrient element (Figure 2.6). The form of the curve is true for many crops and sites, but the effectiveness factors are not absolutely constant (for criticism, see Schilling et al. 2000).

Because the yield formation is affected by many nutrient elements and other growth factors simultaneously, the absolute level of A is variable. The consequence is a variable withdrawal of nutrient elements in dependence on the constellation of all growth factors. Therefore, in fertilizing practice the requirement is defined as the withdrawal by the crop at "sufficient supply" (Table 2.5). This can be calculated from the expected (site-specific) yield and the content of nutrient elements in biomass. Table 2.6 demonstrates examples for N, P, K, and Mg because deficits of these elements are often compensated for by fertilization. The S withdrawal amounts to 20–70 kg ha^{-1} (Sturm et al. 1994), the Ca requirement (Mengel 1991) varies from 15 to 30 kg ha^{-1} (cereals), from 50 to 90 kg ha^{-1} (root crops), and from 150 to 250 kg ha^{-1} (alfalfa). The latter need is mostly met by soil and by liming for correcting the pH value. For the important micronutrient elements, the following withdrawals are given [related to a cereal yield of 60 dt ha^{-1} grain (g ha^{-1}): Fe

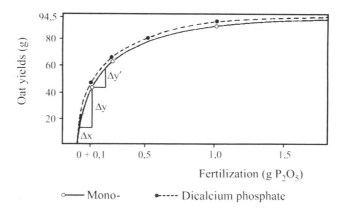

Fig. 2.6   Dependence of oats yield (grain, straw, roots) on phosphate supply in pot experiment using quartz sand as substrate (from Mitscherlich 1909). The dimension (g P$_2$O$_5$) was normal in this time.

Tab. 2.6: Nutrient element withdrawal by fresh weight of crops (as kg dt^{-1}) main product (e.g., grain) and per ha. All data include the withdrawal by the by-product (e.g., straw). (From Kerschberger et al. 1997, enlarged.)

Crop	Ratio main- : by-product	Withdrawal (kg dt^{-1})			
		N	P	K	Mg
Winter wheat (> 12% c.p.)[a]	1:0.8	2.6	0.45	1.43	0.22
Winter wheat (< 12% c.p.)	1:0.8	2.2	0.45	1.43	0.22
Winter barley	1:0.8	2.1	0.45	1.63	0.22
Oats (*Avena sativa* L.)	1:1.1	1.94	0.49	1.78	0.19
Broad bean (*Vicia faba* L.)	1:1	5.6	0.65	3.32	0.36
Maize (*Zea mays* L.)	1:1	2.75	0.51	2.49	0.51
Winter rape (*Brassica napus* L.)	1:1.5	4.42	1.06	4.15	0.45
Potato (*Solanum tuberosum* L.)	1:0.2	0.39	0.06	0.56	0.03
Alfalfa (*Medicago sativa* L.)	–	0.55	0.07	0.54	0.05
Sugar beet (*Beta vulgaris* L. ssp.*vulgaris* var. *altissima* Doell)	1:0.7	0.46	0.08	0.62	0.09
		Withdrawal (kg ha^{-1})			
Winter wheat (70 dt ha^{-1} grain, > 12% c.p.)[a]	1:0.8	182	31.5	100.1	15.4
Banana (450 dt ha^{-1})[b]	–	45	10	100	11

[a] c.p. = crude protein in grain. [b] According to Campbell (1998), without by-product.

< 1000, Mn ~500, Cu ~80, Zn ~300, B ~40, Mo 4–16. Ni and Cl seem to be sufficiently present in European soils, and available Fe is lacking only in calcareous soils.

For land-use management, the question arises as to which dose of fertilizer combines a high economic effect for the agriculturist with harmlessness for the environment. Various soil and plant testing programs are recommended for its estimation (Van Erp and Van Beusichem 1998). Here, only the principle of the German practice is outlined. The procedure is different for nutrient elements which have a noticeable buffer capacity in soil and for those which are easily leached. P, K, Mg, Ca, Mn, B, Cu, Mo, and Zn belong to the first group. Here, the available (diffusible; Schilling et al. 1998) fraction in soil is raised step-by-step to a level that renders production of the desired yield when withdrawal is equalized by fertilizer supply. For estimating the required nutrient amount, the soil content of available nutrient elements is determined by chemical extraction methods (using water, salt solutions or resin exchange reactions). If the detected content is within the sufficient range, the dose of nutrient elements must equalize the expected withdrawal. In other cases, additional or lesser amounts are applied. The data for evaluating the soil contents as well as for correcting withdrawal doses are taken from tables for each individual method (for details, see Schilling et al. 2000). In the case of micronutrient elements, distinction must be made between the decisions of "fertilizing" or "not fertilizing" with a fixed dose (Podlesak et al. 1991).

In the case of N, this procedure is not suitable because N compounds are transformed very rapidly in soil and the available forms are easily leached. Therefore, the requirement is to estimate repeatedly during the vegetation period. Early in spring, the content of NH_4^+ and NO_3^- is determined in

0 to 90 cm or 0 to 60 cm soil depths using an extraction with 0.0125 M $CaCl_2$ solution. The amounts found (kg ha^{-1} N) are taken into account for the first N application rate. Later, the N nutrient status of plants is estimated by use of quick-tests, for instance using the nitrate test in young cereal plants. The results serve for allocating the second (and third) N rate. Recently, so-called N testers have been introduced which determine the N requirement of growing plants during fertilizer application. For instance, an active sensor excites the photosynthesis system of plant by means of a laser beam. Re-emitted fluorescent light allows direct ascertainment of chlorophyll content as a criterion for the N nutrient status. Measurement is carried out from the tractor roof or from the fertilizer spreader to the left and to the right. In this manner, it is possible to bring the N supply into line with requirement of (e.g., cereal) plants.

After balancing all of the nutrient element amounts, it is necessary to choose suitable fertilizers. In every case the contents of all applied organic fertilizers and incorporated plant residues are to be taken into account, after which the gaps in the balance may be closed by use of manufactured fertilizers.

Every arable site is inhomogeneous, which means that the soil-borne available nutrient elements are not distributed homogeneously over the area. As a result, side effects of management are possible, including nutrient element losses to the environment following high application rates, poor crop quality a.o. The new system, which is known as "Precision farming" or site-specific management, aspires to prevent such effects. It uses mapping of soil characteristics and yield monitoring on small areas (< 0.5 ha) in combination with computer data management and global positioning systems (GPS). Thereby, fertilizer application rates, pesticide placement, and other measurements can be varied corresponding to the inhomogeneity of the site.

In comparison with the situation in developed countries, the problems of less- developed countries are much more important. It is expected that the world's population will rise from 5.7 billion in 1995 to 8.5 billion in 2025. Hence, the demand is that food production must increase by over 50% during this period (Byrnes and Bumb 1998). Since scope for extending cultivated or irrigated areas in most regions of the world is limited, more fertilizers will be required to increase the yield potential in less-developed countries. Moreover, fertilization must contribute to the preservation of natural resources. At present, overgrazing and deforestation are the most important factors in enhancing soil degradation in the poor regions of the world, and fertilization may help to stop these dangerous processes. Because the production of organic fertilizers and residues of crops requires the addition of nutrient elements, manufactured fertilizers will play an important role in the future. Consequently, "ecofarming" will not obtain greater dissemination because this approach disclaims the use of most mineral (especially N-) fertilizers, and consequently the crop yields are lower than in conventional agriculture. Therefore, the outlined global problems may not be resolved by using this type of management.

References

Agarie S, Hanaoka N, Ueno O, Miyazaki A, Kubota F, Agata W and Kaufman PB (1998) *Effects of silicon on tolerance to water deficit and heat stress in rice plants (Oryza sativa L.), monitored by electrolyte leakage.* Plant Prod Sci 1:96–103.

ANDERSON DL (1991) *Soil and leaf nutrient interactions following application of calcium silicate slag to sugar cane.* Fertil Res **30**:9–18.

ANKE M, DORN W, MÜLLER R and SCHÄFER U (2000) *Umwelt und Mensch – Langzeitwirkungen und Schlussfolgerungen für die Zukunft.* Abhdl. Sächs. Ak. Wiss. Leipzig, Math.-naturw. Kl. 59, H. **5**:45–61.

ARNON DI and STOUT (1939) *The essentiality of certain elements in minute quantity for plants with special reference to Copper.* Plant Physiol **14**:371–375.

ASHER CJ (1991) *Beneficial elements, functional nutrients and possible new essential elements.* In: Morvedt JJ, Cox FR, Shuman LM and Welch RM, eds., Micronutrients in Agriculture. 2nd ed., pp. 703–723. Soil Sci Soc Amer Book Series No. 4, Madison, WI, USA.

BALKE NE and HODGES TK (1975) *Plasma membrane adenosine triphosphatase of oat roots.* Plant Physiol **55**:83–86.

BERGMANN W (1993) *Ernährungsstörungen bei Kulturpflanzen.* 3rd edn. Gustav Fischer Verlag, Jena-Stuttgart.

BERGMANN W and NEUBERT P (eds) (1976) *Pflanzendiagnose und Pflanzenanalyse zur Ermittlung von Ernährungsstörungen und des Ernährungszustandes von Kulturpflanzen.* VEB Gustav Fischer Verlag, Jena.

BEYERSMANN D (1991) *The significance of interactions in metal essentiality and toxicity.* In: Merian E, ed., Metals and Their Compounds in the Environment, pp. 491–509. Verlag Chemie, Weinheim-New York-Basel-Cambridge.

BOLLARD E (1983) *Involvement of unusual elements in plant growth and nutrition.* In: Läuchli A and Bieleski RL, eds., Inorganic plant nutrition. In: Encyclopedia of Plant Physiology, New Series, Vol. 15 B, pp. 695–755. Springer Verlag, Berlin-Heidelberg-New York-Tokyo.

BOWYER JR and LEEGOOD RC (1997) *Photosynthesis.* In: Dey PM and Herborne JB, eds., Plant Biochemistry, pp. 49–110. Academic Press, San Diego-London-Boston-New York-Sydney-Tokyo-Toronto.

BRISKIN DP (1986) *Plasma membrane H⁺ transporting ATPase: role in potassium ion transport?* Physiol Plant **68**:159–163.

BROWN TA and SHRIFT A (1982) *Selenium: toxicity and tolerance in higher plants.* Biol Rev Camb Philos Soc **57**:59–84.

BYRNES BH and BUMB BL (1998) *Population growth, food production and nutrient requirements.* In: Rengel Z, ed., Nutrient use in crop production,

pp. 1–27. Food Products Press, an Imprint of the Haworth Press, Inc., New York-London.

CAKMAK I (2000) *Possible roles of zinc in protecting plant cells from damage by reactive oxygen species.* New Phytol **146**:185–205.

CAKMAK I and MARSCHNER H (1986) *Mechanism of phosphorus-induced zinc deficiency in cotton. I. Zinc deficiency-enhanced uptake rate of phosphorus.* Physiol Plant **68**:483–490.

CAKMAK I and MARSCHNER H (1988) *Increase in membrane permeability and exsudation of roots of zinc deficient plants.* Plant Physiol **132**:356–361.

CAKMAK I and ROEMHELD V (1997) *Boron deficiency-induced impairments of cellular functions in plants.* Plant Soil **193**:71–83.

CAMPBELL LC (1998) *Managing soil fertility decline.* In: Rengel Z, ed., Nutrient use in crop production, pp. 29–52. Food Products Press, an Imprint of the Haworth Press, Inc., New York-London.

CHANSON A (1991) *A Ca²⁺/H⁺ antiport system driven by the tonoplast pyrophosphate-dependent proton pump from maize roots.* J Plant Physiol **137**:471–476.

CORIA NA, SARQUIS JI, PENALOSA I and URZUA M (1998) *Heat induced damage in potato (Solanum tuberosum) tubers: membrane stability, tissue viability and accumulation of glycoalkaloids.* J Agric Food Chem **46**:4524–4528.

DE VOS CHHR, VONK MJ, VOOIJS R and SCHAT H (1992) *Glutathione depletion due to copper-induced phytochelatin synthesis causes oxidative stress in Silene cucubalus.* Plant Physiol **98**:853–858.

DILWORTH MJ and BISSELING T (1984) *Cobalt and nitrogen fixation in Lupinus angustifolius L. III. DNA and methionine in bacteroids.* New Phytol **98**:311–316.

ELSTNER EF und HIPPELI S (1995) *Schadstoffe aus der Luft.* In: Bock B und Elstner EF, eds., Schadwirkungen auf Pflanzen. 3rd edn., pp. 79–117. Spektrum Verlag, Heidelberg-Berlin-Oxford.

EUSTICE DC, KULL FJ and SHRIFT A (1981) *In vitro incorporation of selenomethionine into protein by Astragalus polysomes.* Plant Physiol **67**:1059–1060.

FERROL N, BELVER A, ROLDAN M, RODRIGUEZ-ROSALES MP and DONAIRE JP (1993) *Effects of boron on proton transport and membrane properties of sunflower (Helianthus annuus L.) cell microsomes.* Plant Physiol **103**:763–769.

FIGUEIRA A, KIDO EA and ALMEIDA RS (2002) *Identifying sugarcane expressed sequences associated* with nutrient transporters and peptide metal chelators. Genet Mol Biol **24**:207–220.

FINCK A (1991) *Düngung*. Verlag Eugen Ulmer, Stuttgart.

GALLEGO SM, BENAVIDES MP and TOMARO ML (1999) *Effect of cadmium ions on antioxidant defense system in sunflower cotyledons*. Biol Plantarum (Prague) **42**:49–55.

GARBARINO J and DU PONT FM (1989) *Rapid induction of Na⁺/H⁺ exchange activity in barley root tonoplast*. Plant Physiol **89**:1–4.

GLADSTONES JS, LONERAGAN JF and GOODCHILD NA (1977) *Field responses to cobalt and molybdenum by different legume species with interferences on the role of cobalt in legume growth*. Aust J Agric Res **28**:619–628.

GOLDBACH HE, RERKASEM B, WIMMER MA, BROWN PH, TELLIER M and BELL RW (2002) *Boron in plant and animal nutrition*. Kluwer Academic/Plenum Publishers, New York.

HAMPE T and MARSCHNER H (1982) *Effect of sodium on morphology, water relations and net photosynthesis in sugar beet leaves*. Z Pflanzenphysiol **108**:151–162.

HAWKESFORD MJ and SMITH FW (1997) *Molecular biology of higher plant sulfate transporters*. In: Cram WJ, De Kok Lj, Stulen I, Brunhold C and Renneberg H, eds., Sulfur metabolism in higher plants, pp. 13–25. Backhuys Publishers, Leiden (Netherlands).

HECHT-BUCHHOLZ C (1979) *Calcium deficiency and plant ultrastructure*. Commun Soil Sci Plan **10**:67–81.

HEISS S, SCHAEFER HJ, HAAG-KERWER A and RAUSCH T (1999) *Cloning sulfur assimilation genes of Brassica juncea L.: Cadmium differentially affects the expression of a putative low-affinity sulfate transporter and isoform of ATP sulfurylase and APS reductase*. Plant Mol Biol **39**:847–857.

HUANG JW, SHAFF JE, GRUNES DL and KOCHIAN LV (1992) *Aluminum effects on calcium fluxes at the root apex of aluminum-tolerant and aluminum-sensitive wheat cultivars*. Plant Physiol **98**:230–237.

JONES DL, PRABOWO AM and KOCHIAN LV (1996) *Aluminium-organic acid interactions in organic soils*. Plant Soil **182**:229–237.

KERSCHBERGER M, FRANKE G and HESS H (1997) *Anleitung und Richtwerte für Nährstoffvergleiche nach Düngeverordnung*. Thüringer Landesanstalt f. Landwirtschaft, Jena.

KIEFFER F (1991) *Metals as essential trace elements for plants, animals and humans*. In: Merian E, ed., Metals and their compounds in the environment,

pp. 481–489. Verlag Chemie, Weinheim-New York-Basel-Cambridge.

KROGMANN DW, JAGENDORF AT and AVRON M (1959) *Uncouplers of spinach chloroplast photosynthetic phosphorylation*. Plant Physiol **34**:272–277.

LÄUCHLI A (1993) *Selenium in plant: uptake, functions, and environment toxicity*. Bot Acta **106**:455–468.

LINDHAUER MG, HAEDER HE and BERINGER H (1990) *Osmotic potentials and solute concentrations in sugar beet plants cultivated with varying potassium/sodium ratios*. Z Pflanz Bodenkunde **153**:25–32.

MA JF, HIRADATE S, NOMOTO K, IWASHITA T and MATSUMOTO H (1997) *Internal detoxification mechanism of Al in Hydrangea*. Plant Pysiol **113**:1033–1039.

MANSOUR MMF, VAN HASSELT PR and KNIPER PJC (1998) *Ca²⁺ and Mg²⁺-ATPase activities in winter wheat root plasma membranes as affected by NaCl stress during growth*. J Plant Physiol **153**:181–187.

MARSCHNER H (1991) *Mechanism of adaptation of plants to acid soils*. Plant Soil **134**:1–20.

MARSCHNER H (1995) *Mineral nutrition of higher plants*. 2nd edn. Academic Press Harcourt Brace & Co., Publishers London-San Diego-New York-Boston-Sydney-Tokyo-Toronto.

MARSCHNER H, KYLIN A and KNIPER PJC (1981) *Genotypic differences in the response of sugar beet plants to replacement of potassium by sodium*. Physiol Plant **51**:239–244.

MATCH T and KOBAYASHI M (1998) *Boron and calcium, essential inorganic constituents of pectin polysaccharides in higher plant cell walls*. J Plant Res **111**:179–190.

MATSUMOTO H (1988) *Repression of proton extrusion from intact cucumber roots and the proton transport rate of microsomal membrane vesicles of the roots due to Ca²⁺ starvation*. Plant Cell Physiol **29**:79–84.

MATYSIK J, ALIA, NACHTEGAAL G, VAN GORKOM HJ, HOFF AJ and DE GROTT-HUUB JM (2000) *Exploring the calcium binding site in photosystem II membranes by solid state 113 Cd NMR*. Biochemistry – US **39**:6751–6755.

MENGEL K (1991) *Ernährung und Stoffwechsel der Pflanze*. 7., überarb. Aufl. Gustav Fischer Verlag, Jena.

MITSCHERLICH EA (1909) *Das Gesetz des Minimums und das Gesetz des abnehmenden Bodenertrages*. Landw Jahrb **38**:537–552.

MITSCHERLICH EA (1956) *Ertragsgesetze*. Akademie-Verlag, Berlin.

MOHAMED AA, KHALIL I, VARANINI Z and PINTON R (2000) *Increase in NAD(P)H-dependent generation of active oxygen species and changes in lipid composition of microsomes isolated from roots of zinc-deficient bean plants.* J Plant Nutr **23**:285–295.

NEUMANN D and ZUR NIEDEN U (2001) *Silicon and heavy metal tolerance of higher plants.* Phytochemistry (Oxford) **56**:685–692.

OLBE M and SOMMARIN M (1991) *ATP-dependent Ca²⁺ transport in wheat root plasma membrane vesicles.* Physiol Plant **83**:535–543.

OHNISHI J, FLÜGGE UI, HELDT HW and KANAI R (1990) *Involvement of Na⁺ in active uptake of pyruvate in mesophyll chloroplasts of some C₄ plants.* Plant Physiol **94**:950–959.

PARRY DW and KELSO M (1975) *The distribution of silicon deposits in the root of Molinia caerulea (L.) Moench and Sorghum bicolor (L.) Moench.* Ann Bot (London) [N. S.] **39**:995–1001.

PODLESAK W, BRUCHLOS P, FALKE H and WERNER TH (1991) *Neue Aspekte der Mikronährstoffdüngung des Getreides.* Feldwirtsch **32**:472–473.

POLZ (1965) *Untersuchungen über die Wirkung der Wasserversorgung auf Substanzbildung und Stoffumsatz bei Futtergräsern.* PhD thesis, Friedrich-Schiller-Universität, Jena.

RAVEN JA (1983) *The transport and function of silicon in plants.* Biol Rev Camb Philos Soc **58**:179–207.

REA PA (1999) *MRP subfamily ABC transporters from plants and yeast.* J Exp Bot **50**:895–913.

REA PA and SANDERS D (1987) *Tonoplast energization: two H⁺ pumps, one membrane.* Physiol Plant **71**:131–141.

RENGEL Z and ROBINSON DL (1989) *Competitive Al³⁺ inhibition of net Mg²⁺ uptake by intact Lolium multiflorum roots. I. Kinetics.* Plant Physiol **91**:1407–1413.

RILEY IT and DILWORTH MJ (1985 a) *Cobalt requirement by nodule development and function in Lupinus angustifolius L.* New Phytol **100**:347–359.

RILEY IT and DILWORTH MJ (1985 b) *Recovery of cobalt deficient nodules in Lupinus angustifolius L.* New Phytol **100**:361–365.

SANTOSH TR, SKREEKALA M and LALITHA K (1999) *Oxidative stress during selenium deficiency in seedlings of Trigonella foenum-graecum and mitigation by mimosine: Part II. Glutathione metabolism.* Biol Trace Elem Res **70**:209–222.

SAURE MC (2001) *Blossom –end rot of tomato (Lycopersicon esculentum Mill.): a calcium- or stress-related disorder?* Scientia hortic – Amsterdam **90**:193–208.

SCHILLING G, GRANSEE A, DEUBEL A, LEZOVIC G and RUPPEL S (1998) *Phosphorus availability, root exudates, and microbial activity in the rhizosphere.* Z Pflanz Bodenkunde **161**:465–478.

SCHILLING G, KERSCHBERGER M, KUMMER K-F und PESCHKE H (2000) *Pflanzenernährung und Düngung.* Verlag Eugen Ulmer, Stuttgart.

SCHUBERT S and LÄUCHLI A (1990) *Sodium exclusion mechanisms at the root surface of two maize cultivars.* Plant Soil **123**:205–209.

SHIMIZU K, DEL AMO Y, BRZEZINSKI MA, STUCKY GD and MORSE DE (2001) *A novel fluorescent silica tracer for biological silicification studies.* Chem Biol (London) **8**:1051–1060.

SMITH FW (1999) *Molecular biology of nutrient transporters in plant membranes.* In: Rengel Z, ed., Mineral nutrition of crops – fundamental mechanisms and implications, pp. 67–89. Food Products Press, an Imprint of the Haworth Press, Inc., New York-London-Oxford.

STURM H, BUCHNER A und ZERULLA W (1994) *Gezielter düngen.* 3., vollk. neu überarb. Aufl. Verlagsunion Agrar, Frankfurt/M.-München-Münster-Hiltrup-Wien-Bern.

SUBBARAO GV, WHEELER RM, STUTTE GW and LEVINE LH (1999) *How far can sodium substitute for potassium in red beet?* J Plant Nutr **22**:1745–1761.

TACHIMOTO M, FUKUTOMI M, MATSUSHIRO H, KOBAYASHI M and TAKAHASHI E (1992) *Role of putrescine in Lemna plants under potassium deficiency.* Soil Sci Plant Nutr (Tokyo) **38**:307–313.

TAYLOR GJ, BASU A, SLASKI JJ, ZANG G and GOOD A (1997) *Al-induced, 51-kilo-dalton, membrane-bound proteins are associated with resistance to Al in a segregating population of wheat.* Plant Physiol **114**:363–372.

TERRY N, CARLSON C, RAAB TK and ZAYED AM (1992) *Rates of selenium volatilization among crop species.* J Environ Qual **21**:341–344.

TUKENDORF A and RAUSER WE (1990) *Changes in glutathione and phytochelatins in roots of maize seedlings exposed to cadmium.* Plant Sci **70**:155–166.

VAN CUTSEM P and GILLET C (1982) *Activity coefficient and selectivity values of Cu²⁺, Zn²⁺ and Ca²⁺ ions adsorbed in the Nitella flexilis L. cell wall during triangular ion exchanges* J Exp Bot **33**:847–853.

VAN ERP PJ VAN BEUSICHEM ML (1998) *Soil and plant testing programs as a tool for optimizing fertilizer strategies.* In: Rengel Z, ed., Nutrient use in crop production, pp. 53–80. Food Products Press, an Imprint of the Haworth Press, Inc., New York-London.

Van Goor (1966) *The role of calcium and cell permeability in disease blossom end rot of tomatoes.* Physiol Plant 21:1110–1121.

Wallace T (1961) *The diagnosis of mineral deficiencies in plants by visual symptoms.* 3rd ed., Her Majesty's Stationery Office, London.

Watson RJ, Heys R, Martin T and Savard M (2001) *Sinorhizobium meliloti cells require biotin and either cobalt or methionine for growth.* Appl Environ Microb 67:3767–3770.

Welch RM, Webb MJ and Loneragan JF (1982) *Zinc in membrane function and its role in phosphorus toxicity.* In: Scaife A, ed., Proceedings of the Ninth Plant Nutrition Colloquium, Warwick, England, pp. 710–715. Commonwealth Agricultural Bureau, Farnham Royal, Bucks.

Wenzel P, Chaves AL, Patino GM, Mayer JE and Rao IM (2002) *Aluminum stress stimulates the accumulation of organic acids in root apices of Brachiaria species* J Plant Nutr Soil Sci 165:582–588.

Wissemeier AH and Horst WJ (1992) *Effect of light intensity on manganese toxicity symptoms and callose formation in cowpea (Vigna unguiculata (L.) Walp.).* Plant Soil 143:299–309.

3

Essential and Toxic Effects of Macro, Trace, and Ultratrace Elements in the Nutrition of Animals

Manfred K. Anke

3.1
Introduction

3.1.1
Essentiality and Toxicity

During the long passage of inorganic components of foodstuffs, water and air through the fauna (and man), which has lasted for several hundred million of years, the majority of these substances have most likely become parts or activators of proteins, enzymes, hormones or other essential components of the body. Consequently, either a deficiency or a toxic excess in supply must be considered for most elements (Figure 3.1).

In the transitional zones between deficient and sufficient supply, as well as between normal and toxic supply, adaptation reactions occur as described for copper in sheep (Wiener and Field 1970) and for manganese in goats and cattle (Anke et al. 1973). As a consequence, breeds of farm animals which adapted themselves to the local trace element offer came into being. Depending upon the species and the elements involved, there is a pharmacodynamic or therapeutic range between the optimum and toxic element offers of some elements. Well-known examples of the effect of inorganic components in the diet are arsenical compounds in the

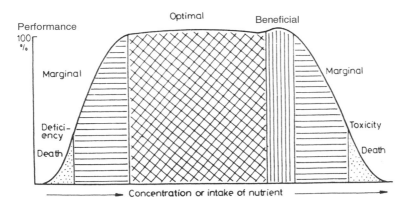

Fig. 3.1 The dependence of performance on trace element supply.

Elements and their Compounds in the Environment. 2nd Edition.
Edited by E. Merian, M. Anke, M. Ihnat, M. Stoeppler
Copyright © 2004 WILEY-VCH Verlag GmbH & Co. KGaA, Weinheim
ISBN: 3-527-30459-2

nutrition of poultry, pigs and man (Anke 1986, Bentley and Chasteen 2002), and copper in pigs (Barber et al. 1955). The toxic effect of elements is species-specific – that is, the results obtained in one species cannot be transferred to another without experimental testing. For example, cattle are extremely sensitive to an exposure of 10 mg Mo kg^{-1} feed dry matter (DM) of the ration and react with molybdenosis, whereas sheep tolerate three times and goats thirty times such molybdenum intake, without difficulties (Falke and Anke 1987).

3.1.2
Groups of Mineral Elements

Three groups of essential mineral elements can be distinguished (Table 3.1). For convention and historical reasons, the elements are divided into macro, trace, and ultratrace elements. The animals' requirement for macro elements is > 100 mg kg^{-1} of food DM (dry matter), while that for trace and ultratrace

elements is in the range of milligrams (mg) or even micrograms (µg) per kg food DM. In the case of adult humans, the same distinguishing units apply when expressed as intake per day. The list of essential mineral elements (Table 3.1) is divided into two categories: (i) those for which essentiality has been confirmed by evidence for an essential biochemical mechanism, involving the element in a catalytic and regulatory role or the syntheses of essential compounds for the fauna by microorganisms (Co → vitamin B$_{12}$; Ni → urease); and (ii) those for which essentiality or a beneficial role is experimentally examined with synthetic rations poor in one element, intrauterine depletion about three to five generations and control groups by the impairment of physiological functions. It is anticipated that future research will shift several ultratrace elements to the category of confirmed trace elements.

Tab. 3.1: Essential mineral elements

Essentiality confirmed by biochemical mechanism			*Essentiality suggested by physiological impairment*		
Macro elements	Calcium	(Ca)	Ultratrace elements	Fluorine	(F)
	Magnesium	(Mg)		Chromium	(Cr)
	Phosphorus	(P)		Silicon	(Si)
	Sulfur	(S)		Arsenic	(As)
	Potassium	(K)		Tungsten	(W)
	Sodium	(Na)		Cadmium	(Cd)
	Chlorine	(Cl)		Lead	(Pb)
				Boron	(B)
Trace elements	Iron	(Fe)		Vanadium	(V)
	Iodine	(I)		Lithium	(Li)
	Copper	(Cu)		Bromine	(Br)
	Manganese	(Mn)		Rubidium	(Rb)
	Zink	(Zn)		Aluminum	(Al)
	Cobalt	(Co)		Titanium	(Ti)
	Molybdenum	(Mo)		Tin	(Sn)
	Selenium	(Se)			
	Nickel	(Ni)			

3.1.3
The Identification of Essential Mineral Elements

The discovery of mineral element essentiality and function has proceeded along multiple roads. Most important during the past thirty years have been: (i) the use of semi-synthetic and synthetic rations, whereby the element to be examined is maintained at poor levels when compared with a control ration of similar composition to element under test (Hennig et al. 1972, 1978, Anke 1974, Anke and Groppel 1989); (ii) the results of parenteral nutrition (i.e., intravenous infusions of highly purified nutrients; Abumrad et al. 1981); (iii) the study of animals living in ecological niches of specific geological origin deficient in elements, such as manganese or cobalt (Werner and Anke 1960); and (iv) the determination of the basis of certain genetic diseases (e.g., molybdenum, copper) (O'Dell and Sunde 1997).

Furthermore, an intrauterine depletion over three or more generations of an animal species has been shown to be highly effective in discovering the essentiality of several elements, especially in case of vanadium, cadmium, lead, lithium, rubidium, aluminum, arsenic, fluorine, and bromine (Anke et al. 1991b, 1998, 2001, 2001a).

Beside the use of synthetic rations and intrauterine depletion over generations, the animal species selected for the study is also very important. The use of ruminant species (goats) has allowed researchers to substitute chemically pure urea for a significant proportion of trace element-containing proteins (Mertz 1986).

An *essential element* is one that is required to support adequate growth, reproduction and health throughout the life cycle if all other nutrients are optimal. Besides the deficiency group, every mineral deficiency experiment needs a control group with identical conditions, feed, and the element tested. In addition, the animals of both groups should live to their natural death.

The synthetic ration, when tested for the essentiality of one element, must be supplemented with all elements contained in the normal feed of the animals. The synthetic ration of the ruminants (goats), in addition to cellulose in the form of purified paper (which was used as litter and nutrient), contained all essential nutrients and all mineral elements present in the normal feed, with the exception of Ra, Fr, Ac, Po, At, Re and the rare earth metals Pr, Nd, Pm, Em, Eu, Gd, Tb, Dys, Ho, Er, Tm, Yb, and Cp) (Anke and Groppel 1989). With the help of synthetic rations, it was possible to obtain extensive data concerning the essentiality of several metals (Cr, W, Cd, Pb, V, Li, Rb, Al, Ti) and also nonmetals (F, Br, Si, As, B). During the 1990 s, the molecular biology paradigms began to offer existing routine potential for bridging the gap between a protein and a function. O'Dell and Sunde (1997) provided four examples for this development:

1. The ability to sequence proteins, cDNAs, and genes rapidly and accurately, in combination with the technological revolution that allows rapid searching of the resulting databases, now permits rapid interconversion of information among these three tiers of gene expression.

2. The ability to discern changes in the regulation of protein expression, not only by the use of antibodies, but also by means of Northern blotting techniques to monitor changes in mRNA levels, and footprinting techniques or gel retardation/mobility shift assays to determine interaction of regulatory proteins with nucleic acids.

3. The use of heterologous expression systems, such as the *Xenopus* oocyte (toad),

baculovirus (insect), yeast or baby kidney hamster cell systems, to clone, characterize, and identify an animal or human gene from a library.

4. The production of transgenic and knockout animals.

The production of transgenic and knockout animals represents a powerful molecular biology technique whereby spontaneous inborn errors of metabolism (e.g., sulfite oxidase deficiency) have provided extensive and valuable insight into the identification of essential elements, for example molybdenum (O'Dell and Sunde 1997).

In future, it seems that it will be necessary to combine the methods of nutrient physiology and molecular biology such that both methods will guarantee the prompt discovery of the essential and toxic effects of inorganic elements.

3.1.4
Requirement and Recommendations of Mineral Elements

For both animals and man, the requirement of inorganic elements is extremely important. Requirement is the lowest continuous level of nutrient intake that, at a specified efficiency of utilization, will maintain the defined level of nutritive in the individual.

At this point a distinction must be made between basal and normative requirements. *Basal requirement* refers to the intake needed to prevent pathologically relevant and clinically detectable signs of impaired function attributable to inadequacy of the nutrient. *Normative requirement* refers to the intake that serves to maintain a level of tissue storage or other reserves that is judged to be desirable.

The essential difference between basal and normative requirement is that the latter usually facilitates the maintenance of a desirable level of tissue stores. For most

trace elements discussed in this chapter, metabolic and tissue-composition studies have indicated the existence of discrete stores which, by undergoing depletion at times of reduced intake or high demand, can provide protection for a certain period against the development of pathological responses to trace element deficiency. As higher levels of intake are needed to maintain this reserve, the normative requirement is necessarily higher than the basal requirement.

Individuals differ in their requirements even if they may have the same general characterization (age, sex, physiological size, body size). One may therefore speak of the average requirements of a group of individuals, or of the level that marks a point in the tail of the requirement distribution curve – the level previously identified as the recommended or safe level of inorganic element level intake.

Normally, the recommended intake of elements is 30 to 100% higher than the normative requirement. In practice, we must distinguish between the (normative) requirement of the elements and the recommendation for the intake of a population. The recommendation is necessarily higher than the normative requirement (Anonymous 1996). Recommended dietary allowances have been adequate to trace element needs of healthy people. They do not delineate a desirable intake against deficiency on one end and against toxicity on the other, or an intake outside the range to increase the risk for deficiency and toxicity, respectively.

3.1.5
Pharmacological Levels of Essential Elements

Numerous studies have been conducted to identify conditions that require nutrient intakes outside the nutritional realm. Examples outside the essential element area include supernutritional levels of copper and zinc in the nutrition of pigs and piglets (Barber et al. 1955, Poulsen 1995). Recently, the pharmacological action of high doses of vanadium which have received the most attention is the element's ability to mimic insulin (Shechter et al. 1990). Another long-standing nutritional example is fluoride, which may not have a prescribed biochemical role, but appears to have a defined level in the diet that protects teeth against decay and bones against premature calcium loss (Cerklevski 1997). Other examples might include the antitumorigenic effect of selenium at levels well above the nutritional levels required (Sill and Dawczynski 1998), and the past intake of very high levels of arsenic by mountaineers and postmen (Benthley and Chasteen 2002). In all of these cases, maximum protective levels occur just at the onset of apparent toxicity.

A second point is that these effects may be antagonized by normal mechanisms that protect animals against toxicity. Also, the manure of pigs supplemented with zinc and/or copper concentration after long-term fertilization of the soil represents a danger for copper- or zinc-sensitive species such as sheep (Davis and Mertz 1987).

3.1.6
Interactions of Mineral Elements

The many mineral interactions which influence the "safe" dietary levels of essential and toxic elements are partly represented in Figure 3.2. While interactions involving dietary elements may be either detrimental or beneficial, the major concern is that an antagonistic element may induce a deficiency of its counterpart nutrient whose concentration in the diet is borderline. The assessment of such in-vivo interactions will be considered here under the limits of bioavailability, which occurs at the site of absorption in the intestinal mucosa or the redistribution from one tissue to another one. Figure 3.2 illustrates the most important, quite different, species-specific interactions of metals, trace and macro elements, respectively, with net requirements in animals and man. Clearly, there are interrelationships in the metabolism of the mineral elements consumed.

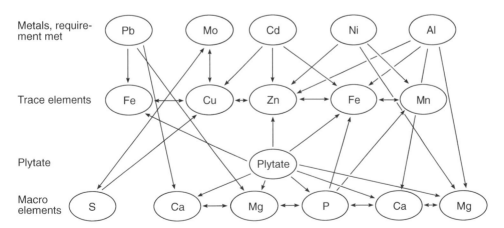

Fig. 3.2 Interactions between and among some macro, trace and ultratrace elements. (Anke et al. 2003)

One ion may be essential for the absorption and utilization of another, and conversely, one ion may adversely affect the absorption and utilization of one or more other ions. The term "interaction" is used to describe such interrelationships among mineral elements, and may be defined as the effect of one element on one or more other elements as revealed by physiological or biochemical consequences. Such interactions must occur at specific sites, on proteins, such as enzyme receptors, or ion channels. There are two major types of interaction: positive and negative. The former is commonly synergistic, while the latter is antagonistic. The many minerals are discussed in later chapters in this book. Figure 3.2 illustrates that the utilization of iron, copper, zinc, and manganese of the trace elements, and calcium, magnesium, phosphorus and sulfur of the macro elements, is specifically influenced by several metals, nonmetals and especially by phytate in case of monogastric animals. Iron utilization is, for example, influenced by high amounts of lead, manganese, zinc, copper, phosphorus and phytate, whereas copper utilization is impaired by high levels of iron, molybdenum, cadmium, sulfur and zinc. Zinc usability is altered by high amounts of copper, cadmium, nickel, iron, calcium and phytate, whereas manganese availability is varied by nickel, iron and phosphorus.

The utilization of the macro element calcium is impaired by a high intake of lead, magnesium, aluminum and phytate, that of magnesium by lead, calcium, nickel, and phytate, and that of phosphorus by magnesium, calcium, and phytate. Several experiments in animals have strikingly illustrated the importance of a trace element dietary balance in determining the "safe" intake of a particular macro, trace or ultratrace element (O'Dell 1997, Grün et al.

1982, Reichlmayer and Kirchgessner 1997, Mertz 1987, Oberleas et al. 1999, Anke et al. 1970, 1988, 1997c, 1997d, Beyersmann et al. 1991).

3.1.7
Toxic Levels of Mineral Elements

In part, the division of the inorganic elements into an essential and a toxic category has tended to confirm the nutritionist's concern with dietary recommendations for the first category of elements, the physiologist with pharmacologically effective levels of the elements (for example, vanadium, Anke 2004) and the toxicologist with establishing tolerances for the "toxic" elements. Lack of communication between the three disciplines has resulted in difficult situations when unenforceable "zero" tolerance levels were applied to selenium – an essential and toxic element with pharmaceutical effects. Arsenic, lead, and cadmium also have long been considered to be "toxic" elements. As a matter of fact, all of the essential elements are toxic if consumed in excess, although the concentration at which toxicity becomes apparent varies widely (O'Dell and Sunde 1997, Mertz 1987, Anke et al. 1991e, Anke and Groppel 1987).

The toxic effects of the mineral elements are extremely element- and species-specific (Hapke 1991). The symptoms of acute poisoning (Geldmacher von Mallinckrodt 1991a) and chronic toxicity of inorganic elements can be completely different (Ewers and Schlipköter 1991). The most common symptoms of acute metal poisoning include the following:

● Gastrointestinal symptoms: Oral ingestion of large quantities of soluble metal salts quickly leads to gastroenteritis. The results are nausea, vomiting, abdominal pain, diarrhea, and possibly shock due to dehydration and loss of electrolytes. Arsenic poisoning is a typical example.

- Damage to the respiratory tract: Inhalation of metals or metal compounds can lead to pulmonary edema. Distinction must also be made between dusts, smoke, and metallic chlorides that lead to the production of hydrochloric acid.
- Cardiovascular effects: Arrhythmia, low blood pressure, and shock.
- Effects on the central nervous system: Cramps, coma, death.
- Kidney damage with oliguria (reduced urine production). Anuria is often the result of tubular necrosis.
- Damage to the blood or blood-producing organs: Hemolytic anemia after inhalation of arsenic hydride or ingestion of copper salts.
- Metal fume fever as an immunological reaction to the inhalation of metallic oxide aerosols (e.g., zinc oxide smoke) (Geldmacher von Mallinckrodt 1991a).

The symptoms of inorganic element poisoning can be found in chapter detailing several elements in volumes II and III of this book. Acute toxicity has been defined as the adverse effect resulting from the administration of a single dose or multiple doses within 24 hours. Acute toxicity tests are conducted to evaluate the relative toxicity of a compound, to investigate its mode of action and its specific toxic effects, and to determine the existence of species differences. Subchronic toxicity tests generally involve daily exposure to an element over a period of about 90 days. These tests are performed to obtain information on the major toxic effects of the test substance, its toxicokinetic behavior, the target organs affected, the reversibility of the observed effects and last, but not least, the amount of the tested element which enables a lifetime of rats of 24 months in chronic exposure studies. The chronic exposure experiments are performed similarly to the subchronic studies, except that the period of exposure in 24 months in rats. Chronic exposure studies are often conducted with the aim of establishing "no observed-effect levels" (NOEL) that may be used in setting acceptable daily intakes (ADI), tolerance limits for chemicals in food or water, or occupational health standards (Ewers and Schlipköter 1991).

Not only arsenic, cadmium, lead, mercury, nickel, aluminum, molybdenum, and bromine, but also zinc, copper and selenium are uniformly toxic to the immune system, although the ultimate effect depends on the species of animals studied and the route and mode of administration. An important feature of inorganic toxicity is the observation that in animals, some metals affect the immune system at doses that are unaccompanied by other clinical manifestations of toxicity. A similar effect also occurs in man (Chowdhury and Chandra 1991)

Any discussion of the mutagenic, cancerogenic, and teratogenic actions of inorganic elements and their compounds must also consider the elements' toxicity. Indeed, in most cases the toxicity may well overwhelm the potential mutagenicity for arsenic, chromium, chromates, nickel, platinum and other metals. With regard to the carcinogenicity of occupational exposure, more detailed epidemiological data point to a possible hazard from arsenic, beryllium, cadmium, chromium and nickel (Gebhart and Rossman 1991). Teratogenic activity so far has been demonstrated for methyl mercury; a teratogenic potential has also been suggested for lithium in animal and man (Puzanova 1983, Anke et al. 2003, Anke 1993).

3.1.8
Ecogenetics

Ecogenetics is a genetic predisposition for an individual reaction, which is not common in the animal or human population. The genetically controlled synthesis of proteins can lead to such variations in all living species.

Certain inbred strains of mice are resistant to cadmium-induced testicular necrosis; other strains of mice show a varying sensitivity to lead poisoning, to a genetic zinc deficiency in milk in mice and minks with the pallid gene. Manganese supplements prevent the appearance of genetically determined ataxia (Geldmacher von Mallinckrodt 1991b). A large proportion of the offspring of goats with manganese-deficient synthetic rations and cows in manganese-deficient areas died, and after three generations of manganese deficiency, only the offspring of some goats and cows with a better manganese utilization are able to survive (Anke et al. 1973). An inheritable defect which leads to copper accumulation in the liver has been found in Bedlington terriers, West Highland white terriers and humans (Morbus Wilson) (Wilson 1912, Sternlieb 1982), while Menkes' disease in babies shows all the symptoms of copper deficiency (Menkes et al. 1962).

Genetically determined variations in the distribution of lithium between plasma and erythrocytes have been found in sheep (Schless et. al. 1975). Likewise, depressive patients react differently to lithium treatment, with heredity a probable contributory factor, as has been shown investigations conducted in twins. A hereditary recessive primary hypomagnesemia that is accompanied by a low level of calcium in the blood leads to tetany. This is presumably caused by a nonfunctional magnesium resorption in the intestine (Lombeck and Bremer 1977, Liebscher and Liebscher 2000, Meij et al. 2002). Meanwhile, four primary hereditary hypomagnesemia disorders have been identified: (i) hypomagnesemia with secondary hypocalcemia (HSH); (ii) familial hypermagnesemia, hypercalciuria and nephrocalcinesis (HHN); and (iii) dominant hypomagnesemia/hypocalciuria and (iv) recessive hypomagnesemia (Meij et al. 2002).

Acrodermatitis enteropathica is an autosomal recessive genetic zinc deficiency which causes parakeratosis of the skin and lowered zinc levels in the plasma (Moynahan 1974).

The common basis for almost all the various genetically determined reactions is that the synthesis of proteins, enzymes or transport proteins is not controlled genetically (Geldmacher von Mallinckrodt 1991b).

3.2
Essentiality and Toxicity of Mineral Elements

3.2.1
Macro Elements

Although the accumulation of empirical knowledge related to mineral nutrition in general is as old as mankind (Table 3.2), perhaps the most recent reports were made during the twentieth by McCollum (1957), by McCay (1973), and by McDowell and Mertz (both 1992). The scientific era of nutrition with macro elements began with their discovery in 1669 when phosphorus was prepared in the free state from urine by Brandt, an alchemist living in Hamburg, Germany. Chlorine was first described by Scheele in 1774, but not named until 1870, by Davy. The nonmetallic element sulfur was first recognized by Antoine Lavoisier, whereas calcium, magnesium, sodium and potassium were discovered some 200 years ago by Sir Humphry Davy, in 1807–1808.

Tab. 3.2: Chronological observations providing essentiality and toxicity of macro elements

Element	Laboratory reference	Essentiality observation	Toxicity symptoms
Ca	Galenus, 2nd century, Paracelsus 1530, Sandström 1880	Prescribed Ca, Ca prevented fragile bones, parathyroid glands are influenced	Benign familial hypercalciuria, polyuria, polydipsia, loss of K and appetite, somnolence, heart-rhythm disturbances
P	Galin 1769 Scheele 1771 Boussingault 1841 Le Vaillant 1796	Essential part of bones as Ca_3PO_4. Ca and P should be part of the diet. Lame sickness and botulism in cattle	Excess of either Ca or P causes bone disorders and reduces feed consumption and gain, calcification of soft tissues in relation with vitamin D in cattle through yellow oat.
Na	Boussingault 1847 Babcock 1905 Orent-Keiles et al. 1937	Abnormal appetite for salt after depletion in cows. Difference between sodium and chlorine	NaCl may be toxic when water intake is limited, anorexia, weight loss, edema, nervousness, paralysis.
Cl	Fettman et al. 1984 Richter et al. 2002	Cl requirement for milking cows and hens, decreased feed intake, laying performance, smaller eggs and growth, alkalosis.	Feed of hens should contain < 1.7 g Cl kg^{-1} feed, high levels of Na and Cl raise blood pressure only in case of gene defects. (Richter and Thieme 2002)
K	Von Liebig 1847 Ringer 1881	Tissues contain primarily K; blood and lymph sodium. K is essential for maintenance of organs.	Maximum dietary tolerable levels of K are 3% in most species, clinical signs of K toxicosis include cardiac insufficiency, edema, muscle weakness.
Mg	Kruse et al. 1932	Essentiality for animals, lactation tetany, grass tetany, vasodilation, hyperirritability in rats	Clinical signs of Mg intoxication in various species are lethargy, disturbance in locomotion, diarrhea, lowered feed intake and performance, drowsiness
S	Henry 1828 Baumann 1876a,b	Two forms of S in tissues and urine, S-deficiency in sheep leads to weight loss, weakness, lacrimation and death	Sulfur toxicity in ruminants occurred through microbially produced H_2S. Sulfide reduces rumen motility and causes nervous and respiratory distress.

In the field of biology, there are three general physiological roles for the macro elements, namely structural, catalytic, and transduction. Of the seven macro elements, calcium and phosphorus play important roles in the skeletal structures of vertebrates. The normative calcium requirements of the different animal species vary between 4.0 g kg^{-1} feed DM in growing ruminants, 35 g kg^{-1} feed DM in hens, and 500–600 mg kg^{-1} in man (Anke et al. 2002, Holtmeier 1995); the normative phosphorus requirement of the same species amounts to between 3.0 and 8.0 g kg^{-1} feed DM (broiler). Phosphorus is also an important component of phospholipids, phosphopro-

growth promoter, whereas the same levels would be toxic for lambs (Anke 1982, Davis and Mertz 1987).

3.2.2.4
Manganese

Manganese deficiency results in distinct pathology including reproductive failure, skeletal defects and ataxia. Manganese is a component of several enzymes, and catalyzes quite different biochemical reactions. The recommendation for manganese intake amounts to 60 mg kg^{-1} feed DM for ruminants and poultry, < 20 mg kg^{-1} for pigs, and 5 mg kg^{-1} for cats. Adverse health effects have not occurred in most animal species fed a dietary manganese concentration of < 1000 mg kg^{-1} feed DM. At 2000 mg kg^{-1} feed DM and above, growth retardation, anemia, gastrointestinal lesions, and (sometimes) also neurological signs have been observed. Swine appear to be more sensitive to high levels of manganese, as 500 mg kg^{-1} feed DM retards both appetite and growth (Anke et al. 1999). The toxicity of excessive manganese appears to be an interaction with iron. Low hemoglobin levels are reported as a result of excessive dietary manganese (McDowell 1992).

3.2.2.5
Zinc

Zinc deficiency includes depressed feed intake, stunted growth, skin lesions, and reproductive difficulties. Zinc is a component of enzymes which catalyze more than fifty different biochemical reactions, as well as a component of proteins involved in gene expression. The recommendation for zinc intake amounts to 30 mg kg^{-1} for calves, heifers and lambs, 40 mg kg^{-1} for dairy cows, piglets, pigs, sows and hens, and 50 mg kg^{-1} feed DM for broilers.

Pigs, poultry, sheep and cattle exhibit considerable tolerance to high intakes of zinc, the extent of the tolerance depending partly on the species, but mainly on the relative content of calcium, copper, iron, nickel, cadmium, and phytate in their feed. In most studies with various species, no adverse physiological effects were observed with < 600 mg kg^{-1} Zn feed DM (Anke 1982, McDowell 1992, O'Dell and Sunde 1997).

3.2.2.6
Cobalt

Cobalt is a component of vitamin B$_{12}$, and is thus classed as an essential element, though there is no evidence that the cobalt ion has any other biochemical function.

The normative cobalt requirement of ruminants amounts to 130 μg kg^{-1} feed DM (Stangl et al. 2000, Stemme et al. 2002).

The toxicity of excessive Co, in part, appears to be an interaction with anemia resulting from decreased iron absorption. In rats, the intestinal absorption of iron is reduced by almost two-thirds in the presence of the ten-fold higher cobalt absorption (Underwood 1984).

3.2.2.7
Molybdenum

The fact that molybdenum is a component of enzymes, such as xanthine oxidase, sulfite oxidase and aldehyde oxidase, provides stronger evidence of its essentiality. On the other hand, the molybdenum intake of animals (and man) is higher than their normative molybdenum requirement of < 100 μg kg^{-1} feed DM in ruminants, < 50 μg kg^{-1} feed DM in monogastric animals, and 25 μg Mo per day in humans (Anke et al. 1985a). Nonruminants are much more resistant to molybdenum toxicity. Ruminants also vary greatly in molybdenum tolerance, from as low as 5 mg kg^{-1} feed DM in cattle to approximately 1000 mg kg^{-1} feed DM in adult goats and mule deer (Graupl 1965, Mills and Davies 1987, Falke and Anke 1987, McDowell 1992).

3.2.2.8
Selenium

Not only does selenium deficiency result in several pathological conditions including cardiomyopathy and skeletal muscle defects, but it is also a component of several proteins, including glutathione peroxidase and the iodothyronine-5′ deiodinases, which are necessary for the conversion of thyroxine (T_4) in cell-active free triiodo-thyronine (f T_3) (Anke et al. 2000a). Selenium toxicity conditions termed "alkali disease" (5–40 mg Se kg^{-1} feed DM) or blind staggers (100 to 9000 mg Se kg^{-1} feed DM) have resulted in extreme losses of livestock. In subacute selenosis, cattle exhibit blindness, abdominal pain, excessive salivation, teeth grating, paralysis, respiratory failure, and death. Death also results from starvation and thirst because, in addition to loss of appetite, the lameness and pain in the hooves are so severe that the animals are unwilling to move about to secure food and water. Swine with selenosis exhibit lameness, hoof malformation, loss of bristles, and emaciation. In hens, the failure of egg hatchability resulted from deformities that prevented hatching, while legs, toes, wings, beaks or eyes were also malformed (McDowell 1992).

3.2.2.9
Nickel

Nickel deficiency reduces growth rate in goats, pigs, and rats. The level of nickel associated with growth depression was significant in the second and following generations. Nickel-deficient goats and their offspring with < 100 μg Ni kg^{-1} feed DM had a higher absorption rate and a decreased viability, a lowered milk production, skin and skeletal lesions, and lower testicle weights.

Baby pigs and kids from the nickel-deficient groups developed a scaly crusty skin similar to that seen in parakeratosis, and lower hematocrit and hemoglobin levels. Nickel deficiency also induces a decreased urease activity in the rumen (Hennig et al. 1978, Anke et al. 1980a, Spears and Hatfield 1978). Urease was the first natural nickel metalloenzyme discovered by Fishbein et al. (1976). Urease is a component of several leguminous plants (jack bean), and is synthesized by the rumen bacteria. Urease catalyzes the reaction:

$$(NH_2)_2\, CO + H_2O \ \rightarrow \ CO_2 + 2NH_3.$$

Binding of the substrate urea to a nickel ion in urease is an integral part of the mechanism in the hydrolysis reaction (Nielsen 1984). Both ruminants and monogastric animals require urease for the decomposition of urea into ammonia, which is needed for the microbial synthesis of ammonia that, in turn, is necessary for amino acid and protein synthesis. This process also takes place in the appendix of monogastric animals and some species of ruminants (roe deer).

The normative nickel requirement of ruminants amounts to 100–350 μg kg^{-1} feed DM, while that of humans amounts to 35–50 μg per day (Anke et al. 1973a, 1974, Anke 1985). The nickel requirement of animals and man is met by all natural diets. Thus, secondary nickel deficiency is to be expected probably because some rumen bacteria use nickel as a part of their enzyme urease. There is clear evidence that nickel is essential for the fauna.

Nickel toxicity is mainly the product of the interactions of this element with zinc, magnesium, and manganese. Approximately 250 mg Ni kg^{-1} feed DM is required to produce a significantly lower feed intake, growth rate and egg production (Anke et al. 1984a, b). On the other hand, a dose of

125 mg Ni kg^{-1} feed DM lowers the zinc, magnesium (and manganese) levels of the indicator organs of hens, broilers and pigs, and also includes a secondary deficiency of these elements that leads to a significant reduction in the hatchability of chickens and their subsequent growth rate and viability (Anke et al. 1997c, 1997d). In humans, nickel is stored in the ribs, with significantly higher levels in women than in men. This fact is of particular importance in connection with nickel allergy, which typically is a disease of women. The nickel-storing capacity of the skeleton is apparently limited and, when all nickel depots have been filled, symptoms of nickel allergy are induced after skin contact with nickel-containing objects of everyday use. The limit value of nickel intake is 600 µg Ni per day (Cronin et al. 1980, Anonymous 1996a, Nielsen et al. 1999, Anke et al. 2000d, 2003a).

3.2.3
Essentiality and Toxicity of Ultratrace Elements (see Table 3.4)

3.2.3.1
Fluorine
Erhard (1874) detected the densifying effect of fluorine on the dental enamel in dogs, and recommended that pregnant women and children during their second dentition should take fluorine pastilles. During the late 1930 s, the correlation between a low degree of caries and the occurrence of dental fluorosis (mottle enamel) was reported (Dean 1938). The essentiality of fluorine has been repeatedly investigated. Schroeder et al. (1968) reported a slightly decreased growth and a reduced life expectancy of mice fed a fluoride-poor diet; Schwarz and Milne (1972) also found a limited influence of fluoride deficiency on the growth of rats. These results were confirmed neither by Messer et al. (1972,

Tab. 3.4: Chronological observations providing evidence of essentiality and toxicity of ultratrace elements

Element	Laboratory reference	Observation	Toxicology, symptoms
F	Erhard 1874, Dean 1938, Schroeder et al. 1968, Schwarz and Milne 1972	Dental caries in dogs, children, mice and rats	Chronic fluorosis is endemic worldwide in farm animals. Major clinical signs of fluorosis found in teeth and bones
Cr	Mertz and Schwarz 1995	Cr (III) is a factor involved in the maintenance of glucose tolerance	Less than three-valent Cr, six-valent Cr induces health injuries. Six-valent Cr taken in via food is quickly reduced to three-valent Cr
Si	Carlisle 1972, Schwarz and Milne 1972b	Retarded growth in chicken and rats	Oral silicon is non-toxic for monogastric species; ruminants consuming high silicon plants may develop silicon renal calculi
As	Nielsen et al. 1975, Anke et al. 1976	Low As intake decreased fertility, birth weight and survival of goats minipigs and rats	Chronic As toxicosis in cattle developed hair coat changes, weight loss, inflamed eyes, diarrhea, incoordination of gait

Tab. 3.4: (Continued)

Element	Laboratory reference	Observation	Toxicology, symptoms
Cd	Anke 1977, Schwarz and Spallholz 1977, Anke et al. 1977a	$<20~\mu g$ Cd kg^{-1} feed DM retarded growth and repro-duction in goats; 4 μg Cd kg^{-1} feed DM lowered growth in rats	Cd has an antagonistic activity to the metabolism of Cu, Zn and Fe. Reduced growth, infertility, abortions (Anke et al. 1970, 1975)
Li	Anke et al. 1981	Low lithium, decreased fertility, pre-and postnatal growth, and increased mortality	In pigs, chicken, cattle, sheep and rats, Li supplements of >100 mg kg^{-1} ration DM reduced feed consumption and induced thirst
B	Hunt and Nielsen 1981, 1983, Nielsen 1984	B stimulated growth and prevented leg abnormalities in chicken and rats	>100 mg B kg^{-1} feed is toxic to animals. 150 mg B L^{-1} in drinking water of cows caused inflammation and edema in the legs and around the claws
Pb	Kirchgessner and Reichl-mayr-Lais 1981a	Low levels of Pb (20 μg kg^{-1} feed DM) produced anemia and decreased growth in second generation	Early symptoms of intoxication include anorexia, fatigue, nervousness, tremor, colli, tremor. Clinical signs: anemia, encephalopathy, renal dysfunction.
V	Anke et al. 1983a, 1984c, Nielsen et al. 1983	<10 V kg^{-1} feed DM decreased feed intake, repro-duction, milk production and increased mortality; in rats, hematocrit values changed	Dietary concentrations of 25 mg V kg^{-1} in rats and up to 50 mg V kg^{-1} in other animals depressed growth and increased mortality
W	M. Anke et al. 1983, 1983c, Groppel et al. 1985	Synthetic rations of 60 μg W kg^{-1} diet DM did not influence growth and reproduction, but increased mor-tality	15 g Na$_2$ WO$_4 \cdot$2H$_2$O per day decreased the hemoglobin levels of cows significantly (Graupl et al. 1965)
Br	Anke et al. 1988a	<800 μg Br kg^{-1} diet DM decreased growth reproduc-tion, hemoglobin level and life time of kids	Cattle, goats and horses with 5 g Br kg^{-1} feed DM developed ataxia and muscle weakness
Al	Anke et al 1990a, Carlisle and Curran 1993, Müller et al. 1995a,b	Al intake of 2.5–6.5 mg kg^{-1} diet DM led to significantly increased mortality and impaired reproduction of chicken	High levels of Al induced P-deficiency signs and decreased the Ca, Mg, Fe, Zn and Fe of several tissues of wild and domestic animals
Rb	Anke et al. 1993b, Angelow and Anke 1994	$>80\%$ of rubidium deficiency goats aborted their fetuses	1000 mg Rb kg^{-1} diet DM depressed growth, reproduction and survival time in rats, LD$_{50}$ in rats was 900 mg kg^{-1} for rubi-dium hydroxide
Ti	Anke 2000	Synthetic ration with 170 μg Ti kg^{-1} DM increased mor-tality of the offspring	Toxicity of TiO$_2$ is very low for animals

1973) nor by Tao and Suttie (1976). Fluoride-deficiency experiments with growing pregnant and lactating goats, though repeated twice, also failed to show any significant influence of fluoride-poor nutrition using a semisynthetic ration that contained all other necessary components (Anke and Groppel 1989) on growth and reproduction performance. In contrast, life expectancy and milk performance declined very significantly, particularly after intrauterine fluorine depletion (Anke 1991, Anke et al. 1991d, 1995). Intrauterine fluorine depletion over a total of 10 generations of goats with 2000 µg F kg^{-1} feed DM in controls and 300 µg F kg^{-1} feed DM in fluorine-deficient goats highlighted the essentiality of fluorine in well-defined form. Goats fed a fluorine-deficient semisynthetic diet consumed significantly more feed, suffered from significant intrauterine and post-natal growth retardation, had a significantly higher kid mortality rate, developed skeletal and joint deformities in old animals (Figure 3.3), possessed a significantly higher phosphorus content in the blood plasma, and had a reduced calcium content and alkaline phosphatase activity (Anke et al. 1995, 1997d). An intrauterine fluorine-depleted kid suffered from thymus hypoplasia and hemosiderosis of the liver (Avtsyn et al. 1992). The normative fluorine requirement of animals (goats) is < 650 µg kg^{-1} feed DM, while that of humans is 250 µg per day. Animals and man store much fluorine in the skeleton, and this can satisfy the fluorine requirement over a long period of time. The fluorine content of foodstuffs in Central Europe considerably exceeds the assumed fluorine requirement of animals (and humans). Therefore, fluorine supplementation of the mineral mixtures of farm animals does not seem necessary (Anke et al. 2001a).

The main problem of fluorine in the food chain of animals is its toxicity. Fluorine is

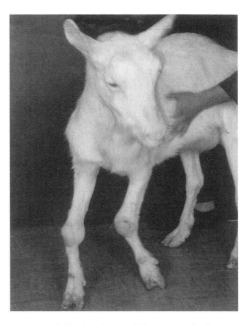

Fig. 3.3 Skeletal and joint deformities of a fluorine-deficient goat.

not equally toxic to all species of animals. Those animals which exhibit the greatest tolerance to fluorosis include poultry, followed by swine (with a short lifetime), horses, sheep, and cattle. Fluorine is a cumulative poison, and when the bone tissue has become saturated then continued intakes are deposited in the soft tissues, with resultant metabolic disturbances and death. Sheep being raised for lamb or wool production can tolerate 60 mg F kg^{-1} in their diet, while finishing lambs can tolerate up to 150 mg kg^{-1}. For cattle, a level of 20–30 mg kg^{-1} F in their feed DM will cause dental mottling; about 50 mg kg^{-1} will cause lameness and decreased milk production. Cattle frequently have a decreased feed intake if dietary fluorine exceeds > 50 mg kg^{-1} DM (Krichnamachari 1987, McDowell 1992).

The impressive damage of the front legs and their joints of older fluorine-deficient goats, together with the significant bio-chemical increase in inorganic phosphorus content and decrease in calcium and alkaline phosphatase activity in the plasma of goats with $< 30 \, \mu g \, kg^{-1}$ feed DM, clearly demonstrates the essentiality of this element (Gürtler et al. 1995, Anke et al. 1991d).

3.2.3.2
Chromium

Although chromium has not yet been finally detected as a component or activator of proteins, enzymes and/or hormones, it belongs to the essential ultratrace elements which must be available to the fauna and humans in minimal amounts.

It has not yet been possible to verify the structure of an insulin-intensifying native and synthetic Cr complex. It was, however, possible to trigger deficiency symptoms in animals and humans with poor rations. These deficiency symptoms of chromium include insufficient glucose tolerance (man, rat, mouse, guinea pig), increased circulating insulin amounts (man, rat) retarded growth (rat, mouse, turkey), reduced life-expectancy (rat, mouse), increased frequency of occurrence of plaques in the blood vessels (rabbit, rat mouse), increased cholesterol and triglyceride serum levels (man, rat, mouse), neuropathies (man), encephalopathies (man), corneal damage (rat, squirrel) as well as reduced reproduction performances and smaller numbers of spermatozoa (rat). The normative chromium requirement of animals is unknown. The individual basic requirement of adults is indicated as $< 20 \, \mu g$ per day by the World Health Organization (WHO). The WHO recommends that the intake of $33 \, \mu g$ Cr per day be averaged over a week. The minimum chromium intake of a population should amount to $25 \, \mu g$ per day (Anonymous 1996a, Anderson 1987, 1998, Offen-

bacher et al. 1997, Barcelaux 1999, Lukaski 1999).

Chromium deficiency is of no major concern in Central Europe in either man or animals. Without exception, the chromium intake in humans of both sexes was $> 25 \, \mu g$ per day, when averaged over a week, with the individual basic requirement indicated as $< 20 \, \mu g$ per day. The chromium intake of German adults is below the WHO limit value of $250 \, \mu g$ per day, but is astonishingly high – at least for chromium-sensitive men (and women) (Anke et al. 1997b, 2000a, 2000b).

In animals, chronic Cr toxicosis results in skin-contact dermatitis, irritation of the respiratory passages, ulceration and perforation of the nasal septum, and lung cancer. Acute systemic Cr intoxication is rare, but was produced with a single oral dose of $700 \, mg \, kg^{-1}$ body weight Cr(VI) in mature cattle, and with $30–40 \, mg \, kg^{-1}$ body weight Cr(VI) in young calves. Signs of acute toxicosis included inflammation and congestion of the rumen and abomasum (Anonymous 1980).

3.2.3.3
Silicon

Silicon (Si) is essential for growth and skeletal development in rats and chicks, and a requirement of sodium silicate is suggested in the range of $100 \, mg \, kg^{-1}$ diet DM or $26–52 \, mg$ per 1000 kcal of an experimental diet. Long-bone joints of deficient chicks were smaller and had reduced strength, while the bones themselves showed an altered chemical composition, with tibia from silicon-deficient chicks containing significantly less glycosaminoglycans and collagen (Carlisle and Alpenfels 1984). Silicon is required for maximum propyl hydroxylase activity, matrix glycoproteins, which together are a measure of collagen biosynthetic rate (Carlisle 1980a, b, Carlisle et al.

1981, Nielsen 2002). In rats, supplemental silicon (250 mg kg^{-1} by DM) had no effect on skeletal development. Egg production in chicken receiving a basal diet containing from 0.6 to 143 mg Si kg^{-1} DM was reduced insignificantly. Supplementation of the hens' feed reduced egg production insignificantly (Vogt 1992). Normally, urinary silicon is readily excreted but, under some conditions in grazing steers and sheep, part of the urinary silicon is deposited in the kidneys, bladder, or urethra to form calculi (McDowell 1992).

Amorphous silicates are considered safe additions to foods, and therefore their use as anticaking agents, for example, is permitted in amounts up to 2% by weight. Water-soluble silicates are also of low toxicity; studies of the effects of feeding various silicon compounds to laboratory animals have generally shown the substances to be innocuous under the test conditions. Likewise, the available data on orally administered silicates in humans substantiate the biological inertness of these compounds (Carlisle 1997).

3.2.3.4
Arsenic

The first suggestions of arsenic's essentiality were made in 1975 and 1976 by two laboratories, each of which was unaware of the other's investigations (Nielsen and Uthus 1980). The arsenic-deficiency experiments in growing, pregnant and lactating goats were started in 1973 and repeated over 12 generations. Likewise, the arsenic-deficiency trials with growing, pregnant and lactating miniature pigs and offspring of both species after intrauterine development were repeated twice (Anke et al 1976, 1985d, 1991c). The signs of arsenic deficiency in minipigs and goats were reviewed by Anke et al. (1977, 1980, 1987a), and those for chicks and rats by Uthus et al.

(1983). The semisynthetic ration of the goats contained 35 µg As kg^{-1} DM, while the control animals received 350 µg As kg^{-1} DM. All animals completed the trial for their natural life-span. In the first year of life, control and arsenic-deficient goats consumed similar amounts of feed (678 and 680 g/day). On average, the surviving adult arsenic-deficient goats ate 7% more semisynthetic ration than did controls (629 and 674 g/day, respectively) (Anke et al. 2001). On average, the arsenic-poor diet reduced intrauterine growth by 6% or 182 g in 133 and 100 kids, respectively (Anke et al. 1998, 1996). Similar effects of arsenic deficiency were demonstrated in minipigs (Anke et al. 1976). During the suckling period, arsenic-deficient kids gained less weight than control kids. Intrauterine arsenic-depleted kids grew more slowly than did kids without intrauterine arsenic depletion and control kids (9 and 17%, respectively) (Anke et al. 1977, 1985e). Arsenic deficiency reduced significantly the success of the first service and the conception rate of the arsenic-deficient goats (11 and 29% respectively of the control and arsenic-deficient goats remained barren). Arsenic deficiency increased the abortion rate and reduced milk production by 20%. The mortality of kids was 6 and 32%, respectively. None of the pregnant, arsenic-deficient goats survived the second lactation.

Barren arsenic-deficient goats achieved an age of 6 years. Death regularly occurred between the 17th and 35th days of lactation; typically, animals suffered spasms and died shortly thereafter.

Arsenic deficiency also led to a significantly reduced ash content in the goat skeleton (Anke et al. 1976), while Uthus and Nielsen (1983) reported damage to the legs of chicks. A systematic investigation of the skeleton and cardiac muscles and liver of arsenic-deficient goats shortly before death showed ultrastructural changes (Schmidt et al. 1983, 1984), with electron-dense material deposited in the mitochondrial membrane of skeletal muscle, cardiac muscle and liver. At an advanced stage, this electron-dense substrate is released from the mitochondrial membrane and is detectable in cytoplasm (Figure 3.4). Schmidt et al. (1984) suggested that this material was insoluble calcium phosphate, and proposed that this change was a form of mitochondrial myopathy. It is possible that cardiomyopathy, in association with a derangement of the cardiac mitochondrial structure, may be caused by arsenic deficiency, though the fundamental mode and site of action of the element are yet to be identified (Anonymous, WHO, 1996).

Arsenic poisoning is commonly an acute clinical syndrome, and death usually occurs rapidly. Clinical signs of acute arsenic toxicity include colicky pain, vomiting, diarrhea, marked depression, and dermatitis usually due to increased capillary permeability and cellular necrosis.

The essential and toxic effects of arsenic are very important and similar to that of selenium as recorded in the past. Hence, in future it may be necessary to examine both faces (essential and toxic effects of arsenic) of this element (Frost 1983, McDowell 1992, Selby et al. 1977, Anke et al. 1990b).

3.2.3.5
Cadmium

Cadmium is both toxic and essential. In 1984, Smith noted that "Although these reports are not sufficient to establish definitely a specific function for cadmium, this metal is a good candidate for essentiality. The history of the trace metal should serve as a warning against taking too pessimistic a view on the possible essentiality of cadmium, because a number of other elements (notably selenium) proved to be essential." More evidence for the essentiality of cadmium is now available after the ten-fold repetition of the cadmium-deficiency experiments with growing, pregnant and lactating goats and their kids, and the discovery of a cadmium-specific carboanhydrase in the diatom *Thallasia sira weissflogii* (Strasdeit 2001).

The essentiality of cadmium was investigated systematically in control goats with 300 µg Cd kg^{-1} ration DM, and in corresponding Cd-deficient animals with <15 µg kg^{-1} ration DM. The cadmium-poor nutrition did not have any effect on feed intake (629 and 644 g/day, respectively). Live weight gain (18.2 and 16.8 kg by day 91 of life, respectively) was not affected by Cd-poor nutrition, whereas Cd deficiency had a significant effect on first insemination, rate of abortion and number of services. The Cd-poor nutrition of mothers affected the activity of the kids. The intra-uterine Cd-depleted kids were often very

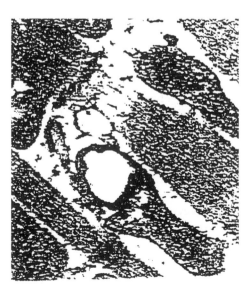

Fig. 3.4 Electron-dense material in the mitochondrial cytoplasm (cardiac muscle) following arsenic intoxication. (Original magnification, ×43 650.)

Fig. 3.5   Left: a cadmium-deficient kid. Right: a control kid.

phlegmatic, moved very little, were too lazy to eat and drink, and had problems holding their head erect (Figure 3.5).

The symptoms of weakness of mobility occurred at different times of the lactation period. Among nine kids (6 males, 3 females) from Cd-deficient goats, all showed clinical deficiency symptoms in the form of muscular weakness at about 6 weeks after weaning. Subsequently, six of the Cd-deficient kids died. The three survivors were then given the ration of control animals, with 300 µg Cd kg^{-1} DM matter, and slowly regained their mobility and achieved their normal body weight. Myasthenia also occurred due to nutrition depletion in lactating goats which, without Cd supplementation, led to death. Feeding cadmium at 65 µg kg^{-1} ration DM prevented the occurrence of cadmium-deficiency symptoms (Anke et al. 1984c, 1986b, 1987a).

The liver, muscles, heart, kidneys and cerebrum of Cd-deficient goats were examined ultrastructurally. Primarily, the mitochondria appeared to be damaged, and their size increased (Figure 3.6), with degenerative changes especially in mitochondria of the liver and kidneys. Christolysis and enlargements were also demonstrated. The mitochondria were only detectable because

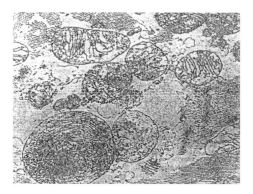

Fig. 3.6   Magnified mitochondria of the cardiac muscle of a 6-month-old goat. (Original magnification, ×20 400.)

of their double membrane and isolated detectable christae – changes which could not be found in other organs. Within these organs the considerable increase in regularly arranged christae was impressive. There were also striking reductions in the contractile apparatus of the cardiac and skeletal muscles. Such findings point to reduced protein synthesis or, in lactating goats, to increased protein mobilization from muscles. This hypothesis might explain the reduction of the contractile system, but not the increased mitochondrial size. Such hyperplasia was found when there was an insufficiency of mitochondria, and might represent compensatory hypertrophy (Anke et al. 1986b, 2000b).

The normative cadmium requirement of goats and animals amounts to 20 µg kg^{-1} DM. Primary cadmium deficiency is not to be expected in animals and humans, as the normal intake is considerably above this range (Kronemann et al. 1982, Anke et al. 1994a).

Ingested or inhaled cadmium is toxic to virtually every system in the animal body. Dietary concentrations of 5 mg kg^{-1} feed DM are always associated with adverse health, but levels as low as 1 mg kg^{-1} feed DM have undesirable effects. A dietary cadmium concentration of 0.5 mg kg^{-1} feed DM is the maximum tolerable level suggested for domestic animals (Anonymous 1980).

Cadmium acts as an antagonist of zinc, copper, and iron. Although cadmium is not transferred to the next generation (not even via milk), it causes a drastic reduction of the copper reserves in the fetus and the copper concentration of the milk (Anke et al. 1970, 1988).

3.2.3.6
Lithium

The essentiality of lithium in goats has been investigated by two research teams in Germany and Hungary, with 15 repetitive experiments over 15 years (Anke et al. 1981, 1983b, 1991a, Arnhold 1989, Szentmihalyi et al. 1985, Szilágyi 1985, 1989). Studies with rats have also been conducted in the United States (Patt et al. 1978, Burt et al. 1982, Pickett et al. 1983, Pickett and Hawkins 1987) and Japan (Ono et al. 1992).

Bach 1990 summarized the results with the comment: "There are animal studies which support lithium's role as a sine qua non for physical health in the case of goats and rats." The kids of lithium-deficient dams (< 1.7 mg Li kg^{-1} and $10-20$ mg Li kg^{-1} DM) had a 9% lower birth weight than those of controls, this difference amounting to 15% by the end of the suckling period of 91 days. The lithium intake of kids is essentially determined by the lithium content of the milk, and Li-deficient kids received only one-third as much lithium. During the following 168 experimental days, the effect of lithium on growth was insignificant (Anke et al. 1991).

The consumption of lithium-poor rations by female goats had no effect on the intensity of estrus behavior. However, the first mating resulted in a significantly lower rate of conception in these animals. The conception rate with repeated services at the following ovulations was improved, but the difference between the groups remained significant. There was also a higher abortion rate among the Li-deficient goats. The effect of lithium deficiency on the sex ratio was most surprising; Li-deficient goats gave birth to significantly more female kids.

Long-term lithium-deficiency experiments with female goats allowed an analysis of the influence of the lithium-poor nutrition on life expectancy. These data showed that 41% of Li-deficient goats and 7% of control animals died during the 2-year experiment. The oldest Li-deficient goats suffered from a disturbed hematopoiesis. Gallicchio et al. (1991) have described the influence of high lithium intake on hematopoiesis.

Lithium deficiency did not affect the biochemical blood profile, but did lower serum lithium levels and the activity of several serum enzymes, mainly those concerned with the citrate cycle [iso-citrate dehydrogenase (ICDH), malate dehydrogenase (MDH), with glycolysis (ALD), and with nitrogen metabolism glutamate dehydrogenase (GLDH). There were significant differences in enzyme activities between control and Li-deficient goats. Owing to the particular role of monoamine oxidase (MAO) in manic-depressive disease, chronic schizophrenia, and unipolar depression, this enzyme was also investigated in the liver of control and Li-deficient goats. MAO activity in the hepatic tissue of the latter group was reduced by 28%; this fall was in good agreement with lithium-poor rations, and disappeared after lithium supplementation. The biochemical effect of lithium on the behavior of rats requires further clarification.

The normative lithium requirement of goats and pigs amounts to < 2.5 mg kg^{-1} DM, while those of adult humans might amount to 200 µg per day (Arnhold 1989, Anke 1991). According to the available data (Regius et al. 1983, Tölgyesi 1983, Lambert 1983, Szentmihalyi et al. 1983, Mertz 1986, Schäfer 1997, Anke et al. 2003), the lithium content of foodstuffs and beverages meets the assumed lithium requirement of the European fauna, including humans.

The effects of lithium intake have been systemically investigated in poultry, cattle, pigs, sheep, and rats (Regius et al. 1983a, Anke et al. 1985d, 1986, Opitz and Schäfer 1976).

Broilers and hens with a supplement of 100 mg Li kg^{-1} ration DM gained 13% less weight than control broilers, while hens laid 14% fewer eggs that were lighter than those of control animals (Anke et al. 1986). Fattening bulls fed 100 mg Li kg^{-1} ration DM gained 18% less weight than corresponding control bulls. The aggression, sexual activity and growth of bulls were significantly decreased and adipose deposition increased by a high lithium intake. In pigs, a supplementation of 500 mg Li kg^{-1} feed DM led to a drastic decrease in feed intake, daily weight gain, and to an enormous water consumption. All pigs fed 1000 mg Li kg^{-1} ration DM died within 92 days (Anke et al. 1984). These findings are important in the context of lithium therapy for

patients with manic depressions insofar as they illustrate the need to minimize the lithium dose (Schäfer 1998).

3.2.3.7
Boron

Boron is essential for higher plants. The initial boron-deficiency experiments in rats were unsuccessful (Orent-Keiles 1941), while boron supplementation in chicks tended to abate signs of vitamin-D deficiency, such as depressed growth and increased plasma alkaline phosphatase (Hunt and Nielsen 1981). Hunt et al. (1983) indicated a relationship between boron, calcium, magnesium and vitamin D$_3$. Signs of boron deficiency may be related to the level of vitamin D, magnesium and possible other nutrients in the diet. Boron deprivation (0.3–0.4 mg boron kg^{-1} diet DM depressed growth, hematocrit, hemoglobin and kidney weight:body weight ratio in rats (Nielsen 1984a). Chicks require 1 mg kg^{-1} dietary boron (Hunt 1989).

Boron apparently has an essential function that somehow regulates parathormone action, and therefore indirectly influences the metabolism of Ca, P, Mg, and cholecalciferol. Boron is needed by the parathyroid and has been shown to prevent loss of Ca and bone demineralization in postmenopausal women (Nielsen et al. 1988).

Boron is similar to the omega-3 fatty acids in that they affect – generally in a beneficial fashion – blood, brain, eye, immune system, and skeletal function, though a specific biochemical function has not been clearly defined. Boron and omega-3 fatty acids most likely interact at the cell membrane level to affect a variety of life processes (Nielsen 2002a).

When boron was administered to rats at 150 mg L^{-1} in drinking water, the animals showed depressed growth, lack of incisor pigmentation, aspermia, and impaired ovarian development (Green et al. 1973). Boron given to rats at 300 mg L^{-1} in drinking water led to depressed triglycerides, protein and alkaline phosphatase, and depressed bone fat (Seal and Weeth 1980). A boron dose level of 8 mg kg^{-1} body weight and day caused osteoporosis associated with a reduction in parathyroid activity (Franke et al. 1985).

3.2.3.8
Lead

Signs of lead deficiency (60 ng Pb kg^{-1} and 1000 ng Pb kg^{-1} food DM, respectively) were seen in the F$_1$ generation from lead-depleted mothers. Besides growth depression, a high mortality of offspring, loss of hair and eczema, microcytic hypochromic anemia disturbances in iron metabolism, lipid metabolism and changes in enzyme activities were observed. Na, K-ATPase, Mg, Ca-ATPase in cell membranes of intrauterine lead-depleted offspring were reduced; the calcium, sodium, potassium, iron, copper, zinc and manganese contents of lead-depleted mothers were also decreased (Kirchgessner and Reichlmayr-Lais 1981a, b, c, 1982, 1986, Reichlmayr-Lais and Kirchgessner 1981a, b, c, d, e, 1986a, b, c, Eder et al. 1990).

Reduced growth rate and disturbances in lipid metabolism were observed in piglets separated from their mother immediately after birth and fed a synthetic lead-poor diet (Kirchgessner et al. 1991, Plass et al. 1991). Lead deficiency in rats and pigs can be prevented or abolished by lead supplementation (Reichlmayr-Lais and Kirchgessner 1997). The practical problem of lead is not a deficiency of this ultratrace element, but rather its toxicology (Grün et al. 1982).

3.2.3.9
Vanadium

The most substantive evidence for vanadium essentiality was provided in the 1980 s from a series of deficiency experiments with goats and rats. In 14 experiments with intrauterine V-depleted goats and $< 10\ \mu g$ V kg^{-1} (range: 1–9 μg V kg^{-1}) in the diet DM of V-deficient goats, these animals ate 20% less feed during lactation than did controls; their pre- and postnatal growth was uninfluenced, the success of first mating and conception rate of she goats with V-poor nutrition were significantly lowered, their abortion rate increased (p < 0.001), and their mortality was 24% compared with 5% for controls. V-deficient goats suffered pain in the extremities, developed swollen forefoot tarsal joints, glandular hyperplasia of the endometrium, and increased size of pancreas, thymus and thyroid compared with controls. V-deficient nutrition of rats also induced increased thyroid weights and thyroid:body weight ratios. V-deficient nanny goats had only 50% of the lifetime of control goats. The normative requirement of vanadium for animals amounts to $> 10\ \mu g$ V kg^{-1} diet DM (Anke 1991, Anke et al. 1983a, 1984c, 1985c, 1986a, 1988b, 1989, 1991b, 2000, 2000a; Avtsyn et al. 1993, Nielsen 1991, 1997, Nielsen et al. 1983, Uthus and Nielsen 1990).

3.2.3.10
Tungsten

Experiments with a tungsten-poor semisynthetic ration (60 μg kg^{-1} ration DM) were repeated six times and over five generations in growing, gravid, and lactating goats, but did not show any tungsten-related effect on growth (even after intrauterine depletion), reproduction, and mortality. Only the life expectancy of adult goats and blood reticulocyte content were significantly changed due to tungsten-poor nutrition.

The normative tunsten-requirement of goats is apparently $< 60\ \mu g$ W kg^{-1} diet DM. Tungsten forms part of the formate dehydrogenase of *Clostridium thermoaceticum* and *C. formicoaceticum* (Ljungdahl and Andreesen 1975). An essentiality of tungsten, especially for ruminants, is possible (Anke et al. 1983, 1983c, 1985b, Groppel et al. 1985).

3.2.3.11
Bromine

Bromine-deficiency experiments with a total of 32 control (20 mg Br kg^{-1} diet DM) and 30 Br-deficient growing, pregnant and lactating goats (800 μg Br kg^{-1} diet DM) were commenced in 1980 and completed in 1993. The Br-deficiency experiments were repeated six times and, if possible, were continued with intrauterine Br-depleted kids (Anke et al. 1988a, 1994). Bromine-deficient adult goats consumed 9% less feed (601 g/ day) than controls (695 g/day) (p < 0.001). Bromine deficiency did have any significant effect on pre- and post-natal growth of kids. After weaning (100th to 268th day of life), the growth of intrauterine Br-depleted kids was reduced by 15% in females, and by 49% in males. Kids whose mothers had been bought (not Br-depleted) showed normal weight gains (Anke et al. 1993c). Milk is relatively rich in bromine and delivered more bromine to the suckling kid than the semisynthetic Br-deficient diet. The growth rate and bromine requirement of male kids were higher than those of females, thereby accounting for the major differences between he- and she-goats. Br-deficiency decreased the success of first insemination and conception rate, and increased the abortion rate significantly. Br-deficient goats produced 7% less milk than controls, but the protein content was almost identical. Surprisingly, the milk of Br-deficient goats contained significantly

more fat than that of controls (Anke et al. 1993c). The mortality of Br-deficient goats during the first and second year of life (14% and 38%, respectively; p < 0.005) as well as that of kids was significantly reduced (3% and 29%, p < 0.001) (Anke et al. 1989a, 1993c, 1994a).

As reported previously, Br-poor nutrition led to a significantly reduced hemoglobin level rapidly after bromine depletion, accompanied by an increasingly lower hematocrit value, whereas the mean corpuscular hemoglobin concentration of control and Br-deficient goats remained unchanged. Bromine deficiency apparently reduces the cell content of blood (Anke et al. 1993c, 2001a, Gürtler and Anke 1993). Bromine seems directly or indirectly necessary for hemoglobin synthesis. The Br-deficient goats passed significantly higher triglyceride levels and gamma-glutamyl transferase (GGT) activities to the blood plasma. The abnormal metabolism of lipids is a symptom of hypobromosis, which manifests itself in fatty hepatosis, increased accumulation of fatty tissue in the mediastinum, and abdominal activity as well as lipomatosis of somatic muscles (Zhavaronkov et al. 1996).

Feed and foodstuffs in Europe contain bromine concentrations which exceed the normative requirement of animals and man. The normative requirement of goats was calculated to be > 1000 to 1500 $\mu g\ kg^{-1}$ feed DM. The bromine requirement of animals and man is met by feed, foodstuff, and water (Anke et al. 2001); hence, bromine-deficiency experiments with rats, mice and chicks were generally unsuccessful (Winnek and Smith 1937, Huff et al. 1956, Bosshardt et al. 1956).

3.2.3.12
Aluminum

Between 1986 and 1994, nine generations of growing, pregnant and lactating goats receiving 2.5–6.5 mg Al kg^{-1} feed DM matter in the Al-deficient groups and 38 mg Al kg^{-1} feed DM in the control groups, were tested for deficiency symptoms. Aluminum deficiency led to significantly increased mortality, impaired success of first insemination, increased abortions and services per gravidity, mortality of the kids, and mortality in the first year of life (0% and 35%, respectively). In the fourth generation, however, a distinct weakness of the hind legs was registered (Figure 3.7) in Al-deficient kids which could walk forward without difficulty but twisted when trying to turn. The difficulties experienced by these animals in coordinating movements was most obvious when they tried to get up. It cannot be excluded that hind-leg weakness in Al-deficient kids is the cause of higher mortality during the first year of life. An analysis of 20 blood plasma components and the activity of selected enzymes showed that, with the exception of a significantly higher urea content, the Al-deficient goats did not differ from controls. The Al-poor nutrition led to significant variations in the aluminum content of the aorta, spleen, ribs, and carpal bones. Feed intake

Fig. 3.7 An aluminum-deficient goat.

and growth rate of the kids, both pre- and postnatal, was not influenced by Al-poor nutrition (Müller et al. 1995a, b, Anke et al. 1990a, 2001a, Angelow et al. 1993). This effect of Al-poor nutrition was also evident in chickens (Carlisle and Curran 1993).

3.2.3.13
Rubidium
Rubidium-deficiency experiments were carried out in seven-fold repetition between 1990 and 1997 with female goats and their intrauterine Rb-depleted offspring. The semisynthetic rations of the Rb-deficient goats contained < 250 µg Rb kg^{-1} feed DM, while control goats received 10 mg Rb kg^{-1} feed DM. Following exhaustion of the body's rubidium stores, Rb-poor nutrition had a considerable effect on feed consumption (−16%). The birth weight of Rb-deficient kids was 14% lower than that of controls. The growth rate decreased significantly only after intrauterine Rb depletion. The conception rate of Rb-deficient goats was significantly less than that of control. The most important finding was the extremely high abortion rate of Rb-deficient goats, wherein goats with abortions had a progestin level which was only 7% of normal values. The plasma estradiol level in goats aborting ranged from 37 to 280 mmol L^{-1}. Rats receiving 540 µg Rb kg^{-1} feed DM had a decreased rubidium content in all tissues tested (as did goats) when compared with controls (Yokai et al. 1994, 1997). Rubidium is probably an essential element that is available in sufficient amounts within the food chain to prevent deficiency from occurring in animals (and man). The normative requirement of goats and animals might reach 300–400 µg kg^{-1} feed DM (Anke et al. 1993b, 1997c, Anke and Angelow 1995, Angelow and Anke 1994, Gürtler et al. 1999).

3.2.3.14

Titanium

The essentiality of titanium was only examined over two generations of growing, gravid and lactating goats that received 170 μg Ti kg^{-1} ration DM, and a ten-fold amount for controls. The low titanium intake did not affect feed intake, growth and reproductive performance. On the other hand, the titanium content of milk from Ti-poor fed goats was reduced in comparison to the milk of control animals to 50% (Anke 2000). Titanium-poor feeding reduced not only titanium incorporation into the fetus (Anke et al. 1996), but also titanium transfer into the milk. It is possible that the high mortality of offspring of Ti-poor fed mothers is caused by titanium depletion and the low titanium content of their milk. The essentiality of titanium requires further investigation. The natural titanium offer in the food chain of animals and man is sufficiently rich to prevent titanium deficiency in animals and man (Anke et al. 1996).

3.3

Summary

The ultratrace element requirements of animals (and men) are, in part, extremely low and are reliably met everywhere. In practice, symptoms of ultratrace element deficiency in animals and man do not occur, apart from genetic defects which prevent utilization of the ultratrace elements. Hints as to the biological essentiality of these elements were only obtained in experiments with semisynthetic rations that where extremely poor in the element(s) to be tested. These conditions led to depressed performances, deficiency diseases, and reduced life expectancy. Deficiency symptoms of ultratrace elements were not registered in real life, as the natural offers meet the requirements or, in part, exceed them considerably. The biological functions of the above-mentioned ultratrace elements as components of essential parts of the body are unknown, and require clarification (Anke et al. 1998). To date, ultratrace elements are only known as being toxic, and this clearly is not true.

References

ABUMRAD NN, SCHNEIDER AJ, STEEL D and ROGERS LS (1981) *Amino acid intolerance during prolonged total parenteral nutrition reversed by molybdate therapy.* J Clin Nutr **34**:2551–2559.

ANDERSON RA (1987) *Chromium.* In: Metz W, ed. Trace elements in human and animal nutrition. Academic Press Inc, San Diego, New York, pp. 225–244.

ANDERSON RA (1998) *Recent Advances in the Clinical and Biochemical Manifestation of Chromium Deficiency in Human and Animal Nutrition.* J Trace Elem Exp Med **11**:241–250.

ANGELOW L (1994) *Rubidium in der Nahrungskette.* Qualification for lectureship. University of Jena, Germany.

ANGELOW L and ANKE M (1994) *Rubidium in der Nahrungskette.* Mengen- und Spurenelemente **14**:285–300.

ANGELOW L, ANKE M, GROPPEL B, GLEI M and MÜLLER M (1993) *Aluminium: an essential element for goats.* In: Anke M, Meissner D and Mills CF, eds. Trace Element in Man and Animals – TEMA 8, Verlag Media Touristik, Gersdorf, Germany, pp. 699–704.

ANKE M, HENNIG A, SCHNEIDER HJ, LÜDKE H, GAGERN VON W and SCHLEGELM H (1970) *The interrelations between cadmium, zinc, copper and iron in metabolism of hens, ruminants and man.* In: Mills CF and Livingstone S, eds. Trace Element Metabolism in Animals. E & S Livingstone, Edinburgh-London, pp. 317–320.

ANKE M (1973) *Die Bedeutung der Spurenelemente für die tierischen Leistungen.* Tag-Ber, Akad Landwirtsch-Wiss, Berlin **132**:197–218.

ANKE M, GROPPEL B, REISSIG W, LÜDKE H, GRÜN M and DITTRICH G (1973) *Manganmangel beim Wiederkäuer. 3. Mitteilung. Manganmangelbedingte Fortpflanzungs-, Skelett- und Nervenstörun-*

gen bei weiblichen Wiederkäuern und ihren Nach-kommen. Arch Tierernährung **23**:197–211.

ANKE M, GRÜN M, DITTRICH G, GROPPEL B and HENNIG A (1973a) *Low nickel ration for growth and reproduction in pigs.* In: Hoekstra WG, Suttie JW, Ganter HE and Mertz W, eds. Trace Element Metabolism in Animals-2. Baltimore, University Park Press, pp. 715–718.

ANKE M, GROPPEL B, KRONEMANN H and GRÜN M (1974) *Nickel – an essential element.* In: Sunderman FW JR., ed. Nickel in the Human Environment. Oxford University Press, pp. 339–365.

ANKE M, HENNIG A, GRÜN M, GROPPEL B and LÜDKE H (1975) *Cadmium and its influence on plants, animals and man with regard to geological and industrial conditions.* In: Hemphill DD, ed. Trace Substances in Environment Health, 10. University of Missouri, Columbia, pp. 105–111.

ANKE M, GRUEN M and PARTSCHEFELD M (1976) *The essentiality of arsenic for animals.* In: Hemphill DD, ed. Trace Substances in Environmental Health. Vol 10, University of Missouri, Columbia, USA, pp. 403–408.

ANKE M (1977) *Essentiality of cadmium in goats?* In: Anke M and Schneider H-J, eds. Cadmium – Symposium, Friedrich-Schiller University, Jena, Germany.

ANKE M, GRÜN M, PARTSCHEFELD M, GROPPEL B and HENNIG A (1977) *Essentiality and function of arsenic.* In: Kirchgessner M, ed. Trace Element Metabolism in Man and Animals. Techn University of Munich, Freising-Weihenstephan **3**:248–252.

ANKE M, HENNIG A, GROPPEL B, PARTSCHEFELD M and GRÜN M (1977a) *The biochemical role of cadmium.* In: Kirchgessner M, ed. Trace Element Metabolism in Man and Animal 3. Technische University München, Freising-Weihenstephan, Germany, pp. 540–548.

ANKE M, GROPPEL B, GRÜN M, HENNIG A and MEISSNER D (1980) *The Influence of arsenic deficiency on growth, reproductiveness, life expectancy and health of goats.* In: Anke M, Schneider H and Brückner C, eds. 3. Spurenelementsymposium – Arsen, University Leipzig, University Jena, Germany, pp. 25–32.

ANKE M, KRONEMANN H, GROPPEL B, HENNIG A, MEISSNER D and SCHNEIDER HJ (1980a) *The influence of nickel-deficiency on growth, reproduction, longevity and different biochemical parameters of goats.* In: Anke M, Schneider HJ, Brückner CHR, eds. Nickel, 3rd Trace Element Symposium, University of Leipzig and Jena,

Kongress und Werbedruck Oberlungwitz, Germany, pp. 3–10.

ANKE M, GROPPEL B, GRÜN M and KRONEMANN H (1981) *The biological importance of lithium.* Mengen- und Spurenelemente **1**:217–239.

ANKE M (1982) *Anorganische Bausteine.* In: Püschner A and Simon O, eds. Grundlagen der Tierernährung. Gustav Fischer Verlag, Jena, Germany, pp. 46–67.

ANKE M, GROPPEL B, GRÜN M and KRONEMANN H (1983) *Die biologische Bedeutung des Wolframs für die Fauna. 1. Der Einfluss einer wolframarmen Ernährung auf das Wachstum.* Mengen- und Spurenelemente **3**:37–44.

ANKE M, GROPPEL B, KRONEMANN H and FÜHRER E (1983a) *Influence of vanadium deficiency on growth, reproduction and life expectancy of goats.* In: Anke M et al., eds. 4. Spurenelement-Symposium, University Leipzig and Jena, pp. 135–141.

ANKE M, GROPPEL B, KRONEMANN H and GRÜN M (1983b) *Evidence for the essentiality of lithium in goats.* In: Anke M. et al., eds. Lithium 4. Spurenelementsymposium, University Leipzig and Jena, Germany, Kongress- und Werbedruck, Oberlungwitz, Germany, pp. 58–65.

ANKE M, GROPPEL B and PARTSCHEFELD M (1983c) *Die biologische Bedeutung des Wolframs für die Faune. 2. Der Einfluss einer wolframen Ernährung auf die Fortpflanzung und Lebenserwartung.* Mengen- und Spurenelemente **3**:45–51.

ANKE M, GROPPEL B, KUHNERT E and ANGELOW L (1984) *Der Einfluß des Lithiums auf Futterverzehr, Wachstum und Verhalten von Schwein und Rind.* Mengen- und Spurenelemente **4**:537–551.

ANKE M, GROPPEL B and SIEGERT E (1984a) *Auswirkungen einer oralen Nickelbelastung. (3. Mitt.)* Mengen- und Spurenelemente **4**:437–446.

ANKE M, GROPPEL B, BRINSCHWITZ T, KRONEMANN H, RICHTER G and MEIXNER B (1984b) *Auswirkungen einer oralen Nickelbelastung. (1. Mitt.)* Mengen- und Spurenelemente **4**:419–294.

ANKE M, GROPPEL B and KOSLA T (1984c) *Die biologische Bedeutung des Vanadiums für den Wiederkäuer.* Mengen- und Spurenelemente **4**:451–467.

ANKE M, SCHMIDT A and GROPPEL B (1984d) *Die Bedeutung kleinster Kadmiummengen für das Tier.* Mengen- und Spurenelemente **4**:482–489.

ANKE M (1985) *Nickel als essentielles Spurenelement.* In: Gladtke E, ed. Spurenelemente. Georg Thieme Verlag Stuttgart, pp. 106–125.

ANKE M, GROPPEL B and GRÜN M (1985a) *Essentiality, toxicity, requirement and supply of molybde-*

num in human and animals. In: Mills CF, Bremner I and Chesters JK, eds. Trace Element in Man and Animals – TEMA 5, Commonwealth Agricultural Bureaux, Farnham, Royal Slough, UK.

ANKE M, GROPPEL B, KRONEMANN H and GRÜN M (1985b) *On the Biological Importance of Tungsten for Fauna.* In: Mills CF, Bremner I and Chesters JK, eds. Trace Element in Man and Animals TEMA 5, Commonwealth Agricultural Bureaux, Farnham Royal, UK, pp. 157–159.

ANKE M, GROPPEL B, KRONEMANN H and KOSLA T (1985c) *Vanadium deficiency in ruminants.* In: Mills CF et al., eds. Trace Element in Man and Animals TEMA 5, Commonwealth Agricultural Bureaux, London, UK, pp. 275–279.

ANKE M, RICHTER G, MEIXNER B, ARNHOLD W and ANGELOW L (1985d) *Der Einfluss des Lithiums auf den Futterverzehr, das Wachstum bzw. die Eiproduktion des Broilers und der Legehenne.* Mengen- und Spurenelemente **5**:412–419.

ANKE M, SCHMIDT A, KRONEMANN H, KRAUSE U and GRUHN K (1985e) *New data on the essentiality of arsenic.* In: Mills CF, Bremner I and Chesters JK, eds. Trace Elements in Man and Animals, TEMA 5, Commonwealth Agricultural Bureaux, Farnham, Royal Slaught, UK. **5**:151–154.

ANKE M (1986) *Arsenic.* In: Metz W, ed. Trace Elements in Human and Animal Nutrition. Academic Press Inc, Orlando, USA, Volume 2, pp. 347–372.

ANKE M, ARNHOLD W, GROPPEL B, RICHTER G, MEIXNER B and ANGELOW L (1986) *Influence of lithium on feed-intake, growth and egg production of broilers and laying hens.* In: Pais I, ed. New Results in the Research of Hardly Known trace elements. Budapest, University of Horticulture and Food Industry, pp. 41–55.

ANKE M, GROPPEL B, GRUHN K, KOSLA T and SZILÁGYI M (1986a) *New research on vanadium deficiency in ruminants.* In: Anke M et al., eds. 5. Spurenelement-Symposium New Trace Elements, University Leipzig and Jena, Germany, pp. 1266–1275.

ANKE M, GROPPEL B, SCHMIDT A and KRONEMANN H (1986b) *Cadmium deficiency in ruminants.* In: Anke M, Baumann W, Bräunlich H, Brückner CHR and Groppel B, eds. 5. Spurenelement-symposium Univ. Leipzig and Jena, pp. 937–946.

ANKE M and GROPPEL B (1987) *Toxic actions of essential trace elements (Mo, Cu, Zn, Fe, Mn)* In: Brätter P and Schramel P, eds. Trace Element – Analytic Chemistry in Medicine and Biology. Vol

4:201–236. Walter de Gruyter & Co., Berlin-New York.

ANKE M, GROPPEL B and SCHMIDT A (1987) *New results on the essentiality of cadmium in ruminants.* In: Hemphill DD, ed. Trace Substances in Environment Health–21, University of Missouri, USA, pp. 556–566.

ANKE M, KRAUSE U and GROPPEL B (1987a) *The effect of arsenic deficiency on growth, reproduction, life expectancy and disease symptoms in animals.* In: Hemphill DD, ed. Trace Substances in Environment Health – 21. University of Missouri, USA, pp. 533–550.

ANKE M, MASAOKA T, SCHMIDT A and ARNHOLD W (1988) *Antagonistic effects of a high sulphur, molybdenum and cadmium content of diets on copper metabolism and deficiency symptoms in cattle and pigs.* In: Hurley LS, et al., eds. Trace Element in Man and Animals – 6, pp. 317–318.

ANKE M, GROPPEL B, ARNHOLD W and LANGER M (1988a) *Essentiality of the trace element bromine.* In: Brätter P and Schramel P, eds. Trace Element Analytical Chemistry in Medicine and Biology, Vol. 5, Walter de Gruyter, Berlin-New York, pp. 618–626.

ANKE M, GROPPEL B, KOSLA T and GRUHN K (1988b) *Investigations on vanadium deficiency in ruminants.* In: Hurley LS, ed. Trace Elements in Man and Animal, TEMA 6, Plenum Press, New York-London, pp. 659–660.

ANKE M and GROPPEL B (1989) *Fluormangelerscheinungen bei der Ziege.* Mengen- und Spurenelemente **9**:346–364.

ANKE M, GROPPEL B, GRUHN K, LANGER M and ARNHOLD W (1989) *The essentiality of vanadium for animals.* In: Anke M et al., eds. Sixth International Trace Element Symposium, University Leipzig, Vol 1, pp. 17–27.

ANKE M, GROPPEL B and ARNHOLD W (1989a) *Further evidence for the essentiality of bromine in ruminants.* In: Anke M. Baumann W, Bräunlich H. Brückner C, Groppel B and Grün M, eds. Sixth International Trace Element Symposium Karl-Marx-University Leipzig, Friedrich-Schiller-University Jena, Germany, pp. 1120–1131.

ANKE M, GROPPEL B, ARNHOLD W, LANGER M and KRAUSE U (1990) *The influence of the ultra trace element deficiency (Mo, Ni, As, Cd, V) on growth, reproduction performance and life expectancy.* In: Tomita H, ed. Trace Elements in Clinical Medicine. Springer-Verlag, Tokyo, pp. 361–376.

ANKE M, GROPPEL B, MÜLLER M and REGIUS A (1990a) *Effects of aluminium-poor nutrition in animals.* In: Pais I, ed. 4th International Trace element Symposium, University of Horticulture and Food Industry, Budapest, Hungary.

ANKE M, REGIUS A and GROPPEL B (1990b) *Essentiality of the trace element bromine.* Acta Agron Hung **39**:297–307.

ANKE M (1991) *The essentiality of ultra trace elements for reproduction and pre- and postnatal development.* In: Chandra RJ, ed. Trace Elements in Nutrition of Children – II, Nestle' Nutrition Workshop Series, Vol 23, Nestec Ltd, Vevey Raven Press Ltd, New York pp. 119–143.

ANKE M, ARNHOLD W, GROPPEL B and KRAUSE U (1991) *The biological importance of lithium.* In: Schrauzer GH and Klippel GH, eds. Lithium in Biology and Medicine. VCH, Weinheim, pp. 149–167.

ANKE M, ARNHOLD W, GROPPEL B and KRÄUTER U (1991a) *Die biologische Bedeutung des Lithiums als Spurenelement.* Erfahrungsheilkunde **10**:656–664.

ANKE M, ARNHOLD W, GROPPEL B, KRAUSE U and LANGER M (1991b) *Significance of the essentiality of fluorine, molybdenum, vanadium, nickel, arsenic and cadmium.* Acta Agron Hung **40**(1–2):291–215.

ANKE M, GROPPEL B and KRAUSE U (1991c) *The essentiality of the toxic elements cadmium, arsenic and nickel.* In: Momcilovic B, ed. Trace Element in Man and Animals – 7, Zagreb: 11 –6–11–8.

ANKE M, GROPPEL B and KRAUSE U (1991d) *Fluorine deficiency in goats.* In: Momcilovic B, ed. Trace Element in Man and Animals – 7, IMI, Zagreb: 26 –28–26–29.

ANKE M, GROPPEL B and KRAUSE U (1991e) *The essentiality of the toxic elements aluminium and vanadium.* In: Momcilovic B, ed. Trace Elements in Man and Animals 7, IMI, Zagreb, Croatia, pp. 11-9–11-10.

ANKE M (1993) *Lithium.* In: Macrae R, Robinson RK, Sadler MJ, eds. Encyclopaedia of Food Science Food Technology and Nutrition, Vol. 4, Academic Press, Harcourt Brace Jovanowich, Publishers London, pp. 2779–2782.

ANKE M, ANGELOW L, SCHMIDT A and GÜRTLER H (1993b) *Rubidium an essential element for animal and man?* In: Anke, M, Meissner, D and Mills CF, eds. Trace Element in Man and Animals (TEMA 8), Verlag Media Touristik, Gersdorf, Germany, pp. 719–723.

ANKE M, GROPPEL B, ANGELOW L, DORN W and DRUSCH S (1993c) *Bromine: An essential element for goats.* In: Anke M, Meissner D and Mills CF, eds. Trace Element in Man and Animals

(TEMA 8), Verlag Media Touristik, Gersdorf, Germany, pp. 737–738.

ANKE M, GROPPEL B and BAUCH K (1993) *Iodine in the food chain.* In: Delange FJT, Dunn D, Glinoer B, eds. Iodine Deficiency in Europe. A Continuing Concern. Plenum Press, New York-London, pp. 151–158.

ANKE M, KRÄMER K and ILLING H (1994) *Brom: Ein essentielles Element?* Ernährungsumschau **41**:110.

ANKE M, MÜLLER M and KRONEMANN H (1994a) *Cadmium in feed- and foodstuffs.* Proc Sci Nutr Physiol **2**:9–16.

ANKE M and ANGELOW L (1995) *Rubidium in the food chain.* Fresenius J Anal Chem **352**:236–239.

ANKE M, GLEI M, MÜLLER M and ILLING H (1995) *The influence of a fluorine poor nutrition by goat.* Proc Soc Nutr Physiol **4**:67.

ANKE M, ILLING-GÜNTHER H, SCHULZ M, ARNHOLD W, FREYTAG H, ANKE S, GLEI M, JARITZ M and RUNGE A (1996) *The titanium intake of adults in Germany related, sex, living area, age and body weight.* In: Pais I, ed. Proceedings, 7. International Trace Element Symposium Budapest, Hungary, pp. 155–161.

ANKE M, SEIFERT M, ANGELOW L, THOMAS G, DROBNER C, MÜLLER M, GLEI W, FREYTAG H, ARNHOLD W, KÜHNE G, ROTTER C, KRÄUTLER U and HOLZINGER S (1996a) *The biological importance of arsenic–Toxicity, essentiality, intake of adults in Germany.* In: Pais I, ed. 7. International Trace Element Symposium Budapest, Hungary, pp. 103–125.

ANKE M, ANGELOW L, GLEI M, ANKE S, LÖSCH E and GUNSTHEIMER U (1997) *The biological essentiality of rubidium.* In: Ermidou-Pollet E, ed. International Symposium on trace Elements in Human, New Perspectives. G. Morogiannis, Acharnai, pp. 245–263.

ANKE M, ANGELOW L, GLEI M, MÜLLER M, GUNSTHEIMER U, RÖHRIG B, ROTHER C and SCHMIDT P (1997a) *Rubidium in the food chain of humans: Origins and intakes.* In: Fischer PWF, L' Abbe MR, Cockell KA and Gibson RS, eds. Trace Elements in Man and Animals–9. NRC Research Press, Ottawa, Canada, pp. 186–188.

ANKE M, DORN W, MÜLLER M, RÖHRIG B, GLEI M, GONZALES D, ARNHOLD W, ILLING-GÜNTHER H, WOLF S, HOLZINGER S and JARITZ M (1997b) *Der Chromtransfer in der Nahrungskette. 4. Mitteilung: Der Chromverzehr Erwachsener in Abhängigkeit von Zeit, Geschlecht, Alter, Körpermasse, Jahreszeit, Lebensraum, Leistung.* Mengen- und Spurenelemente **17**:912–933.

ANKE M, GÜRTLER H, ANGELOW L, GOTTSCHALK J, DROBNER C, ANKE S, ILLING-GÜNTHER H, MÜLLER M, ARNHOLD W and SCHÄFER U (1997c) *Rubidium – an essential element for animals and humans.* In: Fischer PWF, L' Abbe MR, Cockell KA and Gibson RS, eds. Trace Elements in Man and Animals, NRC Research Press, Ottawa, Canada, pp. 189–192.

ANKE M, GÜRTLER H, NEUBERT E, GLEI M, ANKE S, JARITZ M, FREYTAG H and SCHÄFER U (1997d) *Effects of fluorine-poor diets in 10 Generations of goats.* In: Fischer PWF, L' Abbe MR, Cockell KA, Gibson QS, eds. Trace Element in Man and Animals – 9, NRC Research Press, Ottawa, pp. 192–194.

ANKE M, TRÜPSCHUCH A, ARNHOLD W, ILLING-GÜNTHER H, MÜLLER M, GLEI M, FREYTAG H and BECHSTEDT U (1997e) *Die Auswirkungen einer Nickelbelastung bei Tier und Mensch.* In: Lombeck I, ed. Spurenelemente, Wissenschaftliche Verlagsgesellschaft Stuttgart, Germany, pp. 52–63.

ANKE M, DORN W, GUNSTHEIMER G, ARNHOLD W, GLEI M, ANKE S and LÖSCH E (1998) *Effect of trace and ultratrace elements on the reproduction performance of ruminants.* Vet Med-Czech **43**:272–282

ANKE M, MÓCSÉNYI A, MÜLLER R, ANGELOW L, GÜRTLER H and ANKE S (1999) *The role of manganese in nutrition.* Takarmanyozas, Animal Feeding and Nutrition, 2. évf 4, szám:11–15.

ANKE M (2000) *Ultratrace element intake depending on the geological origin of the habitat, time, sex and form of diet.* In: Seifert M, Langer U, Schäfer U and Anke M, eds. Mengen- und Spurenelemente. Author and Element Index 1981–2000, Schubert-Verlag, Leipzig, pp. 11–19.

ANKE M, DORN W, MÜLLER R and SCHÄFER U (2000) In: Fritsche K and Zerling L, eds. *Umwelt und Mensch Langzeitwirkungen und Schlussfolgerungen für die Zukunft.* Abhandlung der Sächsischen Akademie der Wissenschaften zu Leipzig – Mathematisch-naturwissenschaftliche Klasse-Band 59:45–61 S Hirzel Stuttgart/Leipzig, Germany.

ANKE M, GLEI M, ROTHER C, VORMANN J, SCHÄFER U, RÖHRIG B, DROBNER C, SCHOLZ E, HARTMANN E, MÖLLER E and SÜLZLE A (2000a) *Die Versorgung Erwachsener Deutschlands mit Iod, Selen, Zink bzw. Vanadium und mögliche Interaktionen dieser Elemente mit dem Iodstoffwechsel.* In: Bauch KH, ed. 3. Interdisziplinäres Iodsymposium, Aktuelle Aspekte des Iodmangels und Iodüberschusses, Blackwell Wissenschafts, Verlag Berlin-Wien-Oxford-Edinburgh-London-Kopenhagen-Melbourne-Tokyo, pp. 147–175.

ANKE M, MÜLLER R, DORN W, SEIFERT M, MÜLLER M, GONZALES D, KRONEMANN and SCHÄFER U (2000b) *Toxicity and Essentiality of Cadmium.* In: Ermidou-Pollet S and Pollet S, eds. 2nd International Symposium on Trace Elements in Human: New Perspectives, Athens, Greece, pp. 343–362.

ANKE M, MÜLLER R, TRÜPSCHUCH A, SEIFERT M, JARITZ M, HOLZINGER S and ANKE S (2000c) *Intake of chromium in Germany: risk or normality?* J Trace Microprobe Tech **18(4)**:541–548.

ANKE M, TRÜPSCHUCH A, DORN W, SEIFERT M, PILZ K, VORMANN J and SCHÄFER U (2000d) *Intake of nickel in Germany: risk or normality?* J Trace Microprobe Tech **18(4)**:549–556.

ANKE M, ARNHOLD W, ANGELOW L, LÖSCH E, ANKE S and MÜLLER R (2001) *Essentiality of arsenic, bromine, fluorine, and titanium for animal and man.* In: Ermidou-Pollet S and Pollet S, eds. 3rd International Symposium on Trace Elements in Human. *New Perspectives,* pp. 204–229. G. Morogiannis, Acharnai, Greece, pp. 13671.

ANKE M, MÜLLER M, ANKE S, GÜRTLER H, MÜLLER R, SCHÄFER U and ANGELOW L (2001a) *The biological and toxicological importance of aluminium in the environment and food chain of animals and humans.* In: Ermidou-Pollet S and Pollet S, eds. 3rd International Symposium on Trace Elements in Human: New Perspectives. G. Morogianis, Acharnai, Greece, pp. 230–247.

ANKE M, KRÄMER-BESELIA K, MÜLLER M, MÜLLER R, SCHÄFER U, FRÖBUS K and HOPPE C (2002) *Calcium supply, intake, balance and requirement of man. Fifth information: absorption, balance and requirement.* Mengen- und Spurenelemente **21**:1410–1415.

ANKE M, MÜLLER M, TRÜPSCHUCH A and MÜLLER R (2002a) *Intake and effects of cadmium, chromium and nickel in humans.* J Commodity: 41–63.

ANKE M, SCHÄFER U and ARNHOLD W (2003) *Lithium.* In: Caballero B, Trugo L and Finglas P, eds. Encyclopaedia of Food Sciences and Nutrition. Elsevier Science Ltd, Amsterdam, The Netherlands.

ANKE M, DORN W, SCHÄFER U and MÜLLER R (2003a) *The Biological and Toxicological Importance of Nickel in the Environment and the Food chain of Humans.* In: Romancik V, Koprda V, eds. Industrial Toxicology, University Bratislava, Slovakia, pp. 7–23.

ANKE M (2003) *Personal communications.*

ANKE M (2004) *Vanadium.* In: Merian E, Anke M, Ihnat M and Stoeppler M, eds. Elements and their compounds in the Environment. Part III, Chapter 27. Wiley-VCH, Weinheim, Germany.

ANKE S, ILLING-GÜNTHER H, GÜRTLER H, HOLZINGER S, JARITZ M, ANKE S and SCHÄFER U (2000) *Vanadium – an essential element for animals and humans.* In: Roussel AM, Anderson RA and Favier AE, eds. Trace Elements in Man and Animals – 10. Kluwer Academic/Plenum Publishers, New York-Boston-Dordrecht-London-Moscow, pp. 221–225.

ANONYMOUS (1980) *Mineral Tolerance of Domestic animals.* National Academy of Sciences – National Academy of Sciences – National Research Council, Washington DC, USA.

ANONYMOUS (1996) *Arsenic.* In: Trace elements in Human Nutrition and Health, World Health Organization, Geneva, Switzerland, pp. 217–220.

ANONYMOUS (1996a) *World Health Organization, Geneva.* In: Trace Elements in Human Nutrition and Health, WHO, Geneva, pp. 155–160.

ARNHOLD W (1989) *Die Versorgung von Tier und Mensch mit dem lebensnotwendigen Spurenelement Lithium.* Thesis University Leipzig, Sektion Tierprod und Vet Med, Germany.

AVTSYN AP, ANKE M, ZHAVORONKOW AA, GROPPEL B, KAKTURSKY LV, MICHALJEWA LM and LÖSCH E (1992) *Pathologische Anatomie des intrauterinen Fluormangels bei einem Ziegenlamm: Ein Fallbericht.* Mengen- und Spurenelemente 12:410–417.

AVTSYN AP, ZHAVORONKOV AA, KAKTURSKY LV and GROPPEL B (1993) *Pathological anatomy of the experimental hypovanadiosis of goats.* Arch Pathol 1:121–126.

BABCOCK SM (1905) *The addition of salt to the ration of dairy cows.* Wisconsin Agr Exp Stat Ann Report 22:129.

BACH JO (1990) *Some aspects of lithium in living systems.* In: Bach RO and Gallicchio VS, eds. Lithium and Cell Physiology. Springer-Verlag, New York, pp. 1–15.

BAUMANN E (1876a) *Ueber gepaarte Schwefelsäuren im Organismus.* Arch ges Physiol 13:285.

BAUMANN E (1876b) *Ueber Sulfosäuren im Harn.* Ber deut chem Ges 9:54.

BARBER RS, BRAUDE R, MITCHELL KG and CASSIDY J (1955) *High copper mineral mixture for fattening pigs.* Chem Ind 21:601–609.

BARCELOUX DG (1999) *Chromium.* Clin Toxicol 37(2):173–194.

BENTLEY R and CHASTEEN TG (2002) *Arsenic Curiosa and Humanity.* Chem Educator 7:51–60.

BERTRAND G and BHATTACHERJEE RC (1934) *L'action combinee du zinc et des vitamines dans*

l'alimentation des animaux. Compt Rend Acad Sci 198(21):1823–1827.

BEYERSMANN D (1991) *The significance of interactions in metal essentiality and toxicity.* In: Merian E, ed. Metals and Their Compounds in the Environment. VCH Weinheim, New York-Basel-Cambridge, pp. 491–509.

BOSSHARDT DK, HUFF JW and BARNES RH (1956) *Effect of bromine in chick growth.* Proc Soc Exp Biol Med 92:219.

BOUSSINGAULT JB (1872) *Du fer contenu dans le sang et dans les aliments.* CR Acad Sci Paris 74:1353–1359.

BURT J, DOWDY RP, PICKETT EE and O'DELL BL (1982) *Effects of low dietary lithium on tissue lithium content in rats.* Fed Proc 41:460.

CARLISLE EM (1972) *Silicon: An essential element for the chick.* Science 178:619–621.

CARLISLE EM (1980a) *A silicon requirement for normal skull formation in chicks.* J Nutr 110:352–359.

CARLISLE EM (1980b) *Biochemical and morphological changes associated with long bone abnormalities in silicon deficiency.* J Nutr 110:1046–1056.

CARLISLE EM (1997) *Silicon.* In: O'Dell BL and Sunde RA, eds. Handbook of Nutritionally Essential Mineral Elements. Marcel Dekker Inc, New York-Basel-Hong Kong, pp. 603–618.

CARLISLE EM and ALPENFELS WF (1984) *The role of silicon synthesis.* Fed Proc 43:680.

CARLISLE EM, BERGER JW and ALPENFELS WF (1981) *A Silicon requirement for propyl hydroxylase activity.* Fed Proc 40:866.

CARLISLE EM and CURRAN MJ (1993) *Aluminium: an essential element for the chick.* In: Anke M, Meissner D and Mills CF, eds. Trace Elements in Man and Animal – TEMA 8. Verlag Media Touristik, Gersdorf, Germany, pp. 695–698.

CERKLEWSKI FL (1997) *Fluorine.* In: O'Dell BL and Sunde RA, eds. Handbook of Nutritionally Essential Mineral Elements. Marcel Dekker Inc, New York-Basel-Hong Kong, pp. 583–602.

CHATIN A (1851) *Chimie Appliqués Recherche de l'iode dans l'ain les esux, le sel et les produits alimentaires des Alpes, de la France et du Piémont (première partie).* Comptes Rendus Hebdomaires Aeance de L'académie 33:529.

CHOWDHURY BA and CHANDRA RK (1991) *Metal Compounds and Immunotoxicology.* In: Merian E, ed. Metals and Their Compounds in the Environment. VCH Weinheim, New York-Basel-Cambridge, pp. 605–615.

COINDET JR (1820) *Découverte d'un nouveau remède le goitre.* Ann Chem Phys 15:49–59.

CRONIN E, MICHIEL AD and BROWN SS (1980) *Oral challenges in nickel sensitive women with hand eczema.* In: Brown SS and Sunderman FW, eds. Nickel Toxicology. Academic Press, New York, pp. 149–152.

DAVIS GK and MERTZ W (1987) *Copper.* In: Mertz W, ed. Trace Elements in Human and Animal Nutrition. Academic Press Inc, San Diego, pp. 301–364.

DEAN HT (1938) *Endemic fluorosis is relation to dental caries. Public Health Rep* 53:1443–1452.

DE RENZO EC, KALEITA E, HEYTLER P, OLESON JJ, HUTCHINGS BL and WILLIAMS JH (1953) *The nature of the xanthine oxidase factor.* J Am Chem Soc 75(3):753–756.

EDER K, REICHLMAYR-LAIS AM and KIRCHGESSNER M (1990) *Activity of Na-K-ATPase and Ca-Mg-ATPase in red blood cell membranes of lead-depleted rats.* J Trace Element Electrolytes Health Dis 4:21.

ERHARD F (1874) *Mineralien 8,* Heilbronn, Germany, p. 359.

EWERS U and SCHLIPKÖTER HW (1991) *Chronic toxicity of metals and metal compounds.* In: Merian E, ed. Metals and Their Compounds in the Environment. VCH, Weinheim-New York-Basel-Cambridge, pp. 591–603.

FALKE H and ANKE M (1987) *Die Reaktion der Ziege auf Molybdänbelastungen.* Mengen- und Spurenelemente 7:448–452.

FERGUSON WS, LEWIS AH and WATSON SJ (1938) *Action of molybdenum in nutrition of milking cattle.* Nature 141:553.

FISHBEIN WN, SMITH MJ, NAGARJAN K and SCURZI W (1976) *The first natural nickel metalloenzyme: urease.* Fed Proc 35:1680.

FRANKE J, RUNGE H and BECH R (1985) *Boron as an antidote to flourosis? Part 1. Studies on the skeletal system.* Fluoride 18:187–197.

FROST VD (1983) *The unforeseen need for arsenic and selenium for optimum health.* In: Anke M et al., eds. Fourth International Trace Element Symposium. University of Leipzig and Jena, Germany, Kongress- und Webedruck Oberlungwitz, Germany, pp. 89–96.

GALLICCHIO VS, MESSINO MJ, HULETTE BC and HUGHES NK (1991) *Lithium enhances recovery of haematopoiesis and lengthens survival in an allergenic transplant model.* In: Schrauzer GH and Klippell KF, eds. Lithium in Biology and Medicine. VCH, Weinheim, pp. 33–46.

GEBHART E and ROSSMAN G (1991) *Mutagenicity, carcinogenicity, teratogenicity.* In: Merian E, ed. Metals and Their Compounds in the Environ-ment. VCH, Weinheim-New York-Basel-Cambridge, pp. 617–640.

GELDMACHER VON MALLINCKRODT M (1991a) *Acute Metal Toxicity in Humans.* In: Merian E, ed. Metals and Their Compounds in the Environment. VCH, Weinheim-New York-Basel-Cambridge, pp. 585–590.

GELDMACHER VON MALLINCKRODT M(1991b) *Eco-genetics.* In: Merian E, ed. Metals and Their Compounds in the Environment. VCH, Weinheim-New York-Basel-Cambridge, pp. 641–649.

GONZALES D, RAMIREZ A, PEREZ E, SCHÄFER U and ANKE M (1998) *Der Iodverzehr erwachsener Mischköstler Mexikos.* Mengen- und Spurenelemente 19:85–94.

GREEN GH, LOTT MD and WEETH HJ (1973) *Effects of boron-water on rats.* Proc West Sec Anim Sci 24:254–258.

GRAUPE B (1965) *Untersuchungen über die Wirkung einer Molybdändüngung auf Ertrag und Zusammensetzung von Luzerne und der Einfluss Mo-gedüngten Futters auf den Mineralstoffhaushalt.* Thesis, Agriculture Toxicity, University of Jena.

GROPPEL B and ANKE M (1991) *Iodine content in foodstuffs and iodine intake of adults in central Europe.* In: Momcilovich B, ed. Trace Elements in Man and Animals −7. University of Zagreb, Croatia, Yugoslavia 7:7–6.

GROPPEL B, ANKE M and RIEDEL E (1985) Biological essentiality of tungsten. In: Pais I, ed. New Results in the Research Hardly Known Trace Elements. University of Horticulture Budapest, Hungary, pp. 92–103.

GRÜN M, ANKE M, HENNIG A and KRONEMANN H (1982) *Die biologische Bedeutung des Schwermetalls Blei.* Mengen- und Spurenelemente 2:159–178.

GÜRTLER H and ANKE M (1993) *Brom, ein für die Fauna essentielles Element?* Mengen- und Spurenelemente 13:522–530.

GÜRTLER H, ANKE M, GLEI M, ILLING-GÜNTHER H and ANKE S (1999) *Influence of vanadium deficiency on performance, health as well as blood and tissue parameters in goats.* In: Pais I, ed. New Perspectives in the Research of Hardly Known Trace Elements. University of Horticulture and Food Science, Budapest, Hungary, pp. 84–91.

GÜRTLER H, ANKE M, NEUBERT E, ANKE S and JARITZ M (1995) *Die Auswirkungen einer fluorarmen Ernährung bei der Ziege.* Mengen- und Spurenelemente 15:757–764.

HAPKE H-J (1991) *Metal accumulation in the food chain and load of feed and food.* In: Merian E, ed. Metals and their compounds in the Environment. Wiley-VCH, Weinheim, Germany, pp. 460–479.

HART EB, WADELL J and ELVEHJEM CA (1928) *Iron in nutrition. VII. Copper as a supplement to iron for hemoglobin in the rat.* J Biol Chem **77(2)**:797–812.

HENNIG A, JAHREIS G, ANKE M, PARTSCHFELD M and GRÜN M (1978) *Nickel – ein essentielles Spurenelement. 2. Mitt.: Die Ureaseaktivität im Pansensaft als möglicher Beleg für die Lebensnotwendigkeit des Nickel.* Arch Tierern **28**:267–268.

HENNIG A, ANKE M, GROPPEL B, LÜDKE H, REISSIG W, DITTRICH G and GRÜN M (1972) *Manganmangel beim Wiederkäuer.* Arch Tierern **22**:601–614.

HENRY W (1828) *The Element of Experimental Chemistry. 11th edition.* Baldwin and Cradock, London.

HOLTMEIER HJ (1995) *Calcium: Physiologie, Pathophysiologie und Klinik.* In: Holtmeier HJ, ed. Magnesium und Calcium, Wissenschaftliche Verlagsgesellschaft mbH Stuttgart, pp. 87–143.

HUFF JW, BOSSHARDT DK, MILLER OP and BARNES RH (1956) *A nutritional requirement for bromine.* Proc Soc Exp Biol Med **92**:216–219.

HUNT CD, SHULER TR and NIELSEN FH (1983) *Effect of boron on growth and mineral metabolism.* In: Anke M, Bräunlich H, Baumann H and Brückner CHR, eds. 4. Spurenelemente-Symposium: University Leipzig and Jena, Kongressund Werbedruck Oberlungwitz, Germany, pp. 149–155.

HUNT CD (1989) *Dietary boron modified the effects of magnesium and molybdenum on mineral metabolism in the cholecalciferol-deficient chick.* Biol Trace Element Res **22**:201–220.

HUNT CD and NIELSEN FH (1981) *Interaction between boron and cholecalciferol in the chick.* In: Howell JMCC, Gawthorne JM and Withe CL, eds. Trace Element Metabolism in Man and Animals (TEMA-4). Australian Academy of Science, Canberra, pp. 597–600.

KEMMERER AR, ELVEHJEM CA and HART EB (1931) *Studies on the relation of manganese to the nutrition of the mouse.* J Biol Chem **92(3)**:623–630.

KRONEMANN H, ANKE M and GRIS E (1982) *Der Calciumgehalt der Nahrungsmittel in der DDR.* Zbl Pharm **171**:556–558.

KIRCHGESSNER M and REICHLMAYR-LAIS AM (1981a) *Lead deficiency and its effects on growth and metabolism.* In: McHowell JM, Gawthorne E and White CL, eds. Trace Element Metabolism in Man and Animals – 4. Australian Academy of Sciences, Canberra, pp. 390–393.

KIRCHGESSNER M and REICHLMAYR-LAIS AM (1981b) *Changes of iron concentration and iron binding capacity in serum resulting from alimentary lead deficiency.* Biol Trace Element Res **3**:279.

KIRCHGESSNER M and REICHLMAYR-LAIS AM (1981c) *Retention, Absorbierbarkeit und Intermediäre Verfügbarkeit von Eisen bei alimentären Bleimangel.* Int Z Vit Ern Forschung **51**:421.

KIRCHGESSNER M and REICHLMAYR-LAIS AM (1982) *Konzentrationen verschiedener Stoffwechselmetabolismen im experimentellen Bleimangel.* Ann Nutr Metab **26**:50.

KIRCHGESSNER M and REICHLMAYR-LAIS A (1986) *In Vitro-Absorption von Eisen bei Nachkommen von an Blei depletierten Ratten.* J Animal Physiol Animal Nutr **55**:24.

KIRCHGESSNER M, PLASS DL and REICHLMAYR-LAIS A (1991) *Untersuchungen zur Essentialität von Blei an post partum abgesetzten Ferkeln.* J Animal Physiol Animal Nutr **66**:94.

KRISHNAMACHARI KAVR (1987) *Fluorine.* In: Mertz W, ed. Trace Elements in Human and Animal Nutrition – Fifth Edition. Academic Press Inc, San Diego, USA, pp. 365–415.

KRUSE HD, ORENT ER and MCCOLLUM EV (1932) *Studies on magnesium deficient animals.* J Biol Chem **96**:519.

LANGER P (1960) *History of goitre.* In: Endemic Goitre. World Health Organization Monograph, **44**:9–25.

LAMBERT J (1983) Lithium content in the grassland vegetation. In: Anke M, ed. Lithium. 4. Spurenelementesymposium, eds. University Leipzig, Kongress- und Werbedruck Oberlungwitz, Germany, pp. 32–38.

LIEBSCHER DH and LIEBSCHER DE (2000) *Hereditäre Magnesiumangeltetamie – ein übersehenes Krankheitsbild.* Mengen- und Spurenelemente **20**:661–667.

LINES EW (1935) *The effect of the ingestion of minute quantities of cobalt by sheep affected with 'coast disease'.* J Council Sci Ind Res **8**:117–119.

LJUNGDAHL LG and ANDRESEEN JR (1975) *Tungsten, a component of active formate dehydrogenase from Clostridium thermoaceticum.* FEBS Lett **64**:279–282.

LOMBECK I and BREMER HJ (1977) *Primary and Secondary disturbances in trace element. Metabolism connected with genetic metabolic disorders.* Nutr Metab **21**:49–64.

LUKASKI HC (1999) *Chromium as a supplement.* Annu Rev Nutr **19**:279–302.

MARSTON HR (1935) *Problems associated with 'coast disease'.* J Council Sci Ind Res **8**:111–116.

Blei-Mangel. J Animal Physiol Animal Nutr **56**:123.

REICHLMAYR-LAIS AM and KIRCHGESSNER M (1986b) *Effects of lead deficiency on lipid metabolism.* Z Ernährungswiss **25**:165.

REICHLMAYR-LAIS AM and KIRCHGESSNER M (1986c) *Fe-Retention bei Nachkommen von Blei depletierten Ratten.* J Animal Physiol Animal Nutr **55**:77.

REICHLMAYR-LAIS AM and KIRCHGESSNER M (1997) *Lead.* In: O'Dell BL and Sunde R, eds. Handbook of Nutritionally Essential Mineral Elements. Marcel Dekker, Inc, New York-Basel-Hong Kong, pp. 479–492.

RICHERT DA and WESTERFELD WW (1953) *Isolation and identification of xanthine oxidase factor as molybdenum.* J Biol Chem **203**:915–923.

RICHTER G, BARGHOLZ J, LEITERER M and ARNHOLD W (2002) *Untersuchungen zur Chlorunterversorgung von Legehennen.* Mengen- und Spurenelemente **21**:421–426.

RICHTER G and THIEME R (2002) *NaCl-Überversorgung bei Legehennen.* Mengen- und Spurenelemente **21**:73–79.

RINGER ST (1982/83) *An investigation concerning the action of rubidium and cesium salts compared with the action of potassium salts on the ventricles of the frogs heart.* J Physiolog **4**:270–276.

SCHÄFER U (1997) *Essentiality and toxicity of lithium.* J Trace Micoprobe Techn **15**:341–349.

SCHÄFER U (1998) *Past and present conceptions concerning the use of lithium in medicine.* J Trace Micoprobe Techn **16**:535–556.

SCHLESS AP, FRAZER A, MENDELS J, PANDAY GN and THEODORIDES VJ (1975) *Genetic determination of lithium ion metabolism. II. An in vivo study of lithium ion distribution across erythrocyte membranes.* Arch Gen Psychiatry **32**:337–340.

SCHMIDT A, ANKE M, GROPPEL B and KRONEMANN H (1983) *Histochemical and ultra structural findings in As deficiency.* Mengen- und Spurenelemente **3**:424–425.

SCHMIDT A, ANKE M, GROPPEL B and KRONEMANN H (1984) *Effect of As deficiency on skeletal muscle, myocardium, and liver. A histochemical and ultra-structural study.* Exp Pathol **25**:195–197.

SCHNEGG A and KIRCHGESSNER M (1975) *Zur Essentialität von Nickel für das tierische Wachstum.* Z Tierphysiol Tierernaehrg Futtermittelkde **36**:63–74.

SCHROEDER HA, MITCHENER M, BALASSA JJ, KANISOWA M and NASON AP (1968) *Zirconium, niobium, antimony, and fluorine in mice. Effects on growth, survival, and tissue levels.* J Nutr **95**:95–101.

SCHWARZ K and FOLTZ CM (1957) *Selenium as an integral part of factor 3 against necrotic liver degeneration.* J Am Chem Soc **79**:3292–3293.

SCHWARZ K and MILNE DB (1972) *Fluorine requirement for growth in the rat.* Bioinorg Chem **1**:331–338.

SCHWARZ K and MILNE DB (1972b) *Growth-promoting effects of silicon in rats.* Nature **239**:333–334.

SCHWARZ K and SPALLHOLZ (1977) *The potential essentiality of cadmium.* In: Anke M and Schneider H-J, eds. Cadmium – Symposium, Friedrich-Schiller University, Jena, Germany, pp. 188–192.

SEAL BS and WEETH HJ (1980) *Effect of boron in drinking water on the male laboratory rat.* Bull Environ Contam Toxicol **25**:782–789.

SELBY LA, CASE AA, OSWEILER GD and HAYES H-M JR, (1977) *Epidemiology and toxicology of arsenic poisoning in domestic animals.* Environ Health Perspect **10**:183–189.

SHECHTER Y, MEYEROVITSCH J, FARFEL Z, ET AL. (1990) *Insulin mimetic effects of vanadium.* In: Chasteen ND, ed. Vanadium in Biological Systems. Kluwer, Dordrecht, The Netherlands, pp. 129–142.

SILL R and DAWCZYNSKI H (1998) *Stellenwert von Selen in der Prävention und Therapie von Tumorerkrankungen.* Mengen- und Spurenelemente **18**:949–960.

SMITH HA (1984) *Cadmium.* In: Frieden E, ed. Biochemistry of the Essential Ultra Trace Elements. Plenum Press, New York-London, pp. 341–366.

SPEARS JW and HATFIELD EE (1978) *Nickel for ruminants. I. Influence of dietary nickel on ruminal urease activity.* J Anim Sci **47**:1345–1350.

STANGL GI, SCHWARZ FJ, JAHN B and KIRCHGESSNER M (2000) *Cobalt deficiency-induced hyperhomocysteinaemia and oxidative status of cattle.* Br J Nutr **83**:3–6.

STEMME K, MEYER U, FLACHOWSKY G and SCHOLZ H (2002) *Kobalt-Bedarf von Wiederkäuern – sind die derzeitigen Bedarfsempfehlungen für laktierende Milchkühe ausreichend?* Mengen- und Spurenelemente **21**:404–414.

STERNLIEB I (1982) *Wilson's disease.* In: Sorensen JRJ, ed. Inflammatory Disease and Copper. Humana Press, Clifton, NJ, pp. 75–84.

STRASDEIT H (2001) *Das erste cadmium-spezifische Enzym.* Angew Chem **113**:730–732.

SZENTMIHÁLYI S, ANKE M and REGIUS A (1985) *The importance of lithium for plant and animal.* In: Pais I, ed. New Results in the Research of Hardly Known trace Elements. University of Horticulture, Budapest, pp. 136–151.

SZENTMIHÁLYI S, REGIUS A, LOKAY D and ANKE M (1983) *Der Lithiumgehalt der Vegetation in Abhängigkeit von der geologischen Herkunft des Standortes.* In: Anke M, ed. Lithium 4. Spurenelementsymposium, University Leipzig and Jena, Germany, Kongress- und Werbedruck, Oberlungwitz, Germany, pp. 18–24.

SZILÁGYI M, ANKE M, BALOGH I and REGIUS-MÓCSÉNYI A (1989) *Lithium status and animal metabolism.* In: Anke M et al., ed. Sixth International Trace Element Symposium. Vol. 4. University Leipzig and Jena, Germany, Kongress- und Werbedruck, Oberlungwitz, Germany, pp. 1249–1261.

SZILÁGYI M, ANKE M and SZENTMIHÁLYI S (1985) *The effect of lithium deficiency on the metabolism of goats.* In: Pais I, ed. New Results in the Research of hardly known Trace Elements. Horticultural University, Budapest, pp. 167–172.

TAO S and SUTTIE JW (1976) *Evidence for a lack of an effect dietary fluoride level on reproduction in mice.* J Nutr 106:1115–1122.

TODD WR, ELVEHJEM CA and HART EB (1934) *Zinc in the nutrition of the rat.* Am J Physiol 107(1):146–156.

TÖLGYESI G (1983) Die Verbreitung des Lithiums in ungarischen Böden und Pflanzen. In: Anke M, ed. Lithium 4. Spurenelementsymposium, University Leipzig and Jena, Germany, Kongress- und Werbedruck, Oberlungwitz, Germany, pp. 39–44.

TUCKER HF and SALMON WD (1955) *Parakeratosis or zinc deficiency disease in pig.* Proc Soc Expt Biol Med 88:613–616.

UNDERWOOD EJ (1984) *Present knowledge in nutrition.* In: Olson RE, Broquist HP, Chichester CO, Darby WJ, Kolbe AC and Stalvey RM, eds. Nutrition Reviews: 5th edition. The Nutrition Foundation Inc, Washington DC, pp. 528–537.

UNDERWOOD EJ and FILMER JF (1935) *The determination of the biologically potent element (cobalt) in limonite.* Aust Vet J 11(1):84–92.

UTHUS EO, CORNATZER WE and NIELSEN FH (1983) *Consequences of arsenic deprivation in laboratory animals.* In: Lederer WH, ed. Arsenic symposium, production and use, biomedical and environmental perspectives. Van Nostrand Reinhold, New York, pp. 173–189.

UTHUS EO and NIELSEN FH (1990) *Effect of vanadium, iodine and their interaction on growth, blood variables, liver trace elements and thyroid status indices in rats.* Magnesium Trace Elem 9:219–226.

VOGTS H (1992) *Einfluss von Silikaten im Legehennenfutter.* Mengen- und Spurenelemente 12:304–309.

WADDELL J, STEENBOCK H and HART EB (1931) *Growth and reproduction on milk diets.* J Nutr 4:53–62.

WERNER A and ANKE M (1960) *Der Spurenelementgehalt der Rinderhaare als Hilfsmittel zur Erkennung von Mangelerscheinungen.* Archiv für Tierernährung 10:142–153.

WIENER G and FIELD AC (1970) *Genetic variation in copper metabolism of sheep.* In: Mills CF, ed. Trace element metabolism in animals. Livingstone, Edinburgh-London, pp. 92–102.

WILSON SAK (1912) *Progressive lenticular degeneration: a familial nervous disease associated with cirrhosis of the liver.* Brain 34:295–307.

WINNEK PS and SMITH AH (1937) *Studies on the role of bromine in nutrition.* J Biol Chem 121:345–352.

YOKOI K (1997) *A study on possible essentiality nature of tin and rubidium.* J Jpn Soc Nutr Food Sci 50:15–20.

YOKOI K, KIMURA M and ITOKAWA Y (1994) *Effect of low – rubidium diet on macro – mineral levels in rat tissues.* J Jpn Soc Nutr Food Sci 50:295–299.

ZHAVARONKOV AA, KATURSKY LV, ANKE M, GROPPEL B and MIKHALEVA LM (1996) *Pathology of congenital bromide deficiency in experiment.* Arch Pathol Moscow (Russia) N2:62–67.

4

Essential and Toxic Effects of Macro, Trace, and Ultratrace Elements in the Nutrition of Man

Manfred K. Anke

4.1
Introduction

In general, the elements which are vital to the cells of man are the same elements as those needed by cells of other vertebrate species. The several species of mammals differ, however, with regard to the extent of their needs for inorganic cell components. The normative requirement of manganese for adult humans is, for example, 15 µg kg^{-1} body weight (Anke et al. 1999a), whereas that for cattle and other species of ruminants and birds amounts to 1000–1500 µg kg^{-1} body weight (Anke 1982).

In every case, the manganese needs of man are met by manganese levels in the food. A manganese deficiency does not exist in man, but is well known in domestic ruminants and birds (Anke et al. 1999b). In case of molybdenum, the normative requirements of animals and man are met by the natural feeds and foods (Anke et al. 1989). The molybdenum offer to animals and man is much higher than their normative requirements, which amount to 2.5 µg kg^{-1} body weight in animals, and 0.40 µg kg^{-1} in man (Holzinger et al. 1998b). In case of iodine, copper, zinc, iron, selenium, calcium, phosphorus, magnesium, sodium, and potassium, deficiencies are found in both animals and man. The normative requirements are not in every case satisfied by the natural offer of these elements. Beside the requirement of molybdenum, the needs for nickel, chlorine and all ultratrace elements are met by the intake of natural feeds and foods. With regard to these elements, deficiency symptoms in animals and man have not been found, except in case of genetic defects and long-term parenteral nutrition (Anke et al. 2004d).

In contrast to its essentiality, every element of the periodic system may be toxic; it is only a question of the intake quantities and the element specification. An intoxication can induce interactions with essential elements and induce deficiency symptoms; well-known examples are the interactions of nickel with zinc, magnesium, and manganese (Anke et al. 1997f), or cadmium with copper, zinc, and iron (Anke et al 1970).

Factors which are important for the lowering of human health risks due to mineral deficiency and intoxication are the normative requirements of the macro, trace and ultratrace elements, recommendations for their intake, their apparent absorption and excretion rates, their interactions, tolerance limits for chemicals in food and water, and occupational health standards.

Elements and their Compounds in the Environment. 2nd Edition.
Edited by E. Merian, M. Anke, M. Ihnat, M. Stoeppler
Copyright © 2004 WILEY-VCH Verlag GmbH & Co. KGaA, Weinheim
ISBN: 3-527-30459-2

4.2
Macro elements

4.2.1
Normative Requirements and Recommendations for Intake

The normal daily magnesium requirement of women and men is 200 and 250 mg, respectively, on the average of a week (Table 4.1) (Vormann et al. 1999, Vormann and Anke 2002). The magnesium requirement is met by an intake of 3.0 mg Mg kg^{-1} body weight and a magnesium concentration of 650 mg Mg kg^{-1} food dry matter (DM). The recommendation for the daily magnesium intake of adults (women 65 kg, men 75 kg) amounted to 300 mg (Anonymous 2000). Genetic defects of magnesium homeostasis increase the magnesium requirement considerably (Meij et al. 2002), and a daily Mg intake of 1200 mg is necessary in some cases of this genetic disorder (Liebscher 2003).

The normal daily calcium requirement of women and men is < 500 and < 600 mg, respectively, on the average of a week. Adults require 8 mg Ca kg^{-1} body weight, and a calcium concentration of 1600 mg kg^{-1} DM in their diet (Anke et al. 2002a, b).

The normal phosphorus requirement of animals is only 50% that of the calcium requirement, because these elements are needed for bone formation in a ratio of 2 : 1. In humans, the nutritional intake of phosphorus is similar to, or higher than, that of calcium (Anke et al. 2004e). A phosphorus deficiency in humans with mixed diet is not common.

The daily normal potassium requirement is not well known, but is met in women and men by daily intakes of 1300 and 1600 mg, respectively (Anke et al. 2003).

4.2.2
Macro element Intake of Adults with Mixed and Ovolactovegetarian Diets

The macro element intake was determined by the duplicate portion technique over seven consecutive days in a test team comprising seven women and seven men, aged between 20 and 69 years, as well as in 17 test populations in Germany and Mexico. The subjects collected duplicates of all intakes, which were analyzed daily. Fifteen of the test populations lived in Germany, and two in Mexico. The ovolactovegetarians were all Germans (Anke et al. 1997a). On average, the men with a mixed diet consumed 24% more dry matter than the women, while the male ovolactovegetarians ate 23% more dry matter than the female ones.

Tab. 4.1: Normative requirements and recommendations for macro elements

Element	Normative requirement					Recommendation
	Women	Men	mg/kg	mg/kg food dry matter		
				Mixed diet	Vegetarian diet	
	mg/day	mg/day	Body weight			mg/day
Magnesium	200	250	3,0	650	470	300
Calcium	< 500	< 600	< 8,0	1600	1150	1000
Phosphorus	< 400	< 500	< 6,5	1300	950	700
Potassium	< 1500	< 1800	< 24	5000	3600	1900
Sodium	< 1300	< 1600	< 20	< 4000	3000	< 2000

In general, ovolactovegetarians take in 33% more dry matter than people with mixed diets (Anke et al. 1997a). The energy concentration of ovolactovegetarian foods is lower than that of mixed diets.

In 1996, men with mixed and vegetarian diets consumed only 14 and 8% more Ca, respectively, than women (Table 4.2). It seems that women prefer food rich in calcium, such as vegetables, milk and cheese, whereas men favor calcium-poor meats and sausages (Anke et al. 2001a–d). Ovolactovegetarians consume an extremely calcium-rich diet. On average, the normal calcium requirement of both genders and types of nutrition is met. The bioavailability of calcium in vegetarian diets appears to be low (phytic acid).

In contrast to calcium intake, the magnesium consumption of men is higher than that of women by one third (Table 4.2). Ovolactovegetarians take in much more magnesium than people with mixed diets. These differences were all seen to be significant (Glei et al. 1997, Gonzales et al. 1999, Vor-mann et al. 2002, Anke et al. 1998). On average, the daily magnesium intake of people with mixed and vegetable diets meets the normative requirement, but one-quarter of all Germans living on a mixed diet suffer from a magnesium deficit. Ovolactovegetarians take in more magnesium than people with mixed diet; their magnesium requirement is apparently met, despite the bioavailability of this magnesium (phytic acid) being low.

The normal potassium requirement of both sexes and diet forms is met (Table 4.2). Typically, men consume 25–40% more potassium than women, while vegetarians take in much more potassium than people with mixed diets. (Anke et al. 1992, 1992b)

At 6.0 g in women and 8.0 g in men, the daily salt (NaCl) intake of people with mixed and vegetarian diets is lower than postulated. The danger of sodium-induced hypertension is limited to persons with genetic defects of sodium reabsorption in their kidneys (Anke 2004b). The sodium intake of men is only

Tab. 4.2: Ash and macro element intake of adults with mixed and ovolactovegetarian diets as a function of gender and type of diet

Element	Diet	n (w) ; n (m)	Women (w)		Men (m)		p[2]	%[1]
			SD	Mean	Mean [4]	SD [3]		
Ash (g/d)[6]	Mixed	217;217	4. 4	12.8	17.3	6.0	< 0.001	135
	Vegetarian	70;70	5.8	17.6	22.6	7.3	< 0.001	128
Ca (mg/d)	Mixed	217;217	406	619	705	392	< 0.01	114
	Vegetarian	70;70	523	1176	1251	655	> 0.05	106
Mg (mg/d)	Mixed	217;217	72	205	266	92	< 0.001	130
	Vegetarian	70;70	101	376	474	199	< 0.001	126
K (mg/d)	Mixed	217;217	872	2130	2709	1008	< 0.001	127
	Vegetarian	70;70	900	3195	4577	2305	< 0.001	143
Na (mg/d)	Mixed	217;217	991	2314	3239	1289	< 0.001	140
	Vegetarian	70;70	1036	2372	3181	1272	< 0.001	134
P (mg/d)	Mixed	471;504	307	806	1058	393	< 0.001	131

[1] Women = 100%, Men = x%. [2] Significance level in Student's t-test. [3] Standard deviation. [4] Arithmetic mean. [5] Number. [6] Day.

13% higher than that of women, which suggests that women prefer a salt-rich diet. The salt intake of people living on a mixed diet is similar to that of vegetarians (Anke et al. 1992a, 2000a, b).

The phosphorus intake of people with mixed diets is much higher than their requirement. Men prefer phosphorus-rich foodstuffs with a good bioavailability, such as sausages and meats. The intake by adults in Europe and Mexico is similar to the intake worldwide (Anke et al. 2004e, Parr et al. 1992).

4.2.3
Macro element Concentration of the Dry Matter Consumed

Men prefer diets with significantly higher concentrations of ash and sodium (Table 4.3), whereas women tend to favor a diet which is richer in calcium. The magnesium and phosphorus concentrations of mixed and vegetarian diets were similar. The magnesium concentration of the consumed food dry matter was close to the mag-

nesium concentration needed to satisfy the normal magnesium requirement. The calcium, potassium, sodium, and phosphorus concentrations of the consumed dry matter was higher than the desired concentration of these macro elements.

4.2.4
Macro element Intake per kg Body Weight of Humans

With the exception of magnesium (and partly of calcium), macro element intake in Germany and Mexico by people with mixed diets is higher than the normative requirement of adults (Table 4.4). A deficit of potassium and sodium is not to be expected. Ovolactovegetarians take in significantly more (perhaps two-fold) ash, calcium, magnesium and potassium than people with mixed diets. A gender-related influence was seen only in the sodium intake of men in both diets, with men respectively consuming 17 and 10% more sodium than women.

Tab. 4.3: Ash and macro element concentration of the eaten dry matter in dependence of gender and types of diet

Element	Diet	$n^{5)}$ (w) ; n (m)	Women (w)		Men (m)		$p^{2)}$	$\%^{1)}$
			SD	Mean	Mean [4]	SD [3]		
Ash (g/d)[6]	Mixed	217;217	11	43	46	9.5	< 0.05	107
	Vegetarian	70;70	12	45	50	12	< 0.01	111
Ca (mg/d)	Mixed	217;217	1263	2132	1952	831	> 0.05	92
	Vegetarian	70;70	1177	3018	2607	1089	< 0.05	86
Mg (mg/d)	Mixed	217;217	188	713	743	163	> 0.05	104
	Vegetarian	70;70	257	987	1016	333	> 0.05	103
K (mg/d)	Mixed	217;217	2309	7331	7575	2015	> 0.05	103
	Vegetarian	70;70	2570	8459	9956	4523	< 0.001	118
Na (mg/d)	Mixed	217;217	2917	8054	9076	3036	< 0.01	113
	Vegetarian	70;70	2116	6014	6794	1864	< 0.05	113
P (mg/d)	Mixed	471;504	676	2650	2762	749	> 0.05	104

Footnotes see Table 4.2.

Tab. 4.4: Ash and macro element intake per kg body weight in dependence of gender and type of diet (µg/kg body weight BW)

Element	Diet	$n^{5)}$ (w) ; n (m)	Women (65 kg)		Men (75 kg)		$p^{2)}$	$\%^{1)}$
			SD	Mean	Mean [4]	SD [3]		
Ash (mg)	Mixed	217;217	70	192	221	80	< 0.01	115
	Vegetarian	70;70	109	308	327	111	> 0.05	106
Ca (mg)	Mixed	217;217	6.6	9.5	9.0	5.2	> 0.05	95
	Vegetarian	70;70	9.2	20	18	10	> 0.05	90
Mg (mg)	Mixed	217;217	1.2	3.1	3.4	1.2	> 0.05	110
	Vegetarian	70;70	2.0	6.6	6.8	2.8	> 0.05	103
K (mg)	Mixed	217;217	14	32	35	13	> 0.05	109
	Vegetarian	70;70	19	57	66	33	< 0.05	116
Na (mg)	Mixed	217;217	15	35	41	16	< 0.01	117
	Vegetarian	70;70	23	42	46	20	< 0.05	110

Footnotes [1] to [5] see Table 4.2.

4.2.5
Fecal Excretion and Apparent Absorption of Macro elements

The excretion of macro elements occurs via the feces, urine, and sweat, though the first two routes represent the main pathways. Typically, fecal excretion accounts for only 2–7% of the sodium intake. The apparent absorption rate of sodium has been found to vary between 93 and 98%. On average, the apparent absorption rate of macro elements by ovolactovegetarians is lower than that by people with mixed diets. The binding of magnesium and calcium by phytic acid significantly lowers the bioavailability of both elements in the nutrition of ovolactovegetarians. On average, and for both sexes, the apparent absorption rate of magnesium is 34% in people with mixed diets, and 28% in vegetarians. For calcium, the apparent absorption rate was 13% with a mixed diet, but only 6% in vegetarians (Table 4.5). The plentiful consumption of magnesium and calcium by vegetarians did not lead to any greater absorption of

Tab. 4.5: Fecal excretion and apparent absorption rates of ash and macro elements by adults

Element	$n^{2)}$ (w) ; n (m)	Excretion by feces %				Apparent absorption rate, $\%^{1)}$			
		Women (w)		Men (m)		Women		Men	
		Mixed	Veg.[3]	Mixed	Veg.	Mixed	Veg.	Mixed	Veg.
Na	434;140	1.9	2.7	1.4	6.6	98	97	98	93
K	434;140	15	24	15	21	83	78	83	80
Ash	434;140	24	36	20	34	73	61	79	68
P	294;0	39	–	32	–	55	–	67	–
Mg	434;140	65	73	66	72	33	25	34	30
Ca	434;140	81	89	78	89	9.4	6.4	17	9.5

[1] Women = 100%, Men = x%. [2] Number of samples; [3] Vegetarian diet.

either element, mainly due to a phytase deficit.

Fecal excretion of the macro elements was, on average, 3% for sodium, 19% for potassium, 28% for ash, 36% for phosphorus, 69% for magnesium, and 84% for calcium. The apparent absorption rate is negatively correlated with fecal excretion of the macro elements. The real absorption rate of all these elements is higher than the apparent absorption rate, mainly due to the salivary, biliary and pancreatic excretion of absorbed macro elements into the intestine.

4.2.6
Macro element Intake Measured by the Duplicate Portion Technique and the Basket Method

The macro, trace and ultratrace element intake of adults in Germany in relation to time, gender, age, weight, season, habitat, form of diet and performance (pregnancy, breast-feeding) has been systematically investigated. A total number of 19 test populations, each comprising at least seven women and seven men aged between 20 and 69 years, collected visually assessed duplicates of all consumed foods, beverages and sweets on seven successive days. Any consumption was registered in written reports which enabled dietary intake to be

calculated using the basket method, and the results compared with those obtained with the duplicate method (Table 4.6). Comparison of chemical assays with calculations of macro element intake showed that the basket method overestimated intake by 30–50%, with the exception of sodium. This is one of the reasons why recommendations for calcium and magnesium intake are too high (Bergmann 1995, Glei 1995, Krämer 1993).

4.2.7
Intake of Macro elements Through Animal and Vegetable Foodstuffs and Beverages

Almost 60% of the human phosphorus and sodium intakes are supplied by animal foodstuffs, and about 40% by vegetable foodstuffs. Beverages provide only 2–6% of the phosphorus intake (Table 4.7). Animal and vegetarian foodstuffs account for more than 40% of the sum of all inorganic components of the diet as represented by the ash. Surprisingly, beverages deliver 15% of the ash in the diet. Only 30% and 23% respectively of the potassium and magnesium intakes are delivered by animal foodstuffs. Two-thirds of the potassium intake is from vegetable foodstuffs, and only 8% from beverages. In contrast to potassium, more than one-quarter of the consumed magnesium is

Tab. 4.6: Macro element intake of adults with mixed diets determined by the duplicate portion technique and calculated by the basket method

| Element | Women | | Men | | %[1] |
	Duplicate	Basket	Duplicate	Basket	
Na (mg/d)[2]	2513	2363	3253	3626	104
K (mg/d)	2558	3261	2964	3917	130
Mg (mg/d)	211	278	259	360	136
Ca (mg/d)	512	816	660	869	144
P (mg/d)	784	1160	1046	1549	148

[1] Dublicate (women and men) = 100%, Basket = x%; [2] Day.

Tab. 4.7: Intake of macro elements through animal and vegetable foodstuff and beverages by people with mixed diet in percent

Macro element	Animal foodstuffs	Vegetable foodstuffs	Beverages
Calcium	68	29	3
Phosphorus	58	36	6
Sodium	57	41	2
Ash	42	43	15
Potassium	30	62	8
Magnesium	23	49	28

taken in with beverages. In regions of magnesium-rich rocks (dolomite, Keuper sediments), the drinking water contributes a much higher portion of magnesium to the diet (Glei 1995).

4.3

Trace Elements

4.3.1

Normative Requirements and Recommendations for Intake

The normative daily iron requirement of young women amounts to 7.0 mg, and that of men and older women to 6.0 µg kg^{-1} body weight. The iron concentration of the consumed dry matter should be 20 mg Fe kg^{-1} (Anke 2001) (Table 4.8).

The normative daily zinc requirement of people with moderate bioavailability may be < 6.0 and < 8.0 mg, taking the average over a week (Anonymous 1996). This daily intake corresponds to a zinc level of 90 µg kg^{-1} body weight and a zinc concentration of 20 mg kg^{-1} DM consumed (Anke et al. 1999c, Röhrig et al. 1998).

The normative daily manganese requirement of both sexes is < 1.0 mg (Anke et al. 1999a, Schäfer et al. 2001). A manganese intake of 15 µg kg^{-1} body weight may be needed to meet the normative requirement, with a manganese concentration in diet being 2200 µg kg^{-1} DM consumed (Anke et al. 1999a, b). The recommended daily manganese intake of 2–5 mg is high in comparison with the normative requirement (< 1 mg), but harmless (Röhrig et al. 1996, Röhrig 1998).

Tab. 4.8: Normative requirement and recommendations for trace elements

Element	Normative requirement				Recommendation	Tolerable daily intake
	Women (mg/d)ay	Men (mg/d)ay	µg/kg body weight	mg/kg food dry matter	µg/day	µg/kg body weight
Iron	7.0	6.0	0.100	20	10	700
Zinc	< 6.0	< 8.0	0.090	20	10	600
Manganese	< 1.0	< 1.0	0.015	3.0	2–5	Unknown
Copper	0.700	0.800	0.012	2.2	1.0–1.5	175
Molybdenum	< 0.025	0.025	0.0004	0.09	0.05–0.1	150
Nickel	0.025	0.035	0.0005	0.09	0.050	< 8
Iodine	0.065	0.075	0.001	0.22	0.200	15
Selenium	0.020	0.025	0.0004	0.070	0.03–0.05	6

Women and men require 700 and 800 µg Cu per day, respectively. The normative requirements of copper is met by 12 µg Cu kg^{-1} body weight, or by the consumption of dry matter containing 2200 µg Cu kg^{-1}.

The normal daily requirement of molybdenum, at 25 µg (Anke et al. 1989, Turnlund et al. 1995), is very low compared with the daily intake. The normal molybdenum requirement is 0.4 µg kg^{-1} body weight. The diet consumed should contain 90 µg Mo kg^{-1} DM (Holzinger et al. 1998a, b, 1997, Anke et al. 1993a, Anke and Glei 1993, Reiss and Anke 2002).

At 25–35 µg, the normal daily nickel requirement (Nielsen 1987, Anke et al. 1991a) is comparable with that of molybdenum and selenium (Table 4.8). The normative requirement per kg body weight is 0.5 µg Ni (Anke et al. 2000e, Eder and Kirchgessner 1997, Trüpschuch 1997).

The normal daily iodine requirement is 65 to < 75 µg, equivalent to 1 µg kg^{-1} body weight or to 220 µg I per kg consumed DM (Anke 2004a, Anonymous 1996). The daily recommended iodine intake is much higher, at 200 µg.

The required daily selenium intake of women was found to be 20 µg. In case of a lower selenium intake, the women's selenium balances proved negative, and their f T$_3$-levels in blood serum were too low (Anke et al. 2003a). Men require 25 µg Se per day, while both genders require 0.4 µg Se kg^{-1} body weight and a dietary selenium concentration of 70 µg kg^{-1} consumed DM. The recommended daily selenium intake is 30–50 µg, which is far from the toxic limit of 400 mg per day (Drobner 1997, Anke et al. 2002, 2003a).

4.3.2
Trace Element Intake of Adults with Mixed and Ovolactovegetarian Diets

On average, the iron intake of Germans with a mixed diet is within the limit of the normal daily iron requirement for women of 7 mg. Female ovolactovegetarians consume significantly more iron, the bioavailability of which is lower because of its binding to phytic acid. The iron intake of men with mixed and vegetarian diets was found to meet their iron requirements without difficulties (Anke 2001). Men with a mixed diet consumed 22% more iron than women, this difference corresponding with the 24% higher DM intake of males (Table 4.9).

On average, the zinc intake of women and men in Germany approximately equals the normal daily zinc requirement, which is less than 6 and 8 mg, respectively, and varies with the bioavailability of the element. Women in a placebo-controlled, double-blind study with a daily zinc intake of 6.7 mg in the placebo cluster and 17 mg in the zinc cluster, the latter excreted exactly 10 mg zinc more in the feces. A zinc deficit in Germany, especially in women, cannot be excluded. The zinc intake of women in Mexico is significantly higher than that in Germany (Anke et al. 1996, 1999d).

The normative requirement of manganese is completely satisfied by food in both mixed and vegetarian diets. No adult was found to take in < 1 mg per day on the average of a week (Anke et al. 1999a). Women of both diet forms prefer manganese-rich vegetables. Men were found to take in only 7–12% more manganese than women, rather than an expected value of 24% (Table 4.9).

The normal copper requirement of both sexes is met in Germany and Mexico by both mixed and vegetarian diets (Table 4.9). Vegetarians take in significantly more copper than people with mixed diets

Tab. 4.9: Trace element intake of adults with mixed and ovolactovegetarian diets as a function of gender and type of diet

Element	Diet	$n^{5)}$ (w) ; n (m)	Women (w)		Men (m)		$p^{2)}$	$\%^{1)}$
			SD	Mean	Mean [4]	SD [3]		
Fe (mg/d)[6]	Mixed	217;217	2.9	6.3	7.7	3.8	< 0.001	122
	Vegetarian	70;70	5.4	9.5	10	4.4	> 0.05	105
Zn (mg/d)	Mixed	217;217	2.8	6.0	7.5	3.1	< 0.001	125
	Vegetarian	70;70	2.6	8.6	9.5	3.9	< 0.05	110
Mn (mg/d)	Mixed	217;217	1.2	2.4	2.7	1.2	< 0.001	112
	Vegetarian	70;70	2.1	5.5	5.9	3.9	> 0.05	107
Cu (mg/d)	Mixed	217;217	0.86	1.1	1.2	0.67	> 0.05	109
	Vegetarian	70;70	0.57	1.6	2.1	1.0	< 0.001	131
Mo (µg/d)	Mixed	217;217	98	89	100	66	> 0.05	112
	Vegetarian	70;70	131	179	170	92	> 0.05	95
Ni (µg/d)	Mixed	217;217	61	90	97	91	> 0.05	108
	Vegetarian	70;70	122	185	196	154	> 0.05	106
I (µg/d)	Mixed	217;217	47	83	113	59	< 0.001	136
	Vegetarian	70;70	52	80	123	80	< 0.001	154
Se (µg/d)	Mixed	217;217	16	30	41	25	< 0.001	137
	Vegetarian	70;70	21	30	34	28	> 0.05	113

Footnotes see Table 4.2

(Röhrig 1998). Interestingly, the average copper intake of Germans increased significantly after the country's reunification (Anke et al. 1997e).

The daily molybdenum and nickel consumptions in Europe are much higher than the respective normative requirements of 25 and 35 µg. A deficiency of either element in humans is not to be expected (Anke et al. 1993, 1993a, 2000e). Women prefer to eat more vegetables, which are rich in molybdenum and nickel, than men. The molybdenum and nickel intake of men is less than 24%, and reflects the preference of women for molybdenum- and nickel-rich foodstuffs (vegetables, chocolate and cacao products) (Holzinger et al. 1998a, Anke et al. 1993, 2001a).

Before the iodization of mineral mixtures for domestic animals (with 10 mg I kg^{-1}) and of common salt for kitchen and indus-trial use (with 20 mg I kg^{-1}), Germany and most other European countries were iodine-deficient areas, with corresponding iodine-deficiency diseases.

Since 2000, daily iodine intake has increased to more than 100 µg, and only 10% of German adults take in less than this (50–100 µg). Men prefer food that is iodine-rich, such as sausages (bockwurst), and take in 36–56% more iodine than women (Anke 2004a).

The northern parts of Europe produce selenium-poor foods. Selenium consumption in Germany is low, on average, amounting to between 40 and 90 µg per day. The normal selenium requirement of women and men is not met in every case, and selenium deficiency in German is common (Drobner 1997, Anke et al. 2002, 2003). Vegetarians, on average, take in less selenium than people with mixed diets.

4.3.3
Trace Element Concentration of the Dry Matter Consumed

The trace element concentration of dietary dry matter consumed is a good indicator of the trace element supply of humans with mixed or vegetarian diets. The trace element concentration of the consumed dry matter does not vary with dry matter intake, which is influenced by gender, age, season and eating habits (Anke et al. 1997a).

The iron, zinc, iodine and selenium concentrations of the food dry matter consumed is only slightly higher than the normative requirements of these elements (see Table 4.8). The findings show that the intakes of iron, zinc, iodine and selenium in Germany and Europe do not always meet the normative requirements.

In contrast, the manganese, copper, molybdenum and nickel concentrations of dietary dry matter consumed by people with mixed or vegetarian diets meet normative requirements. A deficiency of copper, molybdenum and nickel is not to be expected in Germany and Europe (Anke et al. 1997e), though the possibility of a secondary copper deficiency cannot be completely excluded.

A significant influence of gender on trace element concentrations of consumed food dry matter has only been found for nickel (Table 4.10). This is most likely due to the higher intake of nickel-rich cacao products (Anke et al. 1993).

4.3.4
Trace Element Intake of Humans per kg Body Weight

The iron, zinc, iodine and selenium intakes of adults with mixed diets illustrates the marginal supply of these elements

Tab. 4.10: Trace element concentration of the eaten dry matter in dependence of gender and type of diet (mg/kg dry matter, DM)

Element	Diet	n[5] (w) ; n (m)	Women (w)		Men (m)		p[2]	%[1]
			SD	Mean	Mean [4]	SD [3]		
Fe	Mixed	217;217	11	22	21	9,1	> 0,05	95
	Vegetarian	70;70	13	24	22	7,8	> 0,05	92
Zn	Mixed	217;217	7,4	21	20	5,0	> 0,05	95
	Vegetarian	70;70	5,1	22	20	5,6	> 0,05	91
Mn	Mixed	217;217	3,5	8,2	7,6	2,7	> 0,05	93
	Vegetarian	70;70	5,2	14	12	5,1	< 0,001	86
Cu	Mixed	217;217	2,2	3,7	3,2	1,4	< 0,05	86
	Vegetarian	70;70	1,3	4,1	4,4	1,7	> 0,05	107
Mo	Mixed	217;217	302	305	276	178	> 0,05	90
	Vegetarian	70;70	330	471	385	167	< 0,05	82
Ni	Mixed	217;217	242	320	263	217	< 0,05	82
	Vegetarian	70;70	308	479	419	331	< 0,05	87
I	Mixed	217;217	145	286	312	134	> 0,05	109
	Vegetarian	70;70	137	212	259	140	> 0,05	122
Se	Mixed	217;217	44	98	110	78	> 0,05	112
	Vegetarian	70;70	52	77	75	93	> 0,05	97

Footnotes see Table 4.2.

Tab. 4.11: Trace element intake per kg body weight in dependence of gender and type of diet (μg/kg body weight, BW)

Element	Diet	n[5] (w) ; n (m)	Women (65 kg)		Men (75 kg)		p[2]	%[1]
			SD	Mean	Mean [4]	SD [3]		
Fe	Mixed	217;217	47	96	97	47	> 0,05	101
	Vegetarian	70;70	110	169	151	62	> 0,05	89
Zn	Mixed	217;217	46	91	95	39	> 0,05	104
	Vegetarian	70;70	49	151	138	59	> 0,05	91
Mn	Mixed	217;217	21	36	35	16	> 0,05	97
	Vegetarian	70;70	37	97	84	51	< 0,01	87
Cu	Mixed	217;217	13	16	15	9,2	> 0,05	94
	Vegetarian	70;70	13	28	30	14	> 0,05	107
Mo	Mixed	217;217	1,68	1,36	1,29	0,92	> 0,05	95
	Vegetarian	70;70	2,49	3,17	2,49	1,33	< 0,05	79
Ni	Mixed	217;217	0,90	1,4	1,2	1,1	< 0,05	86
	Vegetarian	70;70	2,1	3,2	2,8	2,1	< 0,05	88
I	Mixed	217;217	0,7	1,3	1,5	0,8	< 0,01	115
	Vegetarian	70;70	0,9	1,4	1,8	1,2	< 0,001	129
Se	Mixed	217;217	0,27	0,46	0,53	0,31	< 0,05	115
	Vegetarian	70;70	0,39	0,52	0,49	0,38	> 0,05	94

Footnotes see Table 4.2.

(Table 4.11). Vegetarians consume significantly more iron and zinc per kg body weight, but the availability of these elements is reduced through binding to phytic acid (Anke 2004d). The intake of iodine and selenium by vegetarians is not different from that of people with mixed diet. Gender had a significant influence only on nickel and iodine intake per kg body weight (Table 4.11). Men consumed less nickel than women, while took in less iodine than men. The danger of a nickel allergy is much greater in women than in men. An iodine deficiency is more frequent in girls and women than in boys or men (Anke et al. 1993, Anke 2004a).

4.3.5
Fecal Excretion and Apparent Absorption of Trace Elements

Fecal excretion of the two nonmetals, iodine and selenium, account for ~19% of iodine intake and 38% of selenium intake; this is significantly less than that of the metals molybdenum, nickel, zinc, iron, and especially manganese (Table 4.12). At approximately 70%, the fecal excretion of molybdenum and nickel is less than that of zinc, iron, and manganese. The apparent absorption rate of the trace elements correlates negatively with the fecal excretion of all trace elements. On average, the highest absorption rates are 81% for iodine and 62% for selenium. Surprisingly, the high absorption rates of nickel and molybdenum (~40%) and the low rates of zinc, iron and

Tab. 4.12: Fecal excretion and apparent absorption rate of trace elements by adults

Element	Excretion by feces, %				Apparent absorption rate, %[1]			
	Women		Men		Women		Men	
	Mixed	Veg.	Mixed	Veg.[3]	Mixed	Veg.	Mixed	Veg.
I	24	15	19	17	75	80	82	86
Se	38	–	39	–	62	–	62	–
Mo	72	70	69	67	37	37	37	38
Ni	73	–	73	–	37	–	46	–
Zn	91	95	90	97	3,0	3,3	3,3	0,5
Fe	94	94	94	94	–	–	–	–
Mn	99,4	99,7	98,9	99,8	–	–	–	1,9

Footnotes see Table 4.5.

manganese were negative in some populations (intake of these elements from dust over the lung) and could not be counted.

4.3.6
Trace Element Intake Measured by the Duplicate Portion Technique and the Basket Method

The calculation of trace element intake using the basket method led to a significant overestimation compared with direct analysis and the duplicate portion technique (Table 4.13). The overestimation is lowest in the case of copper and manganese (10–20%), medium for iodine, manganese and iron, and high for molybdenum, selenium,

and nickel. This overrating of trace element intakes led to an increased recommendation figures for trace element supply.

4.3.7
Intake of Trace Elements Through Animal and Vegetarian Foodstuffs and Beverages

Animal foodstuffs supply two-thirds and more of the iodine, selenium, and zinc intakes (Table 4.14). Beverages contribute a relatively high share (10%) to iodine consumption, but only 5% to selenium and zinc intakes. The majority of the heavy metals (iron, nickel, molybdenum, copper, manganese) is provided by vegetable foodstuffs, and partly by beverages. Approxi-

Tab. 4.13: Trace element intake of adults with mixed diet determined by the duplicate portion technique and calculated by the basket method

Element	Women		Men		
	Duplicate	Basket	Duplicate	Basket	% [1]
Cu	1,1	1,2	1,2	1,4	113
Mn	2,4	3,1	2,7	3,9	118
I	51	57	57	83	130
Zn	7,5	10,6	9,7	13,4	140
Fe	9,5	12,8	12,0	18,1	144
Mo	69	102	81	125	151
Se	24,8	36,6	31,0	50,6	156
Ni	90	135	97	169	163

Tab. 4.14: Intake of inorganic body component through animal and vegetable foodstuff respectively and beverages by people with mixed diets in percent

Element	Animal foodstuff	Vegetable foodstuff	Beverages
Iodine	74	16	10
Selenium	72	26	2
Zinc	62	33	5
Iron	40	56	4
Nickel	24	49	27
Molybdenium	22	70	8
Copper	21	55	24
Manganese	5	68	27

mately one-quarter of the nickel, copper and manganese consumptions are supplied by beverages. Animal foodstuffs, especially sausages and meat, are an important source of available iron.

4.4
Ultratrace Elements

4.4.1
Normative Requirements and Recommendations for Intake

The normative requirement of ultratrace elements is unknown, because their essentiality in humans has not yet been investigated. In experiments with animals, it was shown that a nutrition which is very poor in several ultratrace elements induces element-specific deficiency symptoms, including decreased feed intake, lowered growth rate, lowered reproduction performance, and increased mortality (Anke 2004d). Based on the results of these animal experiments and on the lowest daily intake of these elements by humans, potential requirements were deduced (Table 4.15). A potential daily aluminum requirement, provided that the element is really essential, should be < 2000 μg, on the average of a week, or < 30 μg Al kg^{-1} body weight. In addition, the aluminum concentration of the dry matter consumed by people with a mixed diet should contain < 7000 μg Al kg^{-1} DM, and by vegetarians 5000 μg Al kg^{-1} DM.

Tab. 4.15: Requirement and tolerable daily intake of ultratrace elements

Element	Normative requirement				Tolerable daily intake
	Day	Body weight 70 kg	Food dry matter		
			Mixed	Vegetarian	kg body weight
Aluminium (μg)	$< 2,000$	< 30	< 7000	5000	1 mg
Rubidium (μg)	< 100	$< 1,5$	350	225	2 mg
Lithium (μg)	< 100	$< 1,5$	350	225	2 mg
Titanium (μg)	< 50	0,7	150	120	5 mg
Arsenic (μg)	< 20	0,30	70	50	0,1 μg
Chromium (μg)	< 20	0,30	70	50	3,5 μg
Lead (μg)	< 10	$< 0,15$	35	25	3,5 μg
Vanadium (μg)	< 10	$< 0,15$	35	25	100 μg
Cadmium (μg)	< 5	$< 0,08$	18	12	1 μg

The tolerable daily intake of aluminum by adults may be $1 \, \text{mg kg}^{-1}$ body weight (Anke et al. 1990, 1991, 2001a, Anonymous 1996, Müller et al. 1995, a, b, Müller and Anke 1995, Carlisle and Curron 1993, Williams 1999, Burgess 1996).

The alkali element rubidium, together with lithium, seems most important for human health. A rubidium offer of $< 250 \, \mu\text{g kg}^{-1}$ feed DM was found to induce an abortion rate of approximately 80%. The postulated daily rubidium requirement of humans is $< 100 \, \mu\text{g}$ (Anke and Angelow 1994, 1995, Anke et al. 1997, 2003b). The postulated daily titanium requirement of $< 50 \, \mu\text{g}$ is low, as is the element's toxicity. An intake of $5 \, \text{mg Ti kg}^{-1}$ body weight in the form of titanium dioxide appears to be well tolerated, but titanium tetrachloride is much more toxic (Anke and Seifert 2004d). The daily chromium requirement of adult humans is $< 20 \, \mu\text{g}$ (Anderson 1987), and its tolerable daily intake was $3.5 \, \mu\text{g kg}^{-1}$ body weight (Anke et al. 1997b, d, g, 2000d, Anonymous 1996). The daily arsenic requirement of adults may be $20 \, \mu\text{g}$, and the intoxication threshold level for a person weighing 70 kg is thought to be $150 \, \mu\text{g}$ As per day, or $2.1 \, \mu\text{g kg}^{-1}$ body weight (Anke et al. 1996b, Anonymous 1996). The essentiality of lead was examined by Kirchgessner et al. (1988), and Reichlmayr-Lais and Kirchgessner (1981). If lead were to be an essential element, the daily requirement might be $< 10 \, \mu\text{g}$; the tolerable daily intake was found to be $3.5 \, \mu\text{g kg}^{-1}$ body weight (Anonymous 1996, Müller and Anke 1995, Anke et al. 1996, Seifert and Anke 2000). The proposed daily vanadium requirement is low ($10 \, \mu\text{g}$ for adults), whereas the tolerable daily intake is high ($100 \, \mu\text{g kg}^{-1}$ body weight) (Anke 2004c, Anonymous 1996). A cadmium requirement of humans can no longer be excluded as a cadmium-specific carboanhydrase was

discovered in the diatom *Thallasia sira weissflogii*, and species of cadmium-depleted goats developed deficiency symptoms (myasthenia) which were curable with a control feed containing $300 \, \mu\text{g kg}^{-1}$ DM per proband. The daily cadmium requirement of humans might be $< 5 \, \mu\text{g}$, and the tolerable daily intake $1 \, \mu\text{g Cd kg}^{-1}$ body weight (Anonymous 1996, Anke et al. 2000c, 2002f).

4.4.2
Ultratrace Element Intake of Adults with Mixed and Ovolactovegetarian Diets

In general, the daily intake of ultratrace elements (Table 4.16) by people with mixed or vegetarian diets is much higher in every case than the proposed requirement of these elements. First, it should be noted that, with the exception of aluminum, lead and uranium, men with either a mixed or a vegetarian diet tend to consume significantly greater quantities of ultratrace elements than women (Table 4.16). Men with a mixed diet take in a highly vanadium-rich diet (Anke 2004c). On average, men eat 24% more dry matter than women, and a deviation of approximately $\pm 12\%$ signals a significant preference of special food or food groups by women or men. The gender-specific choice demonstrates best the concentration of elements in the consumed dry matter (Table 4.17). With the exception of arsenic and titanium, vegetarians take in higher amounts of ultratrace elements than people eating a mixed diet. Vegetarians consume between 50% and 250% more barium, strontium, mercury, vanadium, and cadmium than do omnivores, this difference being most striking in the case of mercury. A deficiency of ultratrace elements in humans is not to be expected.

Tab. 4.16: Ultratrace element intake of adults with mixed and ovolactovegetarian diet in dependence of gender and type of diet

Element	Diet	$n^{5)}$ (w) ; n (m)	Women (w)		Men (m)		$p^{2)}$	$\%^{1)}$
			SD	Mean	Mean [4]	SD [3]		
Al (mg/d)	Mixed	168,168	1,9	3,1	3,2	2,2	> 0,05	103
	Vegetarian	70;70	3,2	4,1	4,1	1,9	> 0,05	100
Sr (mg/d)	Mixed	217;217	1,0	1,8	2,2	1,4	< 0,001	122
	Vegetarian	70;70	1,1	2,7	3,4	2,1	< 0,001	126
Rb (mg/d)	Mixed	196;196	0,732	1,6	1,7	0,76	< 0,05	106
Li (µg/d)	Mixed	294;294	724	713	990	1069	< 0,01	139
Ba (mg/d)	Mixed diet	217;217	0,220	0,490	0,570	0,36	< 0,05	116
	Vegetarian	70;70	0,490	0,950	1,00	0,46	> 0,05	105
As (µg/d)	Mixed	217;217	153	107	145	205	< 0,05	136
	Vegetarian	70;70	74	72	101	282	< 0,05	140
Ti (µg/d)	Mixed	217;217	53	80	90	79	< 0,05	112
	Vegetarian	70;70	48	65	102	81	< 0,01	157
Cr (µg/d)	Mixed	217;217	31	61	84	55	< 0,001	138
	Vegetarian	70;70	25	85	99	40	< 0,001	116
V (µg/d)	Mixed	217;217	15	11	33	35	< 0,001	300
	Vegetarian	70;70	103	49	39	34	> 0,05	80
Pb (µg/d)	Mixed	217;217	16	19	19	16	> 0,05	100
Hg (µg/d)	Mixed	217;217	4,5	2,7	4,8	6,8	< 0,001	178
	Vegetarian	49;49	9,1	12	15	12	< 0,05	125
Cd (µg/d)	Mixed	217;217	4,4	7,1	8,8	5,1	< 0,001	124
	Vegetarian	70;70	6,8	11	17	13	< 0,001	155
U (µg/d)	Mixed	168;168	2,0	2,6	2,8	2,4	> 0,05	108

Footnotes see Table 4.2.

4.4.3
Ultratrace Element Concentration of the Dry Matter Consumed

The ultratrace element concentration of the dry matter consumed shows a clear and significant gender-specific preference of special food or food groups (Table 4.17). Men prefer aluminum- and vanadium-rich foodstuffs, whereas women favor rubidium-, lead- and uranium-rich foods and beverages. The reason for the significantly higher aluminum concentration in the dry matter con-sumed by men is their larger consumption of bread, sausages and, especially, beer (Müller et al. 1995b) in comparison with women. The ample consumption of beer by men also accounts for the significantly higher vanadium content in the food dry matter consumed by them (Anke 2004c). The higher concentrations of rubidium, lead, and uranium in the food dry matter consumed by women is due to their preference of tea and coffee, with their high rubidium contents (Angelow 1994, Anke and Angelow 1994), the lead contained in the

Tab. 4.17: Ultratrace element concentration of the eaten dry matter in dependence of gender and type of diet

Element	Diet	$n^{5)}$ (w) ; n (m)	Women (w)		Men (m)		$p^{2)}$	$\%^{1)}$
			SD	Mean	Mean $^{4)}$	SD $^{3)}$		
Al mg/kg DM	1Mixed	168,168	5,2	8,3	10	5,6	< 0,01	120
	Vegetarian	70;70	4,3	9,1	11	8,3	< 0,01	121
Sr mg/kg DM	Mixed	168;168	4,7	6,6	6,5	4,3	> 0,05	98
	Vegetarian	70;70	2,9	7,1	7,6	5,8	> 0,05	107
Rb mg/kg DM	Mixed	294;294	1,9	5,2	4,6	1,7	< 0,001	88
Li mg/kg DM	Mixed	294;294	2,9	2,5	2,6	2,9	> 0,05	104
Ba mg/kg DM	Mixed	217;217	0,59	1,7	1,6	0,76	> 0,05	94
	Vegetarian	70;70	1,1	2,4	2,1	0,74	< 0,05	88
As µg/kg DM	Mixed	168;168	449	335	375	504	> 0,05	112
	Vegetarian	70;70	179	181	196	192	> 0,05	108
Ti µg/kg DM	Mixed	168;168	255	298	257	210	> 0,05	86
	Vegetarian	70;70	120	170	214	153	> 0,05	126
Cr µg/kg DM	Mixed	217;217	91	212	225	102	> 0,05	106
	Vegetarian	70;70	46	221	207	46	> 0,05	94
V µg/kg DM	Mixed	217;217	37	37	91	91	< 0,001	246
	Vegetarian	70;70	257	120	87	76	> 0,05	72
Pb µg/kg DM	Mixed	217;217	58	68	56	40	< 0,05	82
Hg µg/kg DM	Mixed	217;217	15	9,5	14	20	> 0,05	147
	Vegetarian	49;49	22	32	26	19	> 0,05	81
Cd µg/kg DM	Mixed	217;217	17	26	25	13	> 0,05	96
	Vegetarian	70;70	17	28	35	20	> 0,05	125
U µg/kg DM	Mixed	168;168	9,6	10	8,0	6,8	< 0,001	80

Footnotes see Table 4.2.

drinking water used for these beverages (Müller and Anke 1995), and the consumption of uranium-rich mineral waters (Seeber et al. 1997, 1998).

4.4.4
Ultratrace Element Intake per kg Body Weight

The ultratrace element consumption per kg body weight shows that the tolerable daily intakes of humans with mixed and vegetarian diets do not exceed the threshold levels (Table 4.18). There is one exception: the arsenic intake of people with mixed and vegetarian diets is close to the threshold of a potentially harmful level. The burning of arsenic-rich brown coal in parts of Europe, together with the arsenic emissions by power stations and households, formerly enriched the soils with arsenic and introduced considerable quantities of the element into the food chain over wide regions. Besides these sources of arsenic in European environments, there are several areas in Saxony (Germany) in which rocks (mica, gneiss) and their weathering soils are naturally very rich in arsenic and deliver it into the food chain (Anke 1986, Anke et al. 1997c, Risch 1980, Metzner et al. 1991). The

Tab. 4.18: Ultratrace element intake per kg body (BW) weight in dependence of gender and type of diet

Element	Diet	$n^{5)}$ (w) ; n (m)	Women (w)		Men (m)		$p^{2)}$	$\%^{1)}$
			SD	Mean	Mean [4]	SD [3]		
Al (µg/kg BW)	Mixed	168;168	32	47	42	30	> 0,05	89
	Vegetarian	70;70	47	70	60	27	> 0,05	86
Sr (µg/kg BW)	Mixed	168;168	18	27	28	17	> 0,05	104
	Vegetarian	70;70	25	48	49	32	> 0,05	102
Ba (µg/kg BW)	Mixed	217;217	3,8	7,5	7,3	5,0	> 0,05	97
	Vegetarian	70;70	7,7	16	14	6,2	> 0,05	88
Cr (µg/kg BW)	Mixed	217;217	0,50	0,94	1,1	0,88	> 0,05	117
	Vegetarian	70;70	0,48	1,5	1,4	0,59	> 0,05	93
As (µg/kg BW)	Mixed	168;168	2,49	1,68	1,95	2,86	> 0,05	116
	Vegetarian	70;70	1,26	1,24	1,47	1,52	> 0,05	119
Ti (µg/kg BW)	Mixed	217;217	0,93	1,2	1,1	0,99	> 0,05	92
	Vegetarian	70;70	0,77	1,1	1,4	0,89	> 0,05	127
Hg (µg/kg BW)	Mixed	217;217	77	42	58	79	> 0,05	138
	Vegetarian	49;49	144	202	207	169	> 0,05	102
V (ng/kg BW)	Mixed	217;217	220	170	410	434	> 0,05	241
	Vegetarian	70;70	1505	757	552	472	> 0,05	73
Cd (ng/kg BW)	Mixed	217;217	65	107	111	62	> 0,05	104
	Vegetarian	70;70	128	193	237	170	> 0,05	123
Rb (µg/kg BW)	Mixed	294;294	15	26	23	10	< 0,01	88
Li (µg/kg BW)	Mixed	294;294	13	11	12	13	> 0,05	109
Pb (ng/kg BW)	Mixed	217;217	253	279	240	160	> 0,05	86
U (ng/kg BW)	Mixed	168;168	28	39	36	34	> 0,05	92

Footnotes see Table 4.2.

danger of a harmful arsenic load in Europe is quite tangible.

4.4.5
Fecal Excretion and Apparent Absorption of Ultratrace Elements

The majority of ultratrace elements are metals, and their fecal excretion rate is greater than 50% (Table 4.19). Fecal excretion of lead by people with a mixed diet is astonishingly low (50%), while the apparent absorption rate is very high (40% on average). The daily lead intake of test popula-tions was found to vary between 17 and 27 µg, which is low (Müller et al. 1995, 1995a, 1997). The fecal excretion of lead by people with mixed and vegetarian diets varies significantly.

The phytin-rich nutrition of ovolactovegetar-ians lowers the availability of strontium and increases fecal excretion of this alkali metal. The apparent absorption rate of strontium in vegetarians, like that of calcium, is signifi-cantly lower than in people with a mixed diet.

The fecal excretion rates of the ultratrace elements strontium, titanium, uranium and chromium were 90–99%, and the appa-

Tab. 4.19: Fecal excretion and apparent absorption rate of ultratrace elements by adults

Element (n;n)	Excretion by feces, %				Apparent absorption rate, %[1]			
	Women		Men		Women		Men	
	Mixed	Veg.[3]	Mixed	Veg.	Mixed	Veg.	Mixed	Veg.
Pb	52	–	50	–	45	–	36	–
Sr	88	93	87	94	7,4	0,3	21	6,8
Ti	90	74	87	83	–	–	1,6	–
U	91	–	91	–	4,7	–	6,8	–
V	94	98	98	95	–	2,5	–	10
Cr	99	99	99	99	2,0	0,0	4,4	2,4

Footnotes see Table 4.5.

rent absorption rate, if measurable, varied from 10% to 0%.

4.4.6
Ultratrace Element Intake Measured by the Duplicate Portion Technique and the Basket Method

The calculation of ultratrace element intakes by the basket method overestimated the intake, with the exception of strontium and uranium. The calculation of intakes for aluminum, barium, cadmium, arsenic and ash overestimated the consumption by 10–20% (Table 4.20). Calculated intakes of vanadium, lead, titanium, rubidium and chromium were higher by 35 to 90% than quantities measured using the duplicate portion technique. By contrast, mercury and lithium intakes estimated by the basket method

Tab. 4.20: Element intake of adults with mixed diet determined by the duplicate portion technique and calculated by the basket method

Element	Women		Men		%[1]
	Duplicate	Basket	Duplicate	Basket	
Sr (mg/d)[2]	1,8	1,8	2,2	1,9	92
U (µg/d)	2,2	2,1	2,5	2,4	96
Al (mg/d)	5,4	5,8	6,5	7,4	111
Ba (mg/d)	0,94	1,03	1,07	1,29	115
Cd (µg/d)	9,7	10,9	11,5	14,0	117
As (µg/d)	107	127	145	176	120
Ash (mg/d)	12750	16001	17290	20100	120
V (µg/d)	9,4	11,7	19,1	26,8	135
Pb (µg/d)	19	23	19	30	136
Ti (µg/d)	70	96	82	121	143
Rb (µg/d)	1657	2613	1699	2950	166
Cr (µg/d)	61	115	84	158	188
Hg (µg/d)	2,7	7,2	4,6	10,1	237
Li (µg/d)	313	733	383	941	240

Footnotes see Table 4.6.

were twice as high as values determined using the duplicate portion technique. Hence, the basket method should no longer be used to determine macro, trace and ultratrace element intake in humans.

4.4.7
Intake of Ultratrace Elements Through Animal and Vegetable Foodstuffs and Beverages

Lithium is the only one of the ultratrace elements analyzed that is mainly delivered through animal foodstuffs (Table 4.21). All other elements analyzed are mainly supplied by vegetable foodstuffs or, in the case of uranium, rubidium and vanadium, through beverages. It is very surprising that lithium is accumulated in the milk and eggs of animals, as these are main suppliers of this trace element (Anke et al. 2003, Schäfer 1997). The proportions of ultratrace elements analyzed which are delivered via vegetable foodstuffs are highest in the cases of cadmium (74%) and aluminum

Tab. 4.21: Intake of inorganic body component through animal and vegetable foodstuff, respectively, and beverages by people with mixed diets in percent

Element	Animal foodstuffs	Vegetable foodstuffs	Beverages
Lithium	62	25	13
Mercury	45	43	12
Chromium	43	41	16
Arsenic	39	57	4
Barium	34	57	9
Titanium	33	52	15
Lead	28	54	18
Uranium	26	34	40
Rubidium	23	32	45
Aluminium	21	70	9
Cadmium	20	74	6
Vanadium	12	30	58
Strontium	9	58	33

(70%), but lower for lithium (24%). Drinking water and home-made beverages have regional influences on ultratrace element intake. Likewise, the type of foodstuff ingested, and also the composition of drinking and household waters leads to significant variations in ultratrace element intake.

4.5
Summary

The normal macro, trace and ultratrace requirements of adult humans allows, in relation to the recommended and the tolerable daily intakes, a rating of the supply or loading with the inorganic food and body components.

In adult humans, the form of diet and gender cause significant variations in macro, trace and ultratrace element intake. The intake of inorganic food components can be stated daily on the average of a week, related to the body weight in kg, and also in form of the element concentration in the dry matter consumed.

In men, the daily intake of dry matter, whether with a mixed or a vegetarian diet, is 25% higher than in women. Ovolactovegetarians consume, on average, 28% more dry matter than people with mixed diets. In case of a gender-specific preference of special food or food groups, the element concentrations show this influence least; women significantly prefer calcium, molybdenum, nickel, rubidium, and uranium-rich foodstuffs, while men favor sodium-, aluminum- and vanadium-rich foodstuffs and beverages. These gender-specific preferences of elements are related to differences in the eating and drinking habits of women and men.

In general, intake per day, intake per kg body weight and the element concentration of the consumed dry matter showed, in good agreement the marginal supply of magnesium, iron (women), zinc, selenium and iodine. The normal macro, trace and ultratrace element intake, with the exception of arsenic, is far removed from the upper limit or tolerable daily intake.

The apparent absorption rate of the macro elements in humans with a mixed diet varied between 98% for sodium, 83% for potassium, 61% for phosphorus, 35% for magnesium, and 13% for calcium. Ovolacto-vegetarians have a significantly decreased apparent absorption rate for magnesium and calcium, and apparent absorption rates for trace elements varied between 81% for iodine, 62% for selenium, 42% for nickel, 37% for molybdenum, 2.5% for zinc, and 2.0% for manganese.

The calculation of inorganic food component intake using the basket method generally resulted in an overestimation of element consumption when compared with results obtained with the duplicate portion technique. The results ranged from 4% in the case of sodium to 44% for calcium and iron, 55% for selenium, 66% for chromium, and 240% for mercury and lithium. Hence, basket method should no longer be used to determine macro, trace and ultratrace element intake in humans.

Animal foodstuffs deliver 74% of iodine, 72% selenium, and 62% lithium intake in humans, while vegetables supply the highest amounts of cadmium (74%), molybdenum (70%), aluminum (70%), potassium (62%), strontium (58%), barium (57%), arsenic (57%), and the lowest amounts of iodine (16%) and lithium (25%).

Beverages provide humans with 28% of their magnesium intake, 27% of nickel, 24% of copper, 33% of strontium, 40% of uranium, 45% of rubidium, and 58% of vanadium consumption. Drinking water, home-made beverages (tea, coffee) and beer are important suppliers of inorganic body components.

References

ANDERSON RA (1987) *Chromium*. In: Mertz W, ed. Trace Elements in Human and Animal Nutrition, pp 225–244. Academic Press Inc., San Diego, New York.

ANGELOW L (1994) *Rubidium in der Nahrungskette*. Qualification for a lectureship, Friedrich Schiller University, Biological- Pharmaceutical Faculty Jena, Germany.

ANKE M (1982) *Anorganische Bausteine*. In: Püschner A and Simon O, eds. Grundlagen der Tierernährung, pp 46–67. Gustav Fischer Verlag Jena, Germany.

ANKE M (1986) *Arsenic*. In: Mertz, W, ed. Trace Elements in Human and Animal Nutrition. Volume 2, pp. 347–372. Academic Press Inc. Orlando, San Diego, USA.

ANKE M, HENNIG A, SCHNEIDER H-J, LÜDKE H, VON GAGERN W and SCHLEGEL H (1970) *The interrelations between cadmium, zinc, copper and iron an metabolism of hens, ruminants and man*. In: Mills CF, eds. Trace Element Metabolism in Animals 1, pp. 317–320. Livingstone, Edinburgh and London, UK.

ANKE M and RISCH MA (1989) *Importance of molybdenum in animal and man*. In: Anke M et al. (eds) 6th International Trace Element Symposium, Volume 1, Molybdenum, Vanadium, pp 303–321. University Leipzig and Jena , Kongress- und Werbedruck Oberlungwitz, Germany.

ANKE M, GROPPEL B, MÜLLER M and REGIUS A (1990) *Effects of aluminium-poor nutrition in animals*. In: Pais I ed. International Trace Element Symposium, pp 303–324. St. Istvan University, Budapest, Hungary.

ANKE M, GROPPEL B and KRAUSE U (1991) *The essentiality of the toxic elements aluminium and vanadium*. In: Momcilovich B, ed. Trace Elements in Man and Animal–7. pp 11-9–11-10. IMI Zagreb, Croatia.

ANKE M, GROPPEL B, KRAUSE U, ARNHOLD W and LANGER M (1991a) *Trace element intake (Zink, manganese, copper, molybdenium, iodine and nickel) of humans in Thuringia and Brandenburg of the Federal Republic of Germany*. J Trace Elem Electrolytes Health Dis. **5**:69–74.

ANKE M, LÖSCH E, MÜLLER M, KRÄMER K, GLEI M and BUGDOL K (1992) *Der Kaliumgehalt der Lebensmittel bzw. Getränke sowie die Kaliumaufnahme und Kaliumbilanz Erwachsener in Deutschland*. In: Holtmeier H. J, ed. Kalium, pp. 217–231. Wissenschaftl. Verlagsgesellschaft mbH, Stuttgart.

ANKE M, LÖSCH E, MÜLLER M and GROPPEL B (1992a) *Die Natriumaufnahme Erwachsener in den neuen Bundesländern Deutschlands.* In: Holtmeier H-J, eds. Bedeutung von Natrium und Chlorid für den Menschen, pp 194–205. Springer-Verlag, Berlin.

ANKE M, LÖSCH E, MÜLLER M, KRÄMER K, GLEI M and BUGDOL K (1992b) *Potassium in human nutrition.* In: Anonymous, eds. Potassium in Eosystems, pp 187–204. International Potash Institute, Basel, Switzerland.

ANKE M and GLEI M (1993) *Molybdenum.* In: Seiler HGA and Siegel H, eds. handbook on Metals in Clinical Chemistry, pp. 495–501. Marcel Dekker, INC, New York, Basel.

ANKE M, LÖSCH E, ANGELOW L and KRÄMER K (1993) *Die Nickelbelastung der Nahrungskette von Pflanze, Tier und Mensch in Deutschland. 3. Der Nickelgehalt der Lebensmittel und Getränke des Menschen.* Mengen- und Spurenelemente **13**:400–414.

ANKE M, LÖSCH E, GLEI M, MÜLLER M, ILLING H, KRÄMER K (1993a) *Der Molybdängehalt der Lebensmittel und Getränke Deutschlands.* Mengen- und Spurenelemente **13**:537–553.

ANKE M and ANGELOW L (1994) *The biological importance of rubidium.* In: Pais I, ed. 6. Internat. Trace Element Symposium, Budapest, pp. 241–262, St. Istvan University, Hungary.

ANKE M and ANGELOW L (1995) *Rubidium in the food chain.* Fresenius J Anal Chem **352**:236–239.

ANKE M, GLEI M, MÜLLER M, SEIFERT M, ANKE S, RÖHRIG B, ARNHOLD W and FREYTAG H (1996) *Die nutritive Bedeutung des Zinks.* Vitaminspur **11**:125 –135.

ANKE M, SCHEIDT-ILLIG R, MÜLLER M, ERLER M, MOCANU H, NEAGOE A, KRÄMER K, RICHTER D, ANGELOW L and SEIFERT M (1996a) *Blei in der Nahrungskette des Menschen eines teerbelasteten Lebensraumes (Rositz, Thüringen).* Mengen- und Spurenelemente **16**:836–846.

ANKE M, SEIFERT M, ANGELOW L, THOMAS G, DROBNER C, MÜLLER M, GLEI M, FREITAG H, ARNHOLD W, KÜHNE G, ROTHER C, KRÄUTER U and HOLZINGER S (1996b) *The biological importance of arsenic – toxicity, essentiality, intake of adults in Germany.* In: Pais I, ed. 7th International Trace Element Symposium, pp. 103 –125. St. Istvan University, Budapest, Hungary.

ANKE M, ANGELOW L, GLEI M, ANKE S, LÖSCH E and GUNSTHEIMER G (1997) *The biological essentiality of rubidium.* International Symposium on Trace Elements in Human: New Perspectives, pp.245–263. G. Morogianni, Acharnai, Greece.

ANKE M, DORN W, MÜLLER M, RÖHRIG B, GLEI M, GONZALES D, ARNHOLD W, ILLING-GÜNTHER H, WOLF S, HOLZINGER S and JARITZ M (1997a) *Der Chromtransfer in der Nahrungskette. 4. Mitteilung: Der Chromverzehr Erwachsener in Abhängigkeit von Zeit, Geschlecht, Alter, Körpermasse, Jahreszeit, Lebensraum, Leistung.* Mengen- und Spurenelemente **17**:912–927.

ANKE M, GALAMBOS C, HARTMANN E, LÖSCH E, MÖLLER E, SCHOLZ E, GLEI M, ARNHOLD W, SEEBER O and SEIFERT M (1997b) *Der Chromtransfer in der Nahrungskette. 3. Mitteilung: Der Chromgehalt tierischer Lebensmittel und verschiedener Getränke.* Mengen- und Spurenelemente **17**:903–911.

ANKE M, GLEI M, ARNHOLD W, DROBNER C and SEIFERT M (1997c) *Arsenic.* In: O'Dell BL and Sunde RA, eds. Handbook of Nutritionally Essential Mineral Elements, pp. 631–639. Marcell Dekker, Inc. New York, Basel, Hong Kong.

ANKE M, ILLING-GÜNTHER H, HOLZINGER S, JARITZ M, GLEI M, MÜLLER M, ANKE S, TRÜP-SCHUCH A, NEAGOE A, ARNHOLD W and SCHÄFER U (1997d) *Der Chromtransfer in der Nahrungskette. 2. Mitteilung: Der Chromgehalt pflanzlicher Lebensmittel.* Mengen- und Spurenelemente **17**:894–902.

ANKE M, MÜLLER M, ARNHOLD W, GLEI M, ILLING-GÜNTHER H, HARTMANN E, LÖSCH E, DROBNER C, JARITZ M, HOLZINGER S and RÖHRIG B (1997e) *Problems of trace and ultra trace element supply of humans in Europe.* In: Garban Z and Dragan P, eds. Metal Elements in Environment, Medicine and Biology, pp. 15–34. Publishing House Eurobit Timisoara, Romania.

ANKE M, MÜLLER M, GLEI M, JARITZ M, HOLZINGER S, ROTHER C, ARNHOLD W, SCHMIDT P and DROBNER C (1997g) *Der Chromtransfer in der Nahrungskette. 5. Mitteilung: Die Chromausscheidung über Fäzes, Urin bzw. Milch und die Chrombilanz Erwachsener.* Mengen- und Spurenelemente **17**:928–933.

ANKE M, TRÜPSCHUCH A, ARNHOLD W, ILLING-GÜNTHER H, MÜLLER M, GLEI M, FREYTAG H and BECHSTEDT U (1997f) *Die Auswirkungen einer Nickelbelastung bei Tier und Mensch.* In Lombeck I, eds. Spurenelemente, pp 52–63. Wissenschaftliche Verlagsgesellschaft Stuttgart, Germany.

ANKE M, GLEI M, GROPPEK B, ROTHER C and CONZALES D (1998) *Mengen-, Spuren- und Ultra-*

spurenelemente in der Nahrungskette. Nova Acta Leopoldina NF **79**:157–190.

ANKE M, DORN W, MÜLLER M, ROTHER C, LÖSCH E, HARTMANN E, MÖLLER E, NEAGOE A and MOCANU H (1999c) *Mangantransfer in der Nahrungskette des Menschen. 4. Mitteilung: Der Manganverzehr Erwachsener in Abhängigkeit von Geschlecht, Zeit, Lebensraum, Kostform, Alter, Körpergewicht, Jahreszeit und Stillzeit.* Mengen- und Spurenelemente **19**:1030–1037.

ANKE M, GLEI M, WINNEFELD K, ARNHOLD W, VORMANN J, RÖHRIG B, JARITZ M, HOLZINGER S, LATUNDE-DADA O and HARTMANN E (1999d) *Supplementierung und Therapie mit Zink.* In: Meissner D, ed. Spurenelemente, pp. 98–109. Wissenschaftliche Verlagsgesellschaft mbH Stuttgart.

ANKE M, GÜRTLER H, ANKE S, MÜLLER M, ARNHOLD W, SEIFERT M, LÖSCH E and SEEBERG O (1999b) *Der Mangantransfer in der Nahrungskette des Menschen. 1. Mitteilung: Die biologischen Grundlagen des Mangantransportes vom Boden über die Flora und Fauna bis zum Menschen.* Mengen- und Spurenelemente **19**:1002–1012.

ANKE M, GROPPEL B, ANKE S, RÖHRIG B and NEAGOE A (1999e) *Mangan in der Ernährung.* Rekasan-Journal **6**:10–13.

ANKE M, VORMANN J, GLEI M, MÜLLER R and ARNHOLD W (1999a) *Mangantransfer in der Nahrungskette des Menschen. 5. Mitteilung: Manganbilanz und Manganbedarf Erwachsener.* Mengen- und Spurenelemente **19**:1038–1046.

ANKE M, LÖSCH E und ANKE S (2000a) *Natrium in der Nahrungskette des Menschen. 4. Mitteilung: Der Natriumverzehr Erwachsener in Abhängigkeit von Geschlecht, Zeit, Lebensraum, Kostform, Alter, Körpergewicht, Jahreszeit und Stillzeit.* Mengen- und Spurenelemente **20**:1209–1216.

ANKE M, LÖSCH E, HARTMANN E and ANKE S (2000b) *Natrium in der Nahrungskette des Menschen. 5. Mitteilung: Natriumbilanz und Natriumbedarf des Menschen.* Mengen- und Spurenelemente **20**:1193–1200.

ANKE M, MÜLLER R, DORN W, SEIFERT M, MÜLLER M, GONZALES D, KRONEMANN H and SCHÄFER U (2000c) *Toxicity and essenticality of cadmium.* In: Ermidou-Pollet S and Pollet S, eds. International Symposium on Trace Elements in Human. New Perspectives. pp 343–362. G. Morogiannis, Acharnai, Greece.

ANKE M, MÜLLER R, TRÜPSCHUCH A, SEIFERT M, JARITZ M, HOLZINGER S and ANKE S (2000d) *Intake of chromium in Germany: risk or normality?* J Trace and Microprobe Techn **18**:541–548.

ANKE M, TRÜPSCHUCH A, DORN W, SEIFERT W, PILZ K, VORMANN J and SCHÄFER U (2000e) *Intake of Nickel in Germany: risk or normality?* J Trace and Microprobe Techn **18**:549–556.

ANKE M (2001) *Eisen.* In: Praxishandbuch Functional Food, 4. Akt.-Lfg 09. Behrs Verlag, Hamburg, Germany.

ANKE M, TRÜPSCHUCH A, ANKE S, MÜLLER M, MÜLLER R, SCHÄFER U and BLAHA E (2001) *The essentiality of nickel in the food chain, intake and balance of adults from different places and with various eating habits.* In: Pais I, ed. New Perspectives in the Research of Hardly Known Trace Elements and the Importance of the Interdisciplinary Cooperation, pp. 7–53. St Istvan University, Faculty of Food Science Budapest, Hungary.

ANKE M, MÜLLER M, ANKE S, GÜRTLER H, MÜLLER R, SCHÄFER U and ANGELOW L (2001a) *The biological and toxilogical importance of aluminium in the environment and food chain of animals and humans.* In: Ermidou-Pollet S and Pollet S, eds. 3rd International Symposium on Trace Elements in Human: New Perspectives, pp. 230–247. G. Morogianni, Acharnai, Greece.

ANKE M (2002) *Trace element intake depending on the geological origin of the habitat, time, sex and form of diet.* In: Seifert M, Langer U, Schäfer U and Anke M, eds. Mengen- und Spurenelemente. Author and Element Index 1981–2002. pp. 11–19. Mugler Druck-Service GmbH, Wüstenbrand, Germany.

ANKE M, DROBNER C, RÖHRIG B, SCHÄFER U and MÜLLER R (2002) *Der Selenbestand der Flora und der Selengehalt pflanzlicher und tierischer Lebensmittel Deutschlands.* Ernährungsforschung, **47**:67–79.

ANKE M, KRÄMER-BESELIA K, LÖSCH E, SCHÄFER U and SEIFERT M (2002a) *Calcium supply, intake, balance and requirement of man. Fourth information: calcium intake of man in dependence of sex, time, eating habits, age and performance.* Mengen- und Spurenelemente **21**:1404–1409.

ANKE M, KRÄMER-BESELIA K, MÜLLER M, MÜLLER R, SCHÄFER U, FRÖBUS K and HOPPE C (2002b) *Calcium supply, intake, balance and requirement of men. Fifth information: Apparent absorption, balance and requirement.* Mengen- und Spurenelemente **21**:1410–1415.

ANKE M, KRÄMER-BESELIA K, LÖSCH E, MÜLLER R, MÜLLER M and SEIFERT M (2002c) *Calcium supply, intake, balance and requirement of man. First information: Calcium content of plant food.* Mengen- und Spurenelemente **21**:1386–1391.

ANKE M, KRÄMER-BESELIA K, DORN W and HOPPE C (2002d) *Calcium supply, intake, balance and requirement of man. Second information: Calcium content of animal food.* Mengen- und Spurenelemente **21**:1392–1337.

ANKE M, KRÄMER-BESELIA K, LÖSCH E, SCHÄFER U and MÜLLER R (2002e) *Calcium supply, intake, balance and requirement of man. Third information: Calcium content of beverages and the calcium intake via several groups of food stuffs.* Mengen- und Spurenelemente **21**:1398–1403.

ANKE M, MÜLLER M, TRÜPSCHUCH A and MÜLLER R (2002f) *Intake of effects of cadmium, chromium and nickel in humans.* J Commodity Sci **1**:41–63.

ANKE M, BERGMANN K, LÖSCH E and MÜLLER R (2003) *Potassium intake, balance and requirement of adults.* In: Schubert R, Flachowsky G, Jahreis G and Bitsch R, eds. Vitamine und Zusatzstoffe in der Ernährung von Mensch und Tier. 9 th Symposium, pp. 174–181, Friedrich-Schiller University Jena, Germany.

ANKE M, DROBNER C, ANGELOW L, SCHÄFER U and MÜLLER R (2003a) *Die biologische Bedeutung des Selens-Selenverzehr, Selenbilanz und Selenbedarf der Mischköstler und Vegetarier.* In: Schmitt G, ed. Ernährung und Selbstmedikation mit Spurenelementen, pp 1–17. Wissenschaftliche Verlagsgesellschaft mbH Stuttgart.

ANKE M, SCHÄFER U and ARNHOLD W (2003b) *Lithium.* In: Caballero B, Trogo L and Finglers P, eds. Encyclopedia of Food Sciences and Nutrition, pp. 3589–3593. Elsevier Science Ltd.

ANKE M (2004a) *Iodine.* In: Merian E, Anke M, Ihnat M and Stoeppler M, eds. Elements and their Compounds in the Environment, 2nd edn. Part IV, Chapter 9.4, Wiley-VCH, Weinheim, Germany.

ANKE M (2004b) *Sodium.* In: Merian E, Anke M, Ihnat M and Stoeppler M, eds. Elements and their Compounds in the Environment, 2nd edn. Part III, Chapter 1.2, Wiley-VCH, Weinheim, Germany.

ANKE M (2004c) *Vanadium.* In: Merian E, Anke M, Ihnat M and Stoeppler M, eds. Elements and their Compounds in the Environment, 2nd edn. Part III, Chapter 27, Wiley-VCH, Weinheim, Germany.

ANKE M and SEIFERT M (2004d) *Titanium.* Part III, Chapter 24.

ANKE M (2004e) *Essential and Roxic Effects of Macro-Trace and Ultratrace Elements for Animals.* In: Merian E, Anke M, Ihnat M and Stoeppler M, eds. Elements and their Compounds in the Environment, 2nd edn. Part II, Chapter 3. Wiley-VCH, Weinheim, Germany.

ANKE M, BERGMANN K and LÖSCH E (2004e) Personally information

ANONYMOUS (1996) *World Health Organisation, Geneva. Trace Elements in Human Nutrition and Health,* WHO, Geneva, Switzerland.

ANONYMOUS (2000) *Referenzwerte für die Nährstoffzufuhr.* Umschau/Braus, Frankfurt am Main, Germany.

BERGMANN K (1995) *Die Bedeutung tierischer Lebensmittel für die Natrium- und Kaliumversorgung des Menschen.* Doctoral Thesis. Vet.- Med.-Fakulty University Leipzig, Germany.

BURGESS J (1996) *Man and the Elements of Groups 3 and 13.* Chemical Society Reviews **25**:85–92.

CARLISLE EM and CURRAN MJ (1993) *Aluminium: an essential element for the chick.* In: Anke M, Meissner D and Mills CF, eds. Trace Elements in Man and Animals – 8, pp 695–698. Mugler, Kongress- und Werbedruck, Oberlungwitz, Germany.

DROBNER C (1997) *Die Selenversorgung Erwachsener Deutschlands.* Doctoral thesis, Friedrich Schiller University Jena, Biol. Pharm. Faculty, Germany.

EDER K and KIRCHGESSNER M (1997) *Nickel.* In: O'Dell BL and Sunde RA, eds. Handbook of Nutritionally Essential Mineral Elements, pp. 439–451. Marcel Dekker, New York.

GLEI M (1995) *Magnesium in der Nahrungskette unter besonderer Berücksichtigung der Magnesiumversorgung des Menschen.* Qualification for a lectureship. Biol. Pharm. Faculty, Friedrich Schiller University, Jena, Germany.

GLEI M, ANKE M and RÖHRIG B (1997) *Magnesium intake and magnesium balance of adults eating mixed or vegetarian diets.* In: Fischer PWF, L'Abbe MR, Cockell KA and Gibson RS, eds. Trace Elements in Man and Animals – 9: Proceedings of the Ninth International Symposium on Trace Elements in Man and Animals, pp.181–182. NRC Research Press, Ottawa, Canada.

GONZALES D, RAMIREZ A, HERNANDEZ M, MÜLLER R and ANKE M (1999) *Der Magnesiumverzehr erwachsener Mischköstler Mexikos.* Mengen- und Spurenelemente **19**:130–142.

HOLZINGER S, ANKE M, JARITZ M and RÖHRIG B (1997) *Molybdenum transfer in the food chain of humans.* In: Ermidou-Pollet S, ed. Trace Element in Human: New Perspectives, pp. 209–223. G. Morogiannis, Acharnai, Greece.

HOLZINGER S, ANKE M, RÖHRIG B and GONZALES D (1998a) *Molybdenum intake of adults in Germany and Mexico.* Analyst 123:447–450.

HOLZINGER S, ANKE M, SEEBER O and JARITZ M (1998b) *Die Molybdänversorgung von Säuglingen und Erwachsenen.* Mengen- und Spurenelemente 18:916–923.

KIRCHGESSNER M, REICHLMAYR-LAIS AM and STOKL KN (1988) *Retention of lead in growing rats with varying dietary lead supplements.* Journal of Trace Elements and Electrolytes in Health and Disease, 2:149–152.

KRÄMER K (1993) Calcium- und Phosphorausscheidung Erwachsener Deutschlands nach der Duplikat- und Marktkorbmethode. Doctoral thesis. Biol.- Pharm. Fakulty, Friedrich Schiller University, Jena, Germany.

LIEBSCHER D-H (2003) *Selbstmedikation mit hochdosiertem Magnesium.* In: Schmitt Y, ed. Ernährung und Selbstmedikation mit Spurenelementen, pp. 75–86. Wissenschaftliche Verlagsgesellschaft mbH Stuttgart, Germany.

MEIJ JC, LAMBERT PWJ VAN DEN HEUVEL PWJ and NINE VAM (2002) *Genetic disorders of magnesium homeostasis.* Bio Metalls 15:297–307.

METZNER I, VOLAND B and BOBACH G (1991) *Vorkommen und Verteilung von Arsen in Mittelgebirgsböden des Erzgebirges und Vogtlandes.* Mengen- und Spurenelemente 11:152–159.

MÜLLER M, ANKE M (1995) *Investigation into the oral lead exposure of adults in the former German Democratic Republic.* Z Lebensm Unters Forsch 200:38–43.

MÜLLER M, ANKE M, GÜRTLER H and ILLING-GÜNTHER H (1995) *Die Auswirkungen einer aluminiumarmen Ernährung bei der Ziege. 2. Mitteilung: Milchleistung, Lebenserwartung, Plasmaparameter, Aluminiumgehalt ausgewählter Organe.* Mengen- und Spurenelemente 15:613–620.

MÜLLER M, ANKE M, ILLING-GÜNTHER H (1997) *Oral aluminium exposure of adults in Germany – A long-term survey.* In: Fischer PWF et al. eds. Trace Elements in Man and Animals – 9, pp. 177–178. NRC Research Press, Ottawa, Canada.

MÜLLER M, ANKE M, ILLING-GÜNTHER H (1995a) *Die Auswirkungen einer aluminiumarmen Ernährung bei der Ziege. 1. Mitteilung: Methoden, Futterverzehr, Wachstum und Reproduktion.* Mengen- und Spurenelemente 15:605–612.

MÜLLER M, ANKE M, ILLING-GÜNTHER H and HARTMANN E (1995b) *Möglichkeiten und Risiken der oralen Aluminiumbelastung des Menschen.* Mengen- und Spurenelemente 15621–15636.

MILLER RM, ILLING H, ANKE M, DROBNER C and HARTMANN E (1993) *Die Bleiaufnahme Erwachsener in Deutschland.* Mengen- und Spurenelemente 13:475–482.

NIELSEN FH (1987) *Nickel.* In: Mertz W, ed. Trace Elements in Human and Animal Nutrition, 5th edn, Vol 1, pp. 245–273. Academic Press, San Diego.

PARR RM, CRAWLEY H, ABDULLA M, IYENGAR GV and KUMPULAINEN J (1992) *Human dietary intakes of trace elements.* International Atomic Energy Agency, Vienna.

REISS J and ANKE M (2002) *Molybdän.* In: Biesalski, Köhrle J and Schümann K, eds. Vitamine, Spurenelemente und Mineralstoffe, pp 218–222. Georg Thieme Verlag Stuttgart, Germany.

REICHLMAYR-LAIS AM and KIRCHGESSNER M (1981) *Depletion studies on the essential nature of lead in growing rats.* Archiv für Tierernährung 31:731–737.

RISCH RA (1980) *Arsenhaltige biogeochemische Provinzen Usbekistans.* In: Anke M, Schneider H-J and Brückner CHR, eds. 3rd Trace Element-Symposium, pp. 91 –93. University of Leipzig and Jena: Germany,

RÖHRIG B (1998) *Der Zink- und Kupfergehalt von Lebensmitteln aus ökologischem Landbau und der Zink- und Kupferverzehr erwachsener Vegetarier.* Doctoral Thesis, Friedrich Schiller University Biol. Pharm. Faculty, Jena, Germany.

RÖHRIG B, ANKE M and DROBNER C (1996) *Investigation of copper intake with the duplicate portion method in relation of time and kind of diet.* In: Pais I, ed. International Trace Element Symposium Budapest, pp. 171 –178, St. Istvan University, Budapest, Hungary.

RÖHRIG B, ANKE M, DROBNER C, JARITZ M and HOLZINGER S (1998) *Zinc intake of German adults with mixed and vegetarian diets.* Trace Elements and Electrolytes, 15:81–86.

SCHÄFER U (1997) *Essentiality and toxicity of lithium.* J Trace Microbe Techn 15:341–349.

SCHÄFER U, ANKE M and SEIFERT M (2001) *Manganese intake of adults with mixed and vegetarian diets and of breast-feeding and not breast-feeding women determined with the duplicate portion technique.* In: Ermidou S, and Pollet S, eds. 3rd International Symposium on Trace Elements in Human: New Perspectives, pp. 248 –262. Morogiannis Acharnai, Greece.

SEEBER O, ANKE M,HOLZINGER S, LEITERER M and FRANKE K (1998) *Die Uranaufnahme erwachsener Mischkostler in Deutschland.* Mengen- und Spurenelemente 17:924–931.

SEEBER O, ANKE M, HOLZINGER S, LEITERER M and FRANKE K (1997) *Urangehalt deutscher Mineral- und Heilwasser.* Mengen- und Spurenelemente **17**:924–931.

SEIFERT M and ANKE M (1999) *Alimentary nickel intake of adults in Germany.* Trace Elements and Electrolytes **16**:17–21.

SEIFERT M and ANKE M (2000) *Alimentary lead intake of adults in Thuringia/Germany determined with the duplicate portion technique.* Chemosphere **41**:1037–1043.

TURNLUND JR, KEYES WR, PEIFFER GL and CHIANG G (1995) *Molybdenum absorption, excretion, and retention studied with stable isotopes in young men during depletion and repletion.* Am J Clin Nutr **61**:1102–1109

TRÜPSCHUH A (1997) *Die reproduktionstoxikologischen Wirkungen des Nickels und seine Interaktionen mit Zink, Magnesium und Mangan.* Thesis, Biol-Pharm-Faculty, Friedrich-Schiller University, Jena, Germany.

VORMANN J, ANKE M (2002) *Dietary magnesium: supply, requirements and recommendations – Results from duplicate and balance studies in man.* J Clin Basic Cardial **5**:49–53.

VORMANN J, ANKE M, GLEI M, GÜRTLER H, RÖHRIG B, SCHÄFER U and DORN W (1999) *Magnesium: Verzehr, Ausscheidung, Bilanz und Bedarf Erwachsener.* Mengen- und Spurenelemente **19**:971–988.

WILLIAMS RJP (1999) *What is wrong with aluminium? The JD Birchall memorial lecture.* Journal of Inorganic Biochemistry **76**:81–88.

5
Metal and Ceramic Implants

Hartmut F. Hildebrand

5.1
Introduction

Metals have complex effects on the human organism, and four different forms of biological reaction may be distinguished which depend upon the concentration, the exposure time, and the administration route.

- At very low concentrations, some elements such as Co, Cu, Fe, Mn, Zn and even Ni are *essential* or *trace elements* (Anke et al. 1980).
- At high or excessive concentrations, the same substances can induce *toxic reactions* in man and animals; these are well known for As, Co, Ni, Pb and many others (Haguenoer and Furon 1982; Merian 1984). Cytotoxic effects of metal ions have also been demonstrated in cell culture systems (Frazier and Andrews 1979).
- Metals also have an *allergenic potency*. Ni, Co and Cr are recognized to be redoubtable sensitizing agents, whereas only very few cases are known of allergic contact dermatitis to Au, Pd, Pt, Ti, etc. (Marcussen 1957; Dooms-Gossens et al. 1980; Wall and Calnan 1980).
- Finally, numerous metals and/or their compounds are considered – at least in

animals – to be powerful *carcinogenic agents*. At present, the carcinogenic action of these metals is scarcely known, but recent research investigations have suggested that the induction of free radicals by metal compounds is one of the primary factors in the mechanism of metal carcinogenesis (Sunderman 1988, 1989a, b; Shirali et al. 1994). An increasingly evident source of metals within the human organism is the use of metallic biomaterials for dental and orthopedic implants and prostheses; these may be constructed from a large variety of alloys containing between two and eight different metals. A total of more than 30 elements including boron (B), carbon (C) and nitrogen (N) may be contained in different classes of alloys, which are used for external contact with the skin (maintaining exoprostheses), for fixed or mobile medical devices in orifices (mouth, nose, ear, vagina) in contact with mucosal epithelia, and for implantable devices in hard and soft tissues to replace organ functions (total joint prostheses, dental implants), to consolidate failures (orbital floor, rachis, osteosynthesis), and to maintain grafts (trellis membranes, grids).

Elements and their Compounds in the Environment. 2nd Edition.
Edited by E. Merian, M. Anke, M. Ihnat, M. Stoeppler
Copyright © 2004 WILEY-VCH Verlag GmbH & Co. KGaA, Weinheim
ISBN: 3-527-30459-2

5.2
Alloys

5.2.1
Alloys Used for Surgical Implants

The implants used in stomatology and modern orthopedic surgery are manufactured from three different alloy systems, each presenting main characteristic components:

- Fe-base alloys with high Cr-content are summarized as *stainless steel*.
- Co-base alloys with 25–30% Cr, 5–7% Mo and low amounts of other metals such as Ni, Mn, Fe, Si are called *Co-Cr alloys*, while those with about 20% Cr, 10% Ni and up to 15% ungsten (W) are called *wrong Co-Cr alloys*.
- *Ti-base alloys* with 70–90% or more Ti are increasingly used for surgical implants.

They also contain low amounts of other metals such as Al, V; Nb, Ta, Mn, Zr and/or Sn. The only pure metals used for medical devices are Ti and Ta. The only binary alloys applied for biomaterials are Ti-base alloys, for example, Ti30Nb, Ti30Ta Ti(n)Mn, and memory super alloys: NiTi (Bradley 1994; Breme 1994; Breme and Wadewitz 1989).

All of these alloys can also be integrated into medical devices for neurosurgery and cardiovascular, maxillofacial, otologic and visceral surgery, etc. Their most frequent application, however, is for osteosynthesis and partial and total arthroplasties.

During several decades, stainless steel was the most frequently used alloy for joint replacements. At present, Co-base alloys have taken first place, and about 70% of all orthopedic implants are made from Co-Cr alloys. During the past 20 years, titanium and its alloys have become more important due to their bone-like elasticity and their excellent biological behavior.

5.2.2
Dental Alloys

Dental alloys exist in a wide variety of forms, and can be classified with respect to their multiple use in dentistry:

- Crown and bridge casting alloys (conventional alloys)
- Porcelain fused to metal alloys (ceramo-metallic alloys)
- Wires
- Partial denture alloys
- Implant alloys
- Solders
- Dental amalgams

A range of more than 1100 alloys are known on the European Market. With respect to their chemical composition they can be classified into five families:

- Dental amalgams (Hg, Ag, Zn, Sn, Cu)
- Precious alloys (Au, Pt, Ag, Cu, and low amounts of other metals of the platinum group)
- Semi-precious, low gold and Pd-base alloys (Pd, Ag, Au and lower amounts of Pt and Cu)
- Non-precious alloys, i.e., stainless steel, Co-Cr and Ni-Cr alloys, some of which may also contain Cd and Be
- Ti-base alloys.

Progressive dental reconstructions within the life-time of a patient, and in particular the use of different alloys for total or partial dentures, for dental fillings, for porcelain-fused restorations, and transcutaneous implants, generate unavoidable oral polymetallism. Indeed, two alloys of different composition have different electric potentials and inevitably induce corrosion and subsequently the release of metal ions into the human organism (Bundy 1994; Hornez et al. 2000).

5.2.3
Metals Used for Biomedical Alloys

The variation of biomedical alloys is determined by their application. More than 30 different elements are currently used for dental materials and surgical implants. Other metals (e.g., Ce, Cs, Se) are added in alloys for needles and tools applied in acupuncture and hair transplantation.

- Orthopedic and stomatological implants and prostheses: these incorporate Al, Co, Cr, Fe, Mn, Mo, Nb, Ni, Sn, Ta, Ti, V, W, and Zr.
- Dental alloys: these include:
- *Non-precious:* Al, B, Be, Cd, Co, Cr, Fe, Mn, Mo, Ni, Si, Ti, V, and W
- *Precious and semi-precious:* Ag, Au, Cu, Fe, Ga, In, Ir, Pd, Pt, Rh, Ru, Sn, Ti, and Zn
- *Dental amalgams:* Ag, Cu, Hg, Sn, and Zn

5.3
Risks: the State of the Art

Toxicity in humans of the most frequently used metals and their compounds has long been recognized, and many monographs have been published on this subject during the past two decades (Haguenoer and Furon 1982; Brown and Savory 1983; Merian 1984; Michel 1987; Aitio et al. 1991). As all important characteristics of the metals will be emphasized in Part III of this book, a specific chapter of their general toxic effects is not required at this point. It must be underlined, however, that the pure metals are rarely toxic; that the toxic, allergenic and/or carcinogenic effects depend on the concentration and on the nature of compounds (oxides, simple or complex salts); and that two compounds of the same metal may induce strongly different responses and the toxic potency of different ions or compounds may vary by two to three orders of magnitude. Thus, the speciation of compounds is of primary importance; for example, NiCl is toxic without an evident carcinogenic effect, whereas Ni_3S_2 is highly carcinogenic.

Harm caused by the use of metallic implants is essentially due to the release of ions resulting from the corrosion of these alloys. This concerns principally Ni, Cr and Co for any application, Be, Cd, Pd, Ag and Cu for dental alloys, and Ti for stomatological and orthopedic implants (Hildebrand et al. 1995; Hildebrand and Hornez 1998; Hornez et al. 2002).

The second risk factor is wear, in particular produced by articular prostheses and mobile, nonstabilized implants generating wear particles by abrasion (Laffargue et al. 1998).

5.3.1
Ion Release

Ion release from metallic prostheses and implants is the main origin of any unwanted primary or secondary reaction (Black 1988; Hildebrand and Hornez 1998), except for the electro-galvanic phenomena occurring in the oral cavity due to the presence of different dental alloys. In this case, the primary factor may be an electrochemically induced galvano-electric current used by the saliva as a favorable saline electrolyte (Bundy 1994; Hornez et al. 2000).

5.3.1.1
Alloys for Surgical Implants

For several years increasing interest has been shown in the effects and reactions produced by ion release. Nevertheless, thoroughly conducted investigations of these phenomena are rare, and no systematic epidemiological and statistic study exist on this subject, although the high release of Ni, Cr and Co ions has generally been recognized.

In some cases, 200- to 300-fold concentrations of maximum normal values could be demonstrated in body fluids and in implants surrounding tissue (Hildebrand et al. 1988, 1996b; Laffargue et al. 1998).

Cobalt seems to produce similar effects to those of Ni and Cr. Most authors agree that secondary harmful reactions are not directly generated by the presence of ions, but by their still scarcely known metabolites. The degree of oxidation and the formation of metallo-organic complexes may play an essential role, in particular for Cr, the primary ionic form of which after liberation is the trivalent ion. Its active toxic, allergenic and carcinogenic form, however, is the hexavalent ion (Bartolozzi and Black 1985).

The alloy composition has less importance than the physico-chemical structure and characteristics: the amount of ions released from Co-Cr alloys is similar to that liberated from stainless steel (Pazaglia et al. 1987).

5.3.1.2
Dental Alloys

In the case of dental alloys, it is important to consider also other metals, including:

- Cd: the toxic effects of cadmium are generally recognized, and in particular those of the metal's sulfides, oxides and metallo-organic compounds.
- Be: unfortunately, this is still contained in some dental alloys because the mechanical qualities are improved by its presence. Indeed, beryllium is highly allergenic and toxic, and several cases of lung berylliosis have been reported in dental technicians working with alloys containing Be (Lob and Hugonnaud 1977; Choudat 1982; Choudat et al. 1983).
- Pd: this is contained in semi-precious alloys, and some two years ago provoked a major controversy concerning its innoxiousness or harmfulness. At present, there is no serious criterion justifying the anxiety concerning palladium (Hildebrand et al. 1996a), except that Pd may induce a Ni-concomitant allergic sensitization. Only very few cases of clinical manifestations such as allergic contact dermatitis or stomatitis have been reported.
- Dental amalgam is the major source of inorganic mercury (Hg) exposure in the general population (Veron et al. 1984; Sandborgh-Englund et al. 1998). Hg is released in body fluids and in pulpa and dentine after introduction as dental fillings (Hörsted-Bindslev et al. 1997). After removal, there was an initial increase by two orders of magnitude followed by a considerable decline in the Hg levels of saliva, blood, plasma, urine, and feces, which slowly approached those of subjects without any history of amalgam fillings (Ekstrand et al. 1998).

5.3.2
Wear Particles

Wear particles produced by abrasion appear essentially in the vicinity of articular prostheses and of implants with a certain mobility, for example, uncemented total hip replacements. These wear particles may induce multiple tissue reactions: osteolysis, degradation of normal bone structure, severe macrophagic reactions, granuloma, fibrotic capsules, inflammatory and immune reactions which may cause destabilization and loosening of prostheses and implants (Dorr et al. 1990; McKellop et al. 1990; Sarmiento and Gruen 1985; Weissman et al. 1991).

An arthroplasty with different compounds may subsequently produce different wear particles that are metallic, ceramic and polymeric in nature. In general, it seems that:

- polymers and ceramics give rise to fewer problems than metals (Pazaglia et al. 1987);
- the size and form of the particles play an important role: small or irregular particles are more active than larger or regular ones (Black 1988; Dorr et al. 1990);
- alloys containing Co-Cr-Ni raise more concern than Ti alloys (Sarmiento and Gruen 1985; McCutchen et al. 1990).

These differences are due to the physico-chemical characteristics of alloys and their particles. The particles from Co-Cr-Ni alloys are continuously dissolved in the organism and undergo chemical modifications by the formation of precipitates or met-allo-organic complexes which have been shown to bind Ca, and in particular phosphorus or phosphates (Black 1988; Hildebrand et al. 1988; Dorr et al. 1990).

Particles from Ti alloys, however, arise from the passivation layer of the implant. The particles are not Ti ions, but mostly insoluble Ti-oxides or Ti-suboxides which are recognized to be biologically inert. Indeed, the passivation layer is immediately reformed after abrasion because of the high oxidizability of Ti (it occurs in micro-seconds). This behavior protects the alloy and prevents the formation of chemical compounds other than oxides. Some authors, however, do not believe in the innoxiousness of V and Al contained in some Ti alloys with widespread application, for example, Ti_6Al_4V (Hildebrand et al. 1988; Breme 1989).

5.3.3
Ceramics

Ceramic materials cover a very large range of medical applications such as dental porcelain, silicium-based bioglasses as bone and cartilage substitutes, calcium phosphate bone substitutes, and carbon-based implants (Michel 1991):

- Sintered aluminum oxides (Al_2O_3) or zirconium oxides (ZrO_2) are increasingly used as components or as part of an articular endoprosthesis.
- Resorbable or unresorbable Ca–PO$_4$–SiO$_2$-containing compounds have been developed as bone substitutes.
- The use of carbon materials has been attempted for orthopedic implants, and they are still used for heart valves and dermatological applications.

5.3.3.1
Sintered Ceramics

Al_2O_3 and ZrO_2 are considered to be non-bioactive ceramics and are frequently used as the articular heads of total arthroplasties such as total hip prostheses, total shoulder prostheses, and maxillar articular replacements. No unwanted biological effects could be observed in vivo, and no cytotoxic effects have been evidenced in vitro for both compounds except for some formation of granuloma around wear particles of these materials. Oonishi et al. (1997) have also shown small amounts of new bone formation between Al_2O_3 particles in an experimental rabbit femoral defect model.

5.3.3.2
Bioceramics as Bone Substitutes

The development of the so-called "bioceramics" is based on the knowledge that native bone is essentially composed of a more or less carbonated hydroxyapatite (HA): $Ca_{10}(PO_4)_6(OH)_2$. With respect to the need for low solubility or of a controllable resorption, different compounds of the calcium phosphate system $Ca(OH)_2$-H_3PO_4-H_2O have been applied for bone substitutes or bone fillings (Table 5.1).

A good biocompatibility is observed for a Ca:P ratio between 1 and 2, with an opti-

Tab. 5.1: The essentially used calcium phosphates and carbonates

Name	Symbol	Formula	Ca/P	Solubility
Dicalcium phosphate (Brushite)	DCPD	$Ca(HPO_4)·2H_2O$	1.00	1.87×10^{-7}
Dicalcium phosphate (anhydric)	DCPA	$Ca(HPO_4)$	1.00	1.26×10^{-7}
Octacalcium phosphate	OCP	$Ca_8H2(PO_4)_6·5H_2O$	1.33	5.01×10^{-15}
β-Tricalciumphosphate	β-TCP β	$Ca_3(PO_4)_2$	1.5	2.83×10^{-30}
Hydroxyapatite	HA	$Ca_{10}(PO_4)_6(OH)_2$	1.67	2.35×10^{-59}
Tetracalcium phosphate	TCPM	$Ca_4P_2O_7$	2.00	–
Calcium carbonate (Calcite)	CC	$CaCO_3$	–	4.96×10^{-9}

mum for β-tricalcium phosphate (β-TCP). Hydroxyapatite and the β-TCP are without doubt the most frequently studied and most frequently applied ceramic biomaterials for reasons of their optimal biocompatibility, their osteoconductive properties, and their strong bioactivity (Damien and Pearson 1991; Oonishi et al. 1997).

Another bioceramic family is the bioglass system, the definition of which is a "glass product with the aim to produce specific physiological responses – it has a reactive surface containing calcium, silicium and phosphate ions, a alkaline pH at the interface with the tissue". The optimal composition developed by Hench (1991, 1994) is the so-called BIOGLASS(r) 45S5, which is composed of 45% SiO_2, 24.5% Na_2O, 24.5% CaO and 6% P_2O_5. Hence, the name 45S5 indicates the 45% of SiO_2, S is for Si, and 5 is the Ca:P ratio. This same nomenclature system has been proposed for use with other Si-based bioglasses.

Further phosphate-based bioglasses have been developed within the system P_2O_5, CaO, Na_2O, Al_2O_3, Fe_2O_3, ZrO_2. The principal idea was to create silicium-free phosphate glasses which in elemental content are as close as possible to bone constituents, thereby avoiding the unknown long-term behavior of silicate glasses (Wilson and Low 1992; Wilson et al. 1993; Oonishi et al. 1997). These ceramics have been shown to have the same bioactivity and the same mechanical limits as silicate-based bioglasses. A clear advantage is their increased malleability, and their biocompatibility is excellent.

5.3.3.3
Carbon Materials

Carbon materials find widespread medical applications (Haubold et al. 1986) as ligament replacements (Amis et al. 1988; Demner et al. 1991; Reed et al. 1994), carbon tissues for abdominal implants (Morris et al. 1990), bone substitutes and osteosynthesis devices (Morris et al. 1998; Tayton et al. 1982), cardiovascular devices (Tagusari et al. 1998), activated charcoal for wound dressings (Wollina et al. 1996), percutaneous devices (Tagusari et al. 1998) and in various composites (Tayton et al. 1982; Bercovy et al. 1985; Galand and Lynch 1989; Hetherington et al. 1995). These materials have a very good biocompatibility, in addition to an inert and bioactive behavior. The main problem with these carbon materials is their easy wear, which results in the release of black, sharp-edged particles. This may induce either a tattoo effect or the frequent appearance of granuloma as a foreign body reaction with inflammatory cells. This reaction is neither chemical nor physiological, but is a physical and mechanical effect (Oppenheimer effect). The problem has been resolved by the addition of epoxy resins to harden the carbon fiber agglomerate or by vitrifying processes of the implant surface (e.g., pyrolytic carbon; Hetherington et al. 1995) or diamond-like carbon coating (Kornu et al. 1996; Tessier et al. 2003). In the first case, a toxic element is added which is easily detectable after incomplete polymerization. In the second case, the favorable mechanical characteristics are altered by decreasing considerably the elasticity (increasing Young module).

5.3.4
Tissular Reactions

Multiple investigations have been published reporting one or more tissular reactions in the vicinity of implants or prostheses. Also for these reactions, there exists no systematic study allowing a statistical or epidemiological evaluation relative to primary and/or secondary unwanted effects of metallic implants.

The most frequent injuries are certainly granuloma (particularly around Co-Cr-Ni alloys). These are characterized by a high density of collagen fibers and by the presence of multinucleated giant cells (i.e., severe macrophage reaction), fibroblasts, plasmocytes, and histiocytes. Benign granulomas generally contain precipitates that confer a black color to the tissue. The evolution of granuloma may sometimes lead to the blockage of an articular prosthesis, and this leads to a need for surgical reintervention (Griffith et al. 1987; Nasser et al. 1990).

Some granuloma related to Ti implants have also been reported. The tissue contains multiple intra- and extra-cellular particles, but very few inflammatory cells have been observed and most of the tissue reactions to Ti remain without clinical consequences (Griffith et al. 1987; Nasser et al. 1990).

Another type of injury is the structural modification of bone. Osteosynthesis plates have been observed to be totally recovered by newly-formed bone tissue after an exposure period of three to four years. The retrieval of such implants becomes particularly difficult (Hildebrand et al. 1988).

Numerous authors have reported osteolysis, case by case, which is mostly induced by noncemented arthroplasties that always gain a certain mobility by mechanical solicitation during movement. The same phenomenon may arise for craniofacial (after resection of tumors) and otologic unattached prostheses which are potentially mobilizable.

Osteonecrosis producing in some cases sclerosis of the bone/implant interface, the so-called *metallosis*, has also been observed. Two characteristics must be emphasized with respect to metallosis:

- The frequency and importance of necrosis, which in some cases may be total.
- The simultaneous existence of a lymphocyte reaction; in some cases a lymphoid islet with the early stages of a clear center may appear inside the bone.

Thus, different criteria must be observed in order to improve the performance of metallic implants and of craniofacial, otologic and orthopedic prostheses (Bischoff et al. 1994):

- unattached implants must remain immobile;
- articular prostheses should be cemented;
- the frequency of tissue reactions must be less than 5%;
- the performance of an implant must be characterized by the absence of persistent and/or irreversible symptoms such as pain, infection, and neuropathies.

As with dental nonprecious alloys, many reports have been made of lingual lesions, injuries of the oral mucosa in the form of stomatitis, cheilitis, tissular hypertrophy, oral redness, dryness, and angular stomatitis, without precise knowledge of the etiology of such pathologies, whether mechanical, bacterial, inflammatory, immunologic, toxic or electrogalvanic (Hildebrand et al. 1989a, b, 1995; Bundy 1994).

5.3.5
Inflammatory Reactions

The majority of tissue reactions are of inflammatory origin. Indeed, most granuloma contain different cell types with inflammatory characteristics, including multinucleated giant cells, histiocytes, plasmocytes, mast cells, and lymphocytes (Griffith et al. 1987; Nasser et al. 1990). In addition, neutrophilic polymorphonuclear cells frequently exhibit degranulation (Shanbhag et al. 1992).

The density, activity and function of these cells are controlled by endogenous mediators of inflammation such as histamine, prostaglandins, derivatives of complement, lymphokines, cytokines, and leukotrienes. This problem has been approached only in very few investigations, though initial studies revealed a stimulation of prostaglandin E_2, interleukin-1 and collagenesis (Dorr et al. 1990; Cook et al. 1991). These investigations were the first to consider the general effects induced by metals.

5.3.6
Immunologic Reactions

The inflammatory and immunological symptomatologies are often very closely connected. Certain metals, especially Fe, Ni, and Co have a well-known effect on lymphocyte proliferation as they stimulate the direct complements DC2 and DC3 which are implicated in this process. Such studies are rare, but provide precise and highly specific indications on metal actions (Choudat et al. 1983; Bjurholm et al. 1990; Bravo et al. 1990).

5.3.6.1
Sensitization and Allergy

Allergy was first defined as a pronounced reaction of an individual to a substance when that substance is re-introduced into the organism, though this definition has since been modified in line with the progress made in immunological research. The sensitizing substance is called an "antigen"; this is a molecule or a cell which, once introduced into the organism, induces the formation of antibodies or specific defense cells. Coombs and Gell (1975) defined four different classes of allergy.

Allergic reactions caused by stainless steel or by alloys containing Ni-Co-Cr are referred to as "contact dermatitis" and belong to type IV of the above-cited classification. For this form of allergy, the allergen or hapten is a substance with a low allergenic power (Dupuis and Benezra 1982) which initially is bound strongly to certain endogenous proteins to form a stronger antigenic macromolecule. By a very simplified mechanism, the newly formed antigen is captured by macrophages and "memorized" by certain T lymphocytes (Dupuis and Benezra 1982).

After renewed contact, the formerly sensitized lymphocytes produce different substances (e.g., lymphokines) which they liberate into the organism, thus provoking certain tissue reactions. The hypersensitive effect in contact dermatitis appears generally on the skin as eczema. Mucosal reactions and especially stomatitis are possible.

Allergies of type IV thus appear after cellular mediation and without any production of antibodies. For this reason, desensitization is not possible (Dooms-Gossens et al. 1980; Dupuis and Benezra 1982; Hildebrand et al. 1989a, 1989b). The immunological feature is called (hyper-)sensitization, and one speaks of an allergy when clinical manifestations appear.

A large number of statistical and epidemiological investigations of contact dermatitis were performed to establish the frequency of allergies. Tables 5.2 and 5.3 summarize the data of a previous report in which more than 20 statistical studies of allergy in a consultant population (Table 5.2) and five statistical studies of the general population (Table 5.3) were reviewed (Hildebrand et al. 1989a).

5.3.6.2
Alloys for Surgical Implants

Scientific opinion remains divided with regard to the allergic sensitization of metals in patients with implants and endoprostheses. Several authors have demonstrated a direct relationship between metals contained in medical devices (Ni, Cr, Co) and allergic sensitization, and draw particular attention to this concerning the use of these materials as implants in patients with prior sensitization to these metals (Merrit and Brown 1981; Rostoker et al. 1986, Black 1988). Other authors, however, still repudiate any such relationship (Carlsson and Moller 1989; Gawkrodger 1993). By analyzing the references cited by these authors, one can easily observe a tendentious and unscientific behavior regarding this problem. This becomes even more evident from the fact that in the literature several hundred cases have been cited which demonstrate a clear relationship between allergy and orthopedic implants.

5.3.6.3
Dental Alloys

This relationship between allergy and metallic biomaterials has also been confirmed for alloys used in dentistry: Ni, Cr and Co in non-precious alloys (Hildebrand et al. 1989a, 1989b), and Hg and Ag in dental amalgams (Véron et al. 1986; Hörsted-Bindslev et al. 1997; Sandborgh-Englund et al. 1998; Ekstrand et al. 1998). The sensitizing potency of Pd was for a long time an unexplained phenomenon, but most investigations have suggested a real crossed hypersensitivity between Ni and Pd – that is, a patient sensitized to Ni may react positive in a Pd allergy test. This would represent an extremely rare case of nonrecognition of an antigen by a healthy and immunologically intact organism (Hildebrand et al. 1996a).

Tab. 5.2: Statistical assessment of sensitization to Ni, Cr, and Co in the consultant population (%) (Hildebrand et al. 1989a)

Allergen	Reported cases	Male (%)	Female (%)	Total (%)
Nickel	37849	3.1	12.9	9.6
Chromium	36914	12.7	7.1	9.3
Cobalt	31330	4.7	5.3	6;0

Tab. 5.3: Statistical assessment of sensitization to Ni, Cr, and Co in the general population (Hildebrand et al. 1989a)

Allergen	Reported cases	Male (%)	Female (%)	Total (%)
Nickel	3 207	1.5	8.9	4.2
Chromium	822	2.0	1.5	1.7
Cobalt	758	1.0	1.6	1.4

Very recent studies have shown that Pd has its own sensitizing potency, but the cross-reaction between Ni and Pd is still not excluded. In contrast to Ni sensitization, it seems that sensitization to Pd very scarcely leads to clinical symptoms. In the case of Pd-hypersensitivity related to dental alloys, only a dozen cases have been described to date.

5.3.7
Induction of Cancer

Although different cases of cancer in relation to orthopedic implants and endoprostheses have been reported, there exists no statistical or epidemiological evidence of cancer which has been caused by metals contained in implants (Laffargue et al. 1998). Some cases were reported about malignant fibrous histiocytoma of bone arising at the site of metallic implants, plate and screws, hip prostheses, metallic foreign bodies from shrapnel fragments. The causal relationship between metal implants in humans with this kind of tumor and other types such as Ewing's sarcoma, osteosarcoma, chondrosarcoma, fibrosarcoma, rhabdomyosarcoma, hemangiosarcoma, and immunoblastic lymphoma is difficult to assess (Lee et al. 1984; Sunderman 1988, 1989b; Goodfellow 1992; Jacobs and Rosenbaum 1992; Laffargue et al. 2001). The observation of new cases, however, and other associated tumors has led to an increasing interest in establishing whether the association between orthopedic implants and local malignancy is purely coincidental or represents a real carcinogenic risk (Mirra et al. 1994; Troop et al. 1990; Solomon and Sekel 1992; Khurana et al. 1994; Gillespie et al. 1996; Lewis and Sunderman 1996; Visuri et al. 1996; Laffargue et al. 2001).

The carcinogenic feature of some metals is well known, and tumors arising at the site of metal implants have been observed in animals. Heath *et al.* (1971) showed that wear particles from prostheses constructed from Co-Cr alloys were carcinogenic for rat muscles, while Sinibaldi (1976) reported eight cases of bone sarcoma originating in close proximity to various metallic surgical implants used to treat common canine and feline fractures.

In 1988, Sunderman reviewed the clinical and experimental evidence and appraised the carcinogenic hazards from implanted metal alloys containing Ni, Cr or Co. In the same report, the author provided general background information on epidemiological evidence that certain occupational and environmental exposures to metal compounds are associated with excess cancer risks in humans and in experimental animals.

In addition, a total of 24 cases have been reported of sarcomas developing in dogs around implanted orthopedic pins, nails, plates and screws, mostly fabricated from stainless steel. Furthermore, local sarcomas have been observed in rodents after parenteral injection of metallic Ni or Co powders, but not after injection of metallic Cr powder. Since the metal powders release ions which undergo a biological metabolism and oxidation, different Ni, Co and Cr compounds can be formed which have genotoxic and mutagenic effects. This has largely been demonstrated in in-vitro tests and summarized in several reviews (Sunderman 1988, 1989a, b; Anonymous 1990).

The most frequently reported tumors in humans at sites of metal implants are malignant fibrous histiocytoma, fibrosarcoma, osteosarcoma and rhabdomyosarcoma, with descriptions reported of at least 20 cases for each lesion (Mathiesen et al. 1995; Laffargue et al. 2001). The same amount – 80 to 100 cases – can be estimated as the total of the other tumor types cited above (Laffargue et al. 1998). Epidemiological studies (Anonymous 1990) have shown, that latent periods of less than 5 years cannot be associated with chemically induced tumors in humans. In the case of shorter exposure periods, the association of malignancy with metal implants was probably coincidental. Longer periods reported from 7 to 44 years, are consistent with a possible etiologic relationship between the metal implant and subsequent tumor development.

In spite of the few cases of malignant tumors arising with respect to the large number of metal implants used in humans, these arguments deserve serious consideration. As proposed earlier by Aspley (1989) and Jacobs *et al.* (1992), the establishment of an international or world-wide register of such cases would facilitate future knowledge.

5.4
In-Vitro Toxicity Assessment

The biological testing of medical and dental devices is necessary in order to evaluate the biological behavior of biomaterials. Biocompatibility testing includes numerous methods starting with mechanical, physico-chemical and electro-chemical investigations (i.e., corrosion tests), going through in-vitro and in-vivo tests such as implantation in animals and preclinical evaluation in humans, and arriving at the final clinical use in patients. Cytocompatibility is the in-vitro adequate behavior of cells in the presence of biomaterials, whilst cytotoxicity is the harmful or noxious unwanted effect induced by a biomaterial in cell culture systems.

A scheme for in-vitro cytotoxicity testing is defined by the international and European

standards ISO 10993–5 (Standards 1994, 1992) and EN 30993–5 (Standards 1994) which makes available a battery of tests, the choice of which depends on the nature of the sample to be evaluated, the potential site of use, and the nature of the use.

The numerous methods applied and the end-points measured in cytotoxicity determination can be performed by either qualitative or quantitative means. The following examples and results on the cytotoxicity of metals and implantable alloys correspond to 8.5.1.b "Quantitative evaluation" of the above-mentioned standards:

Measure cell death, inhibition of cell growth, cell proliferation or colony formation. The number of cells, amount of protein, release of enzyme, release of vital dye, reduction of vital dye or other measurable parameters may be quantified by objective means. The objective measure and response is recorded in the test report.

In the present authors' laboratory, in-vitro tests have been carried out on numerous alloys and pure metals in order to determine their effect on cell viability and their capacity to induce inflammatory reactions (Table 5.4).

5.4.1
Cell Viability

The viability tests consisted of the establishment of the relative plating efficiency (RPE) and subsequently, the 50% lethal concentration LC_{50} (or RPE_{50}) by using the colony-forming method on human epithelial cells in culture, the L132 cell-line. This test measures quantitatively only one criterion of toxicity which is cell death or cell survival, and consequently is specific, liable, and easily reproducible. It makes possible the ranking of cytotoxic effects of any chemical substance by comparison with the LC_{50} (Puck and Marcus 1955; Frazier and Andrews 1979).

Tab. 5.4: Cytotoxic effects of pure metal and alloy powders on cell cultures of L132 cells: LC_{50} and survival rates (n = 8). Frequency of multinucleated giant cells in cultures exposed to metal and alloy powders (100 mg L^{-1}) during 8 days (n = 5 × 500 cells).

Pure metals or alloys	LC_{50} (mg L^{-1})	Survival rates (% ± SD) at 400 mg L^{-1}	Multinucleated Giant Cells (% ± SD)
Control	NO	100 ± 5	2.6 ± 0.7
Pt	NO	99 ± 3	ND
Sn	NO	91 ± 4	ND
In	NO	88 ± 4	ND
Ti	NO	81 ± 4	2.7 ± 0.7
Au	NO	78 ± 4*	2.7 ± 0.7
Pd	NO	68 ± 7*	2.8 ± 0.5
Cr	600	62 ± 3***	7.8 ± 1.9**
Cu	450	58 ± 6***	ND
Ag	75	32 ± 8***	16.0 ± 2.8***
Zn	25	NO	ND
Ni	25	NO	15.9 ± 2.6***
Co	20	NO	#20.3 ± 4.9***
Al_2O_3	NO	95 ± 2	ND
Stainless steel	NO	72 ± 7	9.1 ± 1.8**
NiCrCo	100	23 ± 8***	16.6 ± 3.3***
NiCrMo	75	14 ± 2***	15.7 ± 2.9***
Dental amalgam	20	NO	17.1 ± 2.1***
TiAl6V4	NO	98 ± 5	2.5 ± 0.8
TiAl5Fe2.5	NO	91 ± 6	2.5 ± 0.7
TiNb30	NO	60 ± 4***	2.6 ± 0.7
Pd79Au10	NO	95 ± 7	2.8 ± 0.5
Au75Pd19	NO	82 ± 3	2.9 ± 0.7
Au61Pd29	NO	82 ± 3	2.8 ± 0.6
Au32Ag31Pd8	NO	77 ± 5*	2.6 ± 0.6
Au50Ag27Pd14	NO	76 ± 4*	2.9 ± 0.7
Au36Pd50	NO	75 ± 9*	2.7 ± 0.5
Au6Ag51Pd18	200	41 ± 8***	2.9 ± 0.6
Ag32Pd57	200	38 ± 10***	14.4 ± 2.1***

Concentration 25 mg L^{-1}. * $p < 0.05$; ** $p < 0.01$; *** $p < 0.001$ with respect to controls. NO, not obtained; ND, not determined.

5.4.1.1
Liability of Tests

The LC_{50} is the concentration expressed in mg L^{-1} which produces 50% cell death in in-vitro tests on different cell culture systems. It should not be confused with the LD_{50}, which is the dose of a substance which induces a 50% death rate of animals or humans exposed to the substance by oral administration. The LD_{50} is expressed in mg kg^{-1} or g kg^{-1} of the living organism, and generally has higher values than the LC_{50}. The reason for this is the capacity of natural defense in a living organism by the interaction with different cell types in an organ, and also by an active immune system preventing primarily cell injury and organ damage (Hildebrand and Hornez 1998).

In cell culture systems, normally only one cell type exists, and this precludes any natural defense by an active immune system and/or the interaction with other cell types. There is a lack of systemic interaction and no provision for circulation. In addition, laboratories mostly use established cell lines which have reduced physiological responses; that is, with decreased enzyme activities and energy production and which may, under certain circumstances, be considered as "ghost cells" (Hildebrand and Hornez 1998).

This inconvenience can be avoided by the use of primary cultures obtained from the in-vitro outgrowth of cells from a fresh biopsy. In these cultures, the physiological response is much more realistic, but they have the disadvantage that they rarely contain only one cell type. Thus, the biological response can hardly be attributed to a specific cell. Moreover, the reproducibility of results is less liable, since the physiological state of fresh primary cells may be different from their origin and also dependent on the sex, age, and other individual parameters. With this in mind, the choice of experiments requires good qualification and long experience on the part of the deciding responsible person, and it is sometimes necessary to perform different experiments for a better correlation of the results (Hildebrand and Hornez 1998; Hornez et al. 2002).

5.4.1.2
Influence of Metals on Cell Survival
Pure powders of Pt, Sn, In, Au and Pd exhibit an identical (i.e., excellent) biological behavior, and for extreme high concentrations (400 mg L^{-1}) they still have a survival rate of 99% to 70% respectively (Figure 5.1). Cr, Cu and Ag have a medium cytotoxic effect, with a survival rate for the same high concentration of 60% to 30% (Figures 5.1 and 5.2), whereas Co, Ni and Zn induce rapid cell death as expressed by a very low LC_{50} (Figure 5.2). Their survival rate for the highest concentration could not be obtained, since total cell death occurred for significantly lower concentrations.

Precious, semi-precious, most Pd-base alloys, Ti-base alloys and stainless steel also produce excellent biological responses with a survival rate of 98% to 70% for the highest concentration. Two semi-precious alloys with high Ag-content reflect the cytotoxic effect of pure silver (Figure 5.1). Ni-Cr alloys (14% and 23%) induce a strong cytotoxic action, and dental amalgams produce total cell death at very low concentrations. The compound Al_2O_3 was added to these test series for comparative purposes in order to emphasize its excellent cytocompatibility (Figure 5.1).

5.4.2
Inflammatory Response

Different tests can be applied to assess the inflammatory response. In cell culture systems, inflammatory response is expressed by the appearance of cytokines or monokines such as interleukin-1, prostaglandin E_2, leukotriene B_4 (LTB_4), and other compounds deriving from the peroxidation cascade (Shirali et al. 1994; Johnson and Organ 1997) or – as we have observed – by the appearance of multinucleated giant cells (Hornez et al. 2002).

5.4.2.1
Liability of Test
Multinucleated giant cells (MGC) normally appear in cultures of macrophages through the fusion of individual cells. Established cell lines generally exhibit a low percentage of 2–5% of binucleated or multinucleated cells. MGC may not only appear in cellular or organotypic cultures (Ziats et al. 1988),

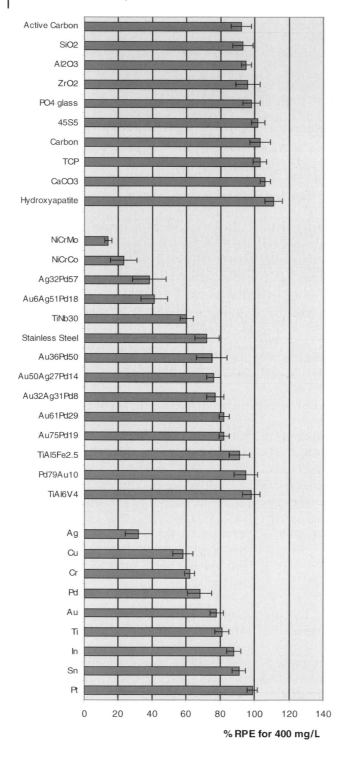

Fig. 5.1 Survival rates (RPE % ± SD) of L132 cells cloned in the continuous presence of an extreme high powder concentration (400 mg L^{-1}) of pure metals and of multiple dental and orthopedic alloys. Al$_2$O$_3$ has been added to this series for comparative purposes. The control culture is 100%.

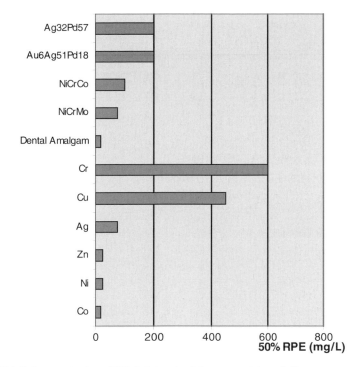

Fig. 5.2   50% lethal concentrations (RPE_{50}) determined for pure metals and alloys.

but also *in vivo* in patients with amalgam tattoos (Buchner and Hansen 1980; Véron et al. 1985, 1986) or with an orthopedic implant (Hildebrand et al. 1988). They are characteristic constituents of granuloma in the vicinity of implants, and their presence is generally considered as a specific inflammatory response (Buchner and Hansen 1980; Hildebrand et al. 1988; Véron et al. 1986; Ziats et al. 1988).

5.4.2.2
Influence of Metals on the Inflammatory Response

The test of inflammatory reactions consisted of quantifying the MGC in monolayer cell cultures of L132 cells. This test reveals morphological modifications in a cell culture by the appearance of MGC, which are directly

related to physiological – that is, functional alterations of the cells (Hildebrand and Hornez 1998; Hornez et al. 2002).

The MGC test confirmed the quasi-perfect cytocompatibility of the of Pd, Au, Ti powders and of alloys containing respectively these metals, since the number of MGC in exposed cultures was identical to that in control cultures (Table 5.4; Figures 5.3 and 5.4). However, cultures exposed to pure Ni, Co, Ag and to Ni-rich alloys and dental amalgams developed eight- to ten-fold increases of MGC with respect to control cultures (Table 5.4; Figures 5.3 and 5.5). The inflammatory effect of silver is confirmed in the Ag32Pd57 alloy. Stainless steel without a major influence on cell viability induces inflammatory reactions comparable to those produced by Cr.

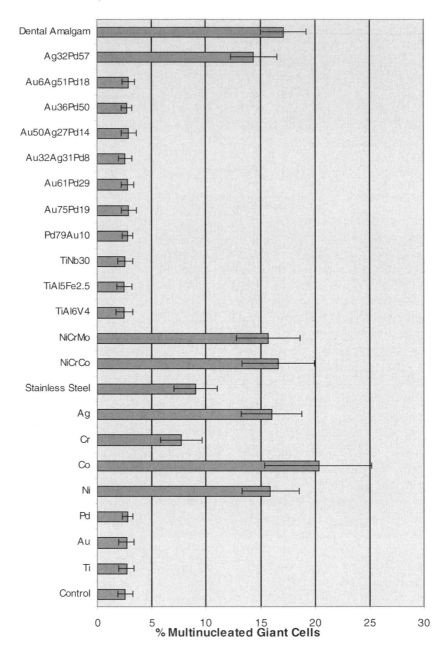

Fig. 5.3   Frequency (% ± SD) of multinucleated giant cells induced in L132 cells exposed to pure metals and alloy powders (100 mg L^{-1}) for 8 days. The values were established by counting 500-cell areas in five different experiments.

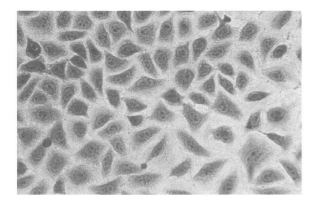

Fig. 5.4 Typical feature of L132 cells corresponding to control cultures or to cultures exposed to Au, Pd, or Ti or to precious, semi-precious and Ti-base alloys.

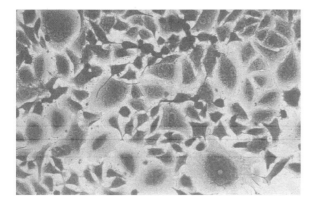

Fig. 5.5 Typical feature of L132 cells grown in the presence of powders of pure Ni or Co, dental amalgam, Ni-Cr containing alloys or alloys with high amounts of Ag. Note the presence of a large number of multinucleated giant cells, indicating an inflammatory reaction.

5.6
Conclusion

Wear particles and metal ions released from dental and surgical alloys can be recovered in the human organism, notably in the urine, blood, plasma, nails, hair, and implant-surrounding tissues. Moreover, these particles and ions may induce not only allergic reactions but also (and in particular) immune and inflammatory reactions, the strength of which at present remains largely underestimated. This finding has been demonstrated in numerous investigations.

Precious and semi-precious dental alloys, with some rare exceptions, are generally recognized as being harmless and perfectly bio-compatible. Ti-base alloys seem to fulfill all mechanical, clinical and biological requirements.

Ceramic materials used for medical devices and bone substitutes generally exhibit a very good biocompatibility, though some observed foreign body reactions have been induced by the presence of sharp-edged wear particles.

There is no evidence for cancer inducement by dental alloys, but orthopedic implants and endoprostheses with evidently higher metal release may cause different types of tumors, as has been reported in several hundred cases.

Some metals should definitely be prohibited; these include beryllium, which is still allowed to be used in some dental alloys, and also nickel, which appears to be one of the worst metals used for biomaterials.

When considering metal-induced pathologies, care must be taken not to confuse the metallic element in its chemical zero state with its salts, oxides and organic complexes: the latter may induce strongly different responses, and the toxic potency of different ions or compounds varies by two to three orders of magnitude. For example, vanadium chloride is extremely toxic, yet low amounts of metallic vanadium do not seem to induce any major undesirable biological effects. Metals such as Ag and Zn no longer exhibit their toxic action in an alloyed state; indeed, a low Zn release may even be used favorably by the human organism.

Particular attention must be given to any metallic biomaterial or medical device, and both known and new devices should be submitted to continuous medical survey in order to avoid as early as possible any potentially unfavorable tissue reactions.

References

AITIO A, ARO A, JÄRVISALO J and VAINIO H (1991) *Trace Elements in Health and Disease*. Royal Society of Chemistry, London.

AMIS AA, KEMPSON AS, CAMPBELL JR and MILLER JH (1988) *Anterior cruciate ligament replacement: Biocompatibility and biomechanics of polyester and carbon fiber in rabbits*. J Bone Joint Surg Br **70**:628.

ANKE M, KRONEMANN H, GROPPEL B, HENNING A, MEISSNER D and SCHNEIDER HJ (1980) *The influence of nickel deficiency on growth reproduction longevity and different biochemical parameters of goats*. In: Anke M, Schneider HJ, Bruckner C, eds. 3. Spurenelement Symposium, Nickel, pp. 3–10. Wiss Beitr Karl-Marx-Univ. Leipzig u Friedrich-Schiller-Univ Jena.

ANONYMOUS (1990) *Report of the international committee on nickel carcinogenesis in man*. Scand J Work Environ Health **16**:1–82.

ARSALANE K, AERTS C, WALLAERT B, VOISIN C and HILDEBRAND HF (1992) *Effects of nickel hydroxy-carbonate on alveolar macrophage functions*. J Appl Toxicol **12**:285–290.

ASPLEY AG (1989) *Editorial: Malignancy and joint replacement: the tip of an iceberg*. J Bone Joint Surg **71B**:1.

BARTOLOZZI A and BLACK J (1985) *Chromium concentrations in serum blood clot and urine from patients following total hip arthroplasty*. Biomaterials **6**:2–8.

BERCOVY M, GOUTALLIER D, VOISIN MC, GEIGER D, BLANQUAERT D, GAUDICHET A and PATTE D (1985) *Carbon-PGLA prostheses for ligament reconstruction*. Clinic Orthop Rel Res **196**:159–168.

BISCHOFF UW, FREEMAN MAR, SMITH D, TUKE MA and GREGSON PJ (1994) *Wear induced by motion between bone and titanium or cobalt-chrome alloys*. J Bone Joint Surg **76-B**:713–716.

BJURHOLM A, AL-TAWIL NA, MARCUSSON JA and NETZ P (1990) *The lymphocyte response to nickel salt in patients with orthopedic implants*. Acta Orthop Scand **61**:248–250.

BLACK J (1988) *In vivo corrosion of a cobalt-base alloy and its biological consequences*. In: Hildebrand HF and Champy M, eds. Biocompatibility of Co-Cr-Ni alloys. NATO-ASI Series 158 A, pp. 83–100. Plenum, London-New York

BRADLEY E (1994) *Ti-8Mn*. In: Material Properties Handbook Titanium Alloys, pp. 755–763. Materials Park, Ohio, ASM Intern.

BRAVO I, CARVALHO GS, BARBOSA MA and DE SOUSA M (1990) *Differential effects of eight metal ions on lymphocyte differentiation antigens in vitro*. J Biomed Materials Res **24**:1059–1068.

BREME J (1989) *Titanium and titanium alloys biomaterials of preference*. Mémoires et Etudes Scientifiques Revue de Métallurgie Octobre 625–638.

BREME J (1994) *Ti-5Al-2,5Fe*. In: Material Properties Handbook. Titanium Alloys, pp. 737–746. Materials Park, Ohio, ASM Intern.

BREME J and WADEWITZ V (1989) *Comparison of titanium-tantalum and titanium-niobium alloys for application as dental implants*. Int J Maxillo-Fac Implant **4**:113–118.

BROWN SS and SAVORY J (1983) *Chemical Toxicology and Clinical Chemistry of Metals*. Academic Press, New York.

BUCHNER A and HANSEN LS (1980) *Amalgam pigmentation (amalgam tattoo) of the oral mucosa. A clinico-pathologic study of 268 cases*. Oral Surg **49**:139–147.

BUNDY KJ (1994) *Corrosion and other electrochemical aspects of biomaterials.* Crit Rev Biomed Eng **22**:139–251.

CARLSSON A and MOLLER H (1989) *Implantation of orthopaedic devices in patients with metal allergy.* Acta Dermatol Venerol (Stockh) **69**:62–66.

CHOUDAT D (1982) *Pathologie pulmonaire et prothésistes dentaires.* Information Dentaire **64**:4157–4160.

CHOUDAT D, BROCHARD P, LEBAS FX, MARSAC J and PHILIBERT H (1983) *Sarcoïdose ou pneumoconiose: Coincidence ou relation.* Arch Mal Prof **44**:339–344.

COOK SD, MCCLUSKEY LC, MARTIN PC and HADDAD RJ (1991) *Inflammatory response in retrieved noncemented porous-coated implants.* Clin Orthop Rel Res **264**:209–222.

COOMBS RR and GELL PHG (1975) *Classification of allergic reactions responsible for clinical hypersensitivity and diseases* In: Gell PHG, Coombs RR and Lachman PJ, eds. Clinical Aspects of Immunology, p. 761. Blackwell Scientific Publications, Oxford, UK.

DAMIEN CJ and PARSONS JR (1991) *Bone graft and bone graft substitutes. A review of current technology and applications.* J Appl Biomater **2**:187–208.

DEMMER P, FOWLER M and MARINO AA (1991) *Use of carbon fibers in the reconstruction of knee ligaments.* Clin Orthop **271**:225.

DOOMS-GOSSENS A, CEUTERICK A, VANMAELE N and DEGREEF H (1980) *Follow-up study of patients with contact dermatitis caused by chromates nickel and cobalt.* Dermatologica (Basel) **160**:249–260.

DORR LD, BLOEBAUM R EMMANUAL J and MELDRUM R (1990) *Histologic biochemical and ion analysis of tissue and fluids retrieved during total hip arthroplasty.* Clin Orthop Rel Res **261**:82–95.

DUPUIS G and BENEZRA C (1982) *Allergic contact dermatitis to simple chemicals.* Marcel Decker Inc, New York-Basel.

EKSTRAND J, BJÖRKMAN L, EDLUND C and SANDBORGH-ENGLUND (1998) *Toxicological aspects on the release and systemic uptake of mercury from dental amalgam.* Eur J Oral Sci **106**:678–686.

FRAZIER ME and ANDREWS TK (1979) *In vitro clonal growth assay for evaluating toxicity of metal salts.* In: Kharasch N, ed. Trace Metals in Health and Disease, pp. 71–81. Raven Press, New York.

GALAN D and LYNCH E (1989) *The effect of reinforcing fibres in denture acrylics.* J Irish Dent Assoc **35**:109–113

GAWKRODGER DJ (1993) *Nickel sensitivity and the implantation of orthopaedic prostheses.* Contact Dermatitis **28**:257–259.

GILLESPIE WJ, HENRY DA, O'CONNEL DL, KENDRICK S, JUSZEZAK E, MCINNENY K and DERBY L (1996) *Development of hematopoietic cancers after implantation of total joint replacement.* Clin Orthop **329S**:290–296.

GIROUX EL and HENKIN RI (1973) *Macromolecular ligands of exchangeable copper zinc and cadmium in human serum.* Bioinorg Chem **2**:125–133.

GOODFELLOW J (1992) *Editorial. Malignancy and joint replacements.* J Bone Joint Surg **74B**:645.

GRIFFITH HJ, BURK EJ and BONFIGLIO TA (1987) *Granulomatous pseudotumors in total joint replacement.* Skeletal Radiol **16**:146–152.

HAGUENOER JM and FURON D (1982) *Toxicologie et Hygiéne Industrielle. Les Dérivés Minéraux.* Technique et Documentation Paris, Vol I and II.

HASAN FM and KAZEMI H (1974) *Chronic beryllium disease: a continuing epidemiological hazard.* Chest **65**:289–293.

HAUBOLD AD, YAPP RA and BOKROS JC (1986) *Carbon for biomedical applications.* In: Bever MB, ed. Encyclopedia of materials science and engineering. Pergamon Press, pp. 513–520.

HEATH JC, FREEMAN MAMR and SWANSON SAV (1971) *Carcinogenic properties of wear particles from prostheses made in cobalt-chromium alloy.* Lancet i, 564–566.

HENCH (1991) *Bioceramics. From concept to clinic.* J Am Ceram Soc **74**:1487–1510.

HENCH (1994) *Bioactive Ceramics Theory and Clinical Application.* In: Andersson OLL, Happonen R-P, Yli-Urpo A, eds. Bioceramics 7, pp. 3–16. Butterworth-Heinemann, Oxford.

HETHERINGTON VJ, LORD CE and BROWN SA (1995) *Mechanical and histological fixation of hydroxylapatite-coated pyrolytic carbon and titanium alloy implants: a report of short-term results.* J Appl Biomaterials **9**:243–248

HILDEBRAND HF and CHAMPY M, eds. *Biocompatibility of Co-Cr-Ni alloys,* pp. 133–153. NATO-ASI Series 158 A. Plenum, London-New York.

HILDEBRAND HF and HORNEZ JC (1998) *Biological response and biocompatibility.* In: Helsen JA and Breme HJ, eds. Metals as Biomaterials, pp. 265–290. John Wiley & Sons, Chichester, UK.

HILDEBRAND HF, OSTAPCZUK P, MERCIER JF, STOEPPLER M, ROUMAZEILLE B and DECOULX J (1988) *Orthopedic implants and corrosion products.* In: Hildebrand HF, Veron C and Martin P (1989a) Nickel Chromium Cobalt dental alloys and allergic reactions: an overview. Biomaterials **10**:545–548.

HILDEBRAND HF, VERON C and MARTIN P (1989b) *Les alliages dentaires en métaux non précieux et l'allergie.* J Biol Buccale **17**:227–243.

HILDEBRAND HF, VERON C, ELAGLI K and DONAZZAN M (1995) *Réactions tissulaires au port des appareils de prothése dentaire partielle ou totale.* Encycl Méd Chir (Paris) Stomatol-Odontol **II**:23–325-P-10.

HILDEBRAND HF, FLOQUET I, LEFÉVRE A and VERON C (1996a) *Biological and hepatotoxic effects of palladium. An overview on experimental investigations and personal studies.* Intern J Risk Safety Med **8**:149–167.

HILDEBRAND HF, LAFFARGUE P, DECOULX J, DUQUENNOY A and MESTDAGH H (1996b) *Retrieval analyses of total hip replacements.* Intern J Risk Safety Med **8**:125–134.

HÖRSTED-BINDSLEV P, DANSCHER GH and HANSEN GH (1997) *Dentinal and pulpal uptake of mercury from lined and unlined amalgam restorations in minipigs.* Eur J Oral Sci **105**:338–343.

HORNEZ JC, ROCHER PH, SPÄTH N, TRAISNEL M and HILDEBRAND HF (2000) *Evaluation électrochimique d'alliages dentaires dans différentes salives artificielles.* In: Mainard D, Merle M, Delagoutte JP and Louis JP, eds. Actualités en Biomatériaux, Vol V, pp. 389–399. Edition Romillat, Paris.

HORNEZ JC, LEFÉVRE A, JOLY D and HILDEBRAND HF (2002) *Multiple parameter cytotoxicity index on dental alloys and pure metals.* Biomol Eng **19**:103–118.

JACOBS JJ, ROSENBAUM DH, HAY RM, GITELIS S and BLACK J (1992) *Early sarcomatous degeneration near a cementless hip replacement: a case report and review.* J Bone Joint Surg **74B**:740–744.

JOHNSON KG and ORGAN CC (1997) *Prostaglandin E2 and interleukin-1 concentrations in nicotine exposed oral keratinocyte cultures.* J Periodontal Res **32**:447–454.

KHURANA JS, ROSENBERG AE, KATTAPURAM SV, FERNANDEZ OS and SHIGERU E (1994) *Malignancy supervening on an intramedullary nail.* Clin Orthop **267**:251–254.

KORNU R, MALONEY WJ, KELLY MA and SMITH RL (1996) *Osteoblast adhesion to orthopaedic implant alloys. Effects of cell adhesion molecules and diamond-like carbon coating.* J Orthop Res **14(6)**:871–877.

LAFFARGUE PH, BREME J, HELSEN JA and HILDEBRAND HF (1998) *Retrieval analyses.* In: Helsen JA and Breme HJ, eds. Metals as Biomaterials, pp. 467–501. John Wiley & Sons, Chichester, UK.

LAFFARGUE PH, HILDEBRAND H.F, LECOMTE-HOUCKE M, BIEHL V, BREME J and DECOULX J (2001) *Histiocytome malin de l'os 20 ans après une fracture du fémur ostéosynthésée Analyse des produits de corrosion et de leur rôle dans la malignité.* Revue de Chirurgie Orthopédique **87**:84–90.

LEE YS, PHO RWH and NATHER A (1984) *Malignant fibrous histiocytoma at site of metal implant.* Cancer **54**:2286–2289.

LEWIS CG and SUNDERMAN FW JR (1996) *Metal carcinogenesis in total joint arthroplasty.* Clin Orthop **329S**:264–268.

LOB M and HUGONNAUD C (1977) *Pathologie pulmonaire.* Arch Mal Prof **38**:543–549.

MARCUSSEN PV (1957) *Occupational nickel dermatitis. Rise in incidence and prevention.* Acta Derm Venereol (Stockh) **2**:289–295.

MATHIESEN EB, AHLBOM A, BERMAN G and LINDGREN JU (1995) *Total hip replacements and cancer: a cohort study.* J Bone Joint Surg **77B**:345–350.

McCUTCHEN JW, COLLIER JP and MAYOR MB (1990) *Osteointegration of titanium implants in total hip arthroplasty.* Clin Orthop Rel Res **261**:114–125.

McKELLOP HA, SARMIENTO A, SCHWINN CP and EBRAMZADEH E (1990) *In vivo wear of titanium-alloy hip prostheses.* J Bone Joint Surgery **72A**:512–517.

MERIAN E (1984) *Metalle in der Umwelt Verteilung Analytik und biologische Relevanz.* Verlag Chemie, Weinheim-Deerfield Beach, Florida-Basel.

MERRIT K. and BROWN SA (1981) *Metal sensitivity reactions to orthopedic implants.* Int J Dermatol **20**:89–94.

MIRRA JM, BULLOUGH PG, MARCOVE RC, JACOBS B and HUVOS AG (1994) *Malignant fibrous histiocytoma and osteosarcoma in association with bone infarcts. Report of four cases two in caisson workers.* J Bone Joint Surg **56A**:932–940.

MICHEL R (1987) *Trace metals in biocompatibility testing.* CRC Critical Reviews in Biocompatibility **3**:235–317.

MICHEL R (1991) *Metal and Ceramic Implants.* In: Merian E, ed. Metals and Their Compounds in the Environment, pp. 557–564 VCH, Weinheim-New York-Basel-Cambridge.

MORRIS DM, HASKINS R, MARINO AA, MISRA RP, ROGERS S, FRONCZAK S and ALBRIGHT JA (1990) *Use of carbon fibers for repair abdominal-wall defects in rats.* Surgery **107**:627–631.

MORRIS DM, HINDMAN J and MARINO AA (1998) *Repair of fascial defects in dogs using carbon fibers.* J Surg Res **80**:300–303.

NASSER S, CAMPBELL PA, KILGUS D, KOSSOVSKY N and AMSTUTZ HC (1990) *Cementless total joint arthroplasty prostheses with titanium-alloy articular surfaces. A human retrieval analysis.* Clin Orthop Rel Res **261**:171–185.

OONISHI H, KUSHITANI S, YASUKAWA E, IWAKI H, HENCH LL, WILSON J and TSUJI E (1997) *Particulate bioglass compared with hydroxyapatite as a bone graft substitute.* Clin Orthop Rel Res **334**:316–325.

PAZAGLIA UE, DELL'ORBO C and WILKINSON MJ (1987) *The foreign body reaction in total hip arthroplasties. A correlated light- microscopy SEM and TEM study.* Arch Orthop Trauma Surg **106**:209–219.

PUCK TT and MARCUS PI (1955) *A rapid method for viable cell titration and clone production with HeLa cells in tissue culture: the use of X-irradiated cells to supply conditioning factors.* Proc Natl Acad Sci USA **41**:432–437.

REED KP, VAN DEN BERG SS, RUDOLPH A, ALBRIGHT JA, CASEY HW and MARINO AA (1994) *Treatment of tendon injuries in thoroughbred racehorses using carbon-fiber implants.* J Equine Vet Sci **14**:371.

ROSTOKER G, ROBIN J, BINET O and PAUPE J (1986) *Dermatoses d'intolérance aux métaux des matériaux d'ostéosynthése et des prothéses (Nickel-Chrome-Cobalt).* Ann Dermatol Venerol **113**:1097–1108.

SANDBORGH-ENGLUND G, ELINDER CG, LANGWORTH S, SCHÜTZ A and EKSTRAND J (1998) *Mercury in biological fluids after amalgam removal.* J Dent Res **77**:615–624.

SARMIENTO A and GRUEN TA (1985) *Radiographic analysis of a low-modulus titanium-alloy femoral total hip component. Two- to six-year follow up.* J Bone Joint Surgery **67A**:48–56.

SHANBHAG A, YANG J, LILIEN J and BLACK J (1992) *Decreased neutrophil respiratory burst on exposure to cobalt-chrome alloy and polystyrene in vitro.* J Biomed Mater Res **26**:185–195.

SHIRALI P, TEISSIER E, MAREZ T, HILDEBRAND HF and HAGUENOER JM (1994) *Effect of Ni_3S_2 on arachidonic acid metabolites in cultured human lung cells.* Carcinogenesis **15**:759–762.

SINIBALDI K, ROSEN H, LIU SK and DEANGELIS M (1976) *Tumors associated with metallic implants in animals.* Clin Orthop **118**:257–266.

SOLOMON MI and SEKEL R (1992) *Total hip arthroplasty complicated by a malignant fibrous histiocytoma. A case report.* J Arthroplasty **7**:549–550.

STANDARDS ISO 10993 –5: 1992 and EN 30993 –5: 1994. *Biological testing of medical and dental devices. Tests for cytotoxicity: in vitro methods.*

SUNDERMAN FW JR (1988) *Carcinogenic risks of metal implant and prostheses.* In: Hildebrand HF and Champy M, eds. Biocompatibility of Co-Cr-Ni alloys, pp. 11–19. NATO-ASI Series 158 A Plenum, London-New York.

SUNDERMAN FW JR (1989a) *Mechanisms of nickel carcinogenesis.* Scand J Work Environ Health **15**:1–2.

SUNDERMAN FW JR (1989b) *Carcinogenicity of metal alloys in orthopedic prostheses Clinical and experimental studies.* Fund Appl Toxicol **13**:205–216.

TAGUSARI O, YAMAZAKI K, LITWAK P, KOJIMA A, KLEIN EC, ANTAKI JF, WATACH M, GORDON LM, KONO K, MORI T, KOYANAGI H, GRIFFITH BP and KORMOS RL (1998) *Fine trabecularized carbon: ideal material and texture for percutaneous device system of permanent left ventricular assist device.* Artif Org **22(6)**:481–487.

TAYTON K, JOHNSON-NURSE C, MCKIBBIN B, BRADLEY J and HASTINGS G (1982) *The use of semi-rigid carbon-fibre-reinforced plastic plates for fixation of human fractures.* J Bone Joint Surg **64B**:105–111.

TESSIER PY, PICHON L., VILLECHAISE P, LINEZ P, ANGLERAUD B, MUBUMBILA N, FOUQUET V, STRABONI A, MILHET X and HILDEBRAND HF (2003) *Carbon nitride thin films as protective coatings for biomaterials synthesis mechanical and biocompatibility characterizations.* Diamond & Related Materials **12**:1066–1069.

TROOP JK, MALLORY TH, FISHER DA and VAUGHN BK (1990) *Malignant fibrous histiocytoma after total hip arthroplasty a case report.* Clin Orthop **253**:297–300.

VÉRON C, HILDEBRAND HF and FERNANDEZ JP (1984) *Les pigmentations gingivales par l'amalgame dentaire Etude ultrastructurale.* J Biol Buccale **12**:273–286.

VÉRON C, HILDEBRAND HF and FERNANDEZ JP (1985) *Les pigmentations gingivales par l'amalgame dentaire Etude ultrastructurale et microanalyse.* J Biomat Dent **1**:47–52.

VÉRON C, HILDEBRAND HF and MARTIN P (1986) *Les amalgames dentaires et l'allergie.* J Biol Buccale **14**:83–100.

VISURI T, PUKKALA E, PAAVOLAINEN P, PULKKINEN P and RISKA EB (1996) *Cancer risk after metal on metal and polyethylene on metal total hip arthroplasty.* Clin Orthop **329S**:280–289.

WALL L and CALNAN CD (1980) *Occupational nickel dermatitis in the electroforming industry.* Contact Derm **6**:414–420.

WEISSMAN BN SCOTT RD BRIDK GW and CORSON JM (1991) *Radiographic detection of metal-induced* synovitis as a complication of arthroplasty of the knee. J Bone Joint Surgery **73A**:1002–1007.

WILLIAMS DF (1981) *Toxicology of implanted metals.* In: Williams DF, ed. Fundamental Aspects of Biocompatibility Vol II, pp. 1–10. CRC Press, Boca Raton, Florida.

WILSON J, CLARK AE, HALL MB and HENCH II (1993) *Tissue response to Bioglass(r) endosseous ridge maintenance implants.* J Oral Implantol **19**:295–302.

WILSON J and LOW S (1992) *Bioactive ceramics for periodontal treatment comparative studies in the patus monkey.* J Appl Biomater **3**:123–129.

WOLLINA U, KNOLL B, PRUFER K, BARTH A, MULLER D and HUSCHENBECK J (1996) *Synthetic wound dressings – evaluation of interactions with epithelial and dermal cells in vitro.* Skin Pharmacol **9**:35–42.

ZIATS NP, MILLER KM and ANDERSON JM (1988) *In vitro and In vivo interactions of cells with biomaterials.* Biomaterials **9**:5–13.

6

Metallothioneins

Bartolome Ribas

6.1

Introduction

During the past 25 years, several well-structured reviews of metallothioneins have been produced, including those by Schäffer and Kägi (1991), Kojima and Kägi (1978), Nordberg and Kojima (1979), Brady (1982), Cousins (1985), Hamer (1986), Kägi (1987), and – most recently – by Hidalgo et al. (2001). In 1957, Margoshes and Vallee first isolated the molecule from equine kidney and proposed the name metallothionein (MT). Although the molecular weight of MT is within the limit of peptides under 10 kDa, it is denominated as a low molecular-weight protein. The name metallothionein reflects the molecule's high metal and sulfur content, both of which are of the order 10% (wt./wt.). Among the composite amino acids, 30% are cysteine, while a number of mineral ions were also seen to be bound to the protein, including Zn(II), Cd(II), Cu(I), and Fe(II) (Kägi and Vallee 1960). More than 10 different isoforms and sub-isoforms of 7 kDa have been detected; this leads to changes in the molecule's affinity and also the degree to which it is induced following exposure to different inorganic and organic cations. Metallothionein-like proteins (MLPs) of molecular weight 10

and 20 kDa have also been characterized, each with four and five isoforms respectively. The binding capacities of MTs are indicative of these proteins' roles in mineral homeostasis and enzymatic catalytic activity. The high redox potential of MTs relates to the etiology of degenerative processes, mainly due to the production of free radicals. The function of MTs as free radical scavengers during oxidative damage is of major value in mineral homeostasis in the central nervous system and in the etiology of neuropathological disorders. Mineral elements are essential in all physiological functions, ranging from cardiorespiratory activity to muscle contraction, hemoglobin synthesis, metal ion absorption, distribution and metabolism, and the electron transport chain through Fe and Cu proteins. MTs are implicated in anemia (Ribas 1983, Ribas et al. 1987, 1988), Alzheimer's disease, Parkinson's disease and other degenerative disorders (Hidalgo et al. 2001). Phylogenic studies have demonstrated that an interrelationship exists between MTs and MLPs, with the number of isoforms higher in more specialized and advanced evolutionarily developed species with a more complex homeostasis (Uchida 1994). MTs are present in all known cells, and occur in vertebrates, invertebrates, plants, and eukary-

Elements and their Compounds in the Environment. 2nd Edition.
Edited by E. Merian, M. Anke, M. Ihnat, M. Stoeppler
Copyright © 2004 WILEY-VCH Verlag GmbH & Co. KGaA, Weinheim
ISBN: 3-527-30459-2

otic microorganisms, as well as in some prokaryotes (Nordberg and Kojima 1979, Hamer 1986, Robinson and Jackson 1986). MLPs have not been detected in some mollusks, however, and only one MT isoform has been identified in certain other species such as crustaceans, teleosteans and mollusks.

6.2
Physico-chemical Characteristics

When, in 1957, Margoshes and Vallee first isolated MT from horse kidney cortex, they found a protein which contained 5.9% Cd, 2.2% Zn, 0.45% Fe, and 0.18% Cu, as well as a high sulfur content. Nordberg et al. (1974) later isolated two isoforms using gel electrophoresis, and these were later designated by Kägi and Bühler (1987) as MT-I and MT-II. Ribas (1981, 1983), used polyacrylamide gel electrophoresis detected four isoforms, while Hunziker and Kägi (1987) used reversed-phase HPLC to resolve six MT-isoforms.

MT have been subdivided into three classes, namely I, II, and III (Fowler et al. 1987), while Binz and Kägi (1999) allocated these compounds to several families. Subsequently, Richards and Beattie (1995), Richards et al. (1996, 1997) and Szpunar (2000) showed capillary zone electrophoresis to be a rapid and reliable method for analysis of this protein. By using a diverse combination of techniques, the metal complexes of metallothionein in rat liver and kidney were first characterized, with two major MT-isoforms (MT-1 and MT-2) being detected in liver, and one MT isoform in kidney (Polec et al. 2002). The order of affinity of metal ions to MTs is Cd > Zn, Cu, Ag, Hg > Bi > Pb, though this may change depending on the tissue involved. MTs are considered to be low molecular-weight proteins (~7 kDa) and to be derived from a wide family of genes. The MT molecule comprises 60–62 amino acids, including ~20 cysteines, but does not contain any aromatic amino acids or histidine. Polymorphically, a minimum of ten isoforms and subisoforms exist. MTs show isoelectric points between 4 and 5.5, are stereochemically and thermodynamically flexible and thermostable up to 70 °C. They are encoded by multiple genes with many alleles, and are inducible both "in vivo" and "in vitro" via gene transcription, with two clusters of α and β domains binding a variable number of mineral elements and with a high redox potential. The reducing potential depends on the 20 cysteine groups ($-SH$) per molecule/60 amino acids (Figure 6.1), this being the same proportion as glutathione, which has only one cysteine for each three amino acids. The positions of the 20 cysteine residues (Cys) in the polypeptide chain is shown in Figure 6.1, and these are highly conserved (Nordberg and Kojima 1979). Their arrangement in the prevailing Cys-X-Cys and Cys-Cys sequences (X = amino acid residue other than Cys) renders the protein a potent metal chelator. Cysteines are implicated in the binding of bivalent metal ions, giving rise to a Cys/metal ratio of about 3. The structures have been established from a large variety of spectroscopic analysis of MTs (Otvos and Armitage 1980).

The maximal UV absorption of MTs occurs at 254 nm, and not at 280 nm as is found with most proteins that contain aromatic amino acids. Specific optical characteristics in terms of the absorption of the metal-thiolate complexes occur at 254 nm with Cd, at 225 nm with Zn, at 275 nm with Cu, and at 300 nm with Hg. MTs bind mineral ions, as both plastic and trace elements, and also toxic heavy metals such as Cd, Hg, and Pb. One of the principal

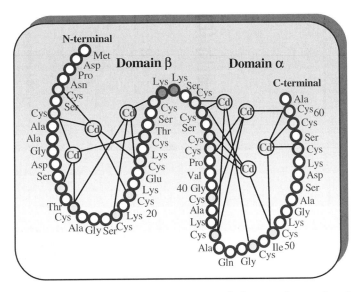

Fig. 6.1   Model of the molecular structure of mammalian metallothionein, showing the primary sequence, and two topologically separate metal-thiolate clusters. (After Otvos and Armitage 1980.)

functions of MTs is the release of trace elements as catalysts for enzymatic activity, being also structural links for requisite specific and catalytic function. MTs are induced by both mineral and organic toxic agents, and act as homeostatic regulators and radical scavengers against oxidative stress in all eukaryotic and prokaryotic cells. Generally, MTs are also induced in inflammatory processes, and by glucocorticoids, hormones, cytokines, endotoxins (Richards and Beattie 1995), antibiotics, vitamins, analgesics (Pountney et al. 1995), and by physical, psychological and pathological stress (Bremner and Beattie 1990, Beattie et al. 1996, Penkowa et al. 2001).

Among the MLPs, all have a very high percentage of thiol groups and a high sulfur content (Uchida 1994, Ferrarello et al. 2002). These proteins are induced principally by cadmium, which is a major environmental problem in public health terms (Piscator 1971). In the tissues, metal ions are bound from water, food and air by metallo-

proteins, MTs, MLPs and metalloenzymes, all of which act as catalysts for enzymatic activity and thus have specific biochemical functions in homeostasis. Cadmium is considered as marker of MTs and, as a biomarker of environmental pollution, has been found in all tissues from mammals, fish, invertebrates, plants, and microorganisms, some of which were prokaryotes (Olafson et al. 1988, Takatera et al. 1994, Ferrarello et al. 2002). All these physico-chemical characteristics suggest that MTs and MLPs form a singular group of proteins, the functions of which can be investigated in a multidisciplinary study covering aspects of clinical benefits and pathological outcome.

The three-dimensional structure of the domains showing the localization of the metal ions is shown in Figure 6.2. The elevated content of −SH groups allows the binding of mineral elements and of toxic heavy metals that are commonly present in situations of environmental pollution, and which are also accumulated in human and animal organs. In this case, MT associates with a certain number of metallic ions, the identity of which induce or inhibit specifically different molecular isoforms of this protein (Ribas 1993). MTs can be detected and quantified for pollution, contamination or accumulation that is principally induced by metallic ions, but also by other toxic agents such as organic compounds (Min et al. 1993). A function of detoxification was proposed for this protein, in order to explain the accumulation of metals or heavy metals from the environment which, when bound to MT, are less toxic than the free metal ions. One suggested use for this protein is as a biomarker for heavy metal exposure or chronic intoxication.

The concentration of this protein varies widely among different tissues, and there are also striking species differences. For example, human liver and kidney are particularly rich in this protein, with levels ranging from 0.01% to 0.1% of the tissue dry weight.

The positions of the 20 cysteine residues (Cys) in the polypeptide (see Figure 6.2) are highly conserved (Nordberg and Kojima 1979). Their arrangement in the pre-

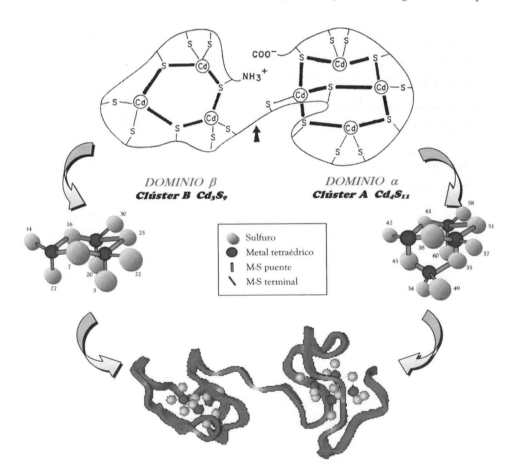

Fig. 6.2 Three-dimensional structure of the metallic domains of the complex of Cd_7MT, and primary model of mammalian metallothionein.

vailing Cys-X-Cys and Cys-Cys sequences (X = amino acid residue other than Cys) renders the protein a potent metal chelator. In fact, all Cys are involved in the binding of bivalent metal ions (usually seven), giving rise to a Cys/metal ratio of about 3. The existence of such clustered structures has also become evident from a large variety of spectroscopic features of MT. Moreover, cross-irradiation studies of ^{113}Cd-containing MT have revealed that the protein harbors two topologically separate metal-thiolate clusters (Otvos and Armitage 1980).

The thermodynamic stabilities of the metal-thiolate clusters follow the order of thiolate model complexes; that is, Zn(II) < Cd(II) < Cu(I), Ag(I), Hg(II), Bi(III) (Kägi and Kojima 1987). Upon acidification, Zn(II) and Cd(II) are released from the protein moiety. At neutral pH, the average apparent stability constants for Zn(II) and Cd(II) have been calculated to be of the order of 10^{12} L mol^{-1} and 10^{16} L mol^{-1}, respectively (Schäffer and Kägi 1991).

6.3
Biological Implications of MT

The physiological and toxicological significance of MT is not well clarified. MT has been suggested to regulate the intake of the essential Fe(II), Zn(II) and Cu(I) metal ions into the cell, as well as participating as a metal donor and acceptor in different biosynthetic and catabolic processes (Beltramini and Lerch 1982, Brady 1982). There is also evidence for a protective role of MT in chronic metal poisoning exposure, especially Cd(II). Through intracellular sequestration, this protein participates effectively in the attenuation of Cd toxicity (Webb 1987), and was also found to be present in metal-tolerant plants (Rauser and Curvetto 1980).

6.3.1
MT Analogues

Mammalian MTs are comprised of an unknown number of MT isoforms. For example, in human systemic organs such as the liver, initial studies of polymorphism suggested the presence of six isoforms (Hunziker and Kägi 1987), but this was later modified to two major and subisoforms. Currently, four major isoforms designated MT-1 to MT-4 are thought to exist, though in the opinion of the present author these should also be subdivided into subisoforms depending upon the heterogeneity of the family of genes. MT-1 and MT-2 are expressed in most tissues (including the brain), whereas MT-3 [which is also known as growth inhibitory factor (GIF) (Uchida 1994)] and MT-4 are expressed predominantly in the central nervous system and in keratinizing epithelia, respectively. MT isoforms have been implicated in very different physiological functions, including zinc, copper and iron metabolism, the protection of enzymatic catalytic activity releasing metal ions, protection against reactive oxygen species to maintain the redox potential in living cells, and adaptation to stress, temperature and inflammatory processes. In the case of the MT-3 isoform, an additional involvement has been reported in neuromodulatory events and in the pathogenesis of Alzheimer's disease, and research in these areas is currently expanding. MLPs, such as the SmtA induced by zinc and with a four-metal cluster, has been shown to protect bacteria against zinc toxicity (Blindauer et al. 2001).

6.3.2
MT Complex with Methotrexate

Induction of MT synthesis after intraperitoneal administration of methotrexate (MTX;

a chemotherapeutic agent used against human cancer) to rats suggested that a complex of MT-MTX-Me might be formed (Gonzalez-Baron et al. 1997, Sanchez et al. 1988). Other chemical agents and organic solvents also induced increases in tissue MT levels, as did other toxic chemical agents (Min et al. 1993). In the case of MTX, the condensation of n molecules of MTX with one molecule of MT goes through the release of several bivalent cations from MT. This could lead to an induction of minor concentrations of metals in the tissues during chemotherapy, and hence to an induced anemic condition (Iniesta et al. 1985).

MT in the thymus was first characterized by Olafson in 1985. The weight and volume of the thymus was seen to diminish in a statistically significant manner in MTX-treated rats, as also occurs with cadmium (Ribas et al. 1987), which acts as a toxic agent for the thymus gland. The toxicity of MTX in humans is well known, and included the development of an anergic condition caused by inhibition of the immune defense mechanisms in which thymus plays an important role. Nevertheless zinc, which is localized in the cellular reticulum of the thymus, is inhibited competitively and substituted by cadmium at its binding sites, which leads to a major regression of the thymus cortical region (Ribas et al. 1987).

6.3.3
MT and Anemia

In anemia, levels of MT, cysteine and transferrin are higher than under normal physiological circumstances (Iniesta et al. 1985, González-Barón et al. 1997), and MT is considered for Fe as also for Zn and Cu, an iron-binding protein (Ribas et al. 1988) and homeostatic regulatory protein (Ribas et al. 1987). It is possible that MTX might condense with native MT and induce the partial

depletion of Zn and Cu, and this may be one way of inhibiting biological activity in living cells, as in the case of tumor growth.

6.3.4
MT and Endocrine Pancreas

MT was monitored in the pancreas of normal mice (Onosaka et al. 1988), while in rats different isoforms were isolated from pancreatic beta cells at MT levels between 5 and 20 ppb Zn μg^{-1}. Isoforms 1 and 5 of the six molecular varieties established in rabbit liver (Ribas and Iniesta 1989) were not detected, though this may be due to the low threshold of sensitivity of the instrumentation used and the low protein concentration in specific organs. MT as a regulatory protein in central intermediary metabolism focuses its main attention on pancreatic secretory activity following a glucose stimulus, mainly because pancreatic MT levels were elevated ~40-fold after an injection of zinc (Onosaka et al. 1988). The human pancreas contains total MT levels of between 50 and 800 $\mu g \ g^{-1}$ pancreatic tissue. Human MTs were shown to contain both zinc and cadmium, indicating that toxic metal elements are concentrated in the pancreas; the subsequent development of human molecular pathologies related to MT isoforms may be linked with zinc-binding proteins in the beta cell (Ribas et al. 1994).

6.4
MT and Tolerance to Ionizing Radiation

Cultured mammalian cells that overexpress MT are unusually resistant to X-ray damage (Bakka and Webb 1981), and this has led some investigators to propose that MTs may function as scavengers for free radicals (Thornalley and Vasak 1985). Glutathione, however, with a much higher overall abundance seems a more available candidate for this purpose (Bremner 1987). Metals and oxidative stress induced MT-1 gene expression in transfected mouse cells and in transgenic mice (Dalton et al. 1997). Oxidative stress after exposure to ionizing γ-irradiation increases lipid peroxidation and subsequent hepatic MT synthesis in damaged tissue (Sato and Bremner 1993) through a variety of mediators, including cytokines, glucocorticoids, or tumor necrosis factor. Interleukin (IL)-6 has also been suggested to play a key role in hepatic MT gene expression during inflammation caused by organic solvents (Min et al. 1993). Irradiation is able to produce a biosynthetic cascade of cytokines and other signal transduction factors such as a protein kinase C pathway in order to stimulate gene expression. This is seen in the fact that MT mRNA expression and MT synthesis are induced by the exposure of cells in vitro or tissues in vivo to either ionizing or UV radiation (Cai et al. 1999). Likewise, Morcillo et al. (2000) have suggested that an enhanced lipid peroxidation is not a prerequisite for the induction of hepatic MT synthesis after γ-irradiation.

6.5
Clinical and Pathologic Involvement of MT

Certain inherited diseases are associated with abnormal trace metal metabolism, and this may be linked to MT synthesis, either directly or indirectly. Zinc, copper, and iron are the target elements in several diseases, and their concentrations may be altered in disorders of the central nervous system or other physiological systems, including Alzheimer-type dementia, amyotrophic lateral sclerosis, acrodermatitis enteropathica, biliary cirrhosis, Wilson's and Menke's diseases (Bremner 1987), epilepsy, Friedreich's ataxia, Guillaine–Barre syndrome, hepatic encephalopathy, multiple sclerosis, Parkinson's disease, Pick's disease, retinitis pigmentosa, retinal dystrophy, schizophrenia, and Wernicke–Korsakoff syndrome (Ebadi et al. 1995). The status of MT isoforms and other MLPs in these pathological conditions, disorders or syndromes are indeed the subjects of ongoing research. Some of these disorders are associated with oxidative stress, and as MT is able to prevent the formation of free radicals, it is believed that cytokine induction of MT provides the redox steady state in human beings. Copper is implicated in Menkes and Wilson's diseases, and iron in various overload and deficiency anemic syndromes (Ribas et al. 1988).

Urinary MT levels have been used to measure Cd exposure of industrial workers (Roels et al. 1983). The side effects of certain metal-containing drugs, such as gold-containing antirheumatic drugs (Glennás and Rugstad 1985) and platinum-containing anticancer agents can also be mitigated by the metal-chelating properties of the protein. MT induction after $Bi(NO_3)_3$ administration was shown to reduce the renal and lethal toxicity of cis-$Pt(NH_3)_2Cl_2$, without compromising its antitumor activity (Naganuma et al. 1987). These experiments point to a potential utility of stimulation MT synthesis in cancer therapy.

References

BAKKA A and WEBB M (1981) *Metabolism of zinc and copper in the neonate: changes in the concentrations and contents of thionein-bound Zn and Cu with age in the livers of the newborns of various mammalian species.* Biochem Pharmacol **30**:721–725.

BELTRAMINI M and LERCH K (1982) *Copper transfer between Neurospora copper metallothionein and type 3 copper apoproteins,* FEBS Lett **142**:219–222.

Binz PA and Kägi JHR (1999) *Metallothionein: Molecular evolution and classification.* In: Klaassen CD, ed. Metallothionein IV, pp. 7–13. Birkhäuser Verlag, Basel.

Blindauer CA, Harrison MD, Parkinson JA, Robinson AK, Cavet JS, Robinson NJ and Sadler PJ (2001) *A metallothionein containing a zinc finger within a four metal cluster protects a bacterium from zinc toxicity.* Proc Natl Acad Sci USA **98**:9593–9598.

Bordin G, Cordeiro F and Rodriguez AR (1998) *Effect of temperature variation on metallothionein sub-isoform separation by reverse phase high performance liquid chromatography.* J Liquid Chrom Rel Technol **21**:2039–2060.

Brady FO (1982) *The physiological function of metallothionein.* Trends Biochem Sci **7**:143–145.

Bremner I (1987) *Nutritional and physiological significance of metallothionein.* In: Kägi JHR and Kojima Y, eds. Metallothionein II, pp. 81–107. Experientia Suppl. 52. Birkhäuser Verlag, Basel.

Bremner I and Beattie JH (1990) *Metallothionein and the trace minerals.* Annu Rev Nutr **10**:63–83.

Beattie JH, Black DJ, Wood AM and Trayhurn P (1996) *Cold-induced expression of the metallothionein-1 gene in brown adipose tissue rats.* Am J Physiol **270**:R971–977.

Cai L, Satoh M, Tohyama C and Cherian MG (1999) *Metallothionein in radiation exposure: its induction and protective role.* Toxicology **132**:85–98.

Cousins RJ (1985) *Absorption, transport and hepatic metabolism of copper and zinc: special reference to metallothionein and ceruloplasmin.* Physiol Rev **65**:238–309.

Dalton TP, Paria BC, Fernando LP, Huet-Hudson YM, Dey SK and Andrews GK (1997) *Activation of the chicken metallothionein promoter by metals and oxidative stress in cultured cells and transgenic mice.* Comp Biochem Physiol B Biochem Mol Biol **116**:75–86.

Durnam DM and Palmiter RD (1981) *Transcriptional regulation of the mouse metallothionein-I-gene by heavy metals.* J Biol Chem **256**:5712–5716.

Durnam DM, Hoffman JS, Quaife CJ, Benditt E P, Chen HY, Brinster RL and Palmiter RD (1984) *Induction of mouse metallothionein b-I mRNA by bacterial endotoxin is independent of metals and glucocorticoid hormones.* Proc Natl Acad Sci USA **81**:1053–1056.

Ebadi M, Iversen PL, Hao R, Cerutis DR, Rojas P, Happe HK, Murrin LC and Pfeiffer RF (1995) *Expression and regulation of brain metallothionein.* Neurochem Int **27**:1–22.

Ferrarello CN, Fernández De La Campa MR, Carrasco JF and Sanz-Medel A (2002) *Speciation of metallothionein-like proteins of the mussel Mytilus edulis by orthogonal separation mechanisms with ICP-MS detection: effect of selenium administration.* Spectrochim Acta B **57**:439–449.

Fowler BA, Hilderbrand CE, Kojima Y and Webb M (1987) *Nomenclature of metallothionein,* In: Kägi JHR and Kojima Y, eds. Metallothionein II, pp. 19–22. Experientia Suppl 52, Birkhäuser Verlag, Basel.

Friedman RL and Stark GR (1985) *Alfa-interferon-induced transcription of HLA and metallothionein genes containing homologous upstream sequences.* Nature **314**:637–639.

Glennás A and Rugstad HE (1985) *Acquired resistance to auranofin in cultures human cells.* Scand J Rheumatol **14**:230–238.

González-Barón M and Casado-Sáenz E (1997) *Cáncer y Medio Ambiente.* Noesis SL, Madrid.

Good M and Vasäk M (1986) *Iron(II)-substituted metallothionein: evidence for the existence of iron-thiolate clusters.* Biochemistry **25**:8353–8356.

Goyer RA (1997) *Toxic and essential metal interactions.* Annu Rev Nutr **17**:37–50.

Hamer DH (1986) *Metallothionein.* Annu Rev Biochem **55**:913–951.

Hidalgo J, Aschner M, Zatta P and Vašak M (2001) *Roles of the metallothionein family of proteins in the central nervous system.* Brain Res Bull **55(2)**:133–145.

Hunziker P and Kägi JHR (1987) *Human hepatic metallothioneins: resolution of six isoforms,* (Proceedings, 2nd International Symposium, Zürich, 1985), Experientia Suppl, vol. 52, pp. 257–264. Metallothionein II, Birkhäuser Verlag, Basel.

Iniesta MP, Rubio MC and Ribas B (1985) *Metallothionein and transferrin concentrations in the rat intestinal mucosa in several anaemic conditions.* 2nd International Meeting on Metallothionein. Abstracts Book, pp. 66, Zürich University, Switzerland.

Iniesta MP, Sánchez-Reus MI, Ribas B, Taxonera C and Diaz Rubio M (1992) *Comparison of metallothionein isoforms induced with cadmium, mercury and lead.* In: Merian E and Haerdi W, eds. Metal Compounds in Environment and Life, Vol. 4, pp. 293–301. Science and Technology, North Wood, UK.

Kägi JHR and Valle BL (1960) *Metallothionein: a cadmium- and zinc-containing protein from equine renal cortex.* J Biol Chem **235**:3460–3465.

KÄGI JHR and KOJIMA Y, eds. *Metallothionein I. (1985) 2nd International Meeting on Metallothionein*, University Zurich, Switzerland.

KÄGI JHR and KOJIMA Y (1987) *Chemistry and biochemistry of metallothionein*, In: Kägi JHR and Kojima Y, eds. Metallothionein II, pp. 25–61. Experientia Suppl 52. Birkhäuser Verlag, Basel.

KARIN M, IMBRA RJ, HEGUY A and WONG G (1985) *Interleukin I regulates human metallothionein gene expression.* Mol Cell Biol **5**:2866–2869.

KOJIMA Y and KÄGI JHR (1978) *Metallothionein.* Trends Biochem Sci **3**:90–93.

LINDE AR, SÁNCHEZ-GALÁN S VALLÉS-MOTA JP and GARCÍA-VAZQUEZ E (2001) *Metallothionein as bioindicator of freshwater pollution: European eel and brown trout.* Ecotoxicol Environ Safety **49**:60–63.

MARGOSHES M and VALLEE BL (1957) *A cadmium protein from equine kidney cortex.* J Am Chem Soc **79**:4813–4814.

MIN KS, ITOH N, OKAMOTO H and TANNAKA K (1993) *Indirect induction of metallothionein by organic compounds.* In: Suzuki KT, Kimura M and Imura N, eds. Metallothionein III: Biological roles and medical implications. pp. 159–174, Birkhäuser, Boston.

MORCILLO MA, RUCANDIO MI and SANTAMARÍA J (2000) *Effect of gamma irradiation on liver metallothionein synthesis and lipid peroxidation in rats.* Cell Mol Biol **46**:435–444.

NAGANUMA A, SATOH M and IMURA N (1987) *Prevention of lethal and renal toxicity of cis-diaminedichloroplatinum(II) by induction of metallothionein synthesis without compromising its antitumour activity in mice.* Cancer Res **47**:983–987.

NORDBERG M, TROJANOWSKA B and NORDBERG GF (1974) *Studies on metal-binding proteins of low molecular weight from renal tissue of rabbits exposed to cadmium or mercury.* Environ Physiol Biochem **4**:149–158.

NORDBERG M and KOJIMA Y (1979) *Metallothionein and other low molecular weight metal binding proteins.* In: Kägi JHR and Nordberg M, eds. Metallothionein, pp. 41–121. Experientia Suppl 34. Birkhäuser Verlag, Basel.

OH SH, DEAGEN, JT WHANGER PD and WESWIG PH (1978) *Biological function of metallothionein. V. Its induction in rats by various stresses.* Am J Physiol **234**:E282–E285.

OLAFSON RW, MC CUBBIN WD and KAY CM (1988) *Primary and secondary structural analysis of a unique prokaryotic metallothionein from a Synechococcus sp. Cyanobacterium.* Biochem J **251**:691–701.

ONOSAKA S, MIN KS, FUJITA Y, TANAKA K, IGUCHI S and OKADA Y (1988) *High concentration of pancreatic metallothionein in normal mice.* Toxicology **50**:27–35.

OTVOS JD and ARMITAGE IM (1980) *Structure of the metal clusters in rabbit liver metallothionein.* Proc Natl Acad Sci USA **77**:7094–7098.

PALMITER RD (1987) *Molecular biology of metallothionein gene expression.* In: Kägi JHR and Kojima Y, eds. Metallothionein II, pp. 63–80. Experientia Suppl. 52. Birkhäuser Verlag, Basel.

PALMITER RD, NORSTEDT G, GELINAS RE, HAMMER RE and BRINSTER RL (1983) *Metallothionein-Human GH fusion genes stimulate growth of mice.* Science **222**:809–814.

PENKOWA M, ESPEJO C, MARTINEZ EM, POULSEN CHR B, MONTALBAN X and HIDALGO JJ (2001) *Altered inflammatory response and increased neurodegeneration in metallothionein I + II deficient mice during experimental autoimmune encephalomyelitis.* J Neuroimmunol **119**:248–260.

POLEC K, PÉREZ-CALVO M, GARCÍA ARRIBAS O, SZPUNAR J, RIBAS B and LOBINSKI R (2002) *Investigation of metal complexes with metallothionein in rat tissues by hyphenated techniques.* J Inorg Biochem **88**:197–206.

POUNTNEY DL, KÄGI JHR and VASAK M (1995) In: Berthon G, ed. Handbook of Metal-Ligand Interactions in Biological Fluids. Vol. 1, pp. 431. CNRS-Marcel Dekker Inc, Toulouse.

PISCATOR M (1971) *Cadmium in the Environment.* In: Friberg L, Piscator M and Nordberg G, eds., pp. 124–135. CRC Press Boca Raton, FL.

RAUSER WE and CURVETTO NR (1980) *Metallothionein occurs in roots of Agrostis tolerant to excess of copper.* Nature **287**:563–564.

RIBAS B. (1981) *Aillament i caracterizació de la miroina cerebral.* An Med **67**:941–953.

RIBAS B (1983) *Isolation of metallothionein isoforms in rat liver.* In: Brätter P and Schramel P, eds. Trace Elements in Analytical Chemistry and Medicinal Biology, Vol. 2, pp. 181–197. Walter de Gruyter Co., Berlin-New York.

RIBAS B, BRENES MA, DE PASCUAL FJ, DEL RIO J and SANCHEZ-REUS MI (1987) *Participation of metallothionein and cerebral structures in iron homeostasis of anaemic rats.* In: Brätter P and Schramel P, eds. Trace Elements in Analytical Chemistry and Medicinal Biology, Vol. 4, pp. 317–324. Walter de Gruyter Co., Berlin-New York.

RIBAS B, DE PASCUAL FJ, DEL RIO J, SANCHEZ REUS MI (1987) *Inhibition of the thymus gland morphology by cadmium.* In: Brätter P and Schramel

P, eds. Trace Elements in Analytical Chemistry and Medicinal Biology, Vol. 4, pp. 325–336. Walter de Gruyter Co., Berlin-New York.

Ribas B, Pelayo JF and Rodrigues NL (1988) *New data on the hypothesis of the brain participation in iron homeostasis.* In: Brätter P and Schramel P, eds. Trace Elements in Analytical Chemistry and Medicinal Biology, Vol. 5, pp. 548–555. Walter de Gruyter Co., Berlin-New York.

Ribas B and Iniesta MP (1989) *Induction of metallothionein 1 with cadmium by high pressure liquid chromatography.* An Real Acad Farm **55**:533–540.

Ribas B (1994) *Heterogeneity of metallothionein isoforms by HPLC under the effect of different metals in rabbit kidney,* In: Brätter P, Ribas B and Schramel P, eds. Trace Elements in Analytical Chemistry and Medicinal Biology, Vol. 6, pp. 303–310. Consejo Superior de Investigaciones Científicas, Madrid, Spain.

Ribas B, Muñoz A, Camblor P, Gomis R and Sarri Y. (1994) *Molecular heterogeneity of pancreatic islet metallothionein.* In: Schramel P, Ribas B and Brätter P, eds. Trace Elements in Analytical Chemistry and Medicinal Biology, Vol. 6, pp. 437–443. Consejo Superior de Investigaciones Científicas, Spain.

Richards MP and Beattie JH (1995) *Comparison of different techniques for the analysis of metallothionein isoforms by capillary electrophoresis.* J Chromatogr B **669**:27–37.

Richards MP, Andrews GK, Winge DR and Beattie J (1996) *Separation of three mouse metallothionein isoforms by free-solution capillary electrophoresis.* J Chromatogr **675**:327–331.

Richards MP and Huang TL (1997) *Metalloprotein analysis by capillary isoelectric focusing.* J Chromatogr B Biomed Sci Appl **690**:43–54.

Robinson NJ and Jackson PJ (1986) *"Metallothionein-like" metal complexes in angiosperms: their structure and function.* Physiol Plant **67**:499–506.

Roels H, Lauwerys R, Buchet JP, Bernard A, Garvey JS and Linton HJ (1983) *Significance of urinary metallothionein in workers exposed to cadmium.* Int Arch Occup Environ Health **52**:159–166.

Sanchez-Reus MI, Iniesta MP and Ribas B (1988) *Metallothionein induction by methotrexate in liver and intestinal mucosa.* In: Schramel P and Brätter P, eds. Trace Elements in Analytical Chemistry and Medicinal Biology, Vol. 5, pp. 437–443. Walter de Gruyter Co., Berlin-New York

Sato M and Bremner I. (1993) *Oxygen free radicals and metallothionein.* Free Rad Biol Med **14**:325–337

Schäffer A and Kägi JHR (1991) *Metallothioneins.* In: Merian E, ed. Metals and Their Compounds in the Environment, pp. 523–530. VCH, Weinheim-New York-Basel-Cambridge.

Szpunar J. (2000) *Bio-inorganic speciation analysis by hyphenated techniques.* Analyst **125**:963–988.

Thornalley PJ and Vasák M (1985) *Possible role for metallothionein in protection against radiation-induced oxidative stress. Kinetics and mechanism of its reaction with superoxide and hydroxyl radicals.* Biochim Biophys Acta **827**:36–44.

Takatera K, Osaki N, Yamaguchi H and Watanabe T (1994) *HPLC/ICP mass spectrometric study of the selenium incorporation into cyanobacterial metallothionein induced under heavy metal stress.* Anal Sci **10**:567–572.

Uchida Y (1994) *Growth inhibitory factor, metallothionein-like protein, and neurodegenerative diseases.* Biol Signals **3**:211–215.

Webb M (1987) *Toxicological Significance of Metallothionein,* In: Kägi JHR and Kojima Y, eds. Metallothionein II, pp. 109–134. Experientia Suppl 52. Birkhäuser Verlag, Basel.

7

Influence of Metals on DNA

Zeno Garban

7.1
Overview on Metals and DNA Interaction

Deoxyribonucleic acid (DNA) is a polyhetero-nucleotic macromolecule present in the genome of both prokaryotic and eukaryotic cells. Its main role is the propagation of genetic information by replication, transcription, and translation. DNA may interact with various physical, chemical, or biological agents which induce changes in its chemical structure with implications on its biological activity; that is, the structure–activity relationship (SAR). Generally, the structure, stability, and reactivity of DNA is governed by the essential cations existing in small amounts in the cells, while their lack can defect or stop the replication.

Metals, as chemical agents, determine specific interactions based on ionization processes occurring at different levels of the DNA macromolecule, followed by the metal ions coordination (Eichhorn 1973, Sissoeff et al. 1976, Gao et al. 1993).

Investigations into the interaction of metal ions (M^{n+}) with DNA have importance in the elucidation of various biochemical and biomedical effects on humans and animals (Eichhorn 1973, Marzilli 1977, Haiduc and Silvestru 1989/1990, Littlefield et al. 1993, McFail-Isom et al. 1998).

The interaction of DNA with metal ions is followed by changes in the double-stranded and even the single-stranded structure of the macromolecule. These changes are caused by destabilization of the biomacro-molecule and the formation of various complexes, initially named "molecular associations", and nowadays "adducts" (Grunberger and Weinstein 1979, Garban et al. 1980, Froystein et al. 1993).

7.2
Steric Parameters of the DNA Macromolecule

Structurally, DNA is formed by the binding of a $C_{3'}$ atom between a phosphate group of one deoxyribonucleotide and a $C_{5'}$ atom of the adjacent deoxyribonucleotide (also noted $-C_{3'}-C_{5'}-C_{3'}-C_{5'}-$). This phospho-diesteric bond is characteristic for the single-stranded DNA, or the "primary structure". Two antiparallely disposed strands with bindings between the complementary nucleobases (NB); that is, adenine-thymine (A-T) and guanine-cytosine (G-C) form the double-stranded DNA, or the "secondary structure" (Figure 7.1). This is considered the classical DNA model of Watson and

Elements and their Compounds in the Environment. 2nd Edition.
Edited by E. Merian, M. Anke, M. Ihnat, M. Stoeppler
Copyright © 2004 WILEY-VCH Verlag GmbH & Co. KGaA, Weinheim
ISBN: 3-527-30459-2

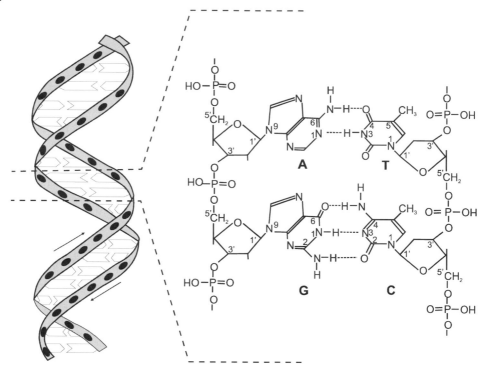

Fig. 7.1 Polyheteronucleotidic macromolecule of DNA (secondary structure). (a) Steric representation (general); (b) binding of complementary and antiparallel strands (details).

Crick (1953), but the Hoogsteen variant (1963) is also possible.

The more complex tertiary structure and quaternary structures of DNA characterize the compacted DNA states present in nucleosomes and microscopically detected in chromosomes.

Studies concerning the interaction of DNA with metal ions – which has as an outcome the formation of DNA-M^{n+} type "adducts" – were performed using various physico-chemical methods, including nuclear magnetic resonance (NMR), optical rotatory dispersion (ORD), and spectroscopy (CD, UV, IR and Raman) (Zimmer 1971, Yang and Samejima 1971, Balasubramanian and Kumar 1976, Marzilli 1977, Arnott 1978, Jack 1979, Theophanides 1976,

Garban et al. 1984, 1988, Froystein et al. 1993).

The dextrogyric rotation of the double-helical DNA may generate the A-DNA, B-DNA, C-DNA, and D-DNA types, while the senestrogyric rotation generates the Z-DNA type. All these types are characterized by linear and angular steric parameters. Details on steric parameters, depending on the residual nucleotides from the macromolecular building and nucleobases pairs (respectively nucleotides) of the DNA double helix are provided below.

7.2.1

Steric Parameters depending on Residual Nucleotides

Defining DNA types implies, among other, the knowledge of the conformational differences given by the torsion angles of the residual deoxyribonucleotide from DNA (Arnott 1978, Jack 1979). The description

of these angles is made relating them to the furanose ring of deoxyribose. Figure 7.2 presents the external and internal intranucleotidic torsion angles ("conformational angles") related to deoxyribose as well as their notation.

The figure shows only the bindings with the nucleobase and the phosphodiester groups. Currently, there are described

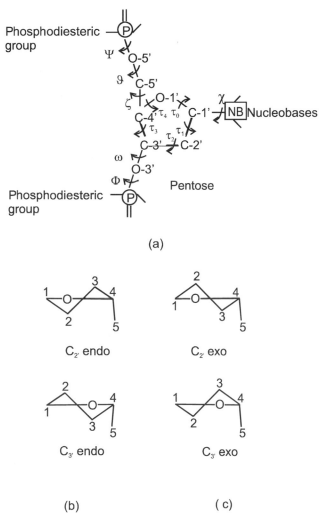

(a)

(b)   (c)

Fig. 7.2 Torison angles of deoxyribonucleotide and conformations of furanose ring in DNA. a) Torsion angles of nucleotide and deoxyribose; b) Conformation of endo furanose; c) Conformation of exo furanose.

Tab. 7.1: Conformational angles and deoxyribose conformations at various dextrogyric DNA types

DNA types	Conformational angles (°)							Conformation of deoxyribose
	Ψ	θ	ζ	τ	ω	ϕ	χ	
A-DNA	−85	−152	45	83	178	−47	86	C-3'-endo
B-DNA	−46	−147	36	157	135	−96	143	C-2'-endo
C-DNA	−39	−160	37	157	161	−106	143	C-2'-endo
D-DNA	−62	−152	69	157	141	−101	144	C-3'-endo

seven external (exocyclic) torsion angles of deoxyribose: ψ, θ, ζ, τ, ω, ϕ, χ characterizing the deoxyribonucleotide and varying in the different DNA types. The nucleobase also shows a free rotation, albeit more limited, around the unique binding between C_1 of the pentose and N_3 purine or N_1 pyrimidine, forming the χ angle. Table 7.1 presents data referring to the angles of dextrogyric DNA types (after Arnott 1978).

The internal torsion angles, which are characteristic for the furanose ring of deoxyribose (noted by $\tau_0 - \tau_4$) determine the appearance of endo- and exo-conformations of deoxyribose. The Watson–Crick model corresponds to B-DNA type, having the deoxyribose in the C-2'-endo conformational form. The secondary structure of B-DNA

was studied using X-ray diffraction to determine the external intranucleotidic torsion angles.

7.2.2
Steric Parameters depending on Nucleobase Pairs

Different conformations of DNA are also characterized by the variation of some steric parameters (linear and angular) established by physico-chemical methods and confirmed by quantum mechanics determinations reported to the nucleobase pairs from the double-helical DNA (Eichhorn 1973, Arnott 1978, Jack 1979, Garban 1996, Howerton et al. 2001). Figure 7.3 presents the modifications of some linear and angu-

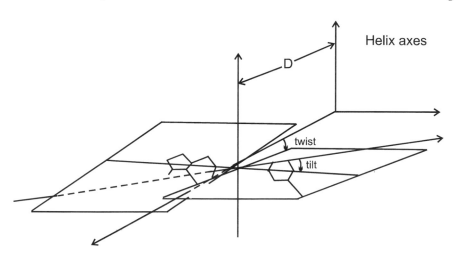

Fig. 7.3 Modifications of some angular parameters of the DNA macromolecule.

lar parameters (e.g., the twisting and the tilting angles).

Data referring to conformational modifications in the DNA macromolecule which lead to variations of linear and angular parameters are presented in Table 7.2.

Linear parameters of secondary structure may be evaluated by: (i) translation (h), which is the interval between two stacked nucleobases; and (ii) distance (D), which is the interval between the weight center of the nucleobases and the axis of the double helix.

Angular parameters refer to the various internal angles between the nucleobases pairs: (i) winding; which is the rotation angle of nucleobase pairs of the double helix; (ii) twisting, which is the inclination angle of vicinal nucleobases in vertical plane; and (iii) tilting, which is the inclination angle of vicinal nucleobases in horizontal plane.

7.3
Interaction of DNA with Divalent Metal Ions

It is considered that B-DNA type is the native form, its occurrence being favored by the increased polarity of the DNA due to the ionic status in DNA/M^{n+} systems.

If the polarity decreases, then under experimental conditions the C-DNA type appears. Then, due to an important number of superposed G-C-type purines (Pu-Pu), the A-DNA may form and, if A-A-type purines are present, then D-DNA may appear (Arnott 1978). In considering the classic type B-DNA, it was estimated that type C-DNA represents a secondary structure closer to that of type B-DNA.

Bindings between DNA and M^{2+} ions depend both on the structure and macromolecular configuration of DNA, as well as on the electronic configuration of the M^{2+}. The interaction of DNA with some M^{n+} destabilizes the polyheteronucleotidic macromolecule, causing changes in its steric parameters. The distinctive electron on the s-orbital of the alkaline-earth metals and that of the d-orbital of transition trace metals provide special peculiarities to the interaction (Eichhorn et al. 1973, Garban 1984, 1998), followed by biological consequences (Hoekstra et al. 1974, Haiduc and Silvestru 1989, Benham 1997).

This chapter deals with the interactions between DNA and divalent metal ions present in the organism ("biometals") as bioconstituents or biochemical effectors (activators/inhibitors) such as alkaline-earth metals (e.g., Ca, Mg) and the biologically active transition metals (e.g., Zn, Fe, Mn, Cu). In addition, the interactions of DNA with potentially toxic metals (Cd, Hg, Ni) or even with some biometals which, when present in excess, may be considered toxic (e.g., Cu, Co), are detailed.

Tab. 7.2: Steric (linear and angular) parameters of dextrogyric DNA types

DNA types	Linear parameters (nm)		Angular parameters (°)		
	Translation [h]	Distance [D]	Winding	Twisting	Tilting
A-DNA	0.256	42.50	32.7	−8.0	+20.0
B-DNA	0.337	6.30	36.0	+ 5.0	−2.0
C-DNA	0.331	21.30	38.6	−5.0	−6.0
D-DNA	0.303	−	45.0	−	−

7.3.1
Interaction with Alkaline-Earth Biometals

The binding between DNA and the alkaline-earth ions is electrostatic in nature, and occurs between the M^{2+} cations and the phosphodiester groups of DNA. The increase in positive charge due to protonization of the bases decreases the DNA affinity for the divalent alkaline-earth cations.

Native DNA binds relatively strong to the alkaline-earth M^{2+} because they stabilize the macromolecule. Experiments have confirmed that Mg^{2+} binds more strongly to native DNA than to denatured DNA (Park and Kohel 1993, Sigel and Sigel 1996).

The melting temperature (T_m), which is defined as the average temperature of transition, increases.

Differences between the interaction of DNA with alkaline-earth and transition divalent metal ions result from investigations made on the binding constants (K_α) using polarography, atomic absorption spectroscopy, and sedimentation analysis. For the alkaline-earth metals, K_α is decreased, and this is confirmed by the lower coordination tendency compared to that of the nucleobases.

The electrostatic nature of the binding between the DNA polyanion and the various cations is explained by the existence of a diffuse electrostatic change-effect, which induces the formation of the DNA-M^{2+} adducts.

Research investigations using NMR and CD and carried out by both others groups (Balasubramanian and Kumar 1976) and by ourselves (Garban et al. 1980, Garban 1994), attest to the modifications induced by DNA interaction with M^{2+} alkaline-earth ions.

7.3.2
Interaction with Transition Biometals

In general, transition metals form covalent bonds with the nitrogen atoms present in the nucleobases.

Investigations into the variation of T_m depending on the molar ratio M^{2+}/DNA-P, offer an overview on the destabilizing effects of divalent transition metals M^{2+}. Such observations were made in case of DNA interaction with Cu^{2+}, Zn^{2+}, and Mn. It was established that destabilization of the double helix by Cu^{2+} depends on the molar ratio M^{2+}/DNA-P, the G-C content of the studied DNA, and the ionic force of the environment.

Measurements made using spin electronic resonance concluded that Cu^{2+} bonds are realized at the outside centers of the double helix (phosphodiester groups) and also at the inside centers (G-C bases groups). Modifications induced by Cu^{2+} interaction with DNA were studied through NMR.

The interaction between Zn^{2+} and DNA leads to DNA-Zn^{2+} complex formation which, by repeated heating-cooling proved the reversibility of the denaturation process (which does not take place in the presence of Mg^{2+}). This is explained by the binding possibility of Zn^{2+} at the G-C base pairs. In the case of DNA which is rich in G-C groups, Zn^{2+} binds preferentially, and this was confirmed by the moving of the maximum of absorption in UV light. Measurements of T_m show that at low concentrations, Zn^{2+} is mostly bound at the phosphodiester centers and at higher concentrations also appears a change effect with the nucleobases.

Experiments with Mn^{2+} also showed in this case a possible binding to phosphodiester groups and to the G-C bases.

In general, it can be affirmed that the peculiarities of DNA-M^{2+} interaction are determined by the nature of the cations, the concentration of the components, ionic strength, pH, and temperature. All of these factors are involved in adduct formation and determine the type of bonds and the chemical structure, and also influence the biological activity.

7.3.3
Interaction with Toxic Transition Metals

Biological effects of some transition metal ions on DNA were associated with their mutagenic and carcinogen actions (Sissoeff et al. 1976, Kazantsis et al. 1979). Various studies in eredopathology and toxicology have shown that some M^{2+} induce malformations and teratogenic effects.

Experimental investigations were mainly convened with metals of group VIII, namely Fe^{2+}, Co^{2+}, and Ni^{2+}. These metals form covalent bonds with the N atoms of DNA, producing pronounced destabilization of the double helix with harmful biological consequences.

There are cases when heavy metals are used in chemotherapy, for example cytostatic chemotherapy with *cis*-platinum (Rosenberg 1969 [cited by Lippert 1999], Garban et al. 1989, Haiduc and Silvestru 1989, Johnsson et al. 1995). Numerous research studies into the interaction of DNA with metal ions also dealt with *cis*-platinum (Lippert 1999, Garban 2000). *Cis*-platinum interacts with nucleosides preferentially through the N$_7$ site of guanine, but may also form bidentate chelates as a result of interaction both with N$_7$ and O$_6$ sites. Modifications induced to *cis*-platinum by the interaction with guanine have been studied using quantum chemistry (Lippert 1999, Chojnacki et al. 2001).

Some metal compounds have been studied in vitro for their antitumoral effects and antiarthritic effects (Gielen et al. 1994, Sadler and Sue 1994), following the interac-

tions of pharmacological interest and indirectly the interactions with proteic macromolecules.

Many carcinogens, which implicitly are metals with toxicogenic potential, bind directly to the DNA chain, forming an adduct and lowering the ionization potential of the DNA.

7.4
Peculiarities of the SAR of DNA-M^{2+} Adducts

A general view concerning the chemical structure–biological activity relationship in the case of DNA-M^{2+} adduct types reveals that studies in this domain, which were conducted between 1970 and 1980, dealt with the physico-chemical aspects regarding the consequences of interaction, as well as with the mechanism of adducts biogenesis.

During the following decade (i.e., 1980–1990), investigations were focused on evaluating the biological activity of adducts using in-vivo experiments in animals, plants, and microorganisms, and well as in-vitro studies in cell culture.

After 1990, the studies were extended to the investigate conformation of adducts, their biomedical and pharmacological aspects, their genotoxicity, and their involvement in biochemical homeostasis.

Experimental investigations showed that the interaction of M^{2+} cations with the DNA polymacroanion affects the phosphodiester backbone of the macromolecule, or the intra-strand nitrogenous atoms from nucleobases (Eichhorn et al. 1973, Sissoeff et al. 1976, Duguid et al. 1995, Garban 1996). A compilation of literature data in Table 7.3 shows the preferential binding sites of various metal ions.

Binding of double-helix DNA at the level of phosphodiester groups is determined by the electrostatic potential, while the

Tab. 7.3: Binding sites of DNA for metal ions

DNA binding sites	Metal ions (M^{n+})
Phosphate	Li^+, Na^+, K^+, Cs^+
	Mg^{2+}, Ca^{2+}, Sr^{2+}, Ba^{2+}
	Cr^{3+}, Fe^{3+}
Phosphate and nucleobases	Co^{2+}, Ni^{2+}, Mn^{2+}, Zn^{2+}, Cd^{2+}, Pb^{2+}, Cu^{2+}, Fe^{2+}, Pt^{2+}, Fe^{3+}
Nucleobases	Ag^+, Hg^{2+}, Sn^{2+}
	Al^{3+}

number of the bidentated cations depends on the ionic strength. The electron donor groups of DNA determine the binding to the level of nitrogenous nucleobases. Guanine is often affected as it has the position N_7 donor electron, characterized as a center with the highest density of electrons reported to the other nucleobases (this fact also accords with the quantum mechanics data).

The existence was also observed of a characteristic affinity of M^{2+} reported to the phosphodiesteric centers and the intra-strand DNA nucleobases. The binding tendency at the intra-strand nucleobases increases in the order:

$$Mg^{2+}, \ Ca^{2+} < Co^{2+}, \ Ni^{2+} < Mn^{2+} < Zn^{2+}$$
$$< Cd^{2+} < Cu^{2+} < Hg^{2+},$$

and the binding tendency at the phosphodiesteric centers increases in the opposite sense.

The complexes formed by DNA-M^{2+} interaction present various binding types:

I. To the phosphodiesteric groups
II. Between a phosphodiesteric group and nucleobases.
III. Between two complementary intra-strand nucleobases
IV. Between two vicinal nucleobases
V. At different positions of the same purine nucleobases.

The binding depending on the absence or the presence of a water molecule (type I) – which is characteristic for the alkaline-earth metals – can present three different structures (Figure 7.4). There is either a direct binding (Figure 7.4(a)) or one which is intermediate with the water molecule (Figure 7.4 (b) and (c)). These bindings are

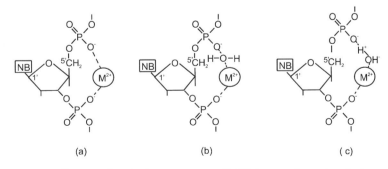

(a) (b) (c)

Fig. 7.4 Binding of M^{2+} to the phosphodiesteric groups of DNA (see text for explanation).

achieved in the DNA samples with either Mg^{2+} or Ca^{2+}.

The location of cations between phosphodiesteric groups (type II) is carried out by chelation of the phosphorous group with N$_7$ of the purine nucleobase from GMP (Figure 7.5). Such a structure characterizes the DNA complexes with Mn^{2+} and Zn^{2+} – cations which present a strong tendency of binding to the phosphodiesteric groups and a low affinity for coordination with purine nucleobases.

Intercalation between complementary intrastrand nucleobases (type III) shows an intrahelical disposition of M^{2+}: (i) in the case of A-T chelation, bindings appear at N$_1$ adenine and N$_3$ thymine; (ii) in the case of the G-C chelation, the bindings appear at N$_1$ guanine and N$_3$ cytosine (Figure 7.6). The first approach is characteristic for Hg^{2+}, and the second for Cu^{2+} and Cd^{2+}. The M^{2+} binding is also possible to N$_7$ guanine, N$_3$ cytosine and to O from C$_6$, respectively C$_2$.

The type IV chelation can be made at the level of two nucleobases situated on the same strand (i.e., on adjacent nucleobases) such as the GpG′ sequence. The binding appears at N$_7$ and O from C$_6$ of the nucleobases (Figure 7.7). Such "sandwich"-type bindings appear in the case of Cu^{2+} and Hg^{2+}.

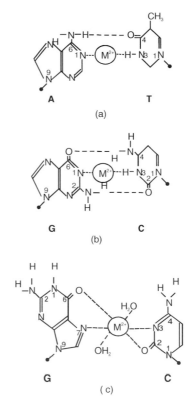

Fig. 7.6 Internal chelatization binding of M^{2+} to adjacent nucleobase pairs of DNA strands: (a) M^{2+} – A-T pair; (b) M^{2+}- G-C pair; (c) M^{2+} – G-C pair (other position). (o— indicates binding to DNA strand).

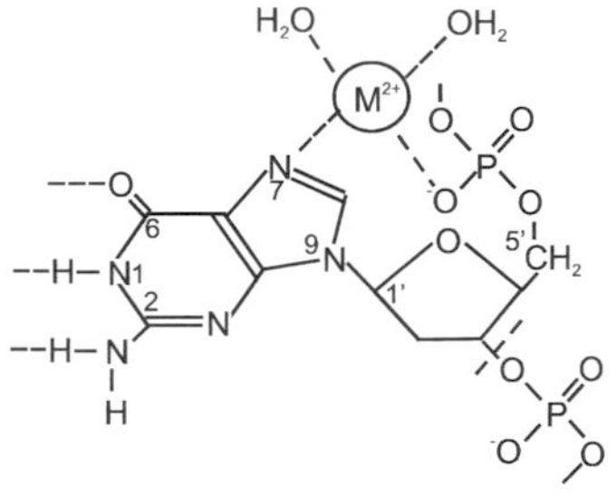

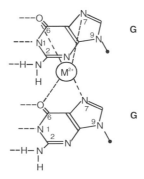

Fig. 7.5 Binding of M^{2+} to guanine and to the phosphodiesteric group of DNA.

Fig. 7.7 Binding of M^{2+} to adjacent guanine nucleobases of DNA.

Binding of the M^{2+} to a purine nucleobase (type V) occurs at N_7 and O from C_6 of adenine or N_7 and O from C_6 of guanine (Figure 7.8). In these cases, water molecules can bind at the chelate. This type of chelation is usually met at transition metals, and affects the conformation of helix causing local denaturations in the macromolecular structure.

Tsukube et al. (1994) investigated the relationship between cytosolic and nuclear calcium content in the ischemic myocardium. Their results indicated that, during ischemia, both cytosolic and nuclear calcium are augmented, with that of nuclear calcium being associated with increased nuclear DNA fragmentation. Treatment with Mg and a combination of Mg-K reduced the nuclear calcium accumulation and decreased the nuclear DNA fragmentation.

Another interesting fact relates to the form of the administered cation. Urlacher and Griep (1995) revealed that magnesium acetate produces conformational changes in the structure of polymerases (e.g., primase from *E. coli*), but no conformational changes were observed in the case of $MgCl_2$, $CaCl_2$ a.o. administration, or they were very reduced by $MgSO_4$ + $MgCl_2$ administration.

In one study, Littlefield et al. (1993) examined the influence of Mg in combination with Ni and Cd in respect of damage of the DNA molecule. These authors proposed to evaluate the influence of Mg on the diminution of the toxic effects of Ni and Cd with respect to sustaining DNA damage. Their conclusions were that: (i) Ni is not directly toxic to DNA; (ii) Cd produces damage directly on the DNA molecule; and (iii) Mg interacts with some of toxic heavy metals and alters the tumorigenic process.

Hartwig and Beyersmann (1989) showed that $NiCl_2$ has mutagenic activity on V79 cells from Chinese hamster, and also had a pronounced co-mutagenic effect towards UV light. Furthermore, $NiCl_2$ was found to enhance the cytotoxicity of *cis*-DDP about 12-fold. All of these observations suggested that the inhibition of DNA repair by Ni^{2+} occurred via the replacement of other divalent metal ions that were essential in repair and regulation processes.

Beyersmann (1994) revealed that the carcinogenicity and genotoxicity of Cd^{2+}, Cr^{2+}, Co^{2+} and Ni^{2+} depended on their chemical ligands, which in turn modulate their bioavailability and reactivity with biochemical targets. The carcinogenic metals Cd, Cr and Ni inhibit the repair of DNA damaged by direct genotoxic agents (UV radiation, alkylating substances) carcinogenic (Williams 1971).

A relatively recent study by Clark and Eichhorn (1995) proposed the use of Cu^{2+} for DNA accessibility in chromatin. It was

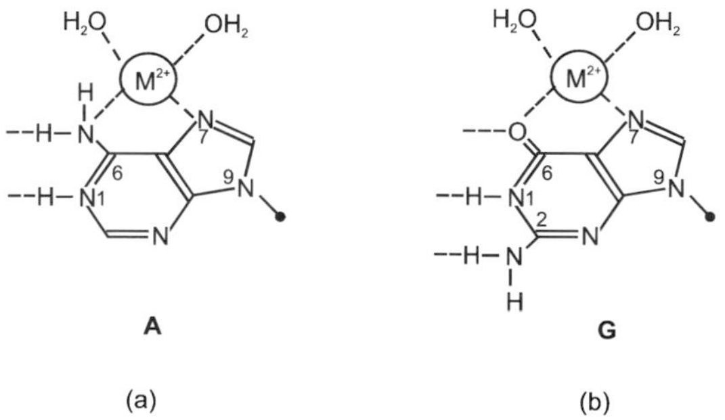

Fig. 7.8 Binding of M^{2+} to purine nucleobases of DNA. (a) M^{2+}-adenine; (b) M^{2+}-guanine.

established that Cu^{2+} binds to the DNA in such a way as to destabilize the double helix and help break the hydrogen bonds between the bases. The process is facilitated by changes in T_m which affect the structure of H-1 histone and the reaction with phosphate binding divalent metal ions (e.g., Mg^{2+}, Mn^{2+}, or Co^{2+}). The removal of H-1, or a decreased affinity of H-1 for DNA, increased the accessibility of the Cu^{2+} to DNA.

Gao et al. (1993) carried out X-ray diffraction analysis of the interactions of two transition ions (Co^{2+} and Cu^{2+}) and an alkaline-earth metal ion (Ba^{2+}) with DNA of different conformations. The findings suggested that: (i) Co^{2+} binds to either B-DNA or A-DNA and may induce significant conformational changes; (ii) Cu^{2+} binds to Z-DNA; and (iii) Ba^{2+} also binds with Z-DNA crystals.

Schiff bases and their complexes with Cu^{2+}, Zn^{2+}, Ni^{2+} and Sn^{2+} as potential antitumor agents, were used in the fluorimetric method for preliminary screening of antitumor agents (Lee et al. 1993). The method was based on the consistency of the in-vivo and in-vitro interactions of drugs with DNA. Studies of Schiff bases-metal complex interaction with DNA using a fluorescence probe [ethidium bromide (EthBr)-DNA system] identified the parallelism between binding constants and antineoplastic

ratios. More recent pharmacological studies have followed aspects of the interaction between chemotherapeutics and DNA in the presence of diverse metal ions such as Zn, Fe, Cu, and Co (Li, 2001).

A number of studies have shown the destabilizing effect on nucleic acids due to the interaction of divalent cations with π-systems of nucleobases (McFail-Isom et al. 1998).

Recent studies on DNA-M^{n+} adducts have also revealed possible semiconductor properties of these structures. In this context, of particular mention are the experiments of Lee et al. (1993), who demonstrated M-DNA formation by coordinative binding of the Zn^{2+} ion to the N_3 imino proton of thymine in the nucelobase pair A-T, and to the N_1 proton of guanine in the C-G nucleobase pair (Figure 7.9). The so-called "engineered DNA" – which is referred to as M-DNA – is able to conduct electricity.

Later on, investigations by Rakitin et al. (2001) on the binding mechanisms of M-DNA and its physical properties, showed that the replacement of imino protons with Zn^{2+} is due to strong hybridization of p-nitrogen and d-zinc electron states. It is thought possible that to use M-DNA as a biosensor might help to reduce adverse drug reaction, improve diagnosis of disease, and also predict the outcome of disease.

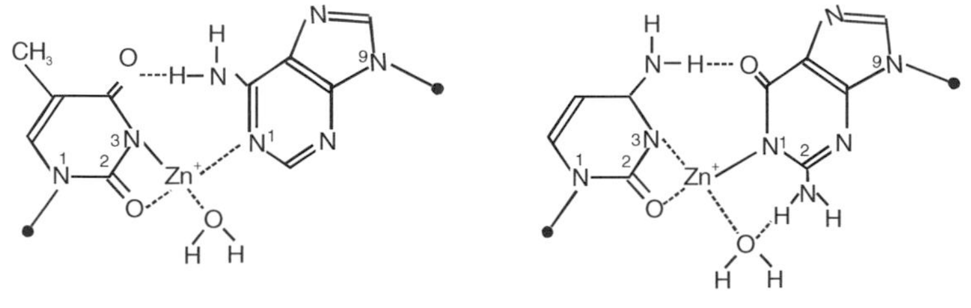

T - A   C - G

Fig. 7.9 Specific bindings for M-DNA formation (coordinative bond of Zn^{2+} ion to the imino protons in N_3 position of thymine and N_1 position of guanine).

Observations on monovalent cations (M^+) attest to their localization in the proximity of phosphate groups, this ranging from 20 to 35% within the major groove, and being about 10% at the sites of the minor groove (Howerton et al. 2001).

In the case of multivalent cations it was observed that the DNA macromolecule condenses in compact structures – a fact which was also proven in an alcohol-water environment.

The study of metal-DNA type adducts was also approached for diverse metal ions from a thermodynamic viewpoint (Duguid et al. 1995). Thus, it was established that no dichotomy was observed between alkaline-earth and transitional metal complexes.

The interaction of DNA with M^{n+} ions produces modifications in SAR, with undesirable biological effects. Consecutive effects of the interactions undergone by some metals with living matter (Woollam 1972, Kazantsis et al 1979) attest to the involvement of DNA, and lead to appearance of chromosomal modifications as well as mutagenic, oncogenic, and teratogenic effects

References

ARNOTT S (1978) *Secondary structure of polynucleotides*. In: First Cleveland Symposium on Macromolecules, pp. 87–104. Elsevier Scientific Publishing Company, Amsterdam.

BALASUBRAMANIAN D and KUMAR C (1976) *Recent studies of the circular dichroism and optical rotatory dispersion of biopolymers*. Appl Spectr Rev 11: 223–286.

BENHAM CJ (1997) *Superhelical duplex destabilization and DNA regulation*. DIMACS/PMMB/MBBC Workshop on DNA topology II, Abstracts Volume, April 3 –4, Rutgers University.

BEYERSMANN D (1994) *Interactions in metal carcinogenicity*. Toxicol Lett 71: 333–338.

CHOJNACKI H, KOLODZIEJCZYK W and PRUCHNIK F (2001) *Quantum chemical studies on molecular and electronic structure of platinum and tin adducts with guanine*. Int J Mol 2: 148–155.

CLARK P and EICHHORN GL (1995) *A simple probe for DNA accessibility in chromatin*. J Inorg Biochem 59: 765–772.

DUGUID JG, BLOOMFIELD VA, BENEVIDES JM and THOMAS G-J, JR (1995) *Raman spectroscopy of DNA-metal complexes. II. The thermal denaturation of DNA in the presence of Sr^{2+}, Ba^{2+}, Mg^{2+}, Ca^{2+}, Mn^{2+}, Co^{2+}, Ni^{2+} and Cd^{2+}*. Biophys J 69: 2623–2641.

EICHHORN GL, ed. (1973) *Inorganic Biochemistry*, Vol. II. Elsevier, Amsterdam-London.

FROYSTEIN NA, DAVIS JT, REID BR and SLETTEN E (1993) *Sequence-selective metal ion binding to DNA oligonucleotides*. Acta Chem Scand 47: 649–657.

GAO YG, SRIRAM M and WANG AH (1993) *Crystallographic studies of metal ion-DNA interactions: different binding modes of cobalt (II), copper (II) and barium (II) to N_7 of guanines in Z-DNA and a drug-DNA complex*. Nucleic Acids Res 21: 4093–4101.

GARBAN Z, MIKLOS J, DARANYI G and SUCIU O (1980) *Investigation of the interaction of deoxyribonucleic acid with divalent metals by circular dichroism II. Interaction with transitional metals*. In: Anuarul "Lucrări ştiinţifice", Vol. XVII, pp. 109–117. Inst. Agr. Timişoara. (In Romanian)

GARBAN Z, EREMIA I and DARYNYI G (1984) *Chronobiochemical aspects of the hepatic DNA biosynthesis in experimental animals under the action of some metals*. J Embryol Exp Morph Suppl Cambridge 1: 6.

GARBAN Z, VĂCĂRESCU G, DARANYI G, POPETI D, EREMIA I and MAURER (1988) *Chemical structure-biological activity relationship in the interaction of DNA with cis-platinum*. 14th International Congress of Biochemistry, Abstracts Vol. V., p. 80, Prague, Czechoslovakia.

GARBAN Z, DANCĂU G, DARANYI G, ERDELEAN R, VĂCĂRESCU G, PRECOB V, EREMIA I and UDRIŞTE C (1989) *Implication of chronobiochemistry-metabolism relationship in the induction of homoeostasis changes. I. The action of cis-platinum on hepatic DNA biosynthesis and on some serum metabolites in rats*. Rev Roum Biochim 26: 107–117.

GARBAN Z (1994) *Zinc and copper effects on deoxyribonucleic acid studied in vitro and in vivo*. In: Pais I, ed. New Perspectives in the research of Hardly Known Trace Elements, pp. 357–362. University Press of U.H.F.S. Budapest.

GARBAN Z (1996) *Interaction of deoxyribonucleic acid with divalent metallic ions and structrual particu-*

larities of the resulted complexes. In: Garban Z, Drăgan P, eds. Metal Elements in Environment, Medicine and Biology, Proceedings of the 2nd International Symposium Vol II, October 27–29, 1996, pp. 99–108. Timişoara, Romania, Publishing House "Eurobit" Timişoara.

GARBAN Z (1998) *Molecular Biology: Fundamental and Applicative Problems.* 3rd edition, Ed. Eurobit Timişoara (In Romanian).

GARBAN Z, CARŢIŞ I, AVACOVICI A and MOLDOVAN I (2000) *Comparative aspects between the interaction of deoxyribonucleic acid with some cytostatic drugs: Particularisation for the interaction with cis-platinum and cyclophosphamide. 1. Investigations in vivo on experimental animals.* In: Anke M et al., eds. Mengen- und Spurenelemente, 20. Arbeitstagung 2000, pp. 1118–1125. Verlag Harald Schubert, Leipzig.

GIELEN M, BOUALAM M, MAHIEU B and TIENKINK ERT (1994) *Crystal structure and in vitro antitumor activity of dibutylbis(5-chloro-2-hydroxybenzoato)-tin(IV).* Appl Organomet Chem **8**:19–23.

GRUNBERGER D and WEINSTEIN IB (1979) *Conformational changes in nucleic acids modified by chemical carcinogens.* In: Grover PL, ed. Chemical Carcinogens and DNA, Vol. 2, pp. 59–93. CRC Press Inc, Boca Raton, Florida.

HAIDUC I and SILVESTRU C (1989/1990) *Organometallics in Cancer Chemotherapy.* Vol. 1 (1989), Vol 2 (1990). CRC Press, Boca Raton, Florida.

HARTWIG A and BEYERSMANN D (1989) *Enhancement of UV-induced mutagenesis and sister-chromatid exchanges by nickel ions in V79 cells: evidence for inhibition of DNA repair.* Mutat Res **217**:65–73.

HOEKSTRA WG, SUTTIE JW, GANTHER HE and MERTZ W (1974) *Proceedings, 2nd International Symposium on Trace Element Metabolism in Animals.* University Park Press, Baltimore-London.

HOOGSTEEN K (1963) *The crystal and molecular structure of hydrogen-bonded complex between 1-methylthimine and 9-methyladenine.* Acta Crystallogr **16**:907–916.

HOWERTON SB, SINES CC, VAN DERVEER D and WILLIAMS LD (2001) *Locating monovalent cations in the grooves of B-DNA* Biochemistry **34**:10023–31.

JACK A (1979) *Secondary and Tertiary Structure of Nucleic acids.* In: Offord RE, ed. International Review of Biochemistry, Chemistry and Macromolecules II A, Vol. 2.24, pp. 211–256. University Park Press, Baltimore.

JOHNSSON A, OLSSON C, NYGREN O, NILSSON M, SEIVING B and CAVALLIN-STAHL E (1995) *Phar-*

macokinetics and tissue distribution of cisplatin in nude mice: platinum levels and cisplatin-DNA adducts. Cancer Chemother Pharmacol **37**:23–31.

KAZANTSIS G, LORNA G and LILL Y (1979) *Mutagenic and carcinogenic effects of metals.* In: Friberg L et al., eds. Handbook on the Toxicology of Metals. Chapter 14, pp. 1–36. Elsevier North Holland Biochemical Press.

LEE JS, LATIMER LJP and REID RS (1993) *A cooperative conformational change in duplex DNA induced by zinc and other divalent metal ions.* Biochem Cell Biol **71**:162–168.

LI W, ZHAO C, XIA C, ANTHOLINE WE and PETERING DH (2001) *Comparative binding properties of metallobleomycins with DNA 10-mers.* Biochemistry **40**:7559–7568.

LIPPERT B (1999) *Cisplatin: Chemistry and Biochemistry of a leading anticancer drug.* Wiley-VCH, Weinheim-New York-Chichester-Brisbane-Singapore-Toronto.

LITTLEFIELD NA, HASS BS, JAMES SJ and POIRIER LA (1993) *Protective effect of magnesium on DNA strand breaks induced by nickel or cadmium.* Cell Biol Toxicol **10**:127–135.

MARZILLI GL (1977) *Metal-ion interactions with nucleic acids and nucleic acids derivatives.* In: Lippard SJ, ed. Progress in Inorganic Chemistry. Vol. 23, pp. 255–378. John Wiley & Sons, Inc.

McFAIL-ISOM L, SHUI X and WILLIAMS LD (1998) *Divalent cations stabilize unstacked conformations of DNA and RNA by interacting with base pi systems.* Biochemistry **37**:17105–17111.

PARK YH and KOHEL RJ (1993) *Effect of concentration of $MgCl_2$ on random-amplified DNA polymorphism.* Biotechniques **16**:652–656.

RAKITIN A, AICH P, PAPADOPOULOS C, KOBZAR YU, VEDENEEV AS, LEE JS and XU JM (2001) *Metallic conduction through engineered DNA: DNA nanoelectronic building blocks.* Phys Rev Lett **86**:3670–3673.

SADLER P-J and SUE RE (1994) *The chemistry of gold drugs.* Met-Based Drugs **1**:107–144.

SIGEL A, SIGEL H, eds. (1996) *Metal Ions in Biological Systems.* Vol. 33, Probing of Nucleic Acids by Metal Ion Complexes of Small Molecules. Marcel Dekker, New York.

SISSOEFF I, GRISVARD J and GUILLE E (1976) *Studies on metal ions-DNA interactions: specific behaviour or reiterative DNA sequences.* In: Progress in Biophysical and Molecular Biology, Vol. 31, pp. 165–199. Pergamon Press, London.

THEOPHANIDES TM (1976) *Interaction of metal ions with nucleic acids.* In: Theophanides TM, ed.

Infrared and Raman Spectroscopy of Biological Molecules, pp. 187–204. D Reidel Publishing Company, Dordrecht, Holland.

TSUKUBE T, McCULLY JD, FAULK EA, FEDERMAN M, LOCICERO J III, KRUKENKAMP IB and LEVITSKY S (1994) *Magnesium cardioplegia reduces cytosolic and nuclear calium and DNA fragmentation in the senescent myocardium.* Ann Thorac Surg **58**:1005–1011.

URLACHER TM and GRIEP MA (1995) *Magnesium acetate induces a conformational change in Escherichia coli primase.* Biochemistry **34**:16708–16714.

WATSON JD and CRICK FHC (1953) *Molecular structure of nucleic acids. A structure for deoxyribonucleic acid.* Nature **171**:737–738.

WILLIAMS RD (1971) *The metals of life (The Solution Chemistry of Metal Ions in Biological Systems).* Van Nostrand Reinhold Company, London.

WOOLLAM DHM, ed. (1972) *Advances in Teratology,* Vol. 5. Logos Press Ltd., London.

YANG JEN TSI and SAMEJIMA TATSUYA (1971) *Optical Rotatory Dispersion and Circular Dichroism of Nucleic Acids.* Ed. University Press, California, San Francisco.

ZIMMER CH (1971) *Bindung von divalenten Metalli-onen an Nucleinsäuren und Wirkungen auf die Konformation der Deoxyribonucleinsäure.* Z Chem **11**:441–458.

8

Acute and Chronic Toxicity of Metals and Metal Compounds for Man

Marika Geldmacher-v. Mallinckrodt and Karl-Heinz Schaller

8.1
Introduction

Metals play a dual role in biological systems. They serve as essential co-factors for a wide range of biochemical reactions, yet these same metals may be extremely toxic to cells. To cope with the stress of increases in environmental metal concentrations, eucaryotic cells have developed sophisticated toxic metal-sensing proteins which respond to elevations in metal concentration. The signal is transmitted to stimulate the cellular transcriptional machinery to active expression of metal detoxification and homeostasis genes. Zhu et al. (1996) summarize the current understanding of the biochemical and genetic mechanisms which underlie cellular responses to toxic metals via metalloregulatory transcription factors. They discuss the molecular mechanisms by which mammalian cells respond to toxic metals by activating the transcription of metallothionein genes. It becomes clear that, although metal-responsive metallothionein gene transcription has been under intensive study for many years, the precise mechanisms by which mammalian cells sense and respond to the large array of chemically distinct metal ions to activate gene expression still remain to be elucidated.

Membrane transport of toxic heavy metals not only controls their access to intracellular target sites but also helps to determine their uptake, distribution, and excretion from the body. The critical role of membranes in the toxicology of metals has attracted the attention of many investigators, and extensive information has been collected on the mechanisms of metal transfer across membranes. Characteristics of metal transport in different cells (see also Part II, Chapter 4), or even on opposite sides of the same cell, or under different physiological conditions, are not identical, and no unitary hypothesis has been formulated until now to explain this process in all cells (Foulkes 2000).

8.2
Mechanisms Responsible for Toxic Effects of Metals

Williams (1981a) provided a general overview on the role of metals in biological systems. These systems were able to alter the relationships of the different metals in their bodies so that they differed from the relationships present in the Earth's crust. The reason for that was the pressure of natural selection, which allows only those species with highly effective and optimized bio-

Elements and their Compounds in the Environment. 2nd Edition.
Edited by E. Merian, M. Anke, M. Ihnat, M. Stoeppler
Copyright © 2004 WILEY-VCH Verlag GmbH & Co. KGaA, Weinheim
ISBN: 3-527-30459-2

chemistry to survive. Natural selection has led to an almost optimal utilization of the various metals. Specialization of metal function was possible in as far as it occurred at the same time as the evolution of proteins. The movement of elements was determined by specific electromechanical and electrochemical events. Williams (1982, 1983, 1985) has extensively discussed the chemical selectivity of protein side-chains and small inorganic molecules for metal ions.

Highly mobile and weakly bound ions such as Na^+ and K^+ serve as charge carriers, while Ca^{2+} and Mg^{2+}, of moderate affinity, act as structure formers. In contrast, static metal ions such as Fe^{2+} and Cu^{2+} – both of which bind with high affinity to biological ligands – often act as redox catalysts. The functions of metal ions in biological systems are governed by their chemical properties such as ion sizes, electron affinities and geometric demands, which also allow biochemical differentiation between similar metal ions.

The regulation mechanisms of biological systems for maintenance of homeostasis are able to adjust to small changes in the concentration of metal compounds for short periods of time. Selection and adaptation to concentration changes occurring over longer periods of time also takes place. However, sudden, significant concentration changes, and their results, caused for example by an acute intake of an overdose of metal ions, cannot be counteracted and may lead to fatal disturbances in the organism. Symptoms of poisoning appear as a result of changes in the molecular structure of proteins, breaking of hydrogen bonds, inhibition of enzymes, changes in potential, and so on. Not only those metal compounds which are "foreign" to the system can have toxic effects, but also those which are essential, if present in large quantities.

If one considers the toxicity of a metal ion with regard to its location in the Periodic System, a pattern can be seen (Luckey et al. 1977). Toxicity decreases with an increase in the stability of the electron configuration. Metal ions of the subgroups IA and IIA are highly electropositive, and these metal ions appear in the biological environment primarily as free cations. The toxicity of the subgroups IA and IIA increases with increasing atomic number:

$$IA : \ Na < K < Rb < Cs$$
$$IIA : \ Mg < Ca < Sr < Ba$$

Also in the subgroups IB, IIB, IIIA, the acute toxicity of the metal ion increases with the electropositivity:

$$IB : \ Cu < Ag < Au$$
$$IIB : \ Zn < Cd < Hg$$
$$IIIA : \ Al < Ga < In < Tl$$

The increase in toxicity can be explained by the increasing affinity of these elements for amino, imino, and sulfhydryl groups, which form the active centers of a number of enzymes. The metals of the sixth Period and their compounds are potentially the most toxic elements of the Periodic System. The generally poor water solubility of their salts, however, often masks their inherent high degree of toxicity. This toxicity becomes apparent in those lead, mercury, and thallium salts that are relatively soluble. The metallic ions of the fourth Period form mostly covalent bonds and complexes with biological ligands, and some form hydroxy acids in which the metal is part of the anion.

In addition to the electrochemical character and the solubility of a metal and its compounds, which influence its bioavailability, the various oxidation states of an element are also important. For example, manganese (VII) compounds such as permanganate are more toxic than manganese (II) compounds, and arsenic (III) oxide is more toxic than arsenic (V) oxide.

Despite these general considerations, until now the unusual diversity of symptoms observed after the uptake of an overdose in the intact organism could not be explained adequately. The effects of an overdose of heavy metals manifest themselves in very different tissues, partially with a rather high specificity for certain metals and their compounds. These organ-specific effects could not yet be explained.

The symptoms of acute and chronic poisoning by the same metal can be completely different. For example, acute mercury poisoning through oral ingestion of large amounts of a soluble mercury salt leads to intense nausea, vomiting and diarrhea, and possibly death from shock within the first 24–36 hours. Chronic poisoning caused by the same compound, however, damages primarily the nervous system, followed by the kidneys (see Moeschlin 1972 and Part III, Chapter 17).

Numerous metals are responsible for immunologically mediated disorders in humans.

The induction of damage by influences in embryonic or fetal development, causing malformations or embryonic death, is a problem of the toxicity rather than mutagenicity of the inducing agent (Gebhart et al. 1991). Mutagenicity as well as carcinogenicity and teratogenicity are discussed in Part II, Chapter 9.

8.3
Role of Speciation and Way of Uptake

The physical or chemical form – the speciation of the metal (see Part II, Chapters 4 and 8) – effects its toxicity. There is also a linguistic problem when speaking of metal toxicity, as these elements almost never exist in the metallic form in living organisms. In fact, we tend to discuss the effects of metal ions, metal complexes, or metal compounds. For instance, metallic mercury, inorganic mercury ions, and organic mercury compounds show completely different spectra of effects. Although orally ingested metallic mercury is largely nontoxic, a one-time inhalation of a high concentration of mercury vapor leads to lethargy, followed by restlessness, nausea, diarrhea, a metallic taste in the mouth, coughing, tachypnea and possibly respiratory arrest. Histological signs are eroded bronchial tubes, bronchiolitis accompanied by interstitial pneumonia, gastroenteritis, colitis, and kidney damage. These symptoms can be accompanied by disturbances of the central nervous system (CNS) such as tremor, and increased excitability.

8.4
Acute Toxicity

8.4.1
General

Acute exposure is defined as a single or multiple exposure occurring within a short time (24 h). Acute metal poisoning is rarely observed in clinical practice, but in the case of an unknown illness a differential diagnosis must take such a situation into account. Metal poisoning produces no specific symptoms, and only rarely tests are carried out to detect toxic metals. The result is that metal poisoning is repeatedly not recognized, and the actual number of cases may be far higher than those reported.

The majority of acute intoxications by metals is usually the result of suicide attempts. Occasionally, errors – for example in the laboratory or clinic, contaminated food, and, in rare cases, medical treatment – can lead to intoxication. Homicide may also be the cause.

The acute toxic effects of metals cannot be considered as isolated phenomena, but rather as a part of the complete spectrum of activity and/or dose–activity relationship of a metal in a biological system (Williams 1981b).

Despite fluid boundaries between subtoxic and toxic amounts, acute metal poisoning due to the intake of a single high dose shows in practice most striking symptoms which develop suddenly. If the poison cannot be removed or inactivated quickly, then irreversible organ and systemic changes (which can be fatal) usually occur. In contrast, chronic poisoning develops gradually as a result of repeated intake of relatively small, but still toxic, doses. This can also lead to irreversible damage.

8.4.2
Uptake and Distribution

Acutely toxic amounts of metals and their compounds are usually taken in through the mouth or lungs. In addition to the dose, the method of uptake determines the intensity as well as the duration of toxic effects, and can lead to very different symptoms (see the above-mentioned example of ingested metallic mercury and inhaled mercury vapor).

Oral uptake often results in vomiting, which reduces the amount of toxin that can be absorbed. Metal compounds can react with either the acid in the stomach or the alkaline environment of the intestine, and this can decidedly influence solubility. Before being distributed throughout the body, the metals pass through the liver, where often detoxification processes begin. The inhalation of fine particles can result in a direct and rapid transfer of soluble metal compounds into the blood, leading to a rapid transfer and onset of symptoms. Toxic quantities of metals can also be absorbed through mucous membranes. Cuta-

neous absorption of acutely toxic amounts of metal compounds seems to be scarce (Guy et al. 1999). Injection has been reported rarely, but happens; for example, Sixel-Dietrich et al. (1985) described a case of acute lead intoxication due to the intravenous injection of lead acetate.

8.4.3
Quantitative Assessment of Acute Toxicity of Individual Metal Compounds

The acute toxicity of a compound can be characterized by its LD_{50}, which is a statistical estimate of the number of milligrams of toxicant per kg body weight required to kill 50% of a large population of animals (e.g., rats). The standard test for acute (short-term) toxicity is to feed animals (e.g., rats), increasing the amounts of a chemical over a period of 14 days until the animals start to die. Alternatively, the chemicals can be applied to the animals' skin until a reaction is observed. The amount of the chemical that kills 50% of the exposed animals is called the lethal dose for 50%, or the LD_{50}.

The LD_{50} may be either "oral" or "dermal", depending on the method of exposure. Lethal doses with respect to inhalation of chemicals in the form of a gas or aerosol can also be tested. In this case, the concentration of gas or vapor that kills half the animals is known as the lethal concentration for 50%, or the LC_{50}. The LD_{50} and LC_{50} are very widely used as indices of toxicity. The criteria shown in Table 8.1 are often used for purposes of classification of acute toxic effects in animals. In order to classify the acute toxicity of chemicals to humans, the scale shown in Table 8.2 can be used (see also IPCS 1996).

It is impossible to assess the health risk posed by a chemical on the basis of its LD_{50} alone. Moreover, the LD_{50} and LC_{50} give no information about the mechanism

Tab. 8.1: Classification of acute toxicity in animals. (After IPCS 1996.)

	Oral LD$_{50}$ rat [mg kg^{-1}]	Dermal LD$_{50}$ rat or rabbit [mg kg^{-1}]	Inhalation LC$_{50}$ rat [mg m^{-3} 4 h^{-1}]
Harmful	200–2000	400–2000	2000–20000
Toxic	25–200	50–400	500–2000
Very toxic	< 25	< 50	< 500

Tab. 8.2: Classification of acute toxicity in humans. (After IPCS 1996.)

Toxicity rating	Dosage	Probable lethal (dose for average human adult)
Practically nontoxic	> 15 g kg^{-1}	> 1 L
Slightly toxic	5–15 g kg^{-1}	0.5–1 L
Moderately toxic	0.5–5 g kg^{-1}	30–500 mL
Very toxic	50–500 mg kg^{-1}	3–30 mL
Extremely toxic	5–50 mg kg^{-1}	7 drops–3 mL
Super toxic	< 5 mg kg^{-1}	< 7 drops

and type of toxicity of a chemical, or its possible long-term or chronic effects. Thus, the LD$_{50}$ and LC$_{50}$ are very crude indices of toxicity. In order to compare the LD$_{50}$ values of two metals, additional information must be available. Important factors are the chemical form in which the element is present (oxidation state, inorganic or organic bonds), the route of uptake (oral, intravenous, intraperitoneal, inhalation, skin adsorption), and the type of animal and its age and state of development. The basic conditions must be comparable.

Luckey et al. (1975) indicated that comparison of the toxicity of metal compounds is more useful when the LD$_{50}$ value is expressed in mmol kg^{-1} instead of the usual mg kg^{-1}. Metals with small differences in atomic mass can show large variances in specific gravity, which influences the toxicity. For example, tungsten and metavanadate are equally toxic if the LD$_{50}$ is expressed in mmol kg^{-1}, but vanadate is three-fold more toxic than tungsten when the LD$_{50}$ is expressed in mg kg^{-1}.

Individual data concerning the LD$_{50}$ of metals and their compounds are published in the Environmental Health Criteria (EHC), edited by IPCS (International Programme on Chemical Safety) since 1976 (IPCS 1976–2003).

The following factors are significant for the toxicity of a compound (Luckey et al. 1975):

- The extent of adsorption, e.g., from the gastrointestinal or respiratory system.
- The particle size of the metal or metal compound (especially important for inhalation toxicity).
- The distribution through the blood to the various organs.
- The extent and the route of excretion, as well as the influence of metabolism and detoxification processes.
- Storage in the cells in the form of harmless particles.

- The efficiency of the mechanisms that control the absorption, excretion, distribution, and retention of toxic metals or compounds.
- The concentration of metal compounds in the organs, which is influenced by the physical form in which the metal is present.
- The influence of the pH of body fluids and organs on the hydrolysis of heavy metal salts as well as their solubility, reactivity, and the toxicity of the hydrolysis products.
- The ability of the metal to chelate ligands of biological macromolecules and other tissue components, as well as the stability of these chelates.
- The ability of the toxic metal to react with other metals, or to suppress or activate essential metals.
- The ability of other metals or body compounds to increase or reduce the toxicity of a metal.

8.4.4
Symptoms of Acute Metal Poisoning

While some groups of toxins (e.g., organophosphorus compounds, digitalis-like drugs) have specific and clearly defined targets in the organism, metals in an acutely toxic amount possess no uniform pattern of action (see Webb 1977). Until now, the usual diversity of symptoms observed in the intact organism could not be explained adequately. The effects of heavy metals and their compounds manifest themselves in very different tissues, partially with a very high specificity for certain metals and their compounds. The most common symptoms of acute metal poisoning can be classified as follows (Kazantzis 1986):

- Gastrointestinal effects: the oral ingestion of large quantities of soluble metal salts leads relatively quickly to gastroenteritis. The result is nausea, vomiting, abdominal pain, diarrhea, and possibly shock due to dehydration and loss of electrolytes. An example is arsenic poisoning.
- Respiratory effects: inhalation of metals or metal compounds can lead to acute chemical pneumonitis and pulmonary edema. One must also distinguish between dusts, smoke, and metallic chlorides that lead to production of hydrochloric acid.
- Cardiovascular effects: arrhythmia, low blood pressure, and shock.
- Effects on the CNS: convulsions, coma, death.
- Renal effects: anuria is often the result of tubular necrosis.
- Hemopoietic effects: acute hemolytic anemia, accompanied by renal failure, e.g., after inhalation of arsenic hydride or ingestion of soluble copper salts.

Acute symptoms of individual metal poisoning can be found in standard reference books, including Luckey et al. (1977), Venugopal et al. (1978), Friberg et al. (1979, 1986), Brown et al. (1987), Seiler et al. (1988), Merian et al. (1991), Goyer et al. (1995), and Zalups et al. (2000), as well as in the individual chapters in Part III of this book.

8.5
Chronic Toxicity

8.5.1
General

Chronic exposure is defined as daily or otherwise repeated exposure over long periods of time, for example, over the working life time or the entire life span. Long-term, low-level exposure usually does not produce immediate toxic effects, but after a certain time signs of chronic toxicity may become apparent. For detailed information on the chronic toxicity of metals and metal compounds, see Ewers and Schlipköter (1991).

In general, chronic toxic effects occur when the agent accumulates in the biological system, when an agent produces irreversible toxic effects, or when there is insufficient time for the system to recover from the toxic effect within the exposure frequency interval. For many agents the toxic effects of acute exposure are quite different from those produced by chronic exposure. According to the site of action, chronic toxic effects can be divided into two groups: local, and systemic effects. For some substances both local and systemic effects can be observed.

8.5.1.1
Local Effects

Local effects occur at the site of first contact between the biological system and the toxic agent. Depending on the route and circumstances of exposure, the gastrointestinal tract, the respiratory organs, the skin, or the eyes can be affected. Gastrointestinal effects such as anorexia, nausea, vomiting, and diarrhea followed by constipation may occur as a result of repeated ingestion of toxic metal compounds over a period of time. Exposure to inhalation of certain metal dusts or aerosols can cause loss of olfactory acuity, atrophy of the nasal mucosa, mucosal ulcers, perforated nasal septum, or sinonasal cancer (see Sunderman 2001). Also chronic pulmonary disorders such as toxic and allergic pulmonary disease may result from inhalation exposure to metals or metal compounds. Other local toxic effects include allergic skin reactions induced for example by nickel or chromates (for details, see Guy et al. 1999).

8.5.1.2
Systemic Effects

Systemic effects require absorption and distribution of the toxic agent to a site distant from its entry point. Metal compounds may produce a variety of systemic effects at different sides of the organism. Usually, the major toxic effects are found to occur in one or two organs. For example, methyl mercury in adult humans is primarily neurotoxic, and the damage is almost exclusively limited to the CNS. The area of damage is highly localized, for example in the visual cortex and the granular layer of the cerebellum (Miura et al. 1995). These organs generally are called the "target organs of toxicity" for that chemical. The most prominent target organs of metals and their compounds include the nervous system, the hemopoietic system, and the kidneys. The target organ of toxicity is not always the site of the highest concentration of the metal, because in the case of some metals inactive complexes of storage depots are formed. Lead, for example, is stored in bones and teeth in an inert form, but its chronic toxicity is mainly directed to the hemopoietic system and the CNS. An example of chronic toxic effects produced by accumulation in a specific organ is the nephrotoxicity of inorganic cadmium and mercury ions, which accumulate in the kidney. At first, the accumulation may be without effect on the functional status of the organ, but when a certain concentration is reached or exceeded the functional status or capacity is affected. The concentration at which functional changes, reversible or irreversible, occur, generally is called the "critical organ concentration". In the case of some other metals, chronic toxicity seems to result from irreversible toxic effects, for example the action of methyl mercury in the CNS. Other examples of apparently irreversible effects are the toxic effects of low-level lead exposure on the immature brain, and the causation of allergic reactions resulting from previous sensitization to certain metals ore metal compounds. Numerous metals and compounds are responsible for systemic immunologically mediated disorders in humans. One of the most common is contact dermatitis, but other manifestations may also be observed (Pelletier et al. 1995, Guy et al. 1999). Cancerogenic and mutagenic effects of metals are discussed in Part II, Chapter 9.

8.5.1.3
Chronic Clinical Effects of Metal Toxicity

The term "chronic effect" is a relative one, signifying that the clinical effect may develop gradually and persist for a longer interval than an acute effect. Only examples of chronic effects can be given here (according to Kazantzis 1986), the great variety of these being described in the individual chapters of Part III of this book:

- Gastrointestinal effects: anorexia, nausea, vomiting, diarrhea followed by constipation, stomatitis, digestive disturbances, intestinal colic.
- Hepatic effects: from abnormalities in hepatic enzyme levels to clinical jaundice, e.g., following exposure to antimony, arsenic, bismuth, copper, chromium, iron, and selenium.
- Respiratory aspects: dyspnea from absorption of metal dust or fume, emphysema, pulmonary fibrosis, formation of granuloma, chronic asthma.
- Effects on the nervous system: tremor, peripheral neuropathy (motor and sensory), paresthesia, ataxia, pyramidal signs, cerebral cortical atrophy, hydrocephalus, convulsions, parkinsonian syndrome, degenerative changes in the nerve cells, personality changes, permanent brain damage.

- Renal effects: tubular proteinuria, glucosuria, aminoaciduria, phosphaturia, hypercalcuria, renal stone formation, uremia, edema.
- Hemopoietic effects: anemia, polycythemia.

Chronic symptoms of individual metal poisoning can be found in standard reference books, including Luckey et al. (1977), Venugopal et al. (1978), Friberg et al. (1979, 1986), Brown et al. (1987), Seiler et al. 1988), Merian (1991), Goyer et al. (1995), and Zalups et al. (2000), as well as in the individual chapters in Part III of this book.

8.5.2
Pathways of Chronic Exposure

Chronic intoxication by metals or metal compounds is usually derived from chronic exposure in the indoor area, the environment, or the workplace. Also, incidences of chronic iatrogenic toxicity have been recognized. Since the placenta provides the route of transfer of both essential and nonessential metals from mother to fetus, prenatal intoxications are also known. (Genetic influence see Part II, Chapter 10.)

8.5.2.1
Chronic Indoor Exposure

Indoor air pollutants include not only biological particles, nonbiological particles, volatile organic compounds, nitrogen oxides, carbon monoxide, various synthetic chemicals, but also metals such as lead from leaded paints, manganese, and cadmium from automobiles exhaust or industrial emission, arsenic from tobacco smoke, and mercury from fungicides, spills or breakage. There are also pollutants generated by human activities; for example, combustion products from stoves with biomass (WHO 1999; see also Part I, Chapter 11). Organic arsenic and mercury compounds mainly derive from marine organisms consumed as food (IPCS 2001; see also Part I, Chapters 6 and 7). Domestic sources further include contamination of food and beverages (e.g., from contact with utensils as earth-glazed pottery) or the use of herbal medicines contaminated with lead or other metals (IPCS 1995, van Vonderen et al. 2000, Ernst et al. 2001).

Chronic lead poisoning seems to remain an important social issue in the United States with regard to certain groups of children (low income, urban, afro-American). The primary lead source for nearly all of the children is leaded paint in deteriorated housing stock (Mushak et al. 1989, Roberts et al. 2001). Indoor problems are often different in developed countries when compared with developing countries.

8.5.2.2
Chronic Environmental Exposure

Itai-itai disease (for details, see Part III, Chapter 6) was first reported in 1955 among the human population of Toyoma, Japan. Urinary cadmium levels were found to be high in affected patients. Studies showed that cadmium content was particularly high in rice, a staple food for this population. The source of cadmium was felt to be via water from the Kamiaha mine upstream from Toyama. It was concluded that cadmium played the most important role in the development of Itai-itai disease (Friberg et al. 1979).

Minamata disease was first discovered in 1956 around Minamata Bay, Japan. A similar epidemic occurred in 1965 along the Agano river, Japan. Minamata disease is methyl mercury poisoning that occurred in humans who ingested fish contaminated with methylmercury discharged in waste water from a chemical plant. Methylmercury is also teratogenic (Ui 1992, Harada 1995, Eto 1997, Schardein 2000; see also Part III, Chapter 17).

A worldwide problem is the chronic environmental exposure to arsenic of geological origin found in groundwater used as drinking water in several parts of the world, for example Bangladesh, India and Taiwan (see NCR 1999, 2001, IPCS 2001, Rahman et al. 2001; see also Part IV, Chapter 6). There seems to be a variation in susceptibility among individuals, and possible reasons for this include age, nutritional status, concurrent exposure to other agents or environmental factors, and genetic polymorphism (Vahter 2000; see also Part II, Chapter 10).

Another problem is drinking water contaminated with lead, an example being that in Philadelphia public schools (Bryant et al. 2001). Pollution of the environment with lead may occur through the smelting and refining of lead, the burning of fuels containing lead additives, the smelting of other metals, and the burning of coal and oil (IPCS 1989, 1995).

8.5.2.3
Chronic Occupational Exposure

Lead is the most widely used nonferrous metal, and a large number of occupations may be associated with risk of exposure. Lead is present in the work atmosphere as fumes, mists (e.g., produced by spray painting) and dust. Inhalation of lead fumes or of fine lead particles is the most important route of absorption in the working atmosphere. Lead poisoning is one of the most common occupational diseases, especially when prevention measures are not established (IPCS 1995). Further metals and their compounds that, among others, are known to cause chronic occupational intoxications at the workplace are antimony, arsenic, beryllium, cadmium, chromium, cobalt, manganese, mercury, nickel, thallium, and vanadium (DFG 1972–2001).

8.5.2.4
Chronic Iatrogenic Exposure

Chronic iatrogenic exposures to metals are rare events. A major iatrogenic poisoning problem of the past few decades was that of aluminum toxicity in patients with chronic renal failure being treated with intermittent hemodialysis. These patients derived aluminum from dialysis solutions and also from treatment with aluminum-containing phosphate-binding agents. For some of the patients the clinical consequences were encephalopathia (a specific form of metabolic bone disease) and a microcytic anemia (Jeffery 1995).

Another source of chronic iatrogenic exposure to metals may be medical implants, which are widely used for dental restorations and prosthetics as well as for orthopedic, restorative, and replacement surgery (see Part II, Chapter 5).

8.5.2.5
Transplacental Transfer

The mechanisms of transfer of toxic metals to the fetus via placenta are different. In humans, there is a placental barrier for cadmium, but not for lead. Epidemiological studies suggest that maternal blood lead levels may affect duration of pregnancy as well as fetal birth weight and outcome. The critical effect of lead on the human fetus is on the CNS, with impairment of cognitive and behavioral development. The second group of substances of particular interest to the teratologist are organic mercury compounds (Mushak et al. 1989, Goyer 1995, Schardein 2000). Until now, there seems to have been no substantial reports which indicate that heavy metals beyond lead and organic mercury compounds have any causal relation to the induction of birth defects in humans, though some 43% of metals tested in animals were teratogenic (Schardein 2000).

Mutagenicity, carcinogenicity and teratogenicity are discussed in detail in Part II, Chapter 9.

8.6
Guidelines and Exposure Limits

This section provides an overview on standards, guidelines and limits relevant in the general and working environment. The listed terms and organizations should be contacted for further information regarding metals, metalloids, and metal compounds.

8.6.1
Guidelines for Drinking Water Quality

WHO has published guidelines for drinking water quality (WHO 1996, 1998), including values for antimony, arsenic, barium, cadmium, chromium, copper, lead, manganese, mercury, molybdenum, nickel, and uranium.

8.6.2
Air Quality Guidelines

WHO has also provided general guidelines for air quality (WHO 1999, last updated 2001). Since the problems are often different in developed and developing countries, WHO will soon offer special Air Quality Guidelines for Europe.

8.6.3
Acceptable Daily Intake (ADI)

The ADI for humans is an estimate of JECFA (Joint FAO/WHO Expert Committee on Food Additives) of the amount of a food additive and contaminants, expressed on a body weight basis, that can be ingested daily over a lifetime without health risk (standard man = 60 kg) (IPCS/JECFA 1987).

JECFA generally sets the ADI of a food additive or food contaminant on the basis of the highest no-observed-effect level (NOEL) in animal studies. In calculating the ADI, a "safety factor" is applied to the NOEL in order to provide a conservative margin of safety on account of the inherent uncertainties in extrapolating animal toxicity data to potential effects in humans and for variation within the human species. NOEL is defined as the greatest concentration or amount of an agent, found by study or observation, that causes no detectable, usually adverse, alteration of morphology, functional capacity, growth, development, or life span of the target. It is intended to provide an adequate margin of safety for humans by assuming that the human being is 10 times more sensitive than the test animal.

The ADI is expressed in a range from 0 to an upper limit, which is considered to be the zone of acceptability of the substance. The ADI is expressed in this way to emphasize that the acceptable level is an upper limit. Substances that accumulate in the body are not suitable for use as food additives. Therefore, ADIs are established only for those compounds that are substantially cleared from the body within 24 h. There was a general consensus that chemicals found to be carcinogenic were not appropriate as food additives at any level whatsoever.

Toxicological evaluation of metals in foods calls for carefully balanced consideration of the following factors:

- Nutritional requirements, including nutritional interactions with other constituents of food (including other metals when the interactions are nutritionally or toxicologically relevant) in respect of, for instance, absorption, storage in the body, and elimination.
- The results of epidemiological surveys and formal toxicological studies, includ-

ing interactions with other constituents of food (including other metals when the interactions are nutritionally or toxicologically relevant), information about pharmaceutical and other medicinal uses, and clinical observations on acute and chronic toxicity in human experience and veterinary practice.
- Total intake on an appropriate time basis (e.g., daily, weekly, yearly or lifetime) from all sources (food, water, air) of metals as normal constituents of the environment, as environmental contaminants, and as food additives of an adventitious or deliberate nature.

The tentative tolerable daily intakes proposed for certain metals provide a guideline for maximum tolerable exposure. In the case of essential elements, these levels exceed the daily requirements, but this should not be construed as an indication of any change in the recommended daily requirements. In the case of both essential and nonessential metals, the tentative tolerable intake reflects permissible human exposures to these substances as a result of natural occurrence in foods or various food processing practices, as well as exposure from drinking water.

8.6.4
Reference Values and Human Biological Monitoring Values for Environmental Toxins

The task of the Commission on Human Biological Monitoring of the German Federal Environmental Agency is to develop scientifically based criteria for the application of human biological monitoring and for the evaluation of human monitoring data in environmental medicine. In principle, two different kinds of criteria are recommended: (i) reference values; and (ii) human biological monitoring values (HBM values). Reference values are intended to indicate the upper margin of the current background exposure of the general population to a given environmental toxin at a given time. Reference values can be used to identify subjects with an increased level of exposure (in relation to background exposure) to a given environmental toxin. Reference values do not represent health-related criteria for the evaluation of human biological monitoring data. HBM values are derived from human toxicology and epidemiology studies and are intended to be used as a basis for a health-related evaluation of human biological monitoring data. Usually, the Commission recommends two different HBM values:

- HBM I, the concentration of an environmental toxin in a human biological material below which there is no risk for adverse health effects in individuals of the general population; and
- HBM II, the concentration of an environmental toxin in a human biological material (usually blood, serum, plasma, or urine) above which there is an increased risk for adverse health effects in susceptible individuals of the general population.

HBM I can be considered a kind of alert value (from the toxicological point of view), whereas HBM II represents a kind of action level, at which attempts should be undertaken to reduce the level of exposure immediately and to carry out further examinations. At present, reference and HBM values are available for lead, cadmium, and mercury (Ewers et al. 1999).

8.6.5
Occupational Exposure Limits

8.6.5.1
Threshold Limit Values (TLV)

The TLVs are provided by the American Conference of Governmental Industrial Hygienists (ACGIH) of the United States (see ACGIH 2002). They refer to airborne concentrations of substances, and represent conditions under which it is believed that nearly all workers may be repeatedly exposed day after day without adverse effects. Because of wide variation in individual susceptibility, however, a small percentage of workers may experience discomfort from some substances at concentrations at or below the threshold limit; a smaller percentage may be affected more seriously by aggravation of a preexisting condition or by development of an occupational illness. The three categories of TLVs are specified as follows:

1. The Threshold Limit Value-Time Weighted Average (TLV-TWA) = the time-weighted average concentration for a normal 8-h working day and a 40-h working week, to which nearly all workers may be repeatedly exposed, day after day, without adverse effect.

2. Threshold Limit Value-Short Term Exposure Limit (TLV-STEL) = the concentration to which the worker can be exposed continuously for a short period of time without suffering from: (i) irritation; (ii) chronic or irreversible tissue damage; or (iii) narcosis of sufficient degree to increase the likelihood of accidental injury, impair self-rescue, or materially reduce work efficiency, and provided that the daily TVL-TWA is not exceeded. It is not a separate independent exposure limit; rather it supplements the TWA limit where there are recognized acute effects from a substance whose toxic effects are primarily of a chronic nature. STELs are recommended only where toxic effects have been reported from high short-term exposures in either humans or animals.

3. Threshold Limit Value-Ceiling (TLV-C) = the concentration that should not be exceeded during any part of the working exposure.

Descriptions of the procedures used by ACGIH in the evaluation of the exposure limits can be found in the appropriate sections of the "Documentation of the Threshold Limit Values and Biological Exposure Indices" (ACGIH 2001).

8.6.5.2
Maximum Allowable Concentration (MAC)

The term MAC is used widely, for example, in the Netherlands and Germany as well as the former Soviet Union and Central and Eastern European countries (IPCS 1996).

In Germany, the MAC values (called MAK = Maximale Arbeitsplatz-Konzentration) are published yearly by the Commission for the Investigation of Health Hazards of Chemical Compounds in the Work Area of the "Deutsche Forschungsgemeinschaft" (DFG 2002). The MAK value is defined as the maximum concentration of a chemical substance (as gas, vapor or particulate matter) in the workplace air which generally does not have known adverse effects on the health of the employee, nor cause unreasonable annoyance (e.g., by nauseous odor), even when the person is repeatedly exposed during long periods, usually for 8 h daily, but assuming on average a 40-h working week. As a rule, the MAK value is given as an average concentration for a period of up to one working day or shift. MAK values are established on the basis of the effects of chemical substances; when possible, practical aspects of the industrial processes and the resulting exposure patterns are also taken into account. Scientific criteria for the prevention of adverse effects on health are decisive, not technical and economical feasibility. For the establishment of a MAK value, the carcinogenicity, sensitizing effects, contribution to systemic toxicity after percutaneous absorption, risks during pregnancy and germ cell mutagenicity of a substance are evaluated, and the substance is classified or designated accordingly. Descriptions of the procedures used by the Commission in the evaluation of these endpoints can be found in the appropriate sections of the "Toxikologische – arbeitsmedizinische Begründungen von MAK-Werten" (DFG 1972 -2001). These justifications are also available in English, and to date 19 volumes have been published (DFG 1991– 2002).

8.6.5.3
Other Terms for Occupational Exposure Limits

Just as the regulations and guidelines of countries are subject to change, so too is the terminology concerning exposure limits. Although TLV and MAC are terms that are well known because of their wide usage over a long time, some countries have developed their own terminology or evolved their own standards, criteria and methods for determining exposure limits. These vary in practice between the stringent concept of MAC and the more elastic approach of TLV, which makes allowances for reversible clinical changes. Some of the expressions currently used in addition to TLV and MAC include (see IPCS 1996):

- Maximum permissible concentration (MPC) is used in Argentina, Finland and Poland, among others. (This term is also used in relation to a chemical's concentration in drinking-water in several countries, including Japan, Germany and the USA.)
- Permissible exposure limit (PEL) is used in the United Kingdom and by the US Occupational Safety and Health Administration, etc.

8.6.6
Biological Monitoring and Biological Limits

Biological monitoring of chemical exposure aims to measure either the amount of a chemical that has been absorbed by the worker, or the effect of that absorbed chemical on the worker. Biological monitoring involves taking a sample of body fluid (usually blood or urine) and measuring the level of the chemical or its metabolite. Alternatively, an effect of that chemical on the body may be determined by measuring the level of an enzyme or other chemical in the blood or urine.

Many chemicals can be assessed by biological monitoring, but the results do not always reflect the level of absorption. Each chemical should be considered separately when deciding whether to perform biological monitoring. Lists of those chemicals for which biological monitoring is recommended are available, and some such chemicals may have a limit value – i.e., a biological exposure index – that should not be exceeded (Lauwerys et al. 2001).

Biological monitoring can assist the occupational health professional to detect and determine absorption via the skin or gastrointestinal system, in addition to that of inhalation; assess body burden; reconstruct past exposure in the absence of other exposure measurements; detect nonoccupational exposure among workers; test the efficacy of personal protective equipment and engineering controls; and monitor work practices.

Scientifically justified threshold limit values in biological material are being compiled and published at present by two institutions (Morgan and Schaller 1999). One of these is the German Senate Commission for the Investigation of Health Hazards of Chemical Compounds in the Work Area of the German Research Foundation (Deutsche Forschungsgemeinschaft). The biological tolerance values (BAT values) are specially drawn up by the working group on "setting of threshold limit values in biological material". The other organization is a committee of the American Conference of Governmental Industrial Hygienists (ACGIH). The biological exposure indices (BEI values) are developed by the BEI Committee working in parallel with the Threshold Limit Values Committee of the ACGIH.

8.6.6.1
BEI Values
BEI values are guidance values for assessing biological monitoring results. BEIs represent the levels of determinants which are most likely to be observed in specimens collected from healthy workers who have been exposed to chemicals to the same extent as workers with inhalation exposure at the TLV. The exceptions are the BEIs for chemicals for which the TLVs are based on protection against nonsystemic effects (e.g., irritation or respiratory impairment) where biological monitoring is desirable because of the potential for significant absorption via an additional route of entry (usually the skin). The BEI generally reflects a concentration below which nearly all workers should not experience adverse health effects. The BEI determinant can be the chemical itself; one or more metabolites; or a characteristic, reversible biochemical change induced by the chemical. In most cases, the specimen used for biological monitoring is urine, blood, or exhaled air. The BEIs are not intended for use as a measure of adverse effects or for diagnosis of occupational illness. BEIs are, for example, given for lead, chromium, arsenic, mercury, and vanadium pentoxide. Descriptions of the procedures used by ACGIH in the evaluation of the exposure limits can be found in the appropriate sections of the "Documentation of the TLVs and BEIs" (ACGIH 2001).

8.6.6.2

BAT Values

The BAT value is defined as the maximum permissible quantity of a chemical substance or its metabolites, or the maximum permissible deviation from the norm of biological parameters induced by these substances in exposed humans. The BAT value is established on the basis of currently available scientific data which indicate that these concentrations generally do not affect the health of the employee adversely, even when they are attained regularly under work place conditions (DFG 1983–2002).

As with MAK values, BAT values are established on the assumption that persons are exposed at work for at most 8 h daily and 40 h weekly. BAT values established on this basis may also be applied without the use of correction factors to other patterns of working hours. BAT values can be defined as concentrations or rates of formation or excretion (quantity per unit time). BAT values are conceived as ceiling values for healthy individuals. They are generally established for blood and/or urine and take into account the effects of the substances and an appropriate safety margin, being based on occupational medical and toxicological criteria for the prevention of adverse effects on health.

To date, the Commission for the Investigation of Health Hazards of Chemical Compounds in the Work Area in Germany (DFG) has published BAT values for the following metals: aluminum, lead, manganese, metallic mercury and inorganic mercury compounds, tetraethyl- and tetramethyl lead, and vanadium pentoxide (see DFG 2002).

Descriptions for the procedures used by this Commission in the evaluation of these endpoints can be found in the appropriate sections of the "Toxikologische Begründungen von BAT-Werten" (DFG 1983–2002).

These justifications are also available in English; to date, three volumes have been published (DFG 1994–1999).

References

ACGIH (American Conference of Governmental Industrial Hygienists) (2001) *Documentation of the Threshold Limit Values and Biological Exposure Indices*, 7th edn. ACGIH Inc, Cincinnati, Ohio.

ACGIH (American Conference of Governmental Industrial Hygienists) (2002) *Threshold limit values for chemical substances and physical agents and biological exposure indices*. ACGIH Inc, Cincinnati, Ohio.

BROWN SS and KODAMA Y, eds. (1987) *Toxicology of metals*. Wiley, New York- Chichester-Brisbane-Toronto.

BRYANT SD, GREENBERG ML and CROF R (2001) *Lead contaminated drinking water in Philadelphia public schools (abstract)*. Clin Toxicol **39**:552.

DFG (Deutsche Forschungsgemeinschaft) (1972–2002) *Gesundheitsschädliche Arbeitsstoffe – Toxikologisch-arbeitsmedizinische Begründungen von MAK-Werten (Maximale Arbeitsplatzkonzentrationen) 1.–35. Lieferung*. Wiley-VCH, Weinheim.

DFG (Deutsche Forschungsgemeinschaft) (1983–2001) *Biologische Arbeitsstoff- Toleranzwerte (BAT-Werte) und Expositionsäquivalente für krebserzeugende Abeitsstoffe (EKA), 1.–10. Lieferung*. Wiley-VCH, Weinheim.

DFG (Deutsche Forschungsgemeinschaft) (1991–2002) *Occupational toxicants*. Critical data evaluation for MAK values and classification of carcinogens, Vol. 1–17. Wiley-VCH, Weinheim-New York-Chichester-Brisbane-Singapore-Toronto.

DFG (Deutsche Forschungsgemeinschaft) (1994–1999) *Biological exposure values for occupational toxicants and carcinogens*. Critical data evaluation for BAT and EKA values, Vol. 1–3, Wiley–VCH, Weinheim, New York, Chichester, Brisbane, Singapore, Toronto.

DFG (Deutsche Forschungsgemeinschaft) (2002) *List of MAK and BAT values 2002; Report No. 38*. Wiley-VCH, Weinheim.

ERNST E and COON JT (2001) *Heavy metals in traditional Chinese medicines: a systematic review*. Clin Pharmacol Ther **7**:497–504.

ETO K (1997) *Pathology of Minamata disease*. Toxicol Pathol **25**:614–623.

EWERS U and SCHLIPKÖTER HW (1991) *Chronic toxicity of metals and metal compounds*. In: Merian

E; ed.: Metals and their compounds in the environment, pp. 591–603. VCH, Weinheim-New York-Basel-Cambridge.

EWERS U, KRAUSE C, SCHULZ C, and WILHELM M (1999) *Reference values and human biological monitoring values for environmental toxins.* Int Arch Occup Environ Health **72**:255–260.

FOULKES EC (2000) *Transport of heavy metals across cell membranes.* Proc Soc Exp Biol Med **223**:234–240.

FRIBERG L, NORDBERG GF and VOUK VB, eds. (1979) *Handbook on the toxicology of metals.* Elsevier/North Holland, Amsterdam-New York-Oxford.

FRIBERG L, NORDBERG GF and VOUK VB, eds. (1986) *Handbook on the toxicology of metals.* 2nd rev. edn. Elsevier/North Holland, Amsterdam-New York-Oxford.

GEBHART E and ROSSMAN TG (1991) *Mutagenicity, carcinogenicity, teratogenicity.* In: Merian E, ed. Metals and their compounds in the environment, pp. 617–640. VCH, Weinheim-New York-Basel-Cambridge.

GOYER LA and CHERIAN MG, eds. (1995) *Toxicology of metals – Biochemical aspects.* Springer-Verlag, Berlin-Heidelberg-New York-London-Paris-Tokyo-Hong Kong-Barcelona-Budapest.

GOYER RA (1995) *Transplacental transfer of lead and cadmium.* In: Goyer RA and Cherian MG, eds. Toxicology of metals, pp. 1–17. Springer-Verlag, Berlin-Heidelberg-New York-London-Paris-Tokyo-Hong Kong-Barcelona-Budapest.

GUY RH, HOSTYNEK JJ, HINZ RS and LORENCE CR (1999) *Metals and the skin. Topical effects and systemic absorption.* Marcel Dekker, New York-Basel.

HARADA M (1995) *Minamata disease: methylmercury poisoning in Japan caused by environmental pollution.* Crit Rev Toxicol **25**:1–24.

IPCS (International Programme on Chemical Safety) (1976–2003) *Environmental Health Criteria (EHC) Monographs.* WHO, Geneva.

IPCS (International Programme on Chemical Safety) (1989) *Environmental Health Criteria Monographs No. 85: Lead: Environmental aspects.* WHO, Geneva.

IPCS (International Programme on Chemical Safety) (1995) *Environmental Health Criteria Monographs No. 165: Lead, inorganic.* WHO, Geneva.

IPCS (International Programme on Chemical Safety)(1996) *Users manual for the IPCS Health and Safety Guides.* WHO, Geneva.

IPCS (International Programme on Chemical Safety) (2001) *Environmental Health Criteria Monographs No. 224: Arsenic* (2nd edn). WHO, Geneva.

IPCS/JECFA (International Programme on Chemical Safety / FAO/WHO Expert Committee on Food Additives) (1987). *Environmental Health Criteria No. 70: Principles for the safety assessment of food additives and contaminants in food.* WHO, Geneva.

JEFFERY EH (1995) *Biochemical mechanisms of aluminum toxicity.* In: Goyer RA and Cherian MG eds. Toxicology of metals, pp. 139–161. Springer-Verlag, Berlin-Heidelberg-New York-London-Paris-Tokyo-Hong Kong-Barcelona-Budapest.

KAZANTZIS G (1986) *Diagnosis and treatment of metal poisoning – general aspects.* In: Friberg L, Nordberg GF and Vouk VB, eds. Handbook on the toxicology of metals, 2nd edn, pp. 294–301. Elsevier/North Holland, Amsterdam-New York-Oxford.

LAUWERYS RR and HOET P (2001) *Industrial chemical exposure, guidelines for biological monitoring.* Lewis Publishers, Boca Raton-London-New York-Washington DC.

LUCKEY TD and VENUGOPAL B (1977) *Metal toxicity in mammals.* Vol. 1, pp. 1, 105. Plenum Press, New York-London.

LUCKEY TD, VENUGOPAL B and HUTCHESON D (1975) *Heavy metal toxicity, safety, homology.* pp. 61–62. Thieme, Stuttgart; Academic Press, New York-San Francisco-London.

MERIAN E, ed. (1991) *Metals and their compounds in the environment.* VCH Weinheim-New York-Basel-Cambridge.

MIURA K; NAGANUMA A; HIMENO S and MIURA N (1995) *Mercury Toxicity.* In: Goyer RA and Cherian MG, eds. Toxicology of metals, pp. 163–187. Springer- Verlag, Berlin-Heidelberg-New York-London-Paris-Tokyo-Hong Kong-Barcelona-Budapest.

MOESCHLIN S (1972) *Klinik und Therapie der Vergiftungen,* 5. Aufl., pp. 72–84. Thieme, Stuttgart.

MORGAN MS and SCHALLER K-H (1999) *An analysis of criteria for biological limit values developed in Germany and in the United States.* Int Arch Occup Environ Health **72**:195–204.

MUSHAK P, DAVIS JM, CROCETTI AF and GRANT LD (1989) *Review. Prenatal and postnatal effects of low-level lead exposure: integrated summary of a report to the U. S. Congress on childhood lead poisoning.* Environ Res **50**:11–36.

NRC (National Research Council) (1999) *Arsenic in drinking water.* National Academy Press, Washington DC.

NRC (National Research Council) (2001) *Arsenic in drinking water: 2001 update.* National Academy Press, Washington DC.

PELLETIER L and DRUET P (1995) *Immunotoxicology of metals.* In: Goyer RA and Cherian MG, eds., Toxicology of metals. Springer-Verlag, Berlin-Heidelberg-New York-London-Paris-Tokyo-Hong Kong-Barcelona-Budapest.

RAHMAN MM, CHOWDHURY UK, MUKHERJEE SC, MONDAL BK, PAUL K, LODH D, BISWAS BK, CHANDA CR, BASU GK, SAHA KC, ROY S, DAS R, PALIT SK, QUAMRUZZAMAN Q and CHAKRABORTI D (2001) *Chronic arsenic toxicity in Bangladesh and West-Bengal, India – A review and commentary.* Clin Toxicol **39**:683–700.

ROBERTS JR, REIGART JR, EBELING M and HULSEY TC (2001) *Time required for blood lead levels to decline in non chelated children.* Clin Toxicol **39**:53–160.

SCHARDEIN JL (2000) *Chemically induced birth defects.* 3rd. edn. pp. 875–909. Marcel Dekker, New York-Basel.

SEILER HG, SIGEL H and SIGEL A, eds. (1988) *Handbook on toxicity of inorganic compounds.* Marcel Dekker, New York-Basel-Hong Kong.

SIXEL-DIETRICH F, DOSS M, PFEIL CH and SOLCHER H (1985) *Acute lead intoxication due to intravenous injection.* Hum Toxicol **4**:301–309.

SUNDERMAN FW JR (2001) *Nasal toxicity, carcinogenicity, and olfactory uptake of metals.* Ann Clin Lab Sci **31**:3–24.

UI J (1992) *Industrial pollution in Japan.* United Nations University Press, Tokyo.

VAHTER M (2000) *Genetic polymorphism in the biotransformation of inorganic arsenic and its role in toxicity.* Toxicol Lett **112/113**:209–217.

VAN VONDEREN MGA, KLINKENBERG-KNOL EC, CRAANEN ME, TOUW DJ, MEUWISSEN SGM and

DE SMET PAGM (2000) *Severe gastrointestinal symptoms due to lead poisoning from Indian traditional medicine.* Am J Gastroenterol **95**:1591–1592.

VENUGOPAL B and LUCKEY TD (1978) *Metal toxicity in mammals.* Vol. 2. Plenum Press, New York-London.

WEBB M. (1977) *Metabolic targets of metal toxicity.* In: Brown SS, ed. Clinical chemistry and chemical toxicology of metals, p. 51. Elsevier/North Holland, Amsterdam- New York-Oxford.

WHO (World Health Organization) (1996, 1998) *Guidelines for drinking-water quality*, 2nd edn. Vol. 2 Health criteria and other supporting information, 1996, pp. 940–949 and Addendum to Vol. 2, 1998, pp. 281–283. WHO, Geneva.

WHO (World Health Organization) (1999, last update 2001) *Air quality guidelines.* WHO, Geneva.

WILLIAMS RJP (1981a) *Natural selection of the chemical elements.* Proc R Soc London B **213**:361–397.

WILLIAMS RJP (1981b) *Physico-chemical aspects of inorganic element transfer through membranes.* Philos Trans R Soc London B **294**; 57–74.

WILLIAMS RJP (1982) *Metal ions in biological catalysts.* Pure Appl Chem **54**:1889–1904.

WILLIAMS RJP (1983) *Inorganic elements in biological space and time.* Pure Appl Chem **55**:1089–1100.

WILLIAMS RJP (1985) *Homeostasis: an outline of the problems.* TEMA **5**:300–306

ZALUPS RK and KOROPATNICK J, eds. (2000) *Molecular biology and toxicology of metals.* Tailor & Francis, London-New York.

ZHU Z and THIELE DJ (1996) *Toxic metal-responsive gene transcription.* EXS **77**:307–320.

9

Mutagenicity, Carcinogenicity, and Teratogenicity

Erich Gebhart

9.1

Introduction

Changes in genetic information (mutations) induced by environmental agents remain one of the intriguing aspects of modern environmental research. Mutations are biological events of sometimes considerable consequences for the affected individual and for the exposed population, but also for the damaged individual cell. The reaction of chemical mutagens with the genetic material yields a broad and variable spectrum of consequences. Beside lethal damage, molecular mutations (gene mutations), and structural chromosome changes (chromosome mutations) may be produced, all of which can cause more or less marked changes of the phenotype. In addition, numerical chromosome alterations (genome mutations) arising from disturbances of the mitotic process may also be caused by chemical mutagens. The progress of molecular genome research (Olden and Guthrie 2001) has generated new aspects of the induction of changes in the genetic information by mutagens. Mutational mechanisms and spectra have now become clearer, although with increasing new insights the complexity of both certainly has not regressed. The consequences of mutations in somatic cells have attracted more attention on the basis of an immense body of knowledge concerning the molecular genetic causes of malignant transformation. The presumed close association of mutagenesis and carcinogenesis, therefore, has increasingly been confirmed. Nonetheless, the meaning of chemically induced mutations of all mentioned types in germ cells with their threatening impact for coming generations must not be neglected in future.

Among chemical mutagens, metals and their compounds have always played a special role because of their wide distribution in our environment and their physiological importance for genome stability (Hartwig 2001a) and enzyme function (Hartwig 2001b) on the one side, and their acute toxicity on the other side which, of course, outweighs their mutagenic potential in several cases. As later chapters will address specifically also the mutagenic, carcinogenic, and teratogenic potential of specific metals and their compounds, this chapter will focus on more generals aspects of metals' mutagenicity and carcinogenicity, but will also discuss toxic action on the developing embryo (teratogenicity). It is understood that, where not specifically mentioned, the data refer to the action of metal compounds

Elements and their Compounds in the Environment. 2nd Edition.
Edited by E. Merian, M. Anke, M. Ihnat, M. Stoeppler
Copyright © 2004 WILEY-VCH Verlag GmbH & Co. KGaA, Weinheim
ISBN: 3-527-30459-2

rather than to elemental metals. In addition, it should be noted that metals are listed in the tables as soon as a report on a positive result in the mentioned organisms has been published, irrespective of its strength. Therefore, the reader's attention is called to the respective references for more details. If only negative results were obtained on a metal or its compounds, this was not registered in the tables but will be mentioned in the text. Mutagenicity mainly caused by emitted ionizing radiation of radioactive metals has not been included.

9.2
Mutagenicity

In 1990, De Flora et al. reviewed the data obtained from 130 short-term tests on 32 chromium compounds, this fact demonstrating the large variety of test assays. The vast range of reliable methods for testing the mutagenicity of chemicals has now been enriched by molecular genetic techniques. Nevertheless, in order to relate all obtained results eventually to the human situation, a comparison of the test agent's metabolism in the test system with that occurring in the human body is an essential component of all tests. For a thorough analysis, all attempts should be directed to detect the whole spectrum of known types of mutations by using a suitable test battery. Each of these test systems may be regarded as a model for the specific demands it fulfils, though none of them can be regarded as an ideal procedure. The original catalogue of mutagenicity test procedures (Kilbey et al. 1984) remains, but has subsequently been extended by a series of molecular methods (Kirkland et al. 2000).

9.2.1
Reactions of Metals with Nucleic Acids and Proteins

The demonstration of direct reactions of an agent with DNA, RNA, or proteins is fundamental in understanding its mutagenicity. However, not each of those reactions actually ends in a persistent mutation, as most of them are eliminated by repair systems or selective factors, including apoptosis.

The numerous nucleophilic centers in nucleic acids are favorite sites for binding metals, but the type and localization of binding apparently depends on the respective metal. Although the basic reactions of metal ions with nucleic acids and proteins have been long been known (for a review, see Gebhart and Rossman 1991), a vast body of data has emerged during the past decade on the grounds of new DNA technologies, and this has also substantially affected our view of the reaction of metals with nucleic acids.

Well-documented reactions of several heavy metals with DNA are:
1. Direct binding to one of the numerous nucleophilic centers in nucleic acids (e.g., N_7 in guanine or the phosphate groups of the DNA backbone).
2. The formation of cross-links and, as a consequence, the induction of strand breaks in DNA.
3. Chelation and formation of other complexes.
4. Interaction with DNA replication and cellular repair mechanisms.

Many of the results in the first three of these categories have been obtained from experiments conducted with isolated (purified) DNA. These do not in all cases represent the situation in the living cell, and in particular the mechanism of mutation induction as a consequence of these reactions occurring. Possible scenarios leading to nucleobase mispairing may include changes in nucleobase selectivity as a consequence of alterations in acid–base properties of nucleobase atoms and groups involved in complementary H bond formation, guanine deprotonation, and stabilization of rare nucleobase tautomers by metal ions (Müller et al. 2000). In addition to the differing affinity of metals towards DNA, the steric structure of DNA bases is crucially important in all reactions with metals. In addition, an indirect mechanism of interaction with DNA is that of reactive oxygen species (ROS) and free radical generation, as well as depurination which, in the cell, is detectable as alkali-labile sites. The existence of apurinic sites eventually can lead to strand breaks via cleavage of AP endonucleases.

In Table 9.1, the various metals are grouped according to their reactions with DNA, and the involvement of several metal ions in more than one of those mechanisms of action are listed. This is particularly evident for Cd(II), Co(II), Cr(III), Ni(II), Zn(II), As(III), Hg(II), and V(IV). Many of these reactions are associated with the valency of the metal ions under consideration (they are particularly well documented for Cr; Cohen et al. 1993), and with their "ionic strength". From a more genetic viewpoint, the interaction also depends on a metal's ability to cross cellular membranes and its availability in an amount which is sufficient for the respective reaction but is not too toxic. In many cases, low concentrations of metal ions are those rendering their reaction with DNA possible.

Reactions of metals with proteins, that is, binding to amino or sulfhydryl groups, can result in an inhibition or at least considerable misfunction of enzymes involved in DNA metabolism and repair (Buchko et al. 2000, Hartwig 2001b). Alterations in genetic information by affecting replication fidelity

may be a consequence of those reactions. However, the replacement by redox metals of zinc in zinc finger proteins which is expected to generate free radicals and thus cause DNA damage (Sarkar 1995), can also lead to considerable changes in cellular transcriptional pathways.

Recently, a method for testing DNA damage in single cells has been developed – the so-called single cell gel electrophoresis (SCGE), now better known as the "comet assay" (Olive et al. 1990, Kassie et al. 2000). In this test, nuclei from mutagen-exposed cells are embedded in an agarose gel and subjected to gel electrophoresis. Using this procedure, pieces of DNA which have arisen from single- or double-strand breaks, or alkali-labile sites are "extracted" from the nucleus and diffuse into the gel according to their length. In this way they form a comet-like tail which can be rendered microscopically visible and measurable by fluorescence staining. A major advantage of this technique is the ability to compare data from different tissues of the same exposed test individual, as was recently shown in experiments with lead and cadmium (Valverde et al. 2000), though the possibility also exists to detect interactions with repair processes (Hartmann and Speit 1996). Using an inhalation model in mice, the former group detected single-strand breaks and alkali-labile sites in several mouse organs. Differences among the organs studied after single and subsequent inhalations of Pb were found, while Cd induced a major effect in all organs studied. A correlation between length of exposure, DNA damage and metal tissue concentration was observed for lung, liver, and kidney. These results show that lead inhalation induces systemic DNA damage, but that some organs (e.g., lung and liver) are special targets of this metal. The damage is dependent in part on the duration of expo-

Tab. 9.1: Interaction of metals (metal compounds) with nucleic acids and proteins. (Data in this table and in Tables 9.2 – 9.5 which were obtained from Gebhart and Rossman (1991) are printed in italics)

Direct binding to nucleophilic centers		Cross-linking	Induction of single-strand breaks	Chelation or other complex formation	Generation of free radicals	Infidelity of DNA synthesis	Action on DNA repair	DNA damage in the comet assay
P*:	G_{N7}*							
Ca(II)[2]	Ag(I)[1]	As(III)[8,21]	As(III)[8,21]	Ag(I,II)[1]	As(III)[16]	Ag(I)	As(III)[18,20,21]	As(III)[21,22,27]
Cd(II)[2.8]	Cd(II)[9]	Cd(II)[9]	Cd(II)[9,11,13,15]	As(III)[21]	Cd(II)[16]	As[21]	Be(II)[8]	Cd(II)[22,27]
Co(II)	Co(II)	Co(II)	Co(II)[14,20]	Cd(II)	Co(II)[17]	Be(II)	Cd(II)[15,18]	Co(II)[23]
Cr(III)[7,10]	Cr(III)[8]	Cr(III)[8]	Cr(III)[8,10,24c]	Co(II)	Cr(III)[16]	Cd(II)[16]	Co(II)[18,20]	Cr(III)[24]
Cu(II)[6]	Cu(II)	Cu(II)	Cu(II)[8]	Cr(III)	Cu(II)[16]	Co(II)[14,16]	Hg(II)	Hg(II)[25]
Eu(III)[4]	Cu(II)[3]	Mn(II)	Hg(II)	Cr(V)	Fe(III)[16,17]	Cr(III)[10,16]	Mn(II)[19]	Mn(VII)[19]
La(III)[4]	Eu(III)[4]	Ni(II)[8]	Mn(II)	Cu(II)	Hg(II)	Cu(II)	Ni(II)[18]	Sb(III)[28]
Mg(II)[2]	Ge(IV)[5]	Pt(II)	Ni(II)[8,13]	Hg(II)	Mg(II)	Mn(II)[19]	Pb(II)[18]	V(III)[26]
Ni(II)	La(III)[4]	V(IV)[12]	Zn(II)	Mg(II)	Mn(II)	Ni(II)[16]	Sb(III)[28]	W(VI)[23]
Pb(II)[6]	Mn(II)[2]	Zn(II)		Mn(II)	Ni(II)[16]	Pb(II)	V(IV)[12]	
Pt(II)	Ni(II)			Ni(II)	Se(II)	V(IV)		
Tb(III)[4]	Tb(III)[4]			Zn(II)	V(IV)[12]	Zn		
Zn(II)	Zn(II)				Zn(II)			

1, Arakawa et al. 2001; 2, Langlais et al. 1990; 3, Theophanides and Anastassopoulou 2002; 4, Taimir-Riahi et al. 1993b; 5, Gerber and Leonard 1997; 6, Tajmir-Riahi et al. 1993a; 7, Zhitkovitch et al. 2001; 8, Hayes 1997; 9, Misra et al. 1998; 10, Snow 1994; 11, Coogan et al. 1992; 12, Leonard and Gerber 1994; 13, Saplakoglu et al. 1997; 14, Beyersmann and Hartwig 1992; 15, Rossman et al. 1992; 16, Galaris and Evangelou 2002; 17, Sarkar 1995; 18, Hartwig 1998; 19, Gerber et al. 2002; 20, Lison et al. 2001; 21, Basu et al. 2001; 22, Hartmann and Speit 1996; 23, van Goethem et al. 1997; 24, Blasiak et al. 1999; Merk et al. 2000; Hodges et al. 2001; 25, Grover et al. 2001; 26, Rojas et al. 1996; 27, Mouron et al. 2001; 28, Schaumlöffel and Gebel 1998.

sure, which suggests that alternative organ processes exist to handle lead intoxication. The results obtained by examination of hexavalent and trivalent chromium salts on human lymphocyte cultures suggest that ROS and hydrogen peroxide may be involved in the formation of DNA lesions by hexavalent chromium. The comet assay did not indicate the involvement of oxidative mechanisms in the DNA-damaging activity of trivalent chromium, and it was speculated that its binding to cellular ligands may play a role in its genotoxicity (Blaziak and Kowalik 2000). A positive effect of hexavalent chromium had also recently been reported in human gastric mucosa cells by use of the comet assay (Blaziak et al. 1999). In addition to these recent and exemplary findings, a series of other metals could be examined with respect to "comet" induction (Table 9.1), and thus been shown to induce DNA damage in living cells. Other assays for detecting DNA damage in living cells or organisms will be presented in the following sections.

9.2.2
Induction of Molecular (Point) Mutations

As mutations are the eventual outcome of a variety of interacting processes governed not only by direct damage to DNA but also by repair and apoptotic mechanisms as well as intracellular transport, not every change in DNA actually leads to a permanent mutation. Therefore, a series of mutagenicity assays have been developed to reveal an utmost realistic picture of the final genetic damage induced by mutagenic agents. A number of short-term in-vitro assays are available which provide indirect evidence of the consequences of damage to DNA.

Over several decades, bacterial mutagenicity assays have played a major role as model test systems in the preliminary assessment of genetic risks of chemicals. They are simple, fast (with automated analysis), and, therefore, rather inexpensive. For instance, the "Ames test" measures reversion from histidine-requiring (his⁻) to histidine-independence (his⁺) in *Salmonella typhimurium*, and includes the possibility to analyze the action of metabolizing enzyme systems (S9-mix). The so-called rec-assay makes use of the repair capacity of strains of *Bacillus subtilis*. Repair-competent (rec⁺) strains are expected to be less affected by the toxic action of a compound than are repair- incompetent (rec⁻) strains. Similar tests use also other bacteria (e.g., the pol-test in *Escherichia coli*; and the umu-test; Yamamoto et al. 2002).

Another assay which indirectly measures DNA damage (or halting of the replication fork) is the induction of prophage in the Microscreen assay; this eventually measures the ability of an agent to induce the SOS system in *E. coli*. Reversion assays depend on mutations of a specific type at specific sites in the DNA, and require different tester strains for each type of reversion.

Although bacterial mutagenicity systems are notoriously insensitive in detecting a mutagenic action of metal compounds, a number of metal compounds yielded positive results in those assays (Table 9.2) if appropriate conditions had been chosen (Pagano and Zeiger 1992). Negative results in the Ames system of alloys used for dental restoration were of practical interest (Wang and Li 1998), but may also be due to the insensitivity of the assay versus titanium compounds. The reasons for that insensitivity are manifold, as pointed out by Gebhart and Rossman (1991). The possible problems included passing the bacterial cell wall, increased toxicity based on a higher sensitivity of bacterial enzymes which masks mutagenicity, a lack of phagocytosis, a mutation spectrum that was not inducible

Tab. 9.2: Metals (metal compounds) inducing point mutations in various short-term tests. (Data from Gebhart and Rossman 1991 in italics, extended by new references.)

Microorganisms			Plants and insects	Mammalian assays		Studies on human cells	
Bacteria	Prophage Induction	other		In-vitro	In-vivo	In-vitro	In-vivo
As(III)[6]	Cr(III,VI)	Al[12]	As[6]	As(III)[6]	As[6]	As[6]	Hg(II)[16]
As(V)	Fe(II,III)	As[6]	Be[13]	Be(II)[13] Cd(II)[17]	Cr[15]	Cr(VI)[22]	
Be(II)[1,13]	Mo(VI)	Be(II)[13]	Co(II)[4,14]	Co[18]	Hg(II)[16]	Hg(II)[16]	
Cd(II)[2,3]	Mn(II)[7]	Co(II)[14]	Hg(II)[16]	Cr(VI)[15]	Tl[23]		
Co(I,II)[2,4]	Ni(II)	Cr[15]	Mn(II)[7]	Hg[16,19]			
Cr(III)[3,5,10]	Pb(II)	Hg(II)[16]	Mo[26]	Mn(II,VII)[7]			
Cu[3,5]	Pt(II)	Mn(II)[7]	Ni[26]	Mo(VI)			
Fe[2]	Se(IV)	V[11]	Zn[26]	Ni(II)[20,21]			
Ga[1]	Sn(II)			Pb(II)[17,19]			
Hg(II)[3,16]	W(VI)			Pt(II,IV)[8,17]			
Ir[5]	Zn(II)			Rh[17]			
Li[24]				Ru[25]			
Mg[5]				V(V)[11]			
Mn(II)[2,7]				W(VI)			
Mo(VI)				Zn(II)			
Ni[3,20,21]							
Os(VIII)							
Pt(II)[8,10]							
Pu[9]							
Rb(I,II)							
Rh[5,10]							
Se(IV)							
Sb(III,V)[1]							
Sn[10]							
Ta[9]							
Te(IV)							
Tl(I)[23]							
U[9]							
V(V)[5,11]							
Zn(II)[2,3]							

1, Kuroda et al. 1991; 2, Pagano and Zeiger 1992; 3, Codina et al. 1995; 4, Ogawa et al. 1994; 5, Yamamoto et al. 2002; 6, Basu et al. 2001; 7, Gerber et al. 2002; 8, Gebel et al. 1997; Uno and Morita 1993; 9, Miller et al. 1998; 10, Lantzsch and Gebel 1997; 11, Leonard and Gerber 1994; 12, Octive et al. 1991; 13, Leonard and Lauwerys 1987; 14, Beyersmann and Hartwig 1992; 15, Cheng et al. 1998; Itoh and Shimada 1997; 16, De Flora et al. 1994; 17, Kanematsu et al. 1990; 18, Kitahara et al. 1996; 19, Ariza and Williams 1996; Schurz et al. 2000; 20, Coogan et al. 1989; 21, Denkhaus and Salnikow 2002; 22, Chen and Thilly 1994; 23, Leonard and Gerber 1997; 24, Leonard et al. 1995; 25, Barca et al. 1999

in bacteria, and highly efficient repair systems.

The mutagenicity of metal compounds in mammalian cells has formerly been assessed using mainly two systems:

- The forward mutation at the thymidine kinase (*tk*) locus of a mouse lymphoma cell line is measured by scoring resistance to trifluorothymidine (see Gebhart and Rossman 1991 for references).

• Chinese hamster cells are used for measuring forward mutations at the hypoxanthine-guanine-phosphoribosyl-transferase (*hprt*) locus by scoring resistance to 6-thioguanine. This test allowed a characterization of nickel-induced mutations in the gene designed CHO variant cellular system AS52 (Rossetto et al. 1994). Both of these mutation assays have now also been established in human cells (Table 9.2). As forward mutation of a non-essential gene is measured by this technique, base pair substitutions, frame shift mutations, deletions, and inactivating rearrangements should all be detectable. The caution which must govern judging data obtained with these test systems has been described in great detail (Gebhart and Rossman 1991). These authors also pointed to overcoming some of the weak points of these assays by using mammalian cells in which bacterial genes have been stably integrated. Nevertheless, a series of metals was found to induce mutations in these demanding classical mammalian cell assays, and the obtained data are in good agreement with the DNA-damaging activity of these metals (Table 9.2 versus 9.1). By contrast, when using chronic exposure some metal compounds did not induce sufficient mutagenic damage (i.e., more than two-fold increase) at high survival levels, including $NaAsO_2$, $BaCl_2$, $CuCl_2$, $HgCl_2$, $MgCl_2$, or NaCl (Gebhart and Rossman 1991). These authors have also emphasized that phagocytosis of insoluble precipitated compounds of Pb(II), Ba(II), Be(II), Ni(II), and Mn(II) may be an important route of entry into mammalian cells. Therefore, to highlight investigations exclusively on soluble metal compounds may be misleading. As an example, the spectrum of mutations induced by metals in these genes have been analyzed for chromium and nickel compounds, and hotspots of mutational changes could be detected within their DNA sequence (Chen and Thilly 1994, Rossetto et al. 1994). Examinations of the induction of point mutations by metals under in-vivo conditions have remained scarce to date (Table 9.2). This may, in part, be due to the toxic action of metal compounds on the test animals, and also on the low sensitivity of those systems towards mutagenic metals, as shown recently for dimethylarsinic acid in the Muta™ mouse (Noda et al. 2002). Only one report on the induction of point mutations in a human population exposed to Hg has been reviewed (De Flora et al. 1994), while another one using three different gene assays (HPRT, TG, glycophorin A) could not detect any significant increase of point mutations related to living near a uranium processing site (Wones et al. 1995). At present, cancer-related genes, such as p53, are becoming the focus of interest (Morris 2002), and the action of metals on these genes will be reported in Section 9.3.

9.2.3
Induction of Chromosome and Genome Mutations

The large body of data from clinical (Schinzel 2001) and oncologic cytogenetic (Heim and Mitelman 1995) investigations clearly illustrates the grave pathologic consequences of structural and numerical chromosomal aberrations. The demonstration of chromosome damaging (clastogenic) activity and the induction of genome mutations has, therefore, become highly significant in modern mutagenicity testing. Although the use of mammalian in-vivo test systems guarantees data of a high relevance for humans, in-vitro tests on cell cultures and other organisms as models allow more extensive experimental investigations to be made. Examinations of metal clastogenicity have, therefore, been performed as in-vitro and in-vivo tests on plants, insects, as well as in a variety of laboratory animals. Beside chromosomal aberrations, the indicators of DNA damage, sister chromatid exchanges (a probable consequence of intrachromosomal rearrangements) and micronuclei (a secondary consequence of clastogenic action) have been the target anomalies of those tests. Although having been considered for a long time, genome mutations (i.e., induced changes of chromosome number) can now be more accurately determined using immunofluorescent kinetochore staining or fluorescence in-situ hybridization (FISH) of the micronuclei of exposed cells. Thus, detection of aneugenic, in addition to clastogenic, activity also of metals and metal compounds (e.g., As, Cd, Co, Cr, Hg, Mn, Ni, Se, Tl, V) has been rendered possible (Table 9.3). The disruption of microtubule assembly and spindle formation has been identified as a mechanism of

Tab. 9.3: Metals (metal compounds) inducing chromosome damage in various organisms

| Non-mammalian test systems | Mammalian assays | | Human cells | | Genome mutations | | |
| | In vitro | In vivo | In vitro | In vivo | Plants: s: | Animals/huma | |
						In vitro	In vivo
Al	$As^{1,6}$	Al^{16}	Al^{23}	$As^{1,6,28}$	As	As^6	Cr^{37}
$As^{1,6}$	Be^7	$As^{1,6,16}$	$As^{1,6,25,28}$	$Cd^{1,29}$	Au	Cd^8	Hg^{38}
Ba	$Cd^{1,8,9}$	$Cd^{16,17}$	$Cd^{1,8,23}$	Co^{30}	Ba	$Co(II)^2$	Mn^{13}
Be	Co^2	Co^{10}	Co^{10}	$Cr^{1,30,31,37}$	Be	Cr	Mo^{18}
Cd^1	$Cr^{1,8,25}$	Cr^1	$Cr^{1,8,25,28}$	Cu^{26}	Cd^8	Hg^1	Se^{21}
Co^2	(Fe)	Fe^1	$Hg^{1,17,23}$	$Hg^{1,11,27,38}$	Co	Mn^{13}	Tl^3
Cr^1	$Hg(II)^{11,12}$	$Hg(II)^{11}$	Mn^{13}	Li	Cr^8	Mo^{18}	
Cu^{39}	Mn^{13}	Li	Mo^{18}	$Ni^{1,30}$	Cu	$Ni^{4,8}$	
Fe^1	$Ni^{1,8,14}$	Mn^{13}	$Ni^{1,8,14}$	$Pb^{1,31}$	Hg^1	$V(V)^5$	
$Hg(II)^{3,11}$	$Pb^{1,9,25}$	Mo^{18}	$Pb^{1,25}$	$Pt(II)^1$	$Ni^{4,8}$		
Mn^{13}	Pt^1	Nd^{19}	Pd^{39}	Tl^{33}	Pa		
Nd	Ru^{15}	$Ni^{1,4}$	$Pt^{1,36}$	$U^{33,35}$	Rb		
Ni^4	Sb^7	Pb^1	Rh^{40}	$Zn^{8,38}$	$Tl^{1,3}$		
Pb^1	Se	Pr^{19}	Sb^{23}		V^5		
Pt	Te	$Pt^{1,32}$	Se				
Th	Ti^{41}	Rh^{40}	Te^{23}				
Tl^1	Tl^1	Sb^{20}	V^5				
V^5	U^{15}	Se^{21}	Zn^1				
Zn^1	$V(III-V)^5$	Tl^3					
	Zn^1	V^{22}					
		Zn^1					

1, Gebhart 1989; 2, Beyersmann and Hartwig 1992; 3, Leonard and Gerber 1997; 4, Coogan et al. 1989; 5, Leonard and Gerber 1994; 6, Basu et al. 2001; 7, Kuroda et al. 1991; 8, Seoane and Dulout 2001; 9, Lin et al. 1994; 10, Lison et al. 2001; 11, De Flora et al. 1994; 12, Akiyama et al. 2001; 13, Gerber et al. 2002; 14, Denkhaus and Salnikow 2002; 15, Lin et al. 1993; 16, Sivikova and Dianovsky 1995; 17, Volkova et al. 1995; 18, Titenko-Holland 1998; 19, Jha and Singh 1995; 20, Gurnani et al. 1992; 21, Biswas et al. 1999; 22, Ciranni et al. 1995; 23, Migliore et al. 1999; 24, Oya Ohta et al. 1996; 25, Wise et al. 1992; 26, Shubber et al. 1998; 27, Schoeny 1996; 28, Gonsebatt et al. 1997; 29, Verougstraete et al. 2002; 30, Gennart et al. 1993; 31, Wu et al. 2000; 32, Adler and El-Tarras 1989; 33, Nikiforov et al. 1999; 34, Sram et al. 1993; 35, McDiarmid et al. 2001; 36, Gebel et al. 1997; 37, Benova et al. 2002; 38, Amorim et al. 1999; 39, Bhunya and Jena 1996; 40, Sadiq et al. 2000; 41, Lu et al. 1998.

aneugenicity induction by As and V (Ramirez et al. 1997).

Previous (in italic print) and more recent studies on the clastogenic action of metals and their compounds in humans are comparatively collected in Table 9.3. A great number of metal compounds have been shown to induce chromosome damage in various test systems, and with differing efficacies. Chromium compounds unfolded clastogenic and aneugenic actions throughout all test categories listed in Table 9.3, while As, Cd, Hg and Ni were present in seven categories, and Co, Mn, and V were in six. However, caution must be taken when weighing-up those data; for example, Leonard and Lauwerys (1990) concluded in their review that "cobalt and its salts appear to be devoid of mutagenic and clastogenic activity in mammalian cells". In considering alloys under practical aspects, recent examinations on extracts of dental amalgam alloys detected the presence of some chromosome damage in SHE cells (Akiyama et al. 2001). These data contrast with a negative outcome of cytogenetic examinations of a nickel-titanium alloy (Wever et al. 1997).

In addition to the positive results collected in Table 9.3, experiments with a series of metals also yielded negative data which are not listed in the table. Li, K, Na, Ca, Al, Cu, Fe, Sb, and Zn were those found to be non-clastogenic in several (most) experimental assays. However, some of those metals presented in Table 9.3 also occasionally yielded negative results, this apparently being dependent upon the type of compound tested, as well as the experimental conditions.

Combinations of certain test assays may provide a clearer view of the mutagenicity of metals if based on reliable comparative pilot studies. A comparison of the activity of metals in the micronucleus test and in

the comet assay was attempted by examining cobalt powder, tungsten carbide and cobalt-tungsten carbide (van Goethem et al. 1997). Both, the comet assay and the micronucleus test were able to detect differences in the genotoxic potential of the compounds studied, but the micronucleus test seemed less sensitive in assessing a synergistic DNA damaging potential of the cobalt-tungsten carbide mixture involved. However, this indirect clastogenicity test system proved to be well suited for screening natural environments (soils, river water, etc.) for metal mutagenicity by using a variety of bioassays, including brown trout (Sanchez-Galan et al. 1999), mussels (Bolognesi et al. 1999), and plants (Minissi and Lombi 1997, Knasmüller et al. 1998).

Sodium arsenite (NaAsO2) and cadmium sulfate (CdSO4) were tested for their ability to induce genotoxic effects in the single cell gel (SCG) assay and the sister chromatid exchange (SCE) test in human blood cultures in vitro (Hartmann and Speit 1994). Treatment of cells for 2 h or 24 h beginning 48 h after the start of the blood cultures did not increase the SCE frequency in the case of cadmium, but did cause a small but significant SCE induction with arsenic at the highest concentration. The metal concentrations which could be investigated in the SCG test were much lower due to a strong toxic effect. Metal concentrations which were toxic in the SCG test were without visible effect in the SCE assay. Thus, the two endpoints for determining genotoxic effects in vitro differed markedly with respect to the detection of genotoxicity induced by metals.

Most of the data presented on chromosomal alterations or secondary aberrations induced by metals and their compounds have been obtained from short-term tests. However, Coen et al. (2001) showed that heavy metals of relevance to human health (e.g., Cd and Ni) may induce a long-term genomic instability which cannot be predicted from acute clastogenicity data.

Cytogenetic examinations of human individuals or groups of individuals exposed to metals and their compounds ("cytogenetic population monitoring") are the most well-documented attempts with respect to practical aspects of a harmful action on human genetic material. The tests revealed a significant increase in chromosome damage in some cases, but in the large majority of cases the increase was not dramatic. As reported previously (see Gebhart and Rossman 1991 for references), combined action of more than one metal may be a feasible means of exposure in heavy metal industries, as well as in individuals involved on metal arc welding (e.g., Jelmert et al. 1994). For several metals, contradictory results have been reported after various examinations, though chromium appeared to be the most consistent human clastogen in those studies. However, as previously pointed out by Leonard and Bernard (1993) for several metals, and recently documented for cadmium by Verougstraete et al. (2002), most of those studies did not fulfill all criteria in a very convincing manner. Therefore in this category, in addition to previous data (Gebhart and Rossman 1991: italic text in Table 9.3), many recent papers have been cited. A very important factor in judging cytogenetic data from mutagen-exposed populations is also the rather high interindividual variation of induced aberration frequency. This phenomenon of heterogeneity reflects the individuality of the human genome, and this must be taken into consideration when judging mutagenic action in humans. Another confounding factor is the evident slightly mutagenic and co-mutagenic action of heavy smoking which could influence the outcome of cytogenetic studies on individuals exposed to metals.

9.2.4
Modulating Effects

As many metal compounds exert a rather weak mutagenic effect, their carcinogenic action cannot be explained by this effect alone. However, in several cases, indirect genotoxic effects – namely synergism with known mutagens – may be one reason for their tumorigenicity.

In contrast, under certain conditions several metal compounds can also display an antagonistic effect if acting together with other mutagens.

Synergistic or antagonistic effects of metals or their compounds versus the action of other mutagens or in interaction with other metals can occur at a variety of levels (see Gebhart and Rossman 1991 for references):

1. Reaction of the metal compound with the genotoxicant or its metabolites, resulting in a stronger or weaker mutagenic action. Co(II), for instance, was shown to form mutagenic complexes with other compounds, Cu(II), Fe(II), and Mn(II), in the presence of ascorbate, generate ROS which are genotoxic. It has long been known that H_2O_2 causes single strand breaks in cellular DNA, but this effect seems to be dependent on a metal-catalyzed Fenton reaction in which a hydroxyl radical is formed. The co-mutagenic effect of Cu(II) with UV light might also be explainable on this basis.

2. Metal compounds can affect the metabolic activation of other compounds, resulting in alterations in the amount or spectrum of metabolites. This effect can arise at the level of the mixed-function oxidases, epoxide hydratase, or formation of sulfate esters or glutathione conjugates. It is thought that some of the antimutagenic/anticlastogenic effects of selenium compounds might be due to effects on metabolic activation.

3. It is possible that some metal compounds also affect the binding of chemicals or other metals to DNA (e.g., Mg; Anastassopoulou and Theophanides 2002).

4. It is also theoretically possible that some metal compounds could affect the integration of viral DNA into the chromosome. This may be an explanation for the formerly shown enhancement of SA7 virus transformation by many metal compounds (e.g., inorganic salts of Ag, Be, Cd, Cr, Cu, Fe, Hg, Mn, Ni, Pb, Pt, Sb, Tl, W, Zn; Casto et al. 1979). This, however, is an indirect rather than a direct co-mutagenic action in its strict sense.

5. A number of metal compounds have been shown to inhibit DNA repair (see Table 9.1), usually by indirect means (Beyersmann 1994). If a metal compound is co-mutagenic with another agent in a repair-proficient cell type, but not in a repair-deficient one, this metal is probably inhibiting DNA repair. Examples are arsenic with UV light, or Ni(II) with methyl methane sulfonate. Weakly mutagenic metals such as cadmium or lead exert a significant co-mutagenic action via repair inhibition (Hartwig 1994).

6. Finally, by affecting the process of DNA replication, metal compounds can alter the genotoxicity of other mutagens. As shown in Table 9.1, several metals induce infidelity of DNA replication, and therefore, although not proven in all details, may act as strong co-mutagens.

A large amount of data has been obtained from examinations of the modifying effect of metals and metal compounds on the mutagenic action of physical and chemical agents (Table 9.4). Most of these findings are certainly attributable to one or more of the mechanisms presented above. Arsenic which per se is toxic rather than mutagenic,

nevertheless has been proven to act antagonistically versus a series of mutagens on different levels of mutation types. Depending on the specific experimental situation (as shown in Table 9.4), a few metals or metal compounds can display a co-mutagenic, but also an anti-mutagenic action (e.g., Co, Ni, Se, Zn). Some of the co-mutagenic activities of certain metals (e.g., Cu, Fe) which have explicitly been attributed to their oxygen radical formation were not included in this table. One long-known anti-mutagen is selenite, but more recently other metal compounds (notably Ge) have also been shown to exert an antagonistic action against a series of mutagens.

A practical aspect of metal-dependent modification of the mutagenic action of X-rays has been examined using comet assay studies in a group of people exposed to lead (Groot de Restrepo et al. 2000). Although the observed effects did not show any significant difference among different lead blood levels, greater effort should be targeted towards this potential of metal exposure.

Finally, the modifying potency of metals and their compounds must be borne in mind when the biological activity of these agents is under consideration. As with other mutagenic agents, the genotoxic activity of metals can be either increased or decreased by co- and anti-mutagens. However, as this is not a specific feature of metals, this aspect is not examined at this point.

9.3
Carcinogenicity

Ongoing developments in the fields of genetics and oncology clearly document an extremely close connection of mutagenic events with the processes involved in malig-

Tab. 9.4: Metals (metal compounds) modifying mutagenic/clastogenic action. (Data taken from Gebhart and Rossman (1991) are printed in italics.)

Comutagenic action

Metal	Mutagens	Test systems (mutation type)
As[26]	UV	E.c.(PM), CHV79 (PM)[1], CHO (CB)
	MMS, MNU, crosslinkers	CHV79 (PM), CHO (CB)
	MNNG[1], DEB[2]	CHV79 (PM)[1], HL (CB,SCE)[2]
	MMS[3], BaP[3]	HL,HF (CA)[3]
Cd	Nitrosamine // aromatics	S.t. (PM) // CH (CT)
	X-rays[4] // UV, MNNG	HL (CB)[4] // CHV79 (SSD,SCE)[11]
Co	Heteroaromatics // UV[5]	S.t. (PM) // CHV79 (PM,SCE)[5]
Cr	Na-azide, 9-aminoacridine	S.t. (PM)
	Aromatics	CH (CT)
Cu	UV // ascorbate, INH	E.c (PM) // CHO (CB)
	Doxorubicin[6] // AgI[7]	S.t.(PM)[6] // plants (PM,CB)[7]
Mn	UV // ascorbate, INH	E.c.(PM) // CHO (CB)
Mo	UV	E.c.(PM)
Ni	MMS, EMS // aromatics	E.c.,S.t.(PM) // CH (CT)
	UV	CHO (SSD,SCE)[8]
Pb	Pt[9] // UV, MNNG[10,11]	CH(SSD,SCE)[9,10], CH(SSD,PM)[11]
Se	MNU, MNNG	S.t.(PM)[12], CH (CB)[12], MBM (MN)[12]
Ti	UV	S.t.(PM)[13], MLY(CA)[13], CHL(CB)[13]
V	Caffeine	HL (CB)[14]

Antimutagenic action

Metal	Mutagens	Test systems (mutation type)
Co	Trp-P2 // UV[5], MNNG[5]	S.t.(PM)[5]
Cr	As(III) // X-rays[5,21]	E.c.(PM) // CHV79(PM)[21]
	BaP	HF (PM)[16]
Ge	Trp-P2[15], AMPI[15], EMS[17]	S.t.(PM)[15,17], CHV79 (MN)[24]
	CdCl2[15], PMA[24]	MBM(CB,MN)[15], MSPA[15]
		HL(SCE)[15,24]
La	UV, nitrogen mustard, cyclo-phosphamide	HF(UDS,CB)[18], MBM(MN)[18]
Mg	Gamma-rays, free radicals	E.c., S. T. (PM); CHO(MN; DNACL)[25]
	NiCl2[25]	CHO/M3T3 (MN,DNACL)[25]
Ni	peroxide[19], UV[23]	CHV79(MN)[19], HL(MN,SCE)[23]
Sb	As	HL (CB,SCE)
Se	As[22], X-rays[22], BaP	CHO(PM)[22]
Zn	MMS, AAF, CdCl2	RHC(SSD)[20]

Abbreviations: UV, ultraviolet irradiation; MMS, methyl methane sulfonate; MNU, methyl nitroso urea; MNNG, N-methyl-N-nitro-N-nitrosoguanidine; INH, isoniazide; EMS ethyl methane sulfonate; BaP Benzo[a]pyrene; Trp-P2, tryptophan pyrolysate P2; AMPI, aminomethyl pyrido indole; PMA, phenyl mercury acetate E.c., *Escherichia coli*; S.t., *Salmonella typhimurium*; CH, Chinese hamster cell lines; HL, human lymphocyte cultures; HF, human fibroblast cultures; MBM mouse bone marrow; RBM rat bone marrow; MLY, mouse lymphoma cell line; RHC, rat hepatocyte cell line. PM, Point mutations; CB, chromosomal breakage; SCE, sister chromatid exchanges; CA, DNA damage in comet assay; CT, cell transformation; MN micronuclei; SSD DNA single strand damage; DNACL, DNA cross links. 1, Rossman et al. 2001; 2, Wiencke and Yager 1992; 3, Hartmann and Speit 1996; 4, Oberheitmann et al. 1999; 5, Beyersmann 1994; 6, Yourtee et al. 1992; 7, Reutova 2001; 8, Lee-Chen et al. 1993; 9, Krueger et al. 1999; 10, Roy and Rossman 1992; 11, Hartwig 1994; 12, Balansky 1991; 13, Nakagawa et al. 1997; 14, Roldan-Reyes et al. 1997; 15, Gerber and Leonard 1997; 16, Tesfai et al. 1998; 17, Schimmer et al. 1997; 18, Zhang and Zhang 1997; 19, Gebel 1998; 20, Coogan et al. 1992; 21, Yokoiyama et al. 1990; 22, Diamond et al. 1996; 23, Katsifis et al. 1998; 24, Lee et al. 1998; 25, Hong et al. 1997; 26, Basu et al. 2001.

nant transformation, tumor progression, and metastatic mechanisms. This relationship between carcinogenicity and mutagenicity is just reflected on the basis of molecular reactions of metals with nucleic acids or proteins. Direct reaction with DNA, infidelity of DNA synthesis, cross-linking, but also generation of radicals and interaction with repair processes are equally valid for the induction of mutations and, in consequence, of malignancy. The pertinent data have just been presented in Section 9.2 (see Table 9.1) and, therefore, are not repeated here.

9.3.1
Genetic and Cellular Mechanisms

Aside those basic mechanisms, the substitution of zinc by lead in several proteins that function as transcriptional regulators has been discussed as a reason for a reduced binding of these proteins to recognition elements in DNA; this suggests an epigenetic involvement of lead in altered gene expression (Silbergeld et al. 2000). These events were suggested by the authors to be of particular relevance in transplacental exposures, and later in cancer. Beyond those direct reactions, influences of metals on variety of molecular and cellular regulation processes can contribute to their carcinogenic action, as there are apoptosis and growth regulation, regulation of transcription, signal transduction, and gene expression (Wang and Shi 2001, Chen and Shi 2002), but also gene silencing by DNA hypermethylation (Costa et al. 2001).

Based on the recent progress of understanding the genetic events of carcinogenesis, of cancer invasion, progression, and metastasis, more recently leading genes of these processes (the so-called proto-oncogenes and tumor suppressor genes) have attracted much attention as targets of muta-

genic/carcinogenic attack. Transcriptional activation of proto-oncogenes (c-fos, c-jun, c-myc) was induced in BALB/c-3T3 and nude mouse tumorigenesis models by cadmium (Joseph et al. 2001), and even amplification of the oncogenes K-ras and c-jun, in addition to genomic instability, was attained in similar experiments using beryllium sulfate (Keshava et al. 2001). Examination of the effects of As, Cd, Cr, and Pb on the gene expression regulated by a battery of 13 different promoters in recombinant HepG2 cells revealed, among others, also induction of fos and the tumor suppressor p53 response element (Tully et al. 2000). From their observation on an alteration of cytosine methylation patterns of the promoter of the tumor suppressor gene p53 in human lung cells, Mass and Wang (1997) deduced a model for a mechanism of carcinogenesis of arsenic. The p53 gene is a central control gene (Morris 2002) which also could be induced by Co in a Syrian hamster embryo (SHE) cell system (Duerksen-Hughes et al. 1999). The direct evidence of metal action on basic genes involved in the process of cell growth, malignancy, and apoptosis (Wang and Shi 2001) thus substantially supports our understanding of the carcinogenic activity of some metals.

In-vitro cell transformation systems, formerly used as indirect mutagenicity tests, examine the ability of an agent to convert nontumorigenic into tumorigenic cells by analyzing their growth pattern (e.g., forming colonies in soft agar) which correlate with tumorigenicity. Most of the studies on metal-induced cell transformations have been carried out using the SHE cell culture system. "Based on 24 metal compounds which have been tested in the SHE and some rodent bioassay, the SHE assay is 92% concordant with rodent bioassay carcinogenicity results, including a sensitivity of 95%." (Kerckaert et al. 1996). All metals known to exert carcinogenic activity in humans have also been able to induce cell transformation in this assay (Table 9.5). Recently, in-vitro transformation of SHE cells into a neoplastic state was shown to be associated with overexpression of c-myc and c-Ha-ras oncogenes (Takahashi et al. 2002). The Balb3T3 mouse cell line which has also been a classical model for cell transformation has recently been used for the assessment of cytotoxicity of a large series of metal compounds in order to predict their carcinogenic potential (Mazzotti et al. 2001, 2002).

9.3.2
Animal Models and Epidemiology

Based on a series of experimental studies on classical animal models (Table 9.5), the IARC Working Group on the Evaluation of the Carcinogenic Risk of Chemicals to

Humans concluded that there was sufficient evidence for the carcinogenicity of soluble calcium chromate and several relatively insoluble hexavalent chromium compounds in laboratory animals. Tumors were mainly induced at the administration site. In addition, experimental exposure to Be, Cd, Ni, and Sb has caused lung tumors in rats, while various beryllium compounds produced osteosarcomas in rabbits by implantation or injection (Hayes 1997). Rossman et al. (2001) could show a co-carcinogenic action of arsenic with solar UV radiation on mouse skin. Apparently strain as well as species differences of the susceptibility to the action of metals may cause variable outcome of carcinogenicity tests; for example, in mice this is caused by higher metallothionein levels (Oberdörster et al. 1994, Waalkes and Rehm 1994).

From an epidemiologic view, the potential carcinogenic consequences due to occupa-

Tab. 9.5: Summarized data on metals (metal compounds) with carcinogenic action

Induction of mutations in cancer-related or reporter genes	In-vitro cell transformation	Carcinogenic action on classical animal models	Epidemiologic data on humans
As(III,V)[1,7]	As(III,V)[7,8]	As[8]	As[7,13,17,.19]
Be(II)[2]	Ba[8]	Be[8,13]	Be[13]
Cd(II)[1,3]	Be[2]	Cd[3,8,13,17]	Cd[13,17,19]
Co[4]	Cd[3,8]	Co[8,9,13]	Co[9,10]
Cr(VI)[1,5]	Co[9]	Cr[8,14,19]	Cr[5,13,14,20]
Ni(II)[6]	Cr[8]	Hg[15]	Hg[15]
Pb(II)[1]	Ga[8]	Mo[8,16]	Ni[10,13,20]
Ti(III)	Mo[8]	Ni[8,10,13,21]	Pb[13]
	Ni[8,10]	Pb[8,13]	Sb[17]
	Pb[8]	Pt[8]	
	Pt(II,IV)[8,11]	Sb[13,18]	
	Ti[8]	Zn[8]	
	U[12]		
	V[8]		
	Zn[8]		

1, Tully et al. 2000; 2, Keshava et al. 2001; 3, Joseph et al. 2001; 4, Duerksen-Hughes et al. 1999; 5, Hirose et al. 2002; 6, Zienolddiny et al. 2000; Costa et al. 2001; 7, Bode and Dong 2002; 8, Kerckaert et al. 1996; 9, Lison et al. 2001; 10, Denkhaus and Salnikow 2002; 11, Chibber and Ord 1989; 12, Miller et al. 1998; 13, Hayes 1997; 14, Cohen et al. 1993; 15, Boffetta et al. 1998; 16, Chan et al. 1998; 17, Waalkes 2000; 18, Leonard and Gerber 1996; 19, Kazantzis et al. 1992; 20, Sunderman 2001; 21, Oller et al. 1997.

tional exposure to metals and/or metal compounds merit greatest attention. Therefore, classical studies on cancer prevalence in certain populations, although not aiming at the genetic basis of these diseases, are of continuing importance. However, biomonitoring exposure to those carcinogenic metal compounds must be critically considered (Leonard and Bernard 1993). Chromium and its compounds have been the subject of a large number of studies on more than 20 000 exposed humans in various industries. In particular, cancers of the respiratory tract (including the nasal cavity and the lung) dominated as the found sites of cancer formation. The exposed groups who were examined comprised those working in chromate-producing industries, in chromate pigment production, in chromate-plating, and in ferrochromium industries (Hayes 1997). Incidental and environmental chromium exposures were also shown to be sources of cancer induction. The IARC Working Group mentioned above, concluded that "there is sufficient evidence of respiratory carcinogenicity in workers occupationally exposed during chromate production, chromate pigment production and chromium plating". Recent examinations of lung cancer samples from chromate-exposed workers detected frequent microsatellite instability (Hirose et al. 2002). In addition, cancers at other sites have also been attributed to hexavalent chromium, including bone, prostate, stomach, genital, renal, and bladder cancer, as well as lymphomas and leukemias (Costa 1997). However, all experimental and epidemiologic data, and the underlying mechanisms of chromium uptake, metabolism and site-specific action, point to the occurrence of thresholds in Cr(VI) carcinogenesis varying between different body compartments (De Flora et al. 1997, De Flora 2000).

The individuality of different nickel species with respect to tumor induction or promotion must be recognized before reaching regulatory decisions (Oller et al. 1997). Sparingly-soluble nickel compounds – and possibly also the soluble compounds – are carcinogens linked to lung and nasal cancers in humans, while evidence on a carcinogenic action in humans of lead is still inconclusive (Hayes 1997). However, new data on the cancer risk of workers exposed to lead (Landrigan et al. 2000) would justify a re-evaluation of the available evidence. Several epidemiological studies suggest an increase in the incidence of respiratory cancers among persons occupationally exposed to arsenic (more than 25 000 exposed; estimated risks 1.4 to 11.9; Hayes 1997), beryllium (more than 10 000 exposed; estimated risk 1.3 to 2.3; Hayes 1997), and also to cadmium (about 15 000 exposed, estimated risks 1.3 to 3.7; Hayes 1997).

The carcinogenic activity of Be, whether administered in the form of the metal, alloys, or other organometallic compounds, has been confirmed in a number of experiments on laboratory animals (Leonard and Bernard 1993), as has been the case of cadmium (see Table 9.5). Epidemiological studies have not provided clear evidence of a carcinogenic hazard of Al (Leonard and Gerber 1988), and many areas of uncertainty also remain on the carcinogenic action of Co (Lison et al. 2001), although experimental data point to such a potential.

In summarizing, the risk estimation from all these data points to a potential carcinogenic action in humans of Cr and Ni, as well as As, Be, and Cd, if the exposure to these metals is sufficiently intense. An estimation of cancerogenic risk from metals for the general population, however, is very difficult. As shown by Merzenich et al. (2001) in a cross-sectional study, a positive association between Ni and the rate of oxidative

DNA lesions could be observed, which – from the authors' viewpoint – "... provides further evidence for the genotoxic effect of nickel in the general population". However, the situation is further complicated by the multiple, and thus possibly interacting, exposures in the actual industrial environment.

9.4
Teratogenicity

The induction of damage by environmental agents in embryonic/fetal development, causing malformations or embryonic death, is a problem of toxicity rather than of mutagenicity of the inducing factor. Teratologic effects, therefore, are not heritable, as in most cases the genetic material remains unaffected by the teratogen. In general, the mechanisms of teratogenicity are different from those of mutagenicity.

The teratogenic action of metals has been tested comprehensively on a large variety of animals (Table 9.6; Schardein 1993). The embryotoxic and teratogenic effects of a number of nickel compounds in mouse, chicken, hamster, and rat has been examined using different routes of exposure (Coogan et al. 1989): "Maternal exposure resulted in a decrease in implantation frequency, increased early and late resorptions, and an increased frequency of stillborn fetuses. In addition, nickel exposure during organogenesis has resulted in a variety of teratogenic effects".

As with the mutagenic and carcinogenic action of metals, the teratogenic effects can be modified by additional influences. For example, the teratogenicity of cadmium in mice was decreased by bismuth-induced metallothionein (Naruse and Hayashi 1989) and caffeine (Lutz and Beck 2000). It is also a well-known fact that species differ-

ences can play a major role (Schardein 1993), as recently confirmed by studies on the developmental toxicity of indium in rats and mice (Nakajima et al. 2000).

Several metals did not produce malformations when tested for their teratogenicity in experimental systems. Cobaltous acetate and cobaltous chloride have not been found to be teratogenic in hamsters and rats respectively (Leonard and Lauwerys 1990). Zinc, in addition, can protect the embryo from damage caused by other teratogenic agents, though zinc deficiency in the mother may be harmful to the embryo (Jankowski et al. 1995). In an experimental rat embryo culture system, abnormal development could also be induced under copper-deficient conditions which were associated with an impaired oxidant defense system (Hawk et al. 1998). Information on the teratogenic effects of vanadium have mainly been obtained from various animal test systems, "but vanadium appears to be only slightly teratogenic, if at all" (Leonard and Gerber 1994, Domingo 1996). Even less is known about beryllium teratogenicity.

Of particular interest are the data from human exposure. Organic mercury compounds were shown to act teratogenically in the human (Schardein 1993), and several reports also pointed to a possible embryotoxic, but also teratogenic, activity of lead. There have, however, been no substantiated reports to indicate that heavy metals other than lead and organic mercury compounds have any causal relationship to the induction of birth defects in humans. The teratogenic potency of some of these metals or their compounds in experimental systems, however, should prompt a further careful examination of the real hazard to man. The use of lithium as an antidepressant during the first trimester, for instance, may be related to an increased incidence (about 10%) of congenital defects, particularly of the cardiovascular

Tab. 9.6: Summarized data on metals (metal compounds) with embryotoxic/teratogenic Action. (In this table data summarized by Schardein (1993) are printed in *italics*; additional literature by citation number.)

Animal studies							Human studies	
Amphi-bians	Chicken	Mouse	Rat	Hamster	Rabbit	Other mammals	Embryo-toxic	Terato-genic
Al(III)[1]	As[6]	*Cd*[7]	*Al*	*Cd*[13]	*Cd*	*Hg*	*Hg*	*Hg*
Cd(II)[1,2,4]	Cd[2,6]	*Cr*[8]	*As*[11]	*Cr*	*Hg*	*Mb*	*Pb*	*Pb*
Co[1,4,5]	Co[6]	*Ga*	*Cd*	*Hg*	In[9]	*Pb*		
Cr(III)[1]	Cu[6]	*Hg*	*Co*[12]	*In*	*Pb*	*Se*		
Cu[3]	Fe[6]	In[9]	*Cr*[8,11]	*Ni*	*Sr*			
Ni[4]	In[6]	*Ni*	*Cu*[11]	*Pb*				
Zn[3,4]	Mn[6]	*Pb*	*Hg*	V[10]				
	Mo[6]	V[10]	In[9,14]	Zn[13]				
			Ni					
			Pb					
			Sr					
			Te					
			V[10]					

1, Calevro et al. 1998; 2, Thompson and Bannigan 2001; 3, Luo et al. 1993a; 4, Luo et al. 1993b; 5, Plowman et al. 1991; 6, Gilani and Alibhai 1990; 7, Mahalik et al. 1995; 8, Kanojia et al. 1996; 9, Nakajima et al. 2000; 10, Domingo 1996; 11, Mason et al. 1989; 12, Paternain et al. 1988; 13, Hartsfield et al. 1992; 14, Ungvary et al. 2000.

system. Taking these facts into account, one cannot regard this light metal as a strong teratogen, but must warn against its intake during the first trimester of pregnancy.

9.5
Concluding Remarks

The evidence obtained from a large number of examinations of the mutagenicity, carcinogenicity, and teratogenicity of metals points to few metals and/or their compounds which are actually hazardous to humans in this regard. Certain chromium compounds evidently create not only mutagenic but also carcinogenic risks for exposed individuals. Nickel, cadmium, beryllium, and arsenic have also been associated with an oncogenic risk, though this may be lower than that for Cr. A large number of other metal compounds have also been shown to exert certain mutagenic, carcinogenic and teratogenic activities in various test systems, but these require further substantiation before their real hazard for humans can be definitely defined. In particular, metals involved in modern technologies (Be, In, Si) must be involved in future examinations on their biological activities.

Of particular interest are recent developments linking the damaging activity of metals to the genetic mechanisms of processes involved in carcinogenesis and cancer progression, and these should increase our knowledge on the genetic activity of metals and their compounds.

References

ADLER ID and EL-TARRAS A (1989) *Clastogenic effects of cis-diamminechloroplatinum. I) Induction of* chromosomal aberrations in somatic germinal cells of mice. Mutat Res 211:131–137.

AKIYAMA M, OSHIMA H and NAKAMURA M (2001) *Genotoxicity of mercury used in chromosome aberration test.* Toxicol In Vitro 25:463–467.

AMORIM MIM, MERGLER D, BAHIA MO, DUBEAU H, MIRANDA D, LEBEL J, BURBANO RR and LUCOTTE M (1999) *Cytogenetic damage related to low levels of methyl mercury contamination in the Brazilian Amazon.* An Acad Bras Ci 72:497–507.

ANASTASSOPOULOU J and THEOPHANIDES T (2002) *Magnesium-DNA interactions and the possible relation of magnesium to carcinogenesis. Irradiation and free radicals.* Crit Rev Oncol Hematol 42:79–91.

ARAKAWA H, NEAULT JF and TAJMIR-RIAHI HA (2001) *Silver(I)complexes with DNA and RNA studied by Fourier transform infrared spectroscopy and capillary electrophoresis.* Biophys J 81:1580–1587.

ARIZA ME and WILLIAMS MV (1996) *Mutagenesis of AS52 cells by low concentrations of lead (II) and mercury (II).* Environ Mol Mutagen 27:30–33.

BALANSKY RM (1991) *Comutagenic and coclastogenic effect of selenium in vitro and in vivo.* Mutat Res 263:231–236.

BARCA A, PANI B, TAMARO M and RUSSO E (1999) *Molecular interactions of ruthenium complexes in isolated mammalian nuclei and cytotoxicity on V79 cells in culture.* Mutat Res 423:171–181.

BASU A, MAHATA J, GUPTA S and GIRI AK (2001) *Genetic toxicology of a paradoxical human carcinogen, arsenic: a review.* Mutat Res 488:171–194.

BENOVA D, HADJIDEKOVA V, HRISTOVA R, NIKOLOVA T, BOULANOVA M, GEORGIEVA I, GRIGOROVA M, POPOV T, PANEV T, GEORGIEVA R, NATARAJAN AT, DARROUDI F and NILSSON R (2002) *Cytogenetic effects of hexavalent chromium in Bulgarian chromium platers.* Mutat Res 514:29–38.

BEYERSMANN D (1994) *Interactions in metal carcinogenicity.* Toxicol Lett 72:333–338.

BEYERSMANN D and HARTWIG A (1992) *The genetic toxicology of cobalt.* Toxicol Appl Pharmacol 115:137–145.

BHUNYA SP and JENA GB (1996) *Clastogenic effect of copper sulphate in chick in vivo test system.* Mutat Res 367:57–63.

BISWAS S, TALUKDER G and SHARMA A (1999) *Comparison of clastogenic effects in inorganic selenium salts in mice in vivo as related to concentrations and duration of exposure.* BioMetals 12:361–368.

BLASIAK J, TRZECIAK A, MALECKA-PANAS E, DRZEWOSKI J, IWANIENKO T, SZUMIEL I and WOJE-

WODZKA M (1999) *DNA damage and repair in human lymphocytes and gastric mucosa cells exposed to chromium and curcumin.* Teratog Carcinog Mutagen 19(1):19–31

BLASIAK J and KOWALIK J (2000) *A comparison of the in vitro genotoxicity of tri- and hexavalent chromium.* Mutat Res 469(1):135–145.

BODE AM and DONG Z (2002) *The paradox of arsenic: molecular mechanisms of cell transformation and chemotherapeutic effects.* Crit Rev Oncol Hematol 42:5–24.

BOFFETTA P, GARCIA-GOMEZ M, POMPE-KIRN V, ZARIDZE D, BOLLANDER T, BULBULYAN M, CABALLERO JD, CECCARELLI F, COLIN D, DIZDAREVIC T, ESPANOL S, KOBAL A, PETROVA N, SALLSTEN G and MERKER E (1998) *Cancer occurrence among European mercury miners.* Cancer Caus Contr 9:591–599.

BOLOGNESI C, LANDINI E, ROGGIERI P, FABBRI R and VIARENGO A (1999) *Genotoxicity biomarkers in the assessment of heavy metal effect in mussels: experimental studies.* Environ Molec Mutagen 33:287–292.

BUCHKO G, HESS NJ and KENNEDY MA (2000) *Cadmium mutagenicity and human nucleotide excision repair protein XPA: CD, EXAFS and $^1H/^{15}N.NMR$ spectroscopic studies on the zinc(II)- and cadmium(II)-associated minimal DNA-binding domain (M98-F219).* Carcinogen 21:1051–1057.

CALEVRO F, CAMPANI S, RAGGHIANTI M, BUCCI S and MANCINO G (1998) *Test of toxicity and teratogenicity in biphasic vertebrates treated with heavy metals (Cr^{3+}, Al^{3+},Cd^{2+})* Chemosphere 37:3011–3017.

CASTO BC, MEYERS J and DI PAOLO JA (1979) *Enhancement of viral transformation for evaluation of the carcinogenic and mutagenic potential of inorganic metal salts.* Cancer Res 39:193–198.

CHAN PC, HERBERT RA, ROYCROFT JH, HASEMAN JK, GRUMBEIN SL, MILLER RA and CHOU BJ (1998) *Lung tumor induction by inhalation exposure to molybdenum trioxide in rats and mice.* Toxicol Sci 45:58–65.

CHEN F and SHI X (2002) *Intracellular signal transduction of cells in response to carcinogenic metals.* Crit Rev Oncol Hematol 42:105–121.

CHEN J and THILLY WG (1994) *Mutational spectrum of chromium (VI) in human cells.* Mutat Res 323:21–27.

CHENG L, LIU S and DIXON K (1998) *Analysis of repair and mutagenesis of chromium-induced DNA damage in yeast, mammalian cells and transgenic mice.* Environ Health Perspect 106 (Suppl 4) 1027–1032.

CHIBBER R and ORD MJ (1989) *The mutagenic and carcinogenic properties of three second generation antitumour platinum compounds: a comparison with cisplatin.* Eur J Cancer Clin Oncol **25**:27–33.

CIRANNI R, ANTONETTI M and MIGLIORE L (1995) *Vanadium salts induce clastogenic effects, in in vivo treated mice.* Mutat Res **343**:53–60.

CODINA JC, PEREZ-TORRENTE C, PEREZ-GARCIA A, CAZORLA FM and DE VICENTE A (1995) *Comparison of microbial tests for the detection of heavy metal genotoxicity.* Arch Environ Contam Toxicol **29**:260–265.

COEN N, MOTHERSILL C, KADHIM M and WRIGHT EG (2001) *Heavy metals of relevance to human health induce genomic instability.* J Pathol **195**:293–299.

COHEN MD, KARGACIN B, KLEIN CB and COSTA M (1993) *Mechanisms of chromium carcinogenicity and toxicity.* Crit Rev Toxicol **23**:255–281.

COOGAN TP, LATTA DM, SNOW ET and COSTA M (1989) *Toxicity and carcinogenicity of Nickel compounds.* CRC Crit Rev Toxicol **19**:341–384.

COOGAN TP, BARE RM and WAALKES MP (1992) *Cadmium-induced DNA strand damage in cultured liver cells: reduction in cadmium genotoxicity following zinc pre-treatment.* Toxicol Appl Pharmacol **113**:227–233.

COSTA M (1997) *Toxicity and carcinogenicity of Cr(VI) in animal models and humans.* Crit Rev Toxicol **27**:431–442.

COSTA M, SUTHERLAND JE, PENG W, SALNIKOW K, BRODAY L and KLUZ T (2001) *Molecular biology of nickel carcinogenesis.* Mol Cell Biochem **222**:205–211.

DE FLORA S, BAGNASCO M, SERRA D and ZANACCHI P (1990) *Genotoxicity of chromium compounds. A review.* Mutat Res **238**:99–172.

DE FLORA S, BENNICELLI C and BAGNASCO M (1994) *Genotoxicity of mercury compounds. A review.* Mutat Res **317**:57–79.

DE FLORA S, CAMOIRANO A, BAGNASCO M, BENICELLI C, CORBETT GE and KERGER BD (1997) *Estimates of the chromium(VI) reducing capacity in human body compartments as a mechanism for attenuating its potential toxicity and carcinogenicity.* Carcinogen **18**:531–537.

DE FLORA S (2000) *Threshold mechanisms and site specificity in chromium(VI) carcinogenesis.* Carcinogen **21**:533–541.

DENKHAUS E and SALNIKOW K (2002) *Nickel essentiality, toxicity, and carcinogenicity.* Crit Rev Oncol Hematol **42**:35–56.

DIAMOND AM, DALE P, MURRAY JL and GRDINA DJ (1996) *The inhibition of radiation-induced mutagenesis by the combined effects of selenium and the aminothiol WR-1065.* Mutat Res **356**:147–154.

DOMINGO JL (1996) *Vanadium: a review of the reproductive and developmental toxicity.* Reprod Toxicol. **10**:175–182.

DUERKSEN-HUGHES PJ, YANG J and OZCAN O (1999) *p53 induction as a genotoxic test for twenty-five chemicals undergoing in vivo carcinogenicity testing.* Environ Health Perspect **107**:805–812.

GALARIS D and EVANGELOU A (2002) *The role of oxidative stress in mechanisms of metal-induced carcinogenesis.* Crit Rev Oncol Hematol **42**:93–103.

GEBEL T, LANTZSCH H, PLESSOW K and DUNKELBERG H (1997) *Genotoxicity of platinum and palladium compounds in human and bacterial cells.* Mutat Res **389**:183–190.

GEBEL T (1998) *Suppression of arsenic-induced chromosome mutagenicity by antimony.* Mutat Res **412**:213–218.

GEBHART E (1989) *Heavy metal induced chromosome damage.* Life Chem Rep **7**:113–148.

GEBHART E and ROSSMAN T (1991) *Mutagenicity, carcinogenicity, teratogenicity.* In: Merian E ed. Metals and their compounds in the environment, pp. 617–640. VCH, Weinheim-New York-Basel-Cambridge.

GENNART JP, BALEUX C, VERELLEN-DUMOULIN C, BUCHET JP, DE MEYER R and LAUWERYS R (1993) *Increased sister chromatid exchanges and tumor markers in workers exposed to elemental chromium-, cobalt- and nickel-containing dusts.* Mutat Res **299**:55–61.

GERBER GB and LEONARD A (1997) *Mutagenicity, carcinogenicity and teratogenicity of thallium compounds.* Mutat Res **387**:47–53

GERBER GB, LEONARD A and HANTSON P (2002) *Carcinogenicity, mutagenicity and teratogenicity of manganese compounds.* Crit Rev Oncol Hematol **42**:25–34.

GILANI SH and ALIBHAI Y (1990) *Teratogenicity of metals to chick embryos.* J Toxicol Environ Health **30**:23–31.

GONSEBATT ME, VEGA L, SALAZAR AM, MONTERO R, GUZMAN P, BLAS J, DEL RAZO LM, GARCIA-VEGAS G, ALBORES A, CEBRIAN ME, KELSH M and OSTROSKY-WEGMAN P (1997) *Cytogenetic effects in human exposure to arsenic.* Mutat Res **386**:219–228.

GROOT DE RESTREPO H, SICARD D and TORRES MM (2000) *DNA damage and repair in cells of lead exposed people.* Am J Ind Med **38**:330–334.

GROVER P, BANU BS, DEVI KD and BEGUM S (2001) *In vivo genotoxic effects of mercury chloride in rat peripheral blood leukocytes using comet assay.* Toxicol 167:191–197.

GURNANI N, SHARMA A and TALUKDER G (1992) *Cytotoxic effects of antimony trichloride on mice in vivo.* Cytobios 70:131–136.

HARTMANN A and SPEIT G (1994) *Comparative investigations of the genotoxic effects of metals in the single cells gel (SCG) assay and the sister chromatid exchange (SCE) test.* Environ Mol Mutagen 23(4):299–305.

HARTMANN A and SPEIT G (1996) *Effect of arsenic and cadmium on the persistence of mutagen-induced DNA-lesions in human cells.* Environ Mol Mutagen 27:98–104.

HARTSFIELD JK, LEE M, MOREL JG and HILBELINK DR (1992) *Statistical analysis of the effect of cadmium and zinc on hamster teratogenesis.* Biochem Med Metab Biol 48:159–173.

HARTWIG A (1994) *Role of DNA repair inhibition in lead and cadmium-induced genotoxicity: a review.* Environ Health Perspect 102, Suppl 3:45–50.

HARTWIG A (1998) *Carcinogenicity of metal compounds: possible role of DNA repair inhibition.* Toxicol Lett 102–103:235–239.

HARTWIG A (2001a) *Role of magnesium in genome stability.* Mutat Res 475:113–121.

HARTWIG A (2001b) *Zinc finger proteins as potential targets for toxic metal ions: differential effects on structure and function.* Antiox Redox Signal 3:625–634.

HAWK SN, URIU-HARE JY, DASTON GP, JANKOWSKI MA, KWIK-URIBE C, RUCKER RB and KEEN CL (1998) *Rat embryos cultured under copper-deficient conditions develop abnormally and are characterized by an impaired oxidant defense system.* Teratology 57:310–320.

HAYES RB (1997) *The carcinogenicity of metals in humans.* Cancer Caus Control 8:371–385.

HEIM S and MITELMAN F (1995) *Cancer cytogenetics,* 2nd edn. Wiley-Liss, New York-Chichester-Brisbane-Toronto-Singapore.

HIROSE T, KONDO K, TAKAHASHI Y, ISHIKURA H, FUJINO H, TSUYUGUCHI H, HASHIMOTO M, YOKOSE T, MUKAI K, KODAMA T and MONDEN Y (2002) *Frequent microsatellite instability in lung cancer from chromate-exposed workers.* Mol Carcinogen 33:172–189.

HODGES NJ, ADAM B, LEE AJ, CROSS HJ and CHIPMAN JK (2001) *Induction of DNA-strand breaks in human peripheral blood lymphocytes and A549 Lung cells by sodium dichromate: association with 8-oxo-2-deoxyguanosine formation and interindividual variability.* Mutagen 16:467–474.

HONG YC, PAIK SR, LEE HJ, LEE KH and JANG SM (1997) *Magnesium inhibits nickel-induced genotoxicity and formation of reactive oxygen.* Environ Health Perspect 105:744–748.

ITOH S and SHIMADA H (1997) *Clastogenicity and mutagenicity of hexavalent chromium in lacZ transgenic mice.* Toxicol Lett 9:229–233.

JANKOWSKI MA, URIU-HARE JY, RUCKER RB, ROGERS JM and KEEN CL (1995) *Maternal zinc deficiency but not copper deficiency or diabetes, results in increased embryonic cell death in the rat: implications for mechanisms underlying abnormal development.* Teratology 51:85–93.

JELMERT Ö, HANSTEEN IL and LANGARD S (1994) *Chromosome damage in lymphocytes of stainless steel welders related to past and current exposure to manual metal arc welding fumes.* Mutat Res 320:223–233.

JHA AM and SINGH AC (1995) *Clastogenicity of lanthanides: induction of chromosomal aberrations in bone marrow cells of mice in vivo.* Mutat Res 341:193–197.

JOSEPH P, MUCHNOK TK, KLISHIS ML, ROBERTS JR, ANTONINI JM WHONG WZ and ONG T (2001) *Cadmium-induced cell transformation and tumorigenesis are associated with transcriptional activation of c-fos, c-jun, and c-myc proto-oncogenes: role of cellular calcium and reactive oxygen species.* Toxicol Sci 61:295–303.

KANEMATSU N, NAKAMINE H, FUKUTA Y, YASUDA JI, KURENUMA S and SHIBATA KL (1990) *Mutagenicity of cadmium, platinum and rhodium compounds in cultured mammalian cells.* Gifu Shika Gakkai Zasshi 17:575–581.

KANOJIA RK, JUNAID M and MURTHY RC (1996) *Chromium induced teratogenicity in female rats.* Toxicol Lett 89:207–213.

KASSIE F, PARZEFALL W and KNASMÜLLER S (2000) *Single cell gel electrophoresis assay: a new technique for human biomonitoring studies.* Mutat Res 463:13–31.

KATSIFIS SP, SHAMY M, KINNEY LP and BURNS FJ (1998) *Interaction of nickel with UV-light in the induction of cytogenetic effects in human peripheral lymphocytes.* Mutat Res 422:331–337.

KAZANTZIS G, BLANKS RG and SULLIVAN KR (1992) *Is cadmium a human carcinogen?* IARC Sci Publ 118:435–446.

KERCKAERT GA, LEBOEUF A and ISFORT RJ (1996) *Use of the Syrian hamster embryo cell transformation assay for determining the carcinogenic potential of heavy metal compounds.* Fund Appl Toxicol 34:67–72.

KESHAVA N, ZHOU G, SPRUILL M, ENSELL M and ONG T (2001) *Carcinogenic potential and genomic instability of beryllium sulphate in BALB/c-3T3 cells.* Mol Cell Biochem **222**:69–76.

KILBEY BJ, LEGATOR M, NICHOLS W and RAMEL C (1984) *Handbook of mutagenicity test procedures.* 2nd edn. Elsevier, Amsterdam-New York-Oxford.

KIRKLAND DJ, HAYASHI M, MACGREGOR JT, MÜLLER L, SCHECHTMAN L and SOFUNI T (2000) *Summary of major conclusions from the International Workshop on Genotoxicity Test Procedures.* Environ Mol Mutagen **35**:162–166.

KITAHARA J, YAMANAKA K, KATO L, LEE YW, KLEIN CB and COSTA M (1996) *Mutagenicity of cobalt and reactive oxygen producers.* Mutat Res **370**:133–140.

KNASMÜLLER S, GOTTMANN E, STEINKELLNER H, FOMIN A, PASCHKE A, GOD R and KUNDI M (1998) *Detection of genotoxic effects of heavy metal contaminated soils with plant bioassays.* Mutat Res **420**:37–48.

KRUEGER I, MULLENDERS LH and HARTWIG A (1999) *Nickel(II) increases the sensitivity of V79 Chinese hamster cells towards cisplatin and trans-platin by interference with distinct steps of DNA repair.* Carcinogen **20**:1177–1184.

KURODA K, ENDO G, OKAMOTO A, YOO YS and HORIGUCHI S (1991) *Genotoxicity of beryllium, gallium, and antimony in short-term assays.* Mutat Res **264**:163–170.

LANDRIGAN PJ, BOFFETTA P and APOSTOLI P (2000) *The reproductive toxicity and carcinogenicity of lead: a critical review.* Am J Ind Med **38**:231–243.

LANGLAIS M, TAJMIR-RIAHI HA and SAVOIC R (1990) *Raman spectroscopic study of the effects of Ca2+, Mg2+, Zn2+, and Cd2+ ions on calf thymus DNA: binding sites and conformational changes.* Biopolymer **30**:743–752.

LANTZSCH H and GEBEL T (1997) *Genotoxicity of selected metal compounds in the SOS chromotest.* Mutat Res **389**:191–197.

LEE CH, LIN RH, LIU SH and LIN-SHIAU SY (1998) *Effects of germanium oxide and other compounds on phenylmercury acetate-induced genotoxicity in cultured lymphocytes.* Environ Mol Mutagen **31**:157–162.

LEE-CHEN SF, WANG MC, YU CT, WU DR and JAN KY (1993) *Nickel chloride inhibits the DNA repair of UV-treated but not methylmethanesulfonate-treated Chinese hamster ovary cells.* Biol Trace Elem Res **37**:39–50.

LEONARD A and BERNARD A (1993) *Biomonitoring exposure to metal compounds with carcinogenic properties.* Exp Health Perspect 101, Suppl 3:127–133.

LEONARD A and GERBER GB (1988) *Mutagenicity, carcinogenicity and teratogenicity of aluminium.* Mutat Res **196**:247–257.

LEONARD A and GERBER GB (1994) *Mutagenicity, carcinogenicity and teratogenicity of vanadium compounds.* Mutat Res **317**:81–88.

LEONARD A and GERBER GB (1996) *Mutagenicity, carcinogenicity and teratogenicity of antimony compounds.* Mutat Res **366**:1–8.

LEONARD A and GERBER GB (1997) *Mutagenicity, carcinogenicity and teratogenicity of thallium compounds.* Mutat Res **387**:47–53.

LEONARD A, HANTSON P and GERBER GB (1995) *Mutagenicity, carcinogenicity and teratogenicity of lithium compounds.* Mutat Res **339**:131–137.

LEONARD A and LAUWERYS R (1987) *Mutagenicity, carcinogenicity and teratogenicity of beryllium.* Mutat Res **186**:35–42.

LEONARD A and LAUWERYS R (1990) *Mutagenicity, carcinogenicity and teratogenicity of cobalt metal and cobalt compounds.* Mutat Res **239**:17–27.

LIN RH, WU LJ, LEE CH and LIN-SHIAU SY (1993) *Cytogenetic toxicity of uranyl nitrate in Chinese hamster ovary cells.* Mutat Res **319**:197–203.

LIN RH, LEE CH, CHEN WK and LIN-SHIAU SY (1994) *Studies on cytotoxic and genotoxic effects of cadmium nitrate and lead nitrate in Chinese hamster ovary cells.* Environ Mol Mutagen **23**:143–149

LISON D, DE BOECK M, VEROUGSTRAETE V and KIRSCH-VOLDERS M (2001) *Update of the genotoxicity and carcinogenicity of cobalt compounds.* Occup Environ Med **58**:619–625.

LU PJ, HO IC and LEE TC (1998) *Induction of sister chromatid exchanges and micronuclei by titanium dioxide in Chinese hamster ovary-K1 cells.* Mutat Res **414**:15–29.

LUO SQ, PLOWMAN MC, HOPFER SM and SUNDERMAN FW JR (1993a) *Embryotoxicity and teratogenicity of Cu2+ and Zn2+ for Xenopus laevis, assayed by the FETAX procedure.* Ann Clin Lab Sci **23**:111–120.

LUO SQ, PLOWMAN MC, HOPFER SM and SUNDERMAN FW JR (1993b) *Mg(2+)-deprivation enhances and Mg(2+)-supplementation diminishes the embryotoxic and teratogenic effects of Ni2+, Co2+, Zn2+, and Cd2+ for frog embryos in the FETAX assay.* Ann Clin Lab Sci **23**:121–129.

LUTZ J and BECK SL (2000) *Caffeine decreases the occurrence of cadmium-induced forelimb ectrodactyly in C57BL/6J mice.* Teratology **62**:325–331.

MAHALIK MP, HITNER HW and PROZIALECK WC (1995) *Teratogenic effects and distribution of cadmium (Cd2+) administered via osmotic mini-pumps to gravid CF-1 mice.* Toxicol Lett 76:195–202.

MASON RW, EDWARDS IR and FISHER IC (1989) *Teratogenicity of combinations of sodium dichromate, sodium arsenate and copper sulphate in the rat.* Comp Biochem Physiol C 93:407–411.

MASS MJ and WANG LJ (1997) *Arsenic alters cytosine methylation patterns of the promoter of the tumor suppressor gene p53 in human lung cells: a model for a mechanism of carcinogenesis.* Mutat Res 386:263–277.

MAZZOTTI F, SABBIONI E, GHIANI M, COCCO B, CECCATELLI R and FORTANER S (2001) *In vitro assessment of cytotoxicity and carcinogenic potential of chemicals: evaluation of the cytotoxicity induced by 58 metal compounds in the Balb/3T3 cell line.* Altern Lab Anim 29:601–611.

MAZZOTTI F, SABBIONI E, PONTI J GHIANI M, FORTANER S and ROSSI GL (2002) *In vitro setting of dose-effect relationships of 32 metal compounds in the Balb/3T3 cell line, as a basis for predicting their carcinogenic potential.* Altern Lab Anim 30:209–217.

MCDIARMID MA, SQUIBB K, ENGELHARDT S, OLIVER M, GUCER P, WILSON PD, KANE R, KABAT M, KAUP B, ANDRSON L, HOOVER D, BROWN L and JACOBSON-KRAM D, Depleted Uranium Follow-Up Program (2001) *Surveillance of depleted uranium exposed Gulf War veterans health effects observed in an enlarged "friendly fire" cohort.* J Occup Environ Med 43:991–1000.

MERK O, REISER K and SPEIT G (2000) *Analysis of chromate-induced DNA-protein crosslinks with the comet assay.* Mutat Res 471:71–80.

MERZENICH H, HARTWIG A, AHRENS W, BEYERSMANN D, SCHLEPEGRELL R, SCHOLZE M, TIMM J and JOCKEL KH (2001) *Biomonitoring on carcinogenic metals and oxidative DNA damage in a cross-sectional study.* Cancer Epidemiol Biomark Prev 10:515–522.

MIGLIORE L, COCCHI L, NESTI C and SABBIONI E (1999) *Micronuclei assay and FISH analysis in human lymphocytes treated with six metal salts.* Environ Mol Mutagen 34:279–284.

MILLER AC, FUCIARELLI AF, JACKSON WE, EJNIK EJ, EMOND C, STROCKO S, HOGAN J, PAGE N and PELLMAR T (1998) *Urinary and serum mutagenicity studies with rats implanted with depleted uranium and tantalum pellets.* Mutagenesis 13:643–648.

MILLER AC, BLAKELY WF, LIVENGOOD D, WHITTAKER T, XU J, EJNIK JW, HAMILTON MF, PAR-

LETTE E, JOHN TS, GERSTENBERG HM and HSU H (1998) *Transformation of human osteoblast cells to the tumorigenic phenotype by depleted uranium-uranyl chloride.* Environ Health Perspect 106:465–471.

MINISSI S and LOMBI E (1997) *Heavy metal content and mutagenic activity, evaluated by Vicia faba micronucleus test, of Tiber river sediments.* Mutat Res 393:17–21.

MISRA RR, SMITH GT and WAALKES MP (1998) *Evaluation of the direct genotoxic potential of cadmium in four different rodent cell lines.* Toxicology 126:103–114.

MORRIS SM (2002) *A role for p53 in the frequency and mechanism of mutation.* Mutat Res 511:45–62.

MOURON SA, GOLIJOW CD and DULOUT FN (2001) *DNA damage by cadmium and arsenic salts assessed by the single cell gel electrophoresis assay.* Mutat Res 498:47–55.

MÜLLER J, SIGEL RK and LIPPERT B (2000) *Heavy metal mutagenicity: insights from bioinorganic model chemistry.* J Inorg Biochem 79:261–265.

NARUSE I and HAYASHI Y (1989) *Amelioration of the teratogenicity of cadmium by the metallothionein induced by bismuth nitrate.* Teratology 40:459–465.

NAKAGAWA Y, WAKURI S, SAKAMOTO K and TANAKA N (1997) *The photogenotoxicity of titanium dioxide particles.* Mutat Res 394:125–132.

NAKAJIMA M, TAKAHASHI H, SASAKI M, KOBAYASHI Y, OHNO Y and USAMI M (2000) *Comparative developmental toxicity study of Indium in rats.* Teratogen Carcinogen Mutagen 20:219–227.

NIKIFOROV A, SLOZINA N, NERONOVA E and KHARCHENKO R (1999) *Cytogenetic investigation of thallium-exposed people: pilot study.* J Toxicol Environ Health Pt A 58:465–468.

NODA Y, SUZUKI T, KOHARA A, HASEGAWA A, YOTSUYANAGI T, HAYASHI M, SOFUNI T, YAMANAKA K and OKADA S (2002) *In vivo genotoxicity of dimethylarsinic acid in the Muta ^{TM} mouse.* Mutat Res 513:205–212.

OBERDÖRSTER G, CHERIAN MG and BAGGS RB (1994) *Importance of species differences in experimental pulmonary carcinogenicity of inhaled cadmium for extrapolation to humans.* Toxicol Lett 72:339–343.

OBERHEITMANN B, SCHÄFER J, DALLY H, GARMS A, FRENTZEL-BEYME R and HOFFMANN W (1999) *The chromosome-based challenge assay using fluorescence in situ hybridisation: a possible test for increased cancer susceptibility.* Mutat Res 428:157–164.

Octive JC, Wood M and Johnson AC (1991) *Mutagenic effects of aluminium.* Mutat Res 264:135–137.

Ogawa HI, Shibahara T, Iwata H, Okada T, Tsuruta S, Kakimoto K, Sakata K, Kato Y, Ryo H and Itoh T (1994) *Genotoxic activities in vivo of cobaltous chloride and other metal chlorides as assayed in the Drosophila wing spot test.* Mutat Res 320:133–140.

Olden K and Guthrie J (2001) *Genomics: implications for toxicology.* Mutat Res 473:3–10.

Olive PL, Banath JP and Durand RE (1990) *Heterogeneity in radiation-induced DNA damage and repair in tumor and normal cells measured using the "comet" assay.* Radiat Res 122:86–94.

Oller AR, Costa M and Oberdörster G (1997) *Carcinogenicity assessment of selected nickel compounds.* Toxicol Appl Pharmacol 143:152–166.

Oya-Ohta Y, Kaise T and Ochi T (1996) *Induction of chromosomal aberrations in cultured human fibroblasts by inorganic and organic arsenic compounds and the different roles of glutathione in such induction.* Mutat Res 357:123–129.

Pagano DA and Zeiger E (1992) *Conditions for detecting the mutagenicity of divalent metals in Salmonella typhimurium.* Environ Mol Mutagen 19:139–146.

Paternain JL, Domingo JL and Corbella J (1988) *Developmental toxicity of cobalt in the rat.* J Toxicol Environ Health 24:193–200.

Plowman MC, Peracha H, Hopfer SM and Sunderman FW Jr (1991) *Teratogenicity of cobalt chloride in Xenopus laevis, assayed by the FETAX procedure.* Teratog Carcinog Mutagen 11:83–92.

Ramirez P, Eastmond DA, Laclette JP and Ostrosky-Wegman P (1997) *Disruption of microtubule assembly and spindle formation as a mechanism for the induction of aneuploid cells by sodium arsenite and vanadium pentoxide.* Mutat Res 386:291–298.

Reutova NV (2001) *Mutagenic potential of copper compound and modification of effects of silver iodide.* Genetika 37:617–623.

Rojas E, Valverde M, Herrera LA, Altamirano-Lozano M and Ostrosky-Wegman P (1996) *Genotoxicity of vanadium pentoxide evaluate by the single cell gel electrophoresis assay in human lymphocytes.* Mutat Res 359:77–84.

Roldan-Reyes E, Aguilar-Morales C, Frias-Vazquez S and Altamirano-Lozano M (1997) *Induction of sister chromatid exchanges in human lymphocytes by vanadium pentoxide in combination with caffeine.* Med Sci Res 25:501–504.

Rossetto FE, Turnbull JD and Nieboer E (1994) *Characterization of nickel-induced mutations.* Sci Total Environ 148:201–206.

Rossman TG, Roy NK and Lin WC (1992) *Is cadmium genotoxic?* IARC Sci Publ 118:367–375.

Rossman TG, Uddin AN, Burns FJ and Bosland MC (2001) *Arsenite is a cocarcinogen with solar ultraviolet radiation for mouse skin: an animal model for arsenic carcinogenesis.* Toxicol Appl Pharmacol 176:64–71.

Roy NK and Rossman TG (1992) *Mutagenesis and comutagenesis of lead compounds.* Mutat Res 298:97–103.

Sadiq MF, Zaghal MH and El-Shanti HE (2000) *Induction of chromosomal aberrations by the rhodium(III) complex cis-[Rh(biq)(2)(Cl)(2)]Cl in cultured human lymphocytes.* Mutagen 15:375–378.

Sanchez-Galan S, Linde AR and Garcia-Vazquez E (1999) *Brown trout and European minnow as target species for genotoxicity tests: differential sensitivity to heavy metals.* Ecotoxicol Environ Saf 43:301–394.

Saplakoglu U, Iscan M and Iscan M (1997) *DNA single-strand breakage in rat lung and kidney after single and combined treatments of nickel and cadmium.* Mutat Res 394:133–140.

Sarkar B (1995) *Metal replacement in DNA-binding zinc finger proteins and its relevance to mutagenicity and carcinogenicity through free radical generation.* Nutrition (Suppl 5):646–649.

Schardein JL (1993) *Chemically induced birth defects.* 2nd edn, pp. 722–750. Marcel Dekker Inc, New York-Basel-Hong Kong.

Schaumlöffel N and Gebel T (1998) *Heterogeneity of the DNA damage provoked by antimony and arsenic.* Mutagen 13:281–286.

Schimmer O, Eschelbach H, Breitinger DK, Grutzner T and Wick H (1997) *Organogermanium compounds as inhibitors of the activity of direct acting mutagens in Salmonella typhimurium.* Arzneim-Forsch/ Drug Res 47:1398–1402.

Schinzel A (2001) *Catalogue of unbalanced chromosome aberrations in man.* 2nd edn. De Gruyter, Berlin-New York.

Schoeny R (1996) *Use of genetic toxicology data in US EPA risk assessment: the mercury study report as an example.* Environ Health Perspect 104, Suppl 3:663–673.

Schurz F, Sabater-Vilar M and Fink-Gremmels J (2000) *Mutagenicity of mercury chloride and mechanisms of cellular defense: the role of metal-binding proteins.* Mutagen 15:525–530.

Seoane AI and Dulout FN (2001) *Genotoxic ability of cadmium, chromium and nickel salts studied by kinetochore staining in the cytokinesis-blocked micronucleus assay.* Mutat Res **490**:99–106.

Shubber E, Amin NS and Eladhami BH (1998) *Cytogenetic effects of copper-containing intrauterine contraceptive device (IUCD) on blood lymphocytes.* Mutat Res **417**:57–63.

Silbergeld EK, Waalkes M and Rice JM (2000) *Lead as a carcinogen: experimental evidence and mechanisms of action.* Am J Ind Med **38**:316–323.

Sivikova K and Dianovsky J (1995) *Sister-chromatid exchanges after exposure to metal-containing emissions.* Mutat Res **327**:17–22.

Snow ET (1994) *Effects of chromium on DNA replication in vitro.* Environ Health Perspect 102, Suppl **3**:41–44.

Sram RJ, Binkova B, Dobias L, Rössner P, Topinka J, Vesela D, Vesely D, Stejskalova J, Bavorova H and Rericha V (1993) *Monitoring genotoxic exposure in uranium miners.* Environ Health Perspect **99**:303–305.

Sunderman FW Jr (2001) *Nasal toxicity, carcinogenicity, and olfactory uptake of metals.* Ann Clin Lab Sci **31**:3–24.

Tajmir-Riahi HA, Naoui M and Ahmad R (1993a) *The effects of Cu2+ and Pb2+ on the solution structure of calf thymus DNA:DNA condensation and denaturation studied by Fourier transform in difference spectroscopy.* Biopolymers **33**:1819–1824.

Tajmir-Riahi HA, Ahmad R and Naoui M (1993b) *Interaction of calf thymus DNA with trivalent La, Eu, and Tb ions: metal ion binding, DNA condensation and structural features.* Biomol Struct Dyn **10**:865–877.

Takahashi M, Barrett C and Tsutsui T (2002) *Transformation by inorganic arsenic compounds of normal Syrian hamster embryo cells into a neoplastic state in which they become anchorage-independent and cause tumors in newborn hamsters.* Int J Cancer **99**:629–634.

Tesfai Y, Davis D and Reinhold D (1998) *Chromium can reduce the mutagenic affects of benzo[a]pyrene diolepoxide in normal human fibroblasts via an oxidative stress mechanism.* Mutat Res **416**:159–168.

Theophanides T and Anastassopoulou J (2002) *Copper and carcinogenesis.* Crit Rev Oncol Hematol **42**:57–64.

Thompson J and Bannigan J (2001) *Effects of cadmium on formation of the ventral body wall in chick embryos and their prevention by zinc pretreatment.* Teratol **64**:87–97.

Titenko-Holland N, Shao J, Zhang L, Xi L, Ngo H, Shang N and Smith MT (1998) *Studies on the genotoxicity of molybdenum salts in human cells in vitro and in mice in vivo.* Environ Mol Mutagen **32**:251–259.

Tully DB, Collins BJ, Overstreet JD, Smith CS, Dinse GE, Mumtaz MM and Chapin RE (2000) *Effects of arsenic, cadmium, chromium, and lead on gene expression regulated by a battery of 13 different promoters in recombinant HepG2 cells.* Toxicol Appl Pharmacol **168**:79–90.

Ungvary G, Szakmary E, Tatrai E, Hudak A, Naray M and Morvai V (2000) *Embryotoxic and teratogenic effects of indium chloride in rats and rabbits.* J Toxicol Environ Health **59**:27–42.

Uno Y and Morita M (1993) *Mutagenic activity of some platinum and palladium complexes.* Mutat Res **298**:269–275.

Valverde M, Fortoul TI, Diaz-Barriga F, Mejia J and Del Castillo ER (2000): *Induction of genotoxicity by cadmium chloride inhalation in several organs of CD-1 mice.* Mutagen **15**:109–114.

Van Goethem F, Lison D and Kirsch-Volders M (1997) *Comparative evaluation of the in vitro micronucleus test and the alkaline single cell gel electrophoresis assay for the detection of DNA damaging agents: genotoxic effects of cobalt powder, tungsten carbide and cobalt-tungsten carbide.* Mutat Res **392**(1–2):31–43.

Verougstraete V, Lison D and Hotz P (2002) *A systematic review of cytogenetic studies conducted in human populations exposed to cadmium compounds.* Mutat Res **511**:15–43.

Volkova NA, Karpliuk IA and Emelianova EV (1995) *Study of mutagenic activity of cadmium by the method of dominant lethal mutation.* Vopr Pitan **2**:24–25.

Waalkes MP and Rehm S (1994) *Chronic toxic and carcinogenic effects of cadmium chloride in male DBA/2NCr and NFS/NCr mice: strain-dependent association with tumors of the hematopoietic system, injection site, liver, and lung.* Fundam Appl Toxicol **23**:21–31.

Waalkes MP (2000) *Cadmium carcinogenesis in review.* J Inorg Biochem **79**:241–244.

Wang RR and Li Y (1998) *In vitro evaluation of biocompatibility of experimental titanium alloys for dental restorations.* J Prosthet Dent **80**:495–500.

Wang S and Shi X (2001) *Molecular mechanisms of metal toxicity and carcinogenesis.* Mol Cell Biochem **222**:3–9.

Weiner ML, Batt KJ, Putman DL, Curren RD and Yang LL (1990) *Genotoxicity evaluation of lithium hypochlorite.* Toxicology **65**:1–22.

WEVER DJ, VELDHUIZEN AG, SANDERS MM, SCHA-KENRAAD JM and VAN HORN JR (1997) *Cytotoxic, allergic and genotoxic activity of a nickel-titanium alloy.* Biomaterials **18**:1115–1120.

WIENCKE JK and YAGER JW (1992) *Specificity of arsenite in potentiating cytogenetic damage induced by the DNA crosslinking agent diepoxybutane.* Environ Mol Mutag **19**:195–200.

WISE JP, LEONARD JC and PATIERNO SR (1992) *Clastogenicity of lead chromate particles in hamster and human cells.* Mutat Res **278**:69–79.

WONES R, RADACK K, MARTIN V, MANDELL K, PINNEY S and BUNCHER R (1995) *Do persons living near a uranium processing site have evidence of increased somatic cell gene mutation? A first study.* Mutat Res **335**:171–184.

WU FY, TSAI FJ, KUO HW, TSAI CH, WU WY, WANG RY and LAI JS (2000) *Cytogenetic study of workers exposed to chromium compounds.* Mutat Res **464**:289–296.

YAMAMOTO A, KOHYAMA Y and HANAWA T (2002) *Mutagenicity evaluation of forty-one metal salts by the umu test.* Biomed Mater Res **59**:176–183.

YOKOIYAMA A, KADA T and KURODA Y (1990) *Anti-mutagenic action of cobaltous chloride on radiation-induced mutations in cultured Chinese hamster cells.* Mutat Res **245**:99–105.

YOURTEE DM, ELKINS LL, NALVARTE EL and SMITH RE (1992) *Amplification of doxorubicin mutagenicity by cupric ions.* Toxicol Appl Pharmacol **116**:57–65.

ZHANG A and ZHANG Q (1997) *Study on antimu-tagenic effect of lanthanum carbonate in CA test, UDS test, and MN test.* Wei Sheng Yan Jiu **26**:306–309.

ZHITKOVICH A, SONG Y, QUIEVRYN G and VOITKUN V (2001) *Non-oxidative mechanisms are responsible for the induction of mutagenesis by reduction of Cr(VI) with cysteine: role of ternary DNA adducts in Cr(III)-dependent mutagenesis.* Biochem **40**:549–560.

ZIENOLDDINY S, SVENDSRUD DH, RYBERG D, MIKALSEN AB and HAUGEN A (2000) *Nickel(II) induces microsatellic mutations in human lung cancer cell lines.* Mutat Res **452**:91–100.

10
Ecogenetics

Marika Geldmacher-von Mallinckrodt

10.1
Introduction

Ecogenetics is understood to be a genetic predisposition for an individual reaction to environmental factors (Brewer 1971, Goedde 1972, Propping 1978, 1980, Kalow 1982, Kalow et al. 1986). Ecogenetic reactions are found in all groups of living organisms. Various reactions to industrial toxins, pesticides, radiation, gaseous emissions, environmental toxins, foodstuffs, pharmaceuticals (pharmacogenetics), and also metals are known. An indication for genetically determined differences in the reactions of organisms to environmental factors is apparent whenever, instead of an unimodal distribution, a bimodal or multimodal distribution is found upon quantification. The common basis for almost all of the various genetically determined reactions is the role, that proteins play – often as enzymes, but also as transport proteins – in almost all life processes. The genetically controlled synthesis of proteins can lead to these variations. As with the findings in pharmacogenetics, some ecogenetic reactions are due to the presence of rare mutant genes, and cause a grossly abnormal response or idiosyncratic reaction. In other instances, the variable response is mediated by a polymorphic system, and a significant proportion (between 2 and 50%) reacts differently. Most frequently, ecogenetic responses involve several genes and lead to unusual responses in a few individuals whose genetic make-up causes them to fall toward one end of the unimodal distribution curve.

In this chapter, examples will be provided of the importance of genetically determined factors relating to the tolerance and sensitivity to toxic metals in bacteria, plants, mammals, and man. Similar effects have also been described, for example, in fishes, insects, nematodes, terrestrial gastropods, crustaceans, and shrimps (Olsson et al. 1997, Forbes 1999).

10.2
Bacteria

Some metal ions, including iron, copper and zinc, are essential in low concentrations for cellular metabolism in bacteria, though at higher concentrations these ions may be toxic. However, other metals such as cadmium, mercury and lead do not play any physiological role and are in fact toxic towards cells. For example, mercury and lead react with sulfhydryl groups of proteins and therefore inhibit their functions. Cad-

Elements and their Compounds in the Environment. 2nd Edition.
Edited by E. Merian, M. Anke, M. Ihnat, M. Stoeppler
Copyright © 2004 WILEY-VCH Verlag GmbH & Co. KGaA, Weinheim
ISBN: 3-527-30459-2

mium is extremely toxic, even in low concentrations, and has been shown to induce DNA breakage (Rossbach et al. 2000).

Many bacteria have specific genes for resistances to the toxic ions of heavy metals and metalloids including Ag^+, TeO_3^-, AsO_4^{3-}, Cd^{2+}, CrO_4^{2-}, Cu^{2+}, Hg^{2+}, Ni^{2+}, Sb^{3+}, TeO_3^{2-}, Tl^+ and Zn^{2+}; there are also reports of resistance to Pb^{2+} and organotin compounds (Silver 1996, 1998, Silver et al. 1996, Rosen 1999, Bruins et al. 2000). Recently, bacteria that are resistant to U(VI) and ^{237}Np have also been found (Lloyd et al. 2001). This leaves out Group IA (e.g., Na, K) and Group IIA (e.g., Ca, Mg) of the Periodic Table, as lacking genes for ion resistance. There are also no resistance genes for Group VIIA halides, although halides are abundant in the environment and toxic in higher concentrations (Ji et al. 1995, Silver 1996).

It is frequently thought that these resistances arose as a result of human pollution in recent centuries. It seems, however, more likely that toxic metal resistance systems arose soon after life began, in a world which was already polluted by volcanic activities and geological sources (Ji et al. 1995).

A number of mechanisms which impart resistance to heavy and soft metals have been identified (Rouch et al. 1995, Osborn et al. 1997), including:

1. Blocking, in which the toxic ion is prevented from entering the cell; e.g. Cu^{2+} (Lutkenhaus 1977).
2. Active efflux of the metal ion from the cell by highly specific systems encoded by resistance genes, e.g. Cd^{2+} (Nies 1992), AsO_2^{1-} and AsO_4^{3-} (Kaur et al. 1992).
3. Intracellular physical sequestration of the metal by binding proteins, e.g. Cd^{2+} and Zn^{2+} (Robinson et al. 1990).
4. Extracellular sequestration, often by extracellular polysaccharides on the cell wall, e.g. Pb^{2+} (Gadd et al. 1978) and Cu^{2+} (Cooksey 1994).
5. Enzymatic conversion of the metal to a form which is less toxic for the bacterium, e.g. CH_3Hg and Hg^{2+} (Misra 1992).

Many of these metal-resistance mechanisms are encoded by genetic systems which have been extensively studied and are well understood. Perhaps the best-studied metal-resistance system is encoded by genes of the *mer*, or mercury resistance, operon. In this system, Hg(II) is transported into the cell via the MerT transporter protein, and detoxified by reduction to less toxic volatile mercury by an intracellular mercury reductase, MerA (see Osborn et al. 1997, Hobman et al. 2000).

In general, these resistance systems have been found in plasmids, but frequently related systems are subsequently found determined by chromosomal genes in other organisms. Examples are mercury resistance in *Bacillus*, and arsenic efflux by chromosomal *E. coli* genes. For some metals (notably mercury and arsenic), the plasmid and chromosomal determinants are basically the same. Other systems, such as copper transport ATPases and metallothionein cation-binding proteins, are only known from chromosomal genes. The largest group of metal-resistance systems function by energy-dependent efflux of toxic ions. Efflux pumps are the major currently known group of plasmid resistance systems, thus reducing their intracellular concentration to subtoxic levels. They can be either ATPases (as is the Cd^{2+} ATPase of Gram-positive and the arsenite ATPase of Gram-negative bacteria) or chemiosmotic (as is the divalent cation efflux system of soil *Alcaligenes* and the arsenite efflux system of the chromosome of Gram-negative bacteria and of plasmids in Gram-positive bacteria). The mechanisms are not precisely the same in all bacterial types: while the mercury resistance systems are highly homologous (but differ in energy coupling), the cadmium resistance involves unrelated ATPases in Gram-positive bacteria and chemiosmotic antiporters in Gram-negative bacteria. These systems appear to be of independent evolutionary origin (Nies et al. 1995, Silver 1996, Silver et al. 1996, Rosen 1999). There is also a well-described bacterial metallothionein, found on the chromosome of some cyanobacteria, and conferring resistances to Cd^{2+} and Zn^{2+} (Turner et al. 1995).

Numerous bacteria have been described and studied in detail for their ability to transform, detoxify or immobilize a variety of metallic and organic pollutants in the environment (Gadd 2000). Like most organisms, however, these bacteria are sensitive to the damaging effects of radiation, and their use in bioremediation will probably be limited to environments where radiation levels are very low. Radiation-resistant bacteria have also been isolated, even from non-extreme environments. Unfortunately, they are often pathogenic, and most lack a developed system for genetic manipulations. However, during the past few years a clean-up technology for environmental biotechnology has been developed based on the radiation-resistant bacterium *Deinococcus radiodurans*, which is being engineered to express bioremediating functions (Daly 2000). Additional advances may be expected with the use of new techniques such as whole-genome transcription (Lloyd et al. 2001).

Microorganisms may have also important roles in the biogeochemical cycling of radionuclides (see Lloyd et al. 2001). New applications of these processes to the detoxification of radionuclide contamination have been developed: U(VI) is a priority pollutant reduced by phylogenetically distinct bacteria. U(VI) can be reduced to U(IV) by certain Fe(III)-dissimulatory microorganisms, for example *Geobacter metallireducens*, and this reduction in solubility can be the basis of U removal from contaminated waters and leachates (Lovley et al. 1997). Biological reduction of U(VI) to U(IV) was stimulated by the addition of ethanol and trimetaphosphate to contaminated groundwaters (Abdelouas et al. 2000), though a detailed analysis of the microbes in this process has not yet been presented.

Another actinide that has attracted attention is the mobile and long-lived α-emitter ^{237}Np which is present in low-activity nuclear wastes. Removal is ineffective when using chemical-based techniques, but biotreatment of ^{237}Np was possible using a combination of the biological reduction of Np(V) by *S. putrefaciens* followed by precipitation of Np(IV) phosphate by a *Citrobacter* sp. (Lloyd et al. 2000).

10.3
Plants

Heavy metal ions such as Cu^{2+}, Zn^{2+}, Mn^{2+}, Fe^{2+}, Ni^{2+}, and Co^{2+} are essential micronutritients for plant metabolism. When these ions are not available to the roots, plants develop specific deficiency symptoms, though when present in excess, these – as well as nonessential metals such as Cd^{2+}, Hg^{2+} and Pb^{2+} – can become extremely toxic. At high concentrations, all these metals may cause symptoms such as chlorosis and necrosis, stunting, leaf discoloration and inhibition of root growth (Marschner 1999). At the cellular level, toxicity may result from binding of metals to sulfhydryl groups in proteins, thereby inhibiting enzyme activity or protein function, or by producing a deficiency of other essential ions. Other possibilities include disruption of cell transport processes and oxidative damage (Williams et al. 2000).

That some plants can evolve tolerance to heavy metal contamination has been recognized for over 60 years (Prat 1934). Over 400 species of plants, algae, and fungi are now known to have evolved tolerance to metals (Guerinot 2000).

Crude estimates can often be made concerning the time scale for tolerance evolution (Shaw 1999). Metaliferous outcrops are especially abundant and extensive in south central Africa and have been available to plants for millions of years. Indeed Brooks et al. (1985) suggest that mineralized outcrops in the Shaban Copper Arc and the Zambian Copperbelt in south-central Africa have been available to plants since the origin of angiosperms during the Mesozoic Era. The significance of such habitats for plant evolution is amply demonstrated by the high percentage of endemics on metal-enriched substrates.

Mine wastes are of more recent origin. The Welsh copper and lead/zinc mines at Drws-y-Coed and Trelogan, respectively, were first worked in the 13th century, and present tailings with tolerant plants probably date from the 19th century (McNeilly et al. 1968). Compared to the African outcrops, a much lower number of endemic metallophyte taxa is known from mine wastes in North America and Europe, where the ages of contaminated sites are measured in decades to centuries, rather than in millions of years. Plant populations growing on mine tailings, from which most of our information about evolutionary processes involved in tolerance evolution has been derived, are relatively recent in origin. Patterns observed in mine populations are not older than decades to a century or two. The fact that the evolution of tolerance can occur very rapidly is not surprising in light of the evidence that tolerant plants can be selected from normal populations in a single generation (Gartside et al. 1974).

Tolerance to metal stress relies on plant capacity to detoxify metals having entered cells. The postulated mechanisms involve biochemical detoxification, for example by binding to organic acids (especially citrate) or proteins like ferritin, metallothioneins and phytochelatins, and finally compartmentalization of the metal within the cell. In most plant cells the vacuole comprises more than 80–90% of the cell volume and is acting as a central storage compartment for ions (Briat et al. 1999).

An extreme case of metal accumulation inside tolerant plants is observed with hyperaccumulating plants. These plants can accumulate a thousand-fold more metal than normal plants without phytotoxic symptoms. Among the 400 metal-hyperaccumulating species of plants which have been reported until now, about 16 are zinc accumulators, containing more than $10000\ \mu g\ g^{-1}$ Zn in shoot dry matter (Brooks et al. 1998). Certain populations of *Thlaspi caerulescens* can tolerate up to $40000\ \mu g$ Zn g^{-1} tissue in their shoots, whereas the normal tissue zinc concentration for most plants is between 20 and 100 $\mu g\,g^{-1}$ (Guerinot 2000). To date, the highest metal content determined in plants concerns nickel accumulation in *Sebertia acuminata* latex, an endemic tree from New Caledonia (Jaffré et al. 1976). Dried latex contains 25% of nickel, with citric acid as the counter-ion for 40% of the metal present (Sagner et al. 1998).

The results of recent studies on several species and metals suggest that tolerance is effected by a limited number of major genes, with modifiers that condition the level of tolerance rather than the presence of tolerance per se (Shaw 1999). Genetic control of metal tolerance by one dominant gene has been observed in many occurrences both in crops and in adapted populations of wild plants, for example, aluminium tolerance in *Zea mays* and shorgum, copper tolerance in *Mimulus guttatus*, and arsenic tolerance in *Agrostis capillaris* (Macnair 1993). According to Macnair et al. (1999), zinc tolerance and zinc accumulation may be genetically independent characters.

A number of genes involved in metal transport in plants have been identified. Many of these belong to previously described transporter families such as the P-type APTases (Axelsen et al. 1998) and the Nramp proteins (natural resistance-associated macrophage protein) (Cellier et al. 1995). Recent studies on metal transport in *Arabidopsis* have identified the founding members of a new family of metal transporters, the ZIP family (Eng et al. 1998), that now has representatives in animals, plants, protists, and fungi (Guerinot 2000). The ZIP family takes its name from the first member to be identified – "ZRT, IRT-like protein". Members of the ZIP gene family are capable of transporting a variety of cations including cadmium, iron, manganese, and zinc. Information on where in the plant each of the ZIP transporters functions and how each is controlled in response to nutritient avaiility may allow the manipulation of plant mineral status with an eye to creating food crops with enhanced mineral content and developing crops that bioaccumulate or exclude toxic metals.

There is now considerable interest in the area of metal transport in plants because of the implications for phytoremediation – the use of plants to extract, sequester, and/or detoxify pollutants such as toxic metals. Phytoremediation strategies for radionuclide and heavy metal pollutants focus on hyperaccumulation above ground. Significant progress has been made in recent years in developing native or genetically modified plants for the remediation of environmental contaminants (Meagher 2000).

Present knowledge of the transport processes for heavy metals across plant membranes at the molecular level remains rudimentary in most cases, however. A comprehensive understanding of metal transport in plants will be essential for developing schemes for the genetic engineering of plants that accumulate specific metals, either for use in phytoremediation or to improve human nutrition (Williams et al. 2000). In order to reach these objectives, a genetic approach would be useful to generate metal-tolerant plants with high biomass efficient for bioextraction (Briat et al. 1999).

10.4
Mammals

Many reports have been made describing mammalian strains which are either sensitive or resistant to toxic metals, and examples of these are provided in the following sections.

10.4.1
Mice

The phenomenon of genetic zinc deficiency in milk has been observed in mice (Piletz et al. 1978). Nursing pups fed only the milk of homozygotic deficient animals do not survive, but if the pups are given additional zinc, then mortality is significantly reduced. It is assumed that the gene responsible for zinc transport from maternal blood to milk is modified. The so-called super mouse, on the other hand, accumulates more zinc from the mother during gestation than do other strains, and excretes it in smaller quantities after birth. In this way it is protected against zinc deficiency (Reis et al. 1977).

"Toxic milk" (tx) is a recessive mutation in mice which causes hepatic accumulation of copper that begins during the third postnatal week. By 6 months of age, the copper concentration can be 100-fold that of the normal adult. This gradual accumulation of copper in the liver resembles that seen in patients with Wilson's disease. In addition, pups are born copper-deficient and the milk produced by mutant mothers is low in copper, resulting in death of the pups (Rauch 1983). Analysis of the morphology of livers from adult tx mice has shown significant differences from the liver damage seen in patients with Wilson's disease. This fact, together with the absence of reports of copper deficiency in infants of mothers with human Wilson's disease, has raised doubts about whether the tx mouse is a valid model for Wilson's disease (Biempica et al. 1988). Recent studies, however, have mapped the toxic milk mutation to the same region of chromosome 8 as the murine homologue of Wilson's disease, consistent with a mutation in this gene causing the tx phenotype (Rauch et al. 1995, Reed et al. 1995). Theophilos et al. (1996) reported the cloning and sequence of the murine homologue of the Wilson's disease gene. These authors demonstrated a point mutation in the tx mouse sequence which resulted in the conversion of a highly conserved methionine to valine in the eigth transmembrane channel, suggesting that the tx mouse could be a valid model for Wilson's disease.

10.4.2
Rats

Long-Evans Cinnamon (LEC) rats, an inbred strain of a mutant rat that was originally isolated from a closed colony of Long-Evans rats, develop hereditary acute hepatitis at about 4 months after birth. Those rats which survive acute hepatitis suffer from chronic hepatitis, and develop hepatocellular carcinoma from one year after birth. The hepatitis is inherited in an autosomal recessive manner. The copper concentration in the liver of LEC rats was over 40-fold that of normal Long-Evans Agouti (LEA) rats, while the serum ceruloplasmin and copper concentrations in LEC rats was significantly decreased (Li et al. 1991). The LEC rat shares many clinical and biochemical features with human Wilson's disease, one of the characteristics being the low level of serum ceruloplasmin. Yet there are some differences: in human Wilson's disease hepatocellular carcinoma are rare, whereas in LEC rats no Kayser-Fleischer ring has been found, and neurological abnormities are rare (Wu et al. 1994). According to Wu et al. (1994), the LEC rat has a deletion in the copper transporting ATPase gene homologous to the human Wilson's disease gene, ATP7B.

10.4.3
Dogs

Copper toxicosis, a hereditary disease in Bedlington terriers, was first described by Hardy et al. (1975). It is an autosomal recessive trait (Johnson et al. 1980) which is characterized by a reduced excretion of copper into the bile. Chronic hepatitis and cirrhosis develop as the amount of accumulated copper in the liver increases, usually leading to clinical signs of liver insufficiency in the middle-aged dog (Twedt et al.1979, van de Sluis et al. 1999). The accumulation of copper becomes histologically evident in liver biopsies at one year of age (Hardy et al. 1975, Twedt et al. 1979). Bedlington terriers with copper toxicosis do not show neurological symptoms and Kayser-Fleischer rings (Twedt et al. 1979, Hunt et al. 1986). A low-copper diet and medical treatment to prevent further accumulation can be established, and may prevent mortality due to hepatic failure (Brewer et al. 1992). The primary genetic defect which underlies the pathogenesis of this form of hepatic copper overload is still unknown. Since an analogous condition, Wilson's disease, exists in humans, for some time the diseases were regarded as equivalent. However, though similarities exist, the result of both linkage studies (Yuzbasiyan-Burkan et al. 1993) and genetic mapping (van de Sluis et al. 1999) suggest that Wilson's disease in humans and copper toxicosis in Bedlington terriers have different biochemical and genetic backgrounds.

Proschowsky et al. (2000) have developed a diagnostic test based on the micro satellite marker C04107 for copper toxicosis in Bedlington terriers described by Yuzbasiyan et al. (1997). Until now, this test has to be considered the best tool that breeders of Bedlington terriers can use in the eradication of copper toxicosis, as surveys carried out so far have indicated that the genetic basis for the disease is similar both in Europe and in the USA.

10.4.3
Humans

10.4.3.1
Arsenic (see also Part IV, Chapter 6)

Arsenic is one of the most important global environmental toxicants (Gebel 2000). Inorganic arsenic is a potent human carcinogen, and it has long been known that occupational exposure (e.g., in copper smelters) increases the risk for lung cancer (IARC 1980), while exposure via the drinking water may cause skin cancer (Tseng 1977). More recent epidemiological studies have demonstrated that there is increased risk for cancer of the urinary bladder and lungs, and possibly also liver and kidneys in persons exposed to arsenic via drinking water (for reviews, see NRC 1999, 2001). Other adverse health effects associated with chronic arsenic exposure are hyperkeratosis, pigmentation changes, and effects on

the circulatory, hepatic and nervous systems. The key mechanism of arsenic's tumorigenicity has still not been elucidated.

Inorganic arsenic is methylated in humans to monomethylarsonic acid, MMA(V), and dimethylarsinic acid, DMA(V), both of which are more rapidly excreted in urine than is inorganic arsenic – especially the trivalent form (As III, arsenite). Absorbed arsenate (As V) is reduced to trivalent arsenic (As III) (for a review, see Vahter 2000). Le et al. (2000a,b) also identified monomethylarsonous acid, MMA(III), in the urine of inhabitants from inner Mongolia after a single oral administration of 300 mg sodium-dimercapto-1-propane sulfonate (DMPS).

It is clear from studies in human volunteers exposed to specified doses of inorganic arsenic that the rate of excretion increases with the methylation efficacy, and that there are large interindividual variations. Yet in many population groups studied with respect to arsenic exposure and metabolism, on average the urine of people exposed to inorganic arsenic either occupationally, experimentally, or in the general environment contains 10–30% inorganic As, 10–20% MMA(V), and 60–80% DMA(V). As might be expected, some studies have indicated a slight decrease in the relative amount of DMA(V) in urine and a corresponding increase in the relative amount of MMA(V) with increasing exposure to arsenic; moreover, in cases of acute intoxication the excretion of methylated metabolites of arsenic was delayed (Vahter 2000).

Investigations on arsenic methylation in different population groups have revealed a low urinary excretion of MMA(V) in native Andean people exposed to arsenic via the drinking water in northern Argentina. On average, about 2% of the urinary arsenic was found to be MMA(V). Later studies on children and pregnant women living in the same area confirmed the results on low MMA(V) excretion (Concha et al. 1998a,b,c). In addition, it was found that people of mixed ethnic origin in north-east Argentina had a similar urinary pattern of arsenic metabolites (Concha et al. 1998a). A study of those exposed to ~600 µg As L^{-1} drinking water in north-eastern Chile showed a few percent of the group to have $<5\%$ MMA(V) in the urine, while the average was 15% (Hopenhayn-Rich et al. 1996). Most of those people with low MMA(V) were Atacamenos. These results indicate a genetic polymorphism of arsenic methylation enzymes. It may be of interest to note that, unlike other population groups studied with respect to arsenic exposure, these people have most likely been exposed to arsenic via drinking water for the hundreds of generations that they have inhabited this area.

Interestingly there is also a report on an unusually high fraction of MMA(V) in the urine (on average, 27% of the total concentration of metabolites of inorganic arsenic) in people exposed to arsenic in drinking water in north-eastern Taiwan. The methylation of arsenic in these people was associated with genetic polymorphism of glutathione-S-transferase (Chiou et al. 1997).

It cannot be ruled out that susceptibility to arsenic carcinogenicity might differ between Andean populations and Taiwanese people according to observed differences in the methylation of arsenic. Some studies have reported that Andean populations do not develop skin cancer after long exposure to As (see Gebel 2000).

10.4.3.2
Calcium (see also Part III, Chapter 2.3)
Numerous reports exist relating to patients with either too-high or too-low concentrations of calcium in the blood and/or urine,

and as yet a reason for this has not been forthcoming. However, in some cases a genetic basis of the disease has been explored.

Dent's disease is a familial renal tubular syndrome characterized by low molecular-weight proteinuria, hypercalciuria, nephrolithiasis and eventual renal failure. Mutations in the CLC-5 chloride channel are the basis for this disease. In humans, this channel is expressed mainly in the kidney, and mutations in the human CLCN5 promoter region might be the cause of Dent's disease (Devuyst et al. 1999, Hayama et al. 2000).

10.4.3.3
Copper (see also Part III, Chapter 8.1)
Wilson's disease is an autosomal recessive disorder of copper metabolism resulting from the absence or dysfunction of a copper-transporting P-type ATPase (ATP7B) encoded on chromosome 13. This ATPase is expressed in hepatocytes where it is localized to the trans-Golgi network and transports copper into the secretory pathway for incorporation into ceruloplasmin and excretion into the bile. Ceruloplasmin contains 95% of the copper found in human serum. Under physiologic circumstances, biliary excretion represents the sole mechanism for copper excretion, and thus affected individuals have progressive copper accumulation in the liver. When the capacity for hepatic storage is exceeded, cell death ensues with copper release into the plasma, hemolysis, and tissue deposition. Presentation in childhood at an average age of 10–13 years may include chronic hepatitis, asymptomatic cirrhosis, or acute liver failure. In young adults, neuropsychiatric symptoms predominate and include dystonia, tremor, personality changes, and cognitive impairments secondary to copper accumulation in the central nervous system. Copper deposition may be detected in

Descemet's membrane in the cornea, producing a Kayser-Fleischer ring and also as azure lunulae in the finger nails. Hepatocellular carcinoma is a rare consequence of Wilson's disease (Loudianos et al. 2000).

Wilson's disease has been detected in all known ethnic groups, and occurs worldwide with an estimated frequency of 1 in 30 000 and a carrier rate of 1 in 90. This frequency is increased in populations where consanguinity was once a common practice (Loudianos et al. 2000).

DNA analysis from patients with Wilson's disease has revealed more than 190 heterogenous mutations (Gu et al. 2000), comprising a very small number of frequent mutations within specific populations and a greater number of rare individual alleles. The H1069Q mutation accounts for more than 40% of the alleles in populations of Northern European origin. An A778L mutation in the fourth transmembrane domain has been observed in 30% of alleles in Oriental patients (Loudianos et al. 2000).

The laboratory diagnosis of Wilson's disease is confirmed by decreased serum ceruloplasmin, increased urinary copper content, and elevated hepatic copper concentration. In most cases, a liver biopsy will be warranted to obtain an accurate measurement of hepatic copper, that will be elevated even in asymptomatic patients. Witt et al. (2001) recently described a DNA-based method for rapid determination of the H1069Q mutation in the ATP7B gene.

Idiopathic (Indian) childhood cirrhosis is a disorder of progressive liver failure in early childhood, with marked accumulation of hepatic copper usually resulting in death from liver failure. Although the etiology of this disorder in unknown, in all such cases the serum ceruloplasmin is elevated, suggesting a defect in biliary copper excretion beyond the point of entry into the secretor pathway. Originally described in infants from rural India and related to excess intake of copper, similar conditions have since been reported in other countries and ethnic groups. The latter cases, plus the occurrence of cases in siblings and a frequent history of consanguinity, support the possibility of an autosomal recessive inherited susceptibility, as did the analysis of a large pedigree (Lockitch 1998).

Menkes' disease is a rare X-linked recessive disease of copper metabolism, the frequency of which has been estimated at between 0.8 and 2 per 100 000 live male births (Tonnesen et al. 1991). Clinical manifestations begin in the first few months of life, or even in the neonatal period. Symptoms include hypothermia, hypotonia, poor weight gain, seizures, and neurodevelopmental delay or regression. The outcome is poor, with death occurring usually by three years of age. Diagnostic characteristics include facial appearance with steely hair, and reduced levels of serum ceruloplasmin and copper (Menkes 1999, Jayawant et al. 2000). In 1993, the Menkes gene was isolated and shown to be a copper transporting P-type APTase protein, ATP7A (Suzuki et al. 1999). The Menkes' protein is expressed in most tissues, except liver. The disease locus was mapped to Xq13.3, and the gene has been isolated by means of positional cloning (Tumer et al. 1999).

Mutations in the Menkes' gene in patients with Menkes' disease show great variety, including missense, nonsense, deletion, and insertion mutations. Mutations in the Menkes' gene have also been identified in patients with mild Menkes' disease or occipital horn syndrome, showing that these diseases are allelic variants of Menkes' disease. In affected cells, copper significantly accumulates as metallothionein-bound copper in the cytosol and copper transport to the organelles – as well as copper efflux – is disturbed. As a result, cuproenzymes cannot receive the copper necessary for their normal function. Thus, the objective in the treatment of Menkes' disease and occipital horn syndrome is to deliver copper to the intracellular compartments where cuproenzymes are synthesized (Kodama et al. 1999, Menkes 1999).

10.4.3.4
Iron (see also Part III, Chapter 13.1)
Remarkable progress is being made in understanding the molecular basis of disorders in human iron metabolism. In a review, Seth et al. (2000) examined the clinical consequences of new insights into the pathophysiology of genetic abnormalities affecting iron metabolism. The most common inherited disorder in individuals of European ancestry is hereditary hemochromatosis, which affects at least 1 in 300 Caucasians. Less frequent or rare disorders include aceruloplasminemia, juvenile hemochromatosis, atransferrinemia, hyperferritinemia with autosomal dominant congenital cataract, Friedreich's ataxia, and X-linked sideroblastic anemia with ataxia.

Hereditary hemochromatosis (HHC) is an autosomal recessive disorder of iron metabolism which is characterized by increased iron absorption and deposition in the liver, pancreas, heart, joints, and pituitary gland. Without treatment, death may occur from cirrhosis, primary liver cancer, diabetes, or cardiomyopathy. Although removal of the excess iron by repeated venesections is an effective therapy, much of the organ damage, once it has occurred, is irreversible (Bothwell et al. 1998). HFE, the gene for hereditary hemochromatosis, was mapped on the short arm of chromosome 6 (6p21.3). Two of the 37 allelic variants described to date (C282Y and H63D) are significantly correlated with the disease. In the study of Hanson et al. (2001), 5% of the probands were found to be compound heterozygotes (C282Y/H63D), and 1.5% were homozygous for the H63D mutation; 3.6% were C282Y heterozygotes, and 5.2% were H63D heterozygotes. In 7% of cases the two mutations were not present. In the general population the frequency of the C282Y/C282Y genotype is 0.4%. C282Y heterozygosity ranges from 9.2% in Europeans to nil in the Asian/Indian subcontinent, African/Middle Eastern and Australasian populations. The H63D carrier frequency is 22% in European populations (Hanson et al. 2001).

Ceruloplasmin is an abundant alpha 2-serum glycoprotein that contains more than 95% of the copper present in human plasma. Aceruloplasminemia is an autosomal recessive disorder characterized by progressive neurodegeneration of the retina and basal ganglia associated with specific inherited mutations in the ceruloplasmin gene. Clinical and pathological studies in patients with aceruloplasminemia revealed a marked accumulation of iron in affected parenchymal tissues – a finding which was consistent with early studies identifying ceruloplasmin as a ferroxidase. The presence of neurologic symptoms in aceruloplasminemia is unique among the known inherited and aquired disorders of iron metabolism. Recent studies have revealed an essential role for astrocyte-specific expression of ceruloplasmin in iron metabolism and neuronal survival in the central nervous system. Recognition of aceruloplasminemia also provides new insides into the genetic and environmental determinants of copper metabolism, and has implications for our understanding of the role of copper in human neurodegenerative diseases (Harris et al. 1998).

10.4.3.5

Lead (see also Part III, Chapter 15)

At least three polymorphic genes have been identified that potentially can influence the bioaccumulation and toxicokinetics of lead in humans: (1) the gene coding for δ-amino-levulinic acid dehydratase (ALAD); (2) the Vitamin D receptor (VDR) gene; and (3) HFE, the gene for hereditary hemochromatosis (Onalaja et al. 2000).

ALAD (E.C. 4.2.1.24) catalyzes the second step in heme synthesis, the asymmetric addition of two molecules of aminolevulinic acid to form porpholobilinogen. The ALAD gene is located on chromosome 9q34. Eight ALAD variants have been described. One polymorphism yields two alleles designated ALAD-1 and ALAD-2. These two alleles determine three isoenzymes, 1–1, 1–2, and 2–2. The prevalence of the ALAD-2 allele ranges from 0 to 20% depending on the population. Generally, Caucasians have the highest frequency of the ALAD-2 allele, with approximately 18% of that population being ALAD 1–2 heterozygotes and 1% being 2–2 homozygotes. In comparison, African and Asian populations have low frequencies of the ALAD-2 allele, with few or no ALAD-2 homozygotes being found. All of these frequencies are in Hardy–Weinberg equilibrium. The rarer ALAD-2 allele has been associated with high blood lead levels, and has been thought to increase the risk of lead toxicity by generating a protein that binds lead more tightly than the ALAD-1 protein. Other evidence suggests that ALAD-2 may confer resistance to the harmful effects of lead by sequestering the metal, making it unavailable for pathophysiologic participation. Recent studies have shown that individuals who are homozygous for the ALAD-1 allele have higher cortical bone levels of lead. This implies that they may be at a greater body lead burden and may be at higher risk of

the long-term effects of lead. No firm evidence exists for an association between ALAD genotype and susceptibility to lead toxicity at background exposure levels (Kelada et al. 2001).

The vitamin D receptor (VDR) gene, located at chromosome 12cen-12 is involved in calcium absorption through the gut and into calcium-rich tissues such as bone. VDR may play a role in susceptibility to lead bioaccumulation (Schwartz et al. 2000). Most studies on the VDR gene have focused on the *Bsm*I polymorphism, defined by the restriction enzyme *Bsm*I (Tokita et al. 1996). This polymorphism results in three genotypes denoted bb when the restriction site is present, BB when it is absent, and Bb when both alleles are present. The study of Schwartz et al. (2000) showed that subjects with the B allele had larger tibial lead concentrations with increasing age, and lower tibial lead concentrations with increasing duration since the last exposure to lead than did subjects without the B allele. In the group of lead workers, ALAD-2 and VDR B were associated with higher blood lead levels; however, only VDR B was associated with higher tibial lead levels.

The third gene that might influence the absorption of lead is HFE, the gene for hereditary hemochromatosis. This gene has been localized on the short arm of chromosome 6 (6p21.3) (see Section 10.4.3.4). Because of associations between iron and lead transport, it is possible that polymorphism in the HFE gene may also influence the absorption of lead. There is evidence to suggest that the hemochromatosis gene may induce susceptibility to increased lead absorption (Onalaja 2000).

10.4.3.6

Magnesium (see also Part III, Chapter 2.2)

The genetic basis and cellular defects of a number of primary magnesium wasting diseases have been elucidated over the past decade. Important disorders are:

- hypomagnesemia with secondary hypocalcemia, an early-onset, autosomal-recessive disease segregating with chromosome 9q12–22.2;
- autosomal-dominant hypomagnesemia caused by isolated renal magnesium wasting, mapped to chromosome 11q23;
- hypomagnesemia with hypercalciuria and nephrocalcinosis, a recessive condition caused by a mutation of the claudin 16 gene (3q27) coding for a tight junctional protein that regulates paracellular Mg^{2+} transport in the loop of Henle;
- autosomal-dominant hypoparathyroidism, a variably hypomagnesemic disorder caused by inactivating mutations of the extracellular Ca^{2+}/Mg^{2+}-sensing receptor (CASR:) gene, at 3q13.3–21 (a significant association between common polymorphism of the CASR: and extracellular Mg^{2+} concentration has been demonstrated in a healthy adult population); and
- Gitelman syndrome, a recessive form of hypomagnesemia caused by mutations in the distal tubular NaCl cotransporter gene, SLC12A3, at 16q13. The basis for renal magnesium wasting in this disease is not known.

These inherited conditions affect different nephron segments and different cell types, and lead to variable but increasingly distinguishable phenotypic presentations (Cole et al. 2000).

10.4.3.7
Molybdenum (see also Part III, Chapter 18)

The molybdenum cofactor (MoCo) is an essential component of a large family of enzymes involved in important transformations in carbon, nitrogen, and sulfur metabolism. The MoCo biosynthetic pathway is evolutionarily conserved and found in archaea, eubacteria, and eucaryotes. In humans,

genetic deficiency in enzymes involved in this pathway trigger an autosomal recessive and usually deadly disease with severe neurological symptoms (Wuebbens et al. 2000).

Molybdenum cofactor deficiency (MoCoD) is an autosomal recessive, fatal neurological disorder, characterized by the combined deficiency of sulphite oxidase, xanthine dehydrogenase, and aldehyde oxidase. No therapy is known for this rare disease, which results in neonatal seizures and other neurological symptoms identical to sulfite oxidase deficiency. Heterozygous carriers of a MoCo deficiency allele do not display any symptoms (Reiss et al. 1999).

An excessive occurrence of this fatal disorder has been found among segments of the Arab population in Northern Israel, suggesting that the true incidence of MoCoD is probably underestimated in this highly inbred population (Shalata et al. 2000). This lethal disease can be diagnosed prenatally by assay of sulphite oxidase activity in chorionic villus samples in pregnancies of couples who have had previously affected children (obligatory carriers). However, to date, there is no biochemical assay for carrier detection among the population at risk.

Mutation of the human molybdenum cofactor sulfurase gene seems to be responsible for classical xanthinuria type II (Ichida et al. 2001). A mutation in the gene for the neurotransmitter receptor-clustering protein gephyrin may cause a novel form of molybdenum cofactor deficiency (Reiss et al. 2001)

10.4.3.8
Zinc (see also Part III, Chapter 29)

Zinc deficiency in humans is widespread throughout the world, though it is more prevalent in areas where the population subsists on cereal proteins. Conditioned Zn deficiency is seen in many disease states. Its deficiency during growth periods results in growth failure and lack of gonadal development in males. Other defects of zinc deficiency include skin changes, poor appetite, mental lethargy, delayed wound healing, neurosensory disorders, and cell-mediated immune disorders (Prasad 1995).

Acrodermatitis enteropathica, however, is a rare autosomal recessive pediatric disease characterized by hypozincemia, dermatitis, diarrhea, alopecia, and growth failure. The disease results from insufficient uptake of zinc by the intestine and can be fatal unless the diet is supplemented with zinc. To map the gene responsible for acrodermatitis enteropathica, Wang et al. (2001) performed a genome-wide screen on 17 individuals (including four affected individuals) in a consanguineous Jordanian family. All four affected individuals, including one who was not genotyped in the genome-wide screen, were found to be homozygous for a common haplotype, spanning approximately 3.5 cM, defined by markers D8S1713 and D8S2334 on chromosomal region 8q24.3. To support these mapping data, seven consanguineous Egyptian families with eight patients with acrodermatitis enteropathica were genotyped using these markers, and six patients from five families were found to be homozygous in this region.

Few reports exist relating to hyperzincemia. Smith et al. (1976) and Smith (1977) were the first to describe an extremely high zinc concentration in plasma in five out of seven members of one family, and in two out of three second-generation individuals with no apparent clinical symptoms or abnormalities. Another three cases of chronic hyperzincemia in three brothers has been described by Failla et al. (1982). Zimmerman (1984) found increased zinc content in umbilical cord serum in eight of nine anencephalics and in three infants with spina bifida. Increased zinc levels were bound to serum albumin or alpha-2 macroglobulin in infants with neural tube defects (NTD). The NTD mothers had normal serum zinc levels, but there was a shift from alpha-2 macroglobulin to albumin. One patient with pyoderma gangrenosum also showed hyperzincemia (Hambidge et al. 1985), as well as an 11-year-old boy with hepatosplenomegaly, rashes, stunted growth, anemia, impaired immune functions, vasculitis, and osteoporosis since infancy (Sampson et al. 1997).

References

ABDELOUAS A, LUTZE W, GONG W, NUTTALL EH, STRIETELMEIER BA and TRAVIS B (2000) *Biological reduction of uranium in groundwater and subsurface soil.* Sci Total Environ **250**:21–35.

AXELSEN KB and PALMGREN MG (1998) *Evolution of substrate specifities in the P-ATPase superfamily.* J Mol Evol **46**:84–101.

BIEMPICA L, RAUCH H, QUINTANA N and STERNLIEB I (1988) *Morphological and chemical studies on a murine mutation (toxic milk) resulting in hepatic copper toxicosis.* Lab Invest **59**:500–508.

BOTHWELL TH and MACPHAIL AP (1998) *Hereditary hemochromatosis: etiologic, pathologic, and clinical aspects.* Semin Hematol **35**:55–71.

BREWER GJ (1971) *Annotation: human ecology, an expanding role for the human geneticist.* Am J Hum Genet **23**:92–94.

BREWER GJ, DICK RD, SCHALL W, YUZBASIYAN – GURKAN V, MULLANEY TP, PACE C, LINDGREN J, THOMAS M and PADGETT G (1992) *Use of zinc acetate to treat copper toxicosis in dogs.* J Am Vet Med Assoc **201**:564–568.

BRIAT J-F and LEBRUN M (1999) *Plant responses to metal toxicity.* Sci R Acad Sci III **322**:43–54.

BROOKS RR and MALAISSE F (1985) *The heavy metal-tolerant flora of south central Africa.* AA Balkema, Rotterdam-Boston.

BROOKS RR, CHAMBERS MF, NICKS LJ and ROBINSON BH (1998) *Phytomining.* Trends Plant Sci **3**:359–362.

BRUINS MR, KAPIL S and OEHME FW (2000) *Microbial resistance to metals in the environment.* Ecotoxicol Environ Safety **45**:198–207.

CELLIER M, PRIVE G, BELOUCHI A, KWAN T, RODRIGUES V, CHIA W and GROS P (1995) *Nramp defines a family of membrane proteins.* Proc Natl Acad Sci USA **92**:10089–10093.

CHIOU HY, HSUEH YM, HSIEH LL, HSU LI, HSU YH, HSIEH FI, WEI M-L, CHEN HC, YANG HT, LEU LC, CHU TH, CHEN-WU C, YANG MH and CHEN CJ (1997) *Arsenic methylation capacity, body retention, and null genotypes of glutathione S-transferase M1 and T1 among current arsenic-exposed residents in Taiwan.* Mutat Res **386**:197–207.

COLE DE and QUAMME GA (2000) *Inherited disorders of renal magnesium handling.* J Am Soc Nephrol **11**:1937–1947.

CONCHA G, VOGLER G, LEZCANO D, NERMELL B and VAHTER M (1998a) *Exposure to inorganic arsenic metabolites during early human development.* Toxicol Sci **44**:185–190.

CONCHA G, VOGLER G, NERMELL B and VAHTER M (1998b) *Metabolism of inorganic arsenic in children with chronic high arsenic exposure in northern Argentina.* Environ Health Perspect **106**:355–359.

CONCHA G, VOGLER G, NERMELL B and VAHTER M (1998c) *Low arsenic excretion in breast milk of native Andean women exposed to arsenic in drinking water.* Int Arch Occup Environ Health **71**:42–46.

COOKSEY DA (1994) *Molecular mechanisms of copper resistance and accumulation in bacteria.* FEMS Microbiol Rev **14**:381–386.

DALY MJ (2000) *Engineering radiation-resistant bacteria for environmental biotechnology.* Curr Opin Biotechnol **11**:280–285.

DEVUYST O, CHRISTIE PT, COURTOY PJ, BEAUWENS R and THAKKER RV (1999) *Intrarenal and subcellular distribution of the human chloride channel, CLC-5, reveals a pathophysiological basis for Dent's disease.* Hum Mol Biol **8**:247–257.

ENG BH, GUERINOT ML, EIDE D and SAIER JR MH (1998) *Sequence analyses and phylogenetic characterization of the ZIP family of metal ion transport proteins.* J Membrane Biol **166**:1–7.

FAILLA ML, VAN DE VEERDONK M, MORGAN WT and SMITH JC (1982) *Characterization of zinc-binding proteins of plasma in familial hyperzincemia.* J Lab Clin Med **100**:943–952.

FORBES VE (1999) *Genetics and ecotoxicology.* Taylor and Francis, Philadelphia.

GADD GM (2000) *Bioremedial potential of microbial mechanisms of metal mobilization and immobilization.* Curr Opin Biotechnol **11**:271–279.

GADD GM and GRIFFITHS AJ (1978) *Microorganisms and heavy metal toxicity.* Microbiol Ecol **4**:303–317.

GARTSIDE DW and McNEILLY T (1974) *The potential for the evolution of heavy metal tolerance in plants II. Copper tolerance in normal populations of different plant species.* Heredity **32**:335–348.

GEBEL T (2000) *Confounding variables in the environmental toxicology of arsenic.* Toxicology **144**:155–162.

GOEDDE HW (1972) *Genetically determined variability in response to drugs.* Pharm Weekbl **107**:437–466.

GU M, COOPER JM, BUTLER P, WALKER AP, MISTRY PK, DOOLEY JS and SCHAPIRA AHV (2000) *Oxidative-phosphorylation defects in liver of patients with Wilson's disease.* Lancet **356**:469–474.

GUERINOT ML (2000) *The ZIP family of metal transporters.* Biochim Biophys Acta **1465**:190–198.

HAMBIDGE KM, NORRIS DA, GITHENS JH, AMBRUSO D and CATALANOTTO FA (1985) *Hyperzincemia in a patient with pyoderma gangrenosum.* J Pediatr **106**:450–451.

HANSON EH, IMPERATORE G and BURKE W (2001) *HFE gene and heredity hemochromatosis: a HuGE review. Human Genome Epidemiology.* Am J Epidemiol **154**:193–206.

HARDY RM, STEVENS JB and STOWE BM (1975) *Chronic progressive hepatitis in Bedlington terriers associated with elevated liver copper concentrations.* Minnesota Veterinarian **15**:13–24.

HARRIS ZL, KLOMP LW and GITLIN JD (1998) *Aceruloplasminemia: an inherited neurodegenerative disease with impairment of iron homeostasis.* Am J Clin Nutr 67 (5 Suppl):972S–977S.

HAYAMA A, UCHIDA S, SASAKI S and MARUMO F (2000) *Isolation and characterisation of the human CLC-5 chloride channel gene promoter.* Gene **261**:355–364.

HOBMAN JL, WILSON JW and BROWN NL (2000) *Microbial mercury reduction.* In: Lovley DR, ed., Environmental microbe–metal interactions, pp. 177–197. ASM Press, Washington DC.

HOPENHAYN-RICH C, BIGGS ML, SMITH AH, KALMAN DA and MOORE LE (1996) *Methylation study of a population environmentally exposed to arsenic in drinking water.* Environ Health Perspect **104**:620–628.

HUNT DM, WAKE SA, MERCER JF and DANKS DM (1986) *A study of the role of metallothionein in the inherited copper toxicosis of dogs.* Biochem J **236**:409–415.

IARC (1980) *Arsenic and arsenic compounds.* In: IARC monographs on the evaluation of the carcinogenic risk of chemicals to humans. Some metals and metallic compounds, Vol. 23. International Agency for Research on Cancer, Lyon, pp. 39–141.

ICHIDA K, MATSUMARA T, SAKUMA R, HOSOYA T and NISHINO T (2001) *Mutation of human molybdenum cofactor sulfurase gene is resonible for classical xanthinuria type II.* Biochem Biophys Res Commun 282:1194–1200.

JAFFRÉ T, BROOKS RR, LEE J and REEVES RD (1976) *Sebertia acuminata: a hyperaccumulator of nickel from New Caledonia.* Science 193:559–580.

JAYAWANT S, HALPIN S and WALLACE S (2000) *Menkes kinky hair disease: an unusual case.* Eur J Paediatr Neurol 4:131–134.

JI G and SILVER S (1995) *Bacterial resistance mechanisms for heavy metals of environmental concern.* J Ind Microbiol 14:61–75.

JOHNSON GF, STERNLIEB I, TWEDT DC, GRUSHOFF PS and SCHEINBERG IH (1980) *Inheritance of copper toxicosis in Bedlington terriers.* Am J Vet Res 41:1865–1866.

KALOW W (1982) *Ethnic differences in drug metabolism.* Clin Pharmacokinet 7:373–400.

KALOW W, GOEDDE HW and AGARWAL DP, eds (1986) *Ethnic differences in reactions to drugs and xenobiotics.* Alan R Liss, New York.

KAUR P and ROSEN BP (1992) *Plasmid encoded resistance to arsenic and antimony.* Plasmid 27:29–40.

KELADA SM, SHELTON E, KAUFMANN RB and KHOURY MJ (2001) *δ-Aminolevulinic acid dehydratase genotype and lead toxicity: A HuGE Review.* Am J Epidemiol 154:1–13.

KODAMA H and MURATA Y (1999) *Molecular genetics and pathophysiology of Menkes disease.* Pediatr Int 41:430–435.

LE XC, LU X, MA M, CULLEN WR, APOSHIAN HV and ZHENG B (2000a) *Speciation of key arsenic metabolic intermediates in human urine.* Anal Chem 72:5172–5177.

LE XC, MA M, CULLEN WR, APOSHIAN HV, LU X and ZHENG B (2000b) *Determination of monomethylarsonous acid, a key arsenic methylation intermediate, in human urine.* Environ Health Perspect 108:1015–1018.

LI Y, TOGASHI Y, SATO S, EMOTO T, KANG J-H, TAKEICHI N, KOBAYASHI H, KOJIMA Y, UNE Y and UCHINO J (1991) *Spontaneous hepatic copper accumulation in Long-Evans Cinnamon rats with hereditary hepatitis.* J Clin Invest 87:1858–1861.

LLOYD JR and LOVLEY DR (2001) *Microbial detoxification of metals and radionuclides.* Curr Opin Biotechnol 12:248–253.

LLOYD , J R, YONG P and MACASKIE LE (2000) *Biological reduction and removal of Np (V) by two microorganisms.* Environ Sci Technol 34:1297–1301.

LOCKITCH G (1998) *Iron and copper-associated cirrhosis in infants.* Clin Lab Med 18:665–671.

LOUDIANOS G and GITLIN JD (2000) *Wilson's disease.* Semin Liv Dis 20:353–364.

LOVLEY DR and COATS JD (1997) *Bioremediation of metal contamination.* Curr Opin Biotechnol 8:285–289.

LUTKENHAUS JF (1977) *Role of a major outer membrane protein in Escherichia coli.* J Bacteriol 131:631–637.

MACNAIR MR (1993) *The genetics of metal tolerance in vascular plants.* New Phytol 124:541–559.

MACNAIR MR, BERT V, HUTSON SB, SAUMITOU-LAPRADE P and PETIT D (1999) *Zinc tolerance and hyperaccumulation are genetically independent characters.* Proc R Soc Lond B Biol Sci 266:2175–2179

MARSCHNER H (1999) *Mineral nutrition of higher plants,* 2nd edn. Academic Press, San Diego-San Francisco-New York-Boston-London-Sydney-Tokyo.

MCNEILLY T and ANTONOVICS J (1968) *Evolution in closely adjacent plant populations IV. Barriers to gene flow.* Heredity 32:205–218.

MEAGHER RB (2000) *Phytoremediation of toxic elemental and organic pollutants.* Curr Opin Plant Biol 3:153–162.

MENKES JH (1999) *Menkes disease and Wilson disease: two sides of the same copper coin. Part I: Menkes disease.* Eur J Paediatr Neurol 3:147–158.

MISRA TK (1992) *Bacterial resistance to inorganic mercury salts and organomercurials.* Plasmid 27:4–16.

NIES DH (1992) *Resistance to cadmium, cobalt, zinc and nickel in microbes.* Plasmid 27:17–28.

NIES DH and SILVER S (1995) *Ion efflux systems involved in bacterial metal resistances.* J Ind Microbiol 14:186–199.

NRC (National Research Council) (1999) *Arsenic in drinking water.* National Academy Press, Washington, DC.

NRC (National Research Council) (2001) *Arsenic in drinking water: 2001 update.* National Academy Press, Washington, DC.

OLLSON PE and KILLE P (1997) *Functional comparison of the metal-regulated transcriptional control regions of metallothionein genes from cadmium-*

sensitive and tolerant fish species. Biochim Biophys Acta **1350**: 325–334.

ONALAJA AO and CLAUDIO L (2000) *Genetic susceptibility to lead poisoning.* Environ Health Perspect **108**: 23–28.

OSBORN AM, BRUCE KD, STRIKE P and RITCHIE DA. (1997) *Distribution, diversity and evolution of the bacterial mercury resistance (mer) operon.* FEMS Microbiol Rev **19**: 239–262.

PILETZ JP and GANSCHOW RE (1978) *Zinc deficiency in murine milk underlies expression of the lethal milk (ml) mutation.* Science **199**: 181–183.

PRASAD AS (1995) *Zinc: an overview.* Nutrition **11**: 93–99.

PRAT S (1934) *Die Erblichkeit der Resistenz gegen Kupfer.* Ber Deutsch Bot. Gesellsch. **52**: 65–67.

PROPPING P (1978) *Pharmacogenetics.* Rev Physiol Biochem Pharmacol **83**: 123–173.

PROPPING P (1980) *Neue Entwicklungen in der Pharmakogenetik.* Klinikarzt **9**: 422–434.

PROSCHOWSKY HF, JEPSEN B, JENSEN HE, JENSEN AL and FREDHOLM M (2000) *Microsatellite marker C04107 as a diagnostic marker for copper toxicosis in the danish population of Bedlington terriers.* Acta Vet Scand **41**: 345–350.

RAUCH H (1983) *Toxic milk: a new mutation affecting copper metabolism in the mouse.* J Hered **74**: 141–144.

RAUCH H and WELLS AJ (1995) *The toxic milk mutation which results in a condition resembling Wilson disease in man is linked to mouse chromosome 8.* Genomics **29**: 551–552.

REED V, WILLIAMSON P, BULL PC, COX DW and BOYD Y (1995) *Mapping of the mouse homologue of the Wilson disease gene to mouse chromosome 8.* Genomics **28**: 573–575.

REIS BL and EVANS GW (1977) *Genetic influence on zinc metabolism in mice.* J Nutr **107**: 1683–1686.

REISS J, CHRISTENSEN E and DORCHE C (1999) *Molybdenum cofactor deficiency: first prenatal genetic analysis.* Prenat Diagn **19**: 386–388.

REISS J, GROSS-HARDT S, CHRISTENSEN E, SCHMIDT P, MENDEL RR and SCHWARZ G (2001) *A mutation in the gene for the neurotransmitter receptorclustering protein gephrin causes a novel form of molybdenum cofactor deficiency.* Am J Hum Genet **68**: 208–213.

ROBINSON NJ, GUPTA A, FORDHAM-SKELTON AP, CROY RRD, WHITTON BA and HUCKLE JW (1990) *Prokaryotic metallothionein gene characterisation and expression: chromosome crawling by ligation mediated PCR.* Proc R Soc Lond B **242**: 241–247.

ROSEN BP (1999) *The role of efflux in bacterial resistance to soft metals and metalloids.* Essays Biochem **34**: 1–15.

ROSSBACH S, KUKUK ML, WILSON TL, FENG SF, PEARSON MM and FISHER MA (2000) *Cadmiumregulated gene fusions in Pseudomonas fluorescens.* Environ Microbiol **2**: 373–382.

ROUCH DA, LEE BTO and MORBY AP (1995) *Understanding cellular responses to toxic agents: a mechanism-choice in bacterial metal resistance.* J Ind Microbiol **14**: 132–141.

SAGNER S, KNEER R, WANNER G, COSSON J-P, DEUS-NEUMANN B and ZENK MH (1998) *Hyperaccumulation, complexation and distribution of nickel in Sebertia acuminata.* Phytochemistry **47**: 339–347.

SAMPSON B, KOVAR IZ, RAUSCHER A, FAIRWEATHER-TAIT S, BEATTIE J, MCARDLE HJ, AHMED R and GREEN C (1997) *A case of hyperzincemia with functional zinc depletion: a new disorder?* Pediatr Res **42**: 219–225.

SCHWARTZ BS, LEE B-K, LEE G-S, STEWART WF, SIMON D, KELSEY K and TODD AC (2000) *Association of blood lead, dimercaptosuccinic acid-chelatable lead, and tibia lead with polymorphism in the Vitamin D receptor and δ-aminolevulinic acid dehydratase genes.* Environ Health Perspect **108**: 949–954.

SETH S and BRITTENHAM GM (2000) *Genetic disorders affecting proteins of iron metabolism: clinical implications.* Annu Rev Med **51**: 443–464.

SHALATA A, MANDEL H, DORCHE C, ZABOT MT, SHALEV S, HUGEIRAT Y, ARIEH D, RONIT Z, REISS J, ANBINDER Y and COHEN N (2000) *Prenatal diagnosis and carrier detection for molybdenum cofactor deficiency type A in northern Israel using polymorphic DNA markers.* Prenat Diagn **20**: 7–11.

SHAW AJ (1999) *The evolution of heavy metal tolerance in plants: Adaptations, limits, and costs.* In: Forbes VE, ed. Genetics and ecotoxicology, pp. 9–30. Taylor and Francis, Philadelphia.

SILVER S (1996) *Bacterial resistances to toxic metal ions – a review.* Gene **179**: 9–19.

SILVER S (1998) *Genes for all metals – a bacterial view of the Periodic Table.* J Ind Microbiol Biotechnol **20**: 1–12.

SILVER S and PHUNG LT (1996) *Bacterial heavy metal resistance: new surprises.* Annu Rev Microbiol **50**: 753–789.

SMITH JC (1977) *Heritable hyperzincemia in humans.* Prog Clin Biol Res **14**: 181–191.

SMITH JC, ZELLER JA, BROWN JA and ONG SC (1976) *Elevated plasma zinc: a heritable anomaly.* Science **193**:496–498.

SUZUKI M and GITLIN JD (1999) *Intracellular localization of the Menkes' and Wilson's disease proteins and their role in intracellular copper transport.* Pediatr Int **41**:436–442.

THEOPHILOS MB, COX DW and MERCER FB (1996) *The toxic milk mouse is a murine model of Wilson disease.* Hum Mol Genet **5**:1619–1624.

TOKITA A, MATSUMOTO H, MORRISON NA, TAWA T, MIURA Y, FUKAMAUCHI K, MITSUHASHI N, IRIMOTO M, YAMAMORI S, MIURA M, WATANABE T, KUWABARA Y, YABUTA K and EISMAN JA (1996) *Vitamin D receptor alleles, bone mineral density and turnover in premenopausal Japanese women.* J Bone Miner Res **11**:1003–1009.

TONNESEN K, KLEIJER WJ and HORN N (1991) *Incidence of Menkes' disease.* Hum Genet **86**:408–410.

TSENG WP (1977) *Effects and dose–response relationships of skin cancer and Blackfoot disease with arsenic.* Environ Health Perspect **19**:109–119.

TUMER Z, MOLLER LB and HORN N (1999) *Mutation spectrum of ATP7A, the gene defective in Menkes' disease.* Adv Exp Med Biol **448**:83–95.

TURNER JS and ROBINSON NJ (1995) *Cyanobacterial metallothioneins: biochemistry and molecular genetics.* J Ind Microbiol **14**:119–125.

TWEDT DC, STERNLIEB I and GILDERSON SR (1979) *Clinical, morphologic and chemical studies on copper toxicosis of Bedlington terriers.* J Am Vet Med Assoc **175**:269–275.

VAHTER M (2000) *Genetic polymorphism in the biotransformation of inorganic arsenic and its role in toxicity.* Toxicol Lett **112–113**:209–217.

VAN DE SLUIS BJA, BREEN M, NANJI M, VAN WOLFEREN M, DE JONG P, BINNS MM, PEARSON PL, KUIPERS J, ROTHUIZEN J, COX DW, WIJMENGA C and VAN OOST BA (1999) *Genetic mapping of the copper toxicosis locus in Bedlington terriers to dog chromosome 10, in a region syntenic to human chromosome region 2p13-p16.* Hum Mol Genet **8**:501–507.

WANG K, PUGH EW, GRIFFEN S, DOHENY KF, MOSTAFA WZ, AL ABOOSI MM, EL SHANTI H and GITSCHIER J (2001) *Homozygosity mapping places the acrodermatitis enteropathica gene on chromosomal region 8q24.3.* Am J Hum Genet **68**:1055–1060.

WILLIAMS LE, PITTMAN JK and HALL JL (2000) *Emerging mechanisms for heavy metal transport in plants.* Biochim Biophys Acta **1465**:104–126.

WITT H and LANDT O (2001) *Rapid detection of the Wilson's disease H1069Q mutation by melting curve analysis with the LightCycler.* Clin Chem Lab Med **39**:953–955.

WU J, FORBES JR, CHEN HS and COX DW (1994) *The LEC rat has a deletion in the copper transporting ATPase gene homologous to the Wilson disease gene.* Nature Genet **7**:541–545.

WUEBBENS MM, LIU MT, RAJAGOPALAN K and SCHINDELIN H (2000) *Insights into molybdenum cofactor deficiency provided by the crystal structure of the molybdenum cofactor biosynthesis protein MoaC.* Structure Fold Des **8**:709–718.

YUZBASIYAN-GURKAN V, BLANTON SH, CAO Y, FERGUSON P, LI J, VENTA PJ and BREWER GJ (1997) *Linkage of a microsatellite marker to the canine copper toxicosis locus in Bedlington terriers.* Am J Vet Res **58**:1–5.

YUZBASIYAN-GURKAN V, WAGNITZ S, BLANTON SH and BREWER GJ (1993) *Linkage studies of the esterase D and retinoblastoma genes to copper toxicosis: a model for Wilson disease.* Genomics **15**:86–90.

ZIMMERMAN AW (1984) *Hyperzincemia in anencephaly and spina bifida: a clue to the pathogenesis of neural tube defects?* Neurology **34**:443–450.